Birkhäuser

John Snygg

A New Approach to Differential Geometry Using Clifford's Geometric Algebra

Birkhäuser

John Snygg
433 Prospect St.
East Orange, NJ 07017
jsnygg@post.harvard.edu

ISBN 978-0-8176-8282-8 e-ISBN 978-0-8176-8283-5
DOI 10.1007/978-0-8176-8283-5
Springer New York Dordrecht Heidelberg London

Library of Congress Control Number: 2011940217

Mathematics Subject Classification (2010): 11E88, 15A66, 53-XX, 53-03

Printed on acid-free paper

Springer is part of Springer Science+Business Media
(www.birkhauser-science.com)

To Pertti Lounesto

Preface

This book was written with the intention that it be used as a text for an undergraduate course. The end result is not only suitable for an undergraduate course but also ideal for a masters level course directed toward future high school math teachers. It is also appropriate for anyone who wants to acquaint himself or herself with the usefulness of Clifford algebra. In that context, instructors teaching Ph.D. students may want to use it as a source book.

Most introductory books on differential geometry are restricted to three dimensions. I use the notation used by geometers in n-dimensions. Admittedly, most of my examples are in three dimensions. However in Chap. 3, I present some aspects of the four-dimensional theory of special relativity. One can present some fun aspects of special relativity without mention of such concepts as "force", "momentum", and "energy." For example, adding speeds possibly near the speed of light results in a sum that is always less than the speed of light.

The capstone topic for this book is Einstein's general theory of relativity. Here again, knowledge of Newtonian physics is not a prerequisite. Because of the geometric nature of Einstein's theory, some interesting aspects can be presented without knowledge of Newtonian physics. In particular, I discuss the possibility of twins aging at different rates, the precession of Mercury, and the bending of light rays passing near the Sun or some other massive body.

The only topic that does require some knowledge of Newtonian physics is Huygen's isochronous pendulum clock, and the relevant section should be considered optional.

My strategy for writing this book had three steps:

Step 1: Steal as many good pedagogical ideas from as many authors as possible.
Step 2: Improve on them if I could.
Step 3: Choose topics that are fun for me and fit together in a coherent manner.

Frequently I was able to improve on the presentation of others using Clifford algebra.

The selection of this book may introduce a hurdle for some instructors. It is likely they will have to learn something new – Clifford algebra. Paradoxically, the use of Clifford algebra will make differential geometry *more* accessible to students who have completed a course in linear algebra. That is because in this book, Clifford algebra replaces the more complicated and less powerful formalism of differential forms. Anyone who is familiar with the concept of non-commutative matrix multiplication will find it easy to master the Clifford algebra presented in this text. Using Clifford algebra, it becomes unnecessary to discuss mappings back and forth between the space of tangent vectors and the space of differential forms. With Clifford algebra, everything takes place in one space.

The fact that Clifford algebra (otherwise known as "geometric algebra") is not deeply embedded in our current curriculum is an accident of history. William Kingdon Clifford wrote two papers on the topic shortly before his early death in 1879 at the age of 33. Although Clifford was recognized worldwide as one of England's most distinguished mathematicians, he chose to have the first paper published in what must have been a very obscure journal at the time. Quite possibly it was a gesture of support for the efforts of James Joseph Sylvester to establish the first American graduate program in mathematics at Johns Hopkins University. As part of his endeavors, Sylvester founded the *American Journal of Mathematics* and Clifford's first paper on what is now known as Clifford algebra appeared in the very first volume of that journal.

The second paper was published after his death in unfinished form as part of his collected papers. Both of these papers were ignored and soon forgotten. As late as 1923, math historian David Eugene Smith discussed Clifford's achievements without mentioning "geometric algebra" (Smith, David Eugene 1923). In 1928, P.A.M. Dirac reinvented Clifford algebra to formulate his equation for the electron. This equation enabled him to predict the discovery of the positron in 1931.

In 1946 and 1958, Marcel Riesz published some results on Clifford algebra that stimulated David Hestenes to investigate the subject. In 1966, David Hestenes published a thin volume entitled *Space-time Algebra* (Hestenes 1966). And 18 years later, with his student Garret Sobezyk, he wrote a more extensive book entitled *Clifford Algebra to Geometric Calculus – A Unified Language for Mathematics and Physics* (Hestenes and Sobczyk 1984). Since then, extensive research has been carried out in Clifford algebra with a multitude of applications.

Had Clifford lived longer, "geometric algebra" would probably have become mainstream mathematics near the beginning of the twentieth century. In the decades following Clifford's death, a battle broke out between those who wanted to use quaternions to do physics and geometry and those who wanted to use vectors. Quaternions were superior for dealing with rotations, but they are useless in dimensions higher than three or four without grafting on some extra structure. Eventually vectors won out.

Since the structure of both quaternions and vectors are contained in the formalism of Clifford algebra, the debate would have taken a different direction had Clifford lived longer. While alive, Clifford was an articulate spokesman and his writing for popular consumption still gets published from time to time. Had Clifford participated in the quaternion–vector debate, "geometric algebra" would have received more serious consideration.

The advantage that quaternions have for dealing with rotations in three dimensions can be generalized to higher dimensions using Clifford algebra. This is important for dealing with the most important feature of a surface in any dimension – namely its curvature.

Suppose you were able to walk from the North Pole along a curve of constant longitude to the equator, then walk east along the equator for 37° and finally return to the North Pole along another curve of constant longitude. In addition, suppose at the start of your trek, you picked up a spear, pointed it in the south direction and then avoided any rotation of the spear with respect to the surface of the earth during your long journey. If you were careful, the spear would remain pointed south during the entire trip. However, on your return to the North Pole, you would discover that your spear had undergone a 37° rotation from its initial position. This rotation is a measure of the curvature of the Earth's surface.

The components of the Riemann tensor, used to measure curvature, are somewhat abstract in the usual formalism. Using Clifford algebra, the components of the Riemann tensor can be interpreted as components of an infinitesimal rotation operator that indicates what happens when a vector is "parallel transported" around an infinitesimal loop in a curved space.

In many courses on differential geometry, the Gauss–Bonnet Formula is the capstone result. Exploiting the power of Clifford algebra, a proof appears slightly less than halfway through this book. If optional intervening historical digressions were eliminated, the proof of the Gauss–Bonnet Formula would appear on approximately p. 115.

This should leave time to cover other topics that interest the instructor or the instructor's students. I hope that instructors endeavor to cover enough of the theory of general relativity to discuss the precession of Mercury. The general theory of relativity is essentially geometric in nature.

Whatever topics are chosen, I hope people have fun.

Acknowledgments

An author would like to have the illusion that he or she has accomplished something without the aid of others. However, a little introspection, at least in my case, makes such an illusion evaporate. To begin with as I prepared to write various sections of this book, I read and reread the relevant sections in Dirk J. Struik's *Lectures in Classical Differential Geometry* 2nd Ed. (1988), and Barrett O'Neill's *Differential Geometry* 2nd Ed. (1997). These two texts are restricted to three dimensions. For higher dimensions, I devoted many hours to Johan C.H. Gerretsen's *Lectures on Tensor Calculus and Differential Geometry* (1962). For my chapter on non-Euclidean geometry, I found a lot of very useful ideas in *Geometry* 1st Ed. (1999) by David A. Brannan, Matthew F. Esplen, and Jeremy J. Gray. From time to time I used other geometry books, which appear in my bibliography.

In addition, I have benefited from those who have developed the tools of Clifford algebra. Perhaps the most relevant book is *Clifford Algebra to Geometric Calculus* (1984) by David Hestenes and Garret Sobczyk. It should be noted that others have demonstrated the usefulness of Clifford algebra at a fairly elementary level. These include Bernard Jancewicz with his *Clifford Algebras in Electrodynamics* (1988), Pertti Lounesto with his *Clifford Algebras and Spinors* (1997), and William E. Baylis with his *Electrodynamics – A Modern Geometric Approach* (1999).

As I was writing the first draft, I found myself frequently contacting Frank Morgan by e-mail and pestering him with questions. As I neared completion of the first draft, I was able to get Hieu Duc Nguyen to go over much of the text. He suggested that I give prospective teachers better guidance on how to use the text. This motivated me to reorganize the book so that the later chapters became more independent of one another. This gives prospective teachers more freedom to design their own course. (At Hieu Duc Nguyen's urging, I also added a section of the Introduction addressed to teachers). I am also grateful to Patrick Girard for his comments. As I was making some finishing touches to the manuscript, I got some valuable comments from G. Stacey Staples.

I also got substantial assistance in my quest to include some historical background. I usually started a search on the Internet that invariably led to items posted by John J. O'Connor and Edmund F. Robertson at the School of Mathematics and Statistics at the University of St. Andrews, Scotland. Due to the ephemeral nature of some web sites, I have been hesitant to cite their work. A few of their entries are mentioned in my bibliography but more often than not, I exploited their bibliographies to find out more about the mathematicians I wanted to discuss.

At a Clifford algebra conference held at Cookeville, Tennessee in 2002, Sergiu Vacaru described to me some of his personal experiences with Dimitri Ivanenkov. This set me off on a grand adventure to discover a little bit about doing physics in Stalinist Russia. Along the way I got help from Gennady Gorelik, Vitaly Ginzburg, Sasha Rozenberg, and Engelbert Schücking.

Engelbert Schücking shared several anecdotes along with a strong sense of history. He also translated a number of German passages that were too difficult for me. I was also able to elicit some comments from Joseph W. Dauben, a noted historian of science. I found myself at odds with George Saliba on one aspect of the impact of Islamic astronomers on the work of Copernicus. Nonetheless I would like to think that most of what I wrote on Islamic science is a reflection of his views.

I benefited from the technical support staff at MacKichan Software in support of their *Scientific WorkPlace* program. And I am also grateful to Alan Bell for keeping my computer working at some critical times.

I wish to thank my daughter, Suzanne,who used her strong editorial skills to clarify the historical sections. I am also grateful to my son, Spencer, who gave me some useful advice on some of my drawings. My brother Charles reviewed a few sections and made some useful comments.

In 1996, I was forced into early retirement when Upsala College (my employer) in East Orange, NJ went bankrupt. In this circumstance, I am grateful to the Rutgers Math Department for granting me status as a "visiting scholar." This has given me valuable library privileges in all the various Rutgers library branches spread out over the state of New Jersey. I am also indebted to the reference librarians in the Newark branch of the Rutgers library system who somehow made sense out of incomplete references and located needed sources.

East Orange, USA John Snygg

Contents

1 **Introduction** ... 1

2 **Clifford Algebra in Euclidean 3-Space** ... 3
- 2.1 Reflections, Rotations, and Quaternions in E^3 ... 3
  - 2.1.1 Using Square Matrices to Represent Vectors ... 3
  - 2.1.2 1-Vectors, 2-Vectors, 3-Vectors, and Clifford Numbers ... 5
  - 2.1.3 Reflection and Rotation Operators ... 6
  - 2.1.4 Quaternions ... 8
- 2.2 The 4π Periodicity of the Rotation Operator ... 13
- 2.3 *The Point Groups for the Regular Polyhedrons ... 15
- 2.4 *Élie Cartan 1869–1951 ... 23
- 2.5 *Suggested Reading ... 25

3 **Clifford Algebra in Minkowski 4-Space** ... 27
- 3.1 A Small Dose of Special Relativity ... 27
- 3.2 *Albert Einstein 1879–1955 ... 37
- 3.3 *Suggested Reading ... 45

4 **Clifford Algebra in Flat n-Space** ... 47
- 4.1 Clifford Algebra ... 47
- 4.2 The Scalar Product and Metric Tensor ... 53
- 4.3 The Exterior Product for p-Vectors ... 61
- 4.4 Some Useful Formulas ... 65
- 4.5 Gram–Schmidt Formulas ... 68
- 4.6 *The Qibla (Kibla) Problem ... 69
- 4.7 *Mathematics of Arab Speaking Muslims ... 74
  - 4.7.1 *Greek Science and Mathematics in Alexandria ... 74
  - 4.7.2 *Hypatia ... 77
  - 4.7.3 *The Rise of Islam and the House of Wisdom ... 79
  - 4.7.4 *The Impact of Al-Ghazālī ... 86

4.7.5 *Claudius Ptolemy, Al-Tūsī, Al-'Urdī, Ibn al-Shātir, Nicholas Copernicus, Tycho Brahe, Johannes Kepler, and Isaac Newton 89
4.8 *Christopher Columbus and Some Bad Geography 109
4.9 *Clifford and Grassmann 110
4.9.1 *William Kingdon Clifford 1845–1879 110
4.9.2 *Hermann Günther Grassmann 1809–1877 116

5 Curved Spaces 121
5.1 Gaussian Curvature (Informal) 121
5.2 n-Dimensional Curved Surfaces (Spaces) 127
5.3 The Intrinsic Derivative ∇_k 132
5.4 Parallel Transport and Geodesics 138
5.5 The Riemann Tensor and the Curvature 2-form 142
5.6 Fock–Ivanenko Coefficients 147
5.6.1 Moving Frames 147
5.6.2 Gauss Curvature via Fock–Ivanenko Coefficients 151
5.6.3 *The Riemann Tensor for Orthonormal Frames 162
5.7 *Doing Physics Under Stalin 165
5.7.1 *George Gamow 1904–1968 166
5.7.2 *Lev Davidovich Landau 1908–1968 167
5.7.3 *Matvei Petrovich Bronstein 1906–1938 170
5.7.4 *Vladimir Alexandrovich Fock 1898–1974 172
5.7.5 *Dmitrii Dmitrievich Ivanenko 1904–1994 174

6 The Gauss–Bonnet Formula 181
6.1 The Exterior Derivative and Stokes' Theorem 181
6.2 *Curvature via Connection 1-Forms 186
6.3 Geodesic Curvature on a 2-dimensional Surface 189
6.4 *Huygens' Pendulum Clock and the Cycloid 195
6.5 The Gauss–Bonnet Formula 202
6.6 The Interpretation of Curvature 2-Forms as Infinitesimal Rotation Operators 209
6.7 *Euler's Theorem for Convex Polyhedrons 210
6.8 *Carl Friedrich Gauss and Bernard Riemann 212
6.8.1 *Carl Friedrich Gauss 1777–1855 212
6.8.2 *Georg Friedrich Bernhard Riemann 1826–1866 223

7 Some Extrinsic Geometry in E^n 227
7.1 The Frenet Frame 227
7.2 *Arbitrary Speed Curves with Formulas 241
7.3 Ruled Surfaces and Developable Surfaces 249
7.4 *Archimedes' Screw 255
7.5 Principal Curvatures 256
7.5.1 The Normal and Geodesic Curvature Vectors 256
7.5.2 The Weingarten Map or Shape Operator 261

7.5.3 Principal Directions and Curvatures 262
7.6 *Leonhard Euler 1707–1783 272
7.7 *Involutes 280
7.8 Theorema Egregium 283
7.9 *The Gauss Map 286
7.10 *Isometries 289
7.10.1 *Isometry of Surfaces with Constant Gaussian Curvature 289
7.10.2 *Isometries for Surfaces with Nonconstant Gaussian Curvature 294

8 *Non-Euclidean (Hyperbolic) Geometry 299
8.1 *Early Developments 299
8.2 *The Poincaré Model and Reflections 303
8.3 *Direct Non-Euclidean Transformations 310
8.4 *Möbius Transformations 314
8.4.1 *The Cross Ratio 317
8.4.2 *Fixed Points 319
8.5 *The Distance Function 322
8.6 *The Law of Cosines and the Law of Sines 326
8.7 *Tessellations 329

9 *Ruled Surfaces Continued 333
9.1 *Lines of Stricture 333
9.2 *Gaussian Curvature of Ruled Surfaces 340
9.3 *The Cusp at the Edge of Regression 342

10 *Lines of Curvature 347
10.1 *Computing Lines of Curvature 347
10.1.1 *Intrinsic Projection Operators for Real Symmetric Linear Operators 347
10.1.2 *The Computation of Principal Directions from Intrinsic Projection Operators 353
10.1.3 *The Computation of Lines of Curvature 355
10.2 *Two Useful Theorems for Lines of Curvature 359
10.3 *Confocal Coordinates in n-Dimensions 364
10.4 *The Nondeformability of a Hypersurface 371

11 *Minimal Surfaces 375
11.1 *Why are Minimal Surfaces Said to be Minimal? 375
11.2 *Minimal Surfaces and Harmonic Functions 379
11.3 *The Enneper–Weierstrass Representations 391

12 Some General Relativity 395
12.1 Einstein's Theory 395
12.2 *Karl Schwarzschild 1873–1916 399
12.3 The Schwarzschild Metric 404

12.4 The Precession of Mercury 412
12.5 The Bending of Light 424

A A Matrix Representation of a Clifford Algebra 431

B Construction of Matrix Representations for Dirac Vectors 437

C A Few Terms of the Taylor's Series for the Urdī-Copernican Model for the Outer Planets 441

D A Few Terms of the Taylor's Series for Kepler's Orbits 445

References 449

Index 459

List of Symbols

Refer to INDEX to get page locations of definitions.

δ^j_k	Kronecker delta symbol
$\delta^{j_1 j_2 \ldots j_n}_{k_1 k_2 \ldots k_n}$	Generalized Kronecker delta symbol
$\mathbf{e}_j$, $\mathbf{e}^k$	Cartesian frame for a Euclidean or pseudo-Euclidean space
$\mathbf{E}_j$, $\mathbf{E}^k$	Orthonormal non-coordinate frame, moving frame, or Frenet frame
$\boldsymbol{\gamma}_j$, $\boldsymbol{\gamma}^k$	Dirac vectors, lower index Dirac vectors, or upper index Dirac vectors
$\mathbf{F} \wedge \mathbf{G}$, $\mathbf{e}_{j_1 j_2 \ldots j_p}$, $\boldsymbol{\gamma}^{k_1 k_2 \ldots k_p}$	Exterior product
Γ^i_{jk}	Christoffel symbol
$\boldsymbol{\Gamma}_k$	Fock–Ivanenko coefficient
g_{jk}	Metric tensor
g^{jk}	Inverse of metric tensor
h^{jk}, h^j_k, and h_{jk}	Tensor components of a shape operator
$\mathbf{A}$, $\mathbf{B}$	Index free Clifford numbers
K or R^{12}_{12}	Gaussian curvature
∇_k	Intrinsic derivative
n_{ij}	Signature matrix
$A^{j_1 j_2 \ldots j_p} \mathbf{e}_{j_1 j_2 \ldots j_p}$, $A_{k_1 k_2 \ldots k_p} \boldsymbol{\gamma}^{k_1 k_2 \ldots k_p}$	p-vectors
R^{jk}_{mn}	Riemann curvature tensor
$\mathbf{R}_{ij}$	Curvature 2-form
$\langle \mathbf{a}, \mathbf{b} \rangle$	Scalar product
$\mathbf{a} \wedge \mathbf{b}$	Wedge product

Chapter 1
Introduction

This text is intended for a one-semester course in differential geometry, although there is enough material for two quarters. The only prerequisite is completion of the calculus – linear algebra sequence. Solutions of differential equations are discussed in the text but do not appear in the problem sets. Thus, completion of a course in differential equations would be useful but not necessary.

Generally, chapters or sections designated with a star in the table of contents are not required for successive sections lacking a star. As a consequence, the starred sections may be considered to be optional.

I have included some historical material which is interesting to me and hopefully both interesting and informative to my readers. Many if not most of the mathematicians and physicists who made differential geometry what it is today did not achieve the total ivory-tower existence that they presumably craved. These historical sections have evoked mixed responses from reviewers. One reviewer wrote, "I think that John Snygg has developed a fascinating historical storyline that is uncommon in undergraduate textbooks. I enjoyed reading the majority of his lengthy historical diversions." Others expressed the opinion that the historical sections are distractions, which have no relevance to the mathematics. Specific suggestions were generally to make omissions which would make the narrative more upbeat and rosy.

It should be observed that for the most part, the historical material is placed in isolated starred sections. That means they are completely optional.

The material contained in the unstarred sections of the first six chapters is sufficient preparation for the material contained in any of the subsequent starred sections with two exceptions. First: Before diving into non-Euclidean geometry in Chap. 8, one should do Problems 58 and 59 in Chap. 4. One should also read Sect. 6.8 in Chap. 6 and do Problem 136a at the end of that section. Problems 58 and 59 in Sect. 4.6 of Chap. 4 are intended to introduce students to a taste of spherical trigonometry. Section 6.8 provides some historical background for the development of non-Euclidean geometry and Problem 136a is an introduction to the metric used for the Poincaré model. Second: The material in Sect. 10.2 requires familiarity with the notion of "lines of stricture", which is covered in Sect. 9.1.

J. Snygg, *A New Approach to Differential Geometry using Clifford's Geometric Algebra*, DOI 10.1007/978-0-8176-8283-5_1, © Springer Science+Business Media, LLC 2012

For a one-quarter course, the unstarred sections in the first seven chapters are probably sufficient. For a basic one-semester course, I hope that readers can go beyond that material and cover the first four sections in Chap. 12 – the chapter on general relativity. If not, there are clearly other options. For example, in Chap. 11 on minimal surfaces, students are invited to invent their own minimal surfaces and then plot them using MAPLE, MATHEMATICA, or other computer programs. (See Problem 278.) Another option is Chap. 8, which is devoted to the Poincaré model for a 2-dimensional surface of constant negative curvature.

Whatever options you choose, I hope you have fun.

Chapter 2
Clifford Algebra in Euclidean 3-Space

2.1 Reflections, Rotations, and Quaternions in E^3

2.1.1 Using Square Matrices to Represent Vectors

One frequently represents a vector $\mathbf{x}$ in the 3-dimensional Euclidean space E^3 by $\mathbf{x} = x\mathbf{i} + y\mathbf{j} + z\mathbf{k}$ or (x, y, z). However, neither of these notations easily generalize to higher dimensions. Alternate notations which do easily generalizes to higher dimensions are $\mathbf{x} = x^1\mathbf{i}_1 + x^2\mathbf{i}_2 + x^3\mathbf{i}_3$ and

$$\mathbf{x} = \left(x^1, x^2, x^3\right) = x^1 \left(1, 0, 0\right) + x^2 \left(0, 1, 0\right) + x^3 \left(0, 0, 1\right). \tag{2.1}$$

These alternate notations have their own problem. In most areas of mathematics, we expect a superscript to designate an exponent. You might think that we could reserve superscripts for exponents and use subscripts to designate different coordinates or other labels. This approach is sometimes used for so-called flat spaces. However, if we accept Einstein's Theory of General Relativity, we live in a space that is curved. To reserve superscripts for exponents in the study of curved spaces is simply too restrictive and inconvenient.

So how can you distinguish a superscript representing an exponent from a superscript representing some kind of label? If you see a superscript outside of some bracket (usually round), you can be confident that it represents an exponent. For example,

$$(a)^2 = aa.$$

On the other hand, if the meaning is clear from the context, the brackets may be omitted. For example, in the next chapter, I will write c to represent the speed of light and c^2 to represent the square of the speed of light.

I now turn to another issue. Usually, one represents a vector as a linear combinations of unit row vectors as in (2.1), or a linear combination of unit column

J. Snygg, *A New Approach to Differential Geometry using Clifford's Geometric Algebra*, DOI 10.1007/978-0-8176-8283-5_2, © Springer Science+Business Media, LLC 2012

vectors. However, as we shall soon see, it is sometimes useful to represent a vector as a linear combination of square matrices. For example, we could write

$$\mathbf{x} = x^1\mathbf{e}_1 + x^2\mathbf{e}_2 + x^3\mathbf{e}_3, \tag{2.2}$$

where

$$\mathbf{e}_1 = \begin{bmatrix} 0 & 0 & 0 & 1 \\ 0 & 0 & 1 & 0 \\ 0 & 1 & 0 & 0 \\ 1 & 0 & 0 & 0 \end{bmatrix}, \quad \mathbf{e}_2 = \begin{bmatrix} 0 & 0 & 1 & 0 \\ 0 & 0 & 0 & -1 \\ 1 & 0 & 0 & 0 \\ 0 & -1 & 0 & 0 \end{bmatrix}, \quad \text{and}$$

$$\mathbf{e}_3 = \begin{bmatrix} 1 & 0 & 0 & 0 \\ 0 & 1 & 0 & 0 \\ 0 & 0 & -1 & 0 \\ 0 & 0 & 0 & -1. \end{bmatrix} \tag{2.3}$$

At first sight, this may seem to be a pointless variation. However, representing a vector in terms of these square matrices enables us to multiply vectors in a way that would not otherwise be possible. We should first note that these matrices have some special algebraic properties. In particular,

$$(\mathbf{e}_1)^2 = (\mathbf{e}_2)^2 = (\mathbf{e}_3)^2 = \mathbf{I}. \tag{2.4}$$

where $\mathbf{I}$ is the identity matrix. Furthermore,

$$\mathbf{e}_2\mathbf{e}_3 + \mathbf{e}_3\mathbf{e}_2 = \mathbf{e}_3\mathbf{e}_1 + \mathbf{e}_1\mathbf{e}_3 = \mathbf{e}_1\mathbf{e}_2 + \mathbf{e}_2\mathbf{e}_1 = 0. \tag{2.5}$$

A set of matrices that satisfy (2.4) and (2.5) is said to form the basis for the *Clifford algebra* associated with Euclidean 3-space. There are matrices other than those presented in (2.3) that satisfy (2.4) and (2.5). (See Prob. 2.) In the formalism of Clifford algebra, one never deals with the components of any specific matrix representation. We have introduced the matrices of (2.3) only to demonstrate that there exist entities that satisfy (2.4) and (2.5).

Now let us consider the product of two vectors. Suppose $\mathbf{y} = y^1\mathbf{e}_1 + y^2\mathbf{e}_2 + y^3\mathbf{e}_3$, then

$$\begin{aligned}\mathbf{xy} = {} & (x^1y^1 + x^2y^2 + x^3y^3)\mathbf{I} + x^2y^3\mathbf{e}_2\mathbf{e}_3 + x^3y^2\mathbf{e}_3\mathbf{e}_2 \\ & + x^3y^1\mathbf{e}_3\mathbf{e}_1 + x^1y^3\mathbf{e}_1\mathbf{e}_3 + x^1y^2\mathbf{e}_1\mathbf{e}_2 + x^2y^1\mathbf{e}_2\mathbf{e}_1.\end{aligned}$$

Using the relations of (2.5), we have

$$\begin{aligned}\mathbf{xy} = {} & (x^1y^1 + x^2y^2 + x^3y^3)\mathbf{I} + (x^2y^3 - x^3y^2)\mathbf{e}_2\mathbf{e}_3 \\ & + (x^3y^1 - x^1y^3)\mathbf{e}_3\mathbf{e}_1 + (x^1y^2 - x^2y^1)\mathbf{e}_1\mathbf{e}_2.\end{aligned} \tag{2.6}$$

(Note $\mathbf{xy} \neq \mathbf{yx}$.)

From (2.6), we can construct formulas for the familiar *scalar product* $\langle \mathbf{x}, \mathbf{y} \rangle$ and the less familiar *wedge product* $\mathbf{x} \wedge \mathbf{y}$. In particular,

$$\langle \mathbf{x}, \mathbf{y} \rangle \mathbf{I} = \frac{1}{2}(\mathbf{xy} + \mathbf{yx}) = \left(x^1 y^1 + x^2 y^2 + x^3 y^3\right) \mathbf{I}, \text{ and} \tag{2.7}$$

$$\mathbf{x} \wedge \mathbf{y} = \frac{1}{2}(\mathbf{xy} - \mathbf{yx}) = (x^2 y^3 - x^3 y^2)\mathbf{e}_2\mathbf{e}_3 + (x^3 y^1 - x^1 y^3)\mathbf{e}_3\mathbf{e}_1 + (x^1 y^2 - x^2 y^1)\mathbf{e}_1\mathbf{e}_2. \tag{2.8}$$

With a slight abuse of notation, we frequently omit the $\mathbf{I}$ that appears in (2.7).

We note that the coefficients of $\mathbf{e}_2\mathbf{e}_3$, $\mathbf{e}_3\mathbf{e}_1$, and $\mathbf{e}_1\mathbf{e}_2$ that appear in the wedge product $\mathbf{x} \wedge \mathbf{y}$ are the three components of the cross product $\mathbf{x} \times \mathbf{y}$.

2.1.2 1-*Vectors*, 2-*Vectors*, 3-*Vectors, and Clifford Numbers*

By considering all possible products of $\mathbf{e}_1$, $\mathbf{e}_2$, and $\mathbf{e}_3$, one obtains an 8-dimensional space spanned by $\{\mathbf{I}, \mathbf{e}_1, \mathbf{e}_2, \mathbf{e}_3, \mathbf{e}_2\mathbf{e}_3, \mathbf{e}_3\mathbf{e}_1, \mathbf{e}_1\mathbf{e}_2, \mathbf{e}_1\mathbf{e}_2\mathbf{e}_3\}$, where

$$\mathbf{e}_2\mathbf{e}_3 = \begin{bmatrix} 0 & 0 & -1 & 0 \\ 0 & 0 & 0 & 1 \\ 1 & 0 & 0 & 0 \\ 0 & -1 & 0 & 0 \end{bmatrix}, \quad \mathbf{e}_3\mathbf{e}_1 = \begin{bmatrix} 0 & 0 & 0 & 1 \\ 0 & 0 & 1 & 0 \\ 0 & -1 & 0 & 0 \\ -1 & 0 & 0 & 0 \end{bmatrix},$$

$$\mathbf{e}_1\mathbf{e}_2 = \begin{bmatrix} 0 & -1 & 0 & 0 \\ 1 & 0 & 0 & 0 \\ 0 & 0 & 0 & -1 \\ 0 & 0 & 1 & 0 \end{bmatrix}, \quad \text{and} \quad \mathbf{e}_1\mathbf{e}_2\mathbf{e}_3 = \begin{bmatrix} 0 & -1 & 0 & 0 \\ 1 & 0 & 0 & 0 \\ 0 & 0 & 0 & 1 \\ 0 & 0 & -1 & 0 \end{bmatrix}.$$

One might think that one could obtain higher order products. However, any such higher order product will collapse to a scalar multiple of one of the eight matrices already listed. For example:

$$\mathbf{e}_1\mathbf{e}_2\mathbf{e}_3\mathbf{e}_2 = \mathbf{e}_1\mathbf{e}_2(\mathbf{e}_3\mathbf{e}_2) = -\mathbf{e}_1\mathbf{e}_2(\mathbf{e}_2\mathbf{e}_3) = -\mathbf{e}_1(\mathbf{e}_2\mathbf{e}_2)\mathbf{e}_3 = -\mathbf{e}_1\mathbf{e}_3 = \mathbf{e}_3\mathbf{e}_1.$$

In this fashion, we have obtained an 8-dimensional vector space that is closed under multiplication. A vector space closed under multiplication is called an *algebra*. An algebra that arises from a vector space with a scalar product in the same manner as this example does from E^3 is called a *Clifford algebra*. (We will give a more formal definition of a Clifford algebra in Chap. 4.)

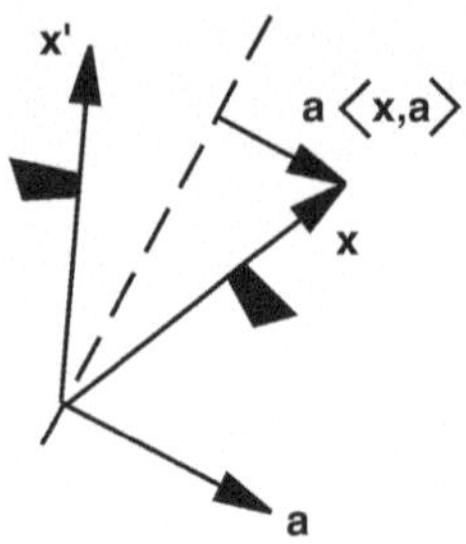

Fig. 2.1 The vector **x**' is the result of reflecting **x** with respect to the plane perpendicular to the unit vector **a**

I label the matrices $\mathbf{e}_1$, $\mathbf{e}_2$, and $\mathbf{e}_3$ to be *Dirac vectors*. Any linear combination of Dirac vectors is a *1-vector*. A linear combination of $\mathbf{e}_2\mathbf{e}_3$, $\mathbf{e}_3\mathbf{e}_1$, and $\mathbf{e}_1\mathbf{e}_2$ is a *2-vector*. In the same vein, a scalar multiple of **I** is a *0-vector* and any scalar multiple of $\mathbf{e}_1\mathbf{e}_2\mathbf{e}_3$ is a *3-vector*. A general linear combination of vectors of possibly differing type is a *Clifford number*.

It will be helpful to use an abbreviated notation for products of Dirac vectors. In particular, let

$$\mathbf{e}_2\mathbf{e}_3 = \mathbf{e}_{23}, \quad \mathbf{e}_3\mathbf{e}_1 = \mathbf{e}_{31}, \quad \mathbf{e}_1\mathbf{e}_2 = \mathbf{e}_{12}, \quad \text{and} \quad \mathbf{e}_1\mathbf{e}_2\mathbf{e}_3 = \mathbf{e}_{123}.$$

2.1.3 Reflection and Rotation Operators

The algebraic properties of Clifford numbers provide us with a convenient way of representing reflections and rotations. Suppose **a** is a vector of unit length perpendicular to a plane passing through the origin and **x** is an arbitrary vector in E^3. (See Fig. 2.1.) In addition, suppose $\acute{\mathbf{x}}$ is the vector obtained from **x** by reflection of **x** with respect to the plane corresponding to **a**. Then

$$\acute{\mathbf{x}} = \mathbf{x} - 2\langle \mathbf{a}, \mathbf{x}\rangle \mathbf{a}. \tag{2.9}$$

From (2.7), it is clear that

$$2\langle \mathbf{a}, \mathbf{x}\rangle \mathbf{a} = (\mathbf{ax} + \mathbf{xa})\,\mathbf{a} = \mathbf{axa} + \mathbf{x}(\mathbf{a})^2 = \mathbf{axa} + \mathbf{x}.$$

So (2.9) becomes

$$\acute{\mathbf{x}} = -\mathbf{axa} \tag{2.10}$$

A rotation is the result of two successive reflections (See Fig. 2.2). From Fig. 2.2, it is clear that $\acute{\acute{\mathbf{x}}}$ is the vector that results from rotating vector **x** through the angle 2ψ about an axis with the direction of the axial vector $\mathbf{a} \times \mathbf{b}$. We can rewrite this relation in the form:

$$\acute{\acute{\mathbf{x}}} = -\mathbf{b}\acute{\mathbf{x}}\mathbf{b} = \mathbf{baxab}, \text{ or}$$

$$\acute{\acute{\mathbf{x}}} = \mathbf{R}^{-1}\mathbf{xR} \quad \text{where} \quad \mathbf{R} = \mathbf{ab}. \tag{2.11}$$

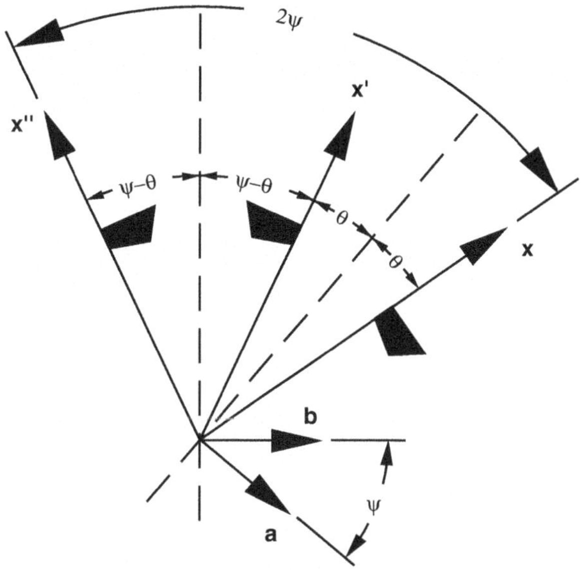

Fig. 2.2 When **x** is subjected to two successive reflections first with respect to a plane perpendicular to **a** and then with respect to a plane perpendicular to **b**, the result is a rotation of **x** about an axis in the direction of $\mathbf{a} \times \mathbf{b}$. The angle of rotation is twice the angle between **a** and **b**

It is useful to explicitly compute the product **ab** and interpret the separate components. If

$$\mathbf{a} = a^1\mathbf{e}_1 + a^2\mathbf{e}_2 + a^3\mathbf{e}_3,$$

and

$$\mathbf{b} = b^1\mathbf{e}_1 + b^2\mathbf{e}_2 + b^3\mathbf{e}_3,$$

then from (2.7) and (2.8):

$$\begin{aligned}\mathbf{R} = \mathbf{ab} &= \frac{1}{2}(\mathbf{ab} + \mathbf{ba}) + \frac{1}{2}(\mathbf{ab} - \mathbf{ba}) \\ &= \mathbf{I}\,\langle \mathbf{a}, \mathbf{b} \rangle + \mathbf{a} \wedge \mathbf{b}.\end{aligned}$$

Since both **a** and **b** are vectors of unit length, $\langle \mathbf{a}, \mathbf{b} \rangle = \cos\psi$. Furthermore, the magnitude of $\mathbf{a} \times \mathbf{b}$ is $\sin\psi$. Although $\mathbf{a} \wedge \mathbf{b}$ unlike $\mathbf{a} \times \mathbf{b}$ is a 2-vector, $\mathbf{a} \wedge \mathbf{b}$ has the same three components as $\mathbf{a} \times \mathbf{b}$. For this reason, we can write

$$\mathbf{a} \wedge \mathbf{b} = \left(n^1\mathbf{e}_{23} + n^2\mathbf{e}_{31} + n^3\mathbf{e}_{12}\right)\sin\psi,$$

where n^1, n^2, and n^3 are the direction cosines of the axial vector $\mathbf{a} \times \mathbf{b}$. With this thought in mind, we have

$$\mathbf{R} = \mathbf{I}\cos\psi + (n^1\mathbf{e}_{23} + n^2\mathbf{e}_{31} + n^3\mathbf{e}_{12})\sin\psi.$$

Note! These ideas can be generalized to higher dimensions. For higher dimensions the entity $\mathbf{a} \wedge \mathbf{b}$ remains well defined, while $\mathbf{a} \times \mathbf{b}$ becomes meaningless. In higher dimensions, you no longer have an axis of rotation; so you must think of the rotation as occurring in the 2-dimensional plane spanned by **a** and **b**.

We should note that ψ represents $\frac{1}{2}$ the angle of rotation. If θ is the actual angle of rotation, we then have

$$\mathbf{R} = \mathbf{I}\cos\frac{\theta}{2} + (n^1\mathbf{e}_{23} + n^2\mathbf{e}_{31} + n^3\mathbf{e}_{12})\sin\frac{\theta}{2}. \tag{2.12}$$

To obtain $\mathbf{R}^{-1}$ from **R**, one can replace θ by $-\theta$ or reverse the order of the Dirac vectors. In either case,

$$\mathbf{R}^{-1} = \mathbf{I}\cos\frac{\theta}{2} - (n^1\mathbf{e}_{23} + n^2\mathbf{e}_{31} + n^3\mathbf{e}_{12})\sin\frac{\theta}{2}. \tag{2.13}$$

Returning to (2.11), we see that there appears to be two representations for the same rotation. In the context of (2.11), **R** is equivalent to $-\mathbf{R}$. From (2.12), we see that changing the sign of **R** is equivalent to replacing θ by $\theta + 2\pi$. Indeed, the operator **R** does not have the expected periodicity of 2π, but it does have a periodicity of 4π. One's first reaction is to think that Clifford algebra has introduced an undesirable complication. In the context of (2.11), this may be the case. However, there are circumstance for which this "complication" corresponds to physical reality. We will discuss this point in the next section.

Meanwhile, we note that for k reflections:

$$\acute{\mathbf{x}} = (-1)^k \mathbf{a}_k\mathbf{a}_{k-1}\ldots\mathbf{a}_1\mathbf{x}\mathbf{a}_1\mathbf{a}_2\ldots\mathbf{a}_k = (-1)^k\mathbf{T}^{-1}\mathbf{x}\mathbf{T}. \tag{2.14}$$

2.1.4 Quaternions

Using quaternions, you can represent a rotation operator in a form essentially identical to that which appears in (2.12). What are quaternions? They were invented (discovered?) by William Rowan Hamilton (1805–1865) in 1843. Before that time, it had been observed that the multiplication of complex numbers could be interpreted as the multiplication of points in a 2-dimensional plane. This was first done by Casper Wessel (1745–1818) in 1797 and then again independently by Jean Robert Argand (1768–1822) in 1806 (Kramer 1981, pp. 72–73). In particular, instead of writing:

$$(a + ib)(c + id) = (ac - bd) + i(ad + bc), \text{ one can write,}$$

$$(a, b)(c, d) = (ac - bc, ad + bc).$$

The question that Hamilton asked himself was, "Could there be a 3-dimensional version of this multiplication that would be useful for the study of physics?" Since his idea was to generalize the notion of complex numbers, he was investigating triples of the form: $a + \mathbf{i}b + \mathbf{j}c$. You can invent all kinds of multiplication rules, but he was looking for a rule that would be meaningful and useful for the study of physics. Starting in 1828, he spent 15 years on this project without success. Finally on October 16, 1843 (a Monday), he had an eureka experience. He was walking along side of the Royal Canal in Dublin with his wife to preside at a Council meeting of the Royal Irish Academy. Then it dawned on him that he should introduce a fourth dimension. In this joyful moment, he carved the formulas for multiplying numbers of the form: $a + \mathbf{i}b + \mathbf{j}c + \mathbf{k}d$ on a stone of the Broome Bridge (or Brougham Bridge as he called it). ((O'Connor and Robertson: Hamilton) and (Boyer 1968, p. 625)).

Time has obliterated the original carving but in 1958, the Royal Irish Academy erected a plaque commemorating the event:

Here as he walked by
on the 16th of October1843
Sir William Rowan Hamilton
in a flash of genius discovered
the fundamental formula for
quaternion multiplication
$\mathbf{i}^2 = \mathbf{j}^2 = \mathbf{k}^2 = \mathbf{ijk} = -1$
and cut it in a stone on this bridge.

From the formula that Hamilton carved in stone, it can be shown that

$$\mathbf{jk} = -\mathbf{kj} = \mathbf{i},\ \ \mathbf{ki} = -\mathbf{ik} = \mathbf{j},\text{ and } \mathbf{ij} = -\mathbf{ji} = \mathbf{k}.$$

(See Prob. 3.)

Due to this achievement, William Hamilton is known as the founder of modern "abstract algebra."

In the theory of quaternions, a rotation operator corresponding to that which appears in (2.12) is written in the form:

$$\mathbf{R} = \mathbf{I}\cos\frac{\theta}{2} - (n^1\mathbf{i} + n^2\mathbf{j} + n^3\mathbf{k})\sin\frac{\theta}{2}. \tag{2.15}$$

Comparison with (2.12) suggests that we can identify identify **i**, **j**, and **k**, respectively, with $-\mathbf{e}_{23}$, $-\mathbf{e}_{31}$, and $-\mathbf{e}_{12}$. As mentioned above, the binary relations for quaternion multiplication are:

$$(\mathbf{i})^2 = (\mathbf{j})^2 = (\mathbf{k})^2 = -1, \tag{2.16}$$

$$\mathbf{jk} = -\mathbf{kj} = \mathbf{i}, \tag{2.17}$$

$$\mathbf{ki} = -\mathbf{ik} = \mathbf{j}, \text{ and} \tag{2.18}$$

$$\mathbf{ij} = -\mathbf{ji} = \mathbf{k}. \tag{2.19}$$

You should check that the same equations hold for the corresponding 2-vectors associated with E^3. Namely:

$$(-\mathbf{e}_{23})^2 = (-\mathbf{e}_{31})^2 = (-\mathbf{e}_{12})^2 = -\mathbf{I}, \tag{2.20}$$

$$(-\mathbf{e}_{31})(-\mathbf{e}_{12}) = -(-\mathbf{e}_{12})(-\mathbf{e}_{31}) = (-\mathbf{e}_{23}), \tag{2.21}$$

$$(-\mathbf{e}_{12})(-\mathbf{e}_{23}) = -(-\mathbf{e}_{23})(-\mathbf{e}_{12}) = (-\mathbf{e}_{31}), \tag{2.22}$$

$$\text{and} \quad (-\mathbf{e}_{23})(-\mathbf{e}_{31}) = -(-\mathbf{e}_{31})(-\mathbf{e}_{23}) = (-\mathbf{e}_{12}). \tag{2.23}$$

In Hamilton's formulation, a vector $\mathbf{x}$ is represented as $x^1\mathbf{i}+x^2\mathbf{j} + x^3\mathbf{k}$ and the rotated vector $\acute{\mathbf{x}}$ is computed by the quaternion version of (2.12).

Neither the usual vector formulation nor the Hamilton approach makes a good distinction between an ordinary vector and an *axial* or *pseudo-vector*.

As we have seen, in the formalism of Clifford algebra, an ordinary vector appears as a 1-vector and a plane of rotation appears as a 2-vector. In three dimensions, a 1-vector and a 2-vector both have three components. In the usual vector formalism, they both appear as 1-vectors. In the quaternion formulation, they both appear as 2-vectors.

The distinction between the two entities arises if we consider a reflection. If, for example, we consider a reflection with respect to the y-z plane, we have

$$\acute{\mathbf{x}} = -\mathbf{e}_1\mathbf{x}\mathbf{e}_1.$$

If

$$\mathbf{x} = x^1\mathbf{e}_1 + x^2\mathbf{e}_2 + x^3\mathbf{e}_3, \text{ then}$$

$$\mathbf{x}' = -x^1\mathbf{e}_1 + x^2\mathbf{e}_2 + x^3\mathbf{e}_3.$$

On the other hand, under the same reflection the 2-vector

$$\mathbf{X} = x^1\mathbf{e}_{23} + x^2\mathbf{e}_{31} + x^3\mathbf{e}_{12} = x^1\mathbf{e}_2\mathbf{e}_3 + x^2\mathbf{e}_3\mathbf{e}_1 + x^3\mathbf{e}_1\mathbf{e}_2$$

becomes

$$\acute{\mathbf{X}} = x^1(-\mathbf{e}_1\mathbf{e}_2\mathbf{e}_1)(-\mathbf{e}_1\mathbf{e}_3\mathbf{e}_1) + x^2(-\mathbf{e}_1\mathbf{e}_3\mathbf{e}_1)(-\mathbf{e}_1\mathbf{e}_1\mathbf{e}_1) + x^3(-\mathbf{e}_1\mathbf{e}_1\mathbf{e}_1)(-\mathbf{e}_1\mathbf{e}_2\mathbf{e}_1) \text{ or}$$

$$\acute{\mathbf{X}} = x^1\mathbf{e}_{23} - x^2\mathbf{e}_{31} - x^3\mathbf{e}_{12}.$$

This same distinction is carried out in the usual vector formulation but in a somewhat awkward fashion. Let us consider the cross product $\mathbf{x} \times \mathbf{y}$. Suppose

$$\mathbf{x} = x^1\mathbf{e}_1 + x^2\mathbf{e}_2 + x^3\mathbf{e}_3, \text{ and}$$

$$\mathbf{y} = y^1\mathbf{e}_1 + y^2\mathbf{e}_2 + y^3\mathbf{e}_3, \text{ then}$$

$$\mathbf{x} \times \mathbf{y} = (x^2y^3 - x^3y^2)\mathbf{e}_1 + (x^3y^1 - x^1y^3)\mathbf{e}_2 + (x^1y^2 - x^2y^1)\mathbf{e}_3.$$

How should the cross product transform under a reflection with respect to the y-z plane? If we treat $\mathbf{x} \times \mathbf{y}$ as an ordinary vector, then

$$(\mathbf{x} \times \mathbf{y})' = -(x^2y^3 - x^3y^2)\mathbf{e}_1 + (x^3y^1 - x^1y^3)\mathbf{e}_2 + (x^1y^2 - x^2y^1)\mathbf{e}_3.$$

On the other hand, if we carry out the same reflection on $\mathbf{x}$ and $\mathbf{y}$ before computing the cross product, we have

$$\begin{aligned}
\acute{\mathbf{x}} &= -\mathbf{x}^1\mathbf{e}_1 + \mathbf{x}^2\mathbf{e}_2 + \mathbf{x}^3\mathbf{e}_3, \\
\acute{\mathbf{y}} &= -\mathbf{y}^1\mathbf{e}_1 + \mathbf{y}^2\mathbf{e}_2 + \mathbf{y}^3\mathbf{e}_3, \text{ and} \\
\acute{\mathbf{x}} \times \acute{\mathbf{y}} &= (x^2y^3 - x^3y^2)\mathbf{e}_1 - (x^3y^1 - x^1y^3)\mathbf{e}_2 - (x^1y^2 - x^2y^1)\mathbf{e}_3.
\end{aligned}$$

When this second interpretation of the impact of a reflection on $\mathbf{x} \times \mathbf{y}$ is applied, $\mathbf{x} \times \mathbf{y}$ is said to be an *axial* or *pseudo-vector.* In the context of Clifford algebra a pseudo-vector is a 2-vector and this awkwardness disappears. Similarly, the entity $\langle \mathbf{x} \times \mathbf{y}, \mathbf{z} \rangle$, which is referred to as a *pseudo-scalar* in the usual vector formulation, becomes a 3-vector in Clifford algebra.

In three dimensions, it is still useful to use the usual *cross product*, when one seeks a vector that is perpendicular to a plane spanned by two vectors such as $\mathbf{x}$ and $\mathbf{y}$. Thus, we will still use the usual definition:

$$\mathbf{x} \times \mathbf{y} = (x^2y^3 - x^3y^2)\mathbf{e}_1 + (x^3y^1 - x^1y^3)\mathbf{e}_2 + (x^1y^2 - x^2y^1)\mathbf{e}_3.$$

However, we will also need the notion of a *wedge product* that we defined in (2.8). Namely:

$$\mathbf{x} \wedge \mathbf{y} = \frac{1}{2}(\mathbf{xy} - \mathbf{yx}) = (x^2y^3 - x^3y^2)\mathbf{e}_{23} + (x^3y^1 - x^1y^3)\mathbf{e}_{31} + (x^1y^2 - x^2y^1)\mathbf{e}_{12}.$$

In closing this section, we wish to bring to your attention the notion of *orthogonal transformations.* An orthogonal transformation is simply a product of reflections. This terminology is chosen when one wishes to focus on the fact that the standard scalar product in E^n is preserved. In this chapter, we have restricted ourselves to E^3. In this context, it is appropriate that you verify the fact that products of reflections do indeed preserve the scalar product (at least in E^3). (See Probs. 6 and 7.)

The product of an even number of reflections (a rotation) is called a *proper orthogonal transformation*, while the product of and odd number of reflections is called an *improper orthogonal transformation.*

Problem 1. From the form of (2.11), it is clear that if the rotation operators $\mathbf{R}$ and $\acute{\mathbf{R}}$ represent two successive rotations, then the combined rotation is represented by the product $\mathbf{R}\acute{\mathbf{R}}$. Use this fact and (2.12) to show that a 90^0 rotation about the y-axis followed by a 90^0 rotation about the x-axis is equivalent to a 120^0 rotation about the axis, which has the direction of the vector $(1, 1, 1)$.

Problem 2. There are many representations that can be used for $\mathbf{e}_1$, $\mathbf{e}_2$, and $\mathbf{e}_3$. One convenient representation is that using *Pauli matrices* $\boldsymbol{\sigma}_1$, $\boldsymbol{\sigma}_2$, and $\boldsymbol{\sigma}_3$. That is, we can let

$$\mathbf{e}_1 = \sigma_1 = \begin{bmatrix} 0 & 1 \\ 1 & 0 \end{bmatrix}, \quad \mathbf{e}_2 = \sigma_2 = \begin{bmatrix} 0 & -i \\ i & 0 \end{bmatrix}, \quad \text{and} \quad \mathbf{e}_3 = \sigma_3 = \begin{bmatrix} 1 & 0 \\ 0 & -1 \end{bmatrix}.$$

Show that in this representation, (2.4) and (2.5) are satisfied.

Problem 3. If you assume associativity for the multiplication of quaternions, then using the equations that appears on Hamilton's plaque, we have

$$\mathbf{ijk} = -1 \Rightarrow (\mathbf{i})^2\mathbf{jk} = -\mathbf{i} \Rightarrow -\mathbf{jk} = -\mathbf{i} \Rightarrow \mathbf{jk} = \mathbf{i}.$$

(a) In a similar fashion, show

$$\mathbf{ki} = \mathbf{j} \text{ and } \mathbf{ij} = \mathbf{k}.$$

(b) Also show that

$$\mathbf{kj} = -\mathbf{i}, \ \mathbf{ik} = -\mathbf{j}, \text{ and } \mathbf{ji} = -\mathbf{k}.$$

Problem 4. In the representation introduced in Prob. 2, the quaternions **i**, **j**, and **k** are represented by complex 2×2 matrices. In particular,

$$\mathbf{i} = -\mathbf{e}_{23} = -i\boldsymbol{\sigma}_1 = \begin{bmatrix} 0 & -i \\ -i & 0 \end{bmatrix}, \quad \mathbf{j} = -\mathbf{e}_{31} = -i\boldsymbol{\sigma}_2 = \begin{bmatrix} 0 & -1 \\ 1 & 0 \end{bmatrix},$$

$$\text{and } \mathbf{k} = -\mathbf{e}_{12} = -i\boldsymbol{\sigma}_3 = \begin{bmatrix} -i & 0 \\ 0 & i \end{bmatrix}.$$

In this representation, the rotation operator

$$\begin{aligned} \mathbf{R} &= \mathbf{I}\cos\frac{\theta}{2} + (\mathbf{e}_{23}n^1 + \mathbf{e}_{31}n^2 + \mathbf{e}_{12}n^3)\sin\frac{\theta}{2} \\ &= \begin{bmatrix} \cos\frac{\theta}{2} + in^3\sin\frac{\theta}{2} & (n^2 + in^1)\sin\frac{\theta}{2} \\ -(n^2 - in^1)\sin\frac{\theta}{2} & \cos\frac{\theta}{2} - in^3\sin\frac{\theta}{2} \end{bmatrix}. \end{aligned}$$

Show that in this representation, the matrix representing **R** is unitary and has determinant equal to 1. (From this result, it is clear that the algebraic properties of the double-valued rotation operators for three dimensions can be ascertained by studying the algebraic properties of 2×2 unitary matrices whose determinant is 1. For this reason, the group of double-valued rotation operators is labeled **SU**(2). The letter **U** indicates "unitary". The letter **S** indicates "special", which in the context of group representation theory means the determinant is 1.)

Problem 5. Suppose

$$\mathbf{R} = \mathbf{I}\cos\frac{\theta}{2} + \hat{\mathbf{n}}\sin\frac{\theta}{2}, \quad \text{where}$$

$$\hat{\mathbf{n}} = n^1\mathbf{e}_{23} + n^2\mathbf{e}_{31} + n^3\mathbf{e}_{12}.$$

(a) Using the fact that

$$(n^1)^2 + (n^2)^2 + (n^3)^2 = 1, \text{ show that}$$

$$(\hat{\mathbf{n}})^2 = -1.$$

(b) Show that $\exp\left[\hat{\mathbf{n}}(\frac{\theta}{2})\right] = \mathbf{R}$. Hint: represent $\exp\left[\hat{\mathbf{n}}(\frac{\theta}{2})\right]$ by a Taylor's series and then separate the odd and even odd and even powers $\hat{\mathbf{n}}$.

Problem 6. Suppose $\acute{\mathbf{x}} = -\mathbf{axa}$ and $\acute{\mathbf{y}} = -\mathbf{aya}$, where $\mathbf{a}$ is a unit vector. Show $\langle \acute{\mathbf{x}}, \acute{\mathbf{y}} \rangle = \langle \mathbf{x}, \mathbf{y} \rangle$. (Remember from (2.7), $\langle \mathbf{x}, \mathbf{y} \rangle \, \mathbf{I} = \frac{1}{2}(\mathbf{xy} + \mathbf{yx})$.)

Problem 7. Suppose $\acute{\mathbf{x}} = (-1)^k \mathbf{a}_k \mathbf{a}_{k-1} \ldots \mathbf{a}_1 \mathbf{x} \mathbf{a}_1 \mathbf{a}_2 \ldots \mathbf{a}_k$ and $\mathbf{y}' = (-1)^k \mathbf{a}_k \mathbf{a}_{k-1} \ldots \mathbf{a}_1 \mathbf{y} \mathbf{a}_1 \mathbf{a}_2 \ldots \mathbf{a}_k$. Show $\langle \acute{\mathbf{x}}, \acute{\mathbf{y}} \rangle = \langle \mathbf{x}, \mathbf{y} \rangle$.

2.2 The 4π Periodicity of the Rotation Operator

From the consequences of the last section, we see that if the vector $\mathbf{x}(\theta)$ represents the result of rotating vector $\mathbf{x}(0)$ through an angle θ, then we can represent the rotation in the form:

$$\mathbf{x}(\theta) = \mathbf{R}^{-1}(\theta)\mathbf{x}(0)\mathbf{R}(\theta), \quad \text{where}$$

$$\mathbf{R}(\theta) = \mathbf{I}\cos\frac{\theta}{2} + \hat{\mathbf{n}}\sin\frac{\theta}{2},$$

$$\hat{\mathbf{n}} = n^1\mathbf{e}_{23} + n^2\mathbf{e}_{31} + n^3\mathbf{e}_{12}, \quad \text{and}$$

n^1, n^2, along with n^3 are the direction cosines for the axis of rotation.

Although $\mathbf{x}(\theta)$ has a period of 2π, $\mathbf{R}(\theta)$ has a period of 4π! With the development of quantum mechanics in the 1920s, it became recognized that a 4π periodicity sometimes occurs in nature. To explain the observed structure of the hydrogen energy spectrum, it was necessary to attribute to the electron a spin of $\frac{1}{2}$ and a periodicity of 4π. Later, it became recognized that some objects larger than electrons also have a 4π periodicity (Bolker 1973). A demonstration of this fact has been put forward by Edgar Riefin (1979).

For an object to display a 4π periodicity, it is necessary that it be in some sense attached to its surroundings.

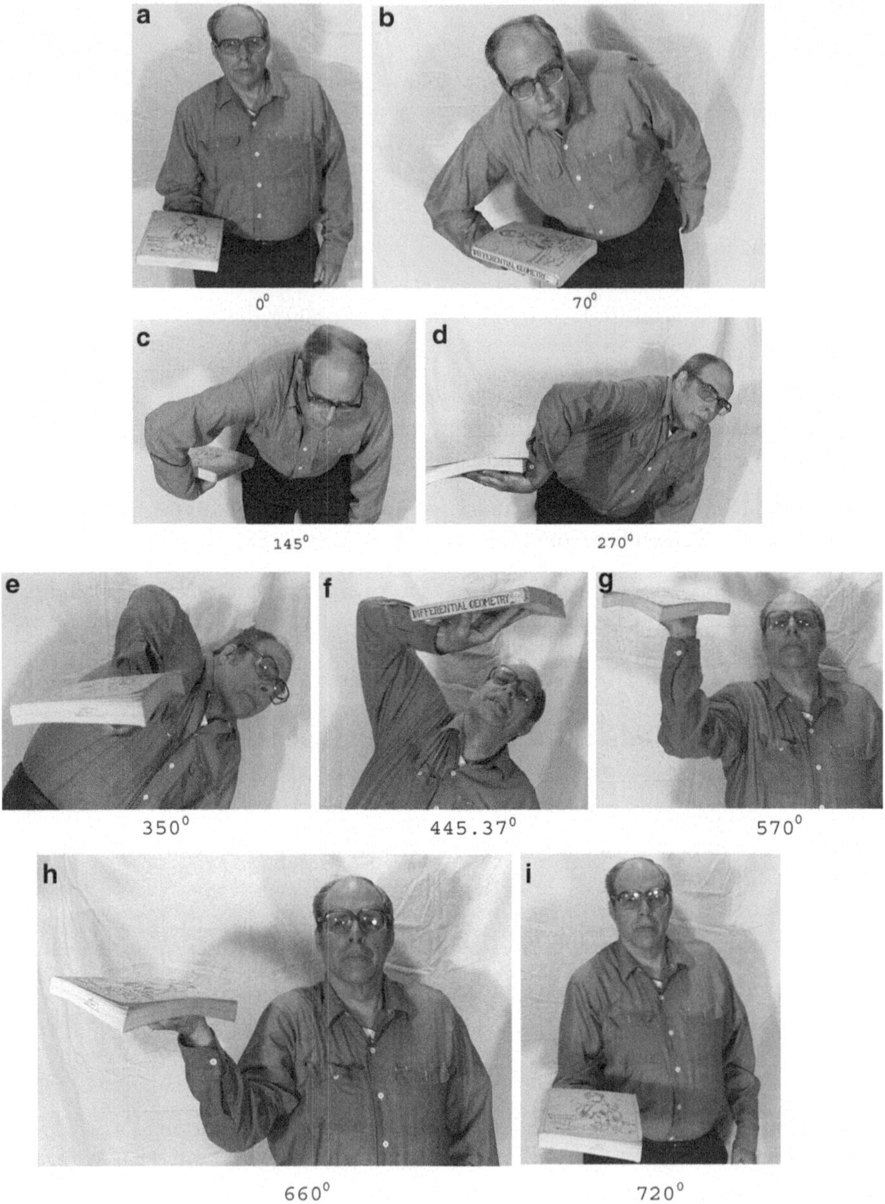

Fig. 2.3 A book with a 4π periodicity

To illustrate this, you may wish to carry out a demonstration. First, hold a glass of water in the palm of your hand. The hand holding the glass may be left or right but it is important that your hand be under the glass with palm up. Then maintain a firm grip on the glass and rotate it 360^0 without moving your feet or spilling any water.

When you have completed this maneuver, you will find yourself in an awkward position with the glass slightly above your head and your elbow pointed upward. Clearly, the relationship of the glass to you is quite different from what it was in its initial position. However, if you continue the rotation, you may be surprised to find that your arm will unwind itself and the glass will return to its initial position with its initial relationship to you. Thus, the glass attached to your arm does not have a 2π periodicity but it does have a 4π periodicity.

This demonstration is shown in Fig. 2.3 where a book is used in place of a glass of water.

2.3 *The Point Groups for the Regular Polyhedrons

One aspect of geometry, which attracts a lot of attention in physics, is symmetry groups. The symmetry of a body can be characterized by the set of transformations that maintain distances between points and bring the body into its original space of occupation. Quite reasonably, these are called *symmetry transformations*. For infinite bodies (for example an infinite crystal lattice), the set of symmetry transformations may contain translations.

But for finite bodies, symmetry transformations are restricted to rotations and products of rotations and reflections. For this reason, Clifford algebra is a good tool to attack the mathematics of symmetry for finite bodies.

Before getting very deep into this topic, it is useful to prove a theorem by Élie Cartan (1938, pp. 13–17; 1966, pp. 10–12). His theorem states that in an n-dimensional space (real or complex), a transformation consisting of any finite number of reflections can also be obtained by a number of reflections that does not exceed n.

In this text, we only need the real 3-dimensional version and that is the only version we will prove.

Theorem 8. *Suppose* $\acute{\mathbf{x}} = (-1)^k \mathbf{a}_k\, \mathbf{a}_{k-1} \ldots \mathbf{a}_1 \mathbf{x} \mathbf{a}_1 \mathbf{a}_2 \ldots \mathbf{a}_k$. *That is we have a transformation consisting of k reflections. Then this same transformation (in E^3) can be achieved by three or fewer reflections.*

Proof. Case 1. The number of reflections k is even. If we multiply an even number of 1-vectors, we get a linear combination of the 0-vector $\mathbf{I}$ and the three 2-vectors $\mathbf{e}_{23}$, $\mathbf{e}_{31}$, and $\mathbf{e}_{21}$. That is

$$\mathbf{a}_1 \mathbf{a}_2 \ldots \mathbf{a}_k = \mathbf{I}\alpha + \mathbf{e}_{23}\beta^1 + \mathbf{e}_{31}\beta^2 + \mathbf{e}_{12}\beta^3.$$

(This already looks like a rotation operator!) The operator $\mathbf{a}_k \mathbf{a}_{k-1} \ldots \mathbf{a}_1$ is essentially the same as $\mathbf{a}_1 \mathbf{a}_2 \ldots \mathbf{a}_k$ except for the fact that the underlying Dirac vectors are in reverse order. Thus,

$$\mathbf{a}_k \mathbf{a}_{k-1} \ldots \mathbf{a}_1 = \mathbf{I}\alpha + \mathbf{e}_{32}\beta^1 + \mathbf{e}_{13}\beta^2 + \mathbf{e}_{21}\beta^3 = \mathbf{I}\alpha - \mathbf{e}_{23}\beta^1 - \mathbf{e}_{31}\beta^2 - \mathbf{e}_{12}\beta^3.$$

Since $(\mathbf{a}_1)^2 = (\mathbf{a}_2)^2 = \ldots = (\mathbf{a}_k)^2 = \mathbf{I}$, $(\mathbf{a}_k\mathbf{a}_{k-1}\ldots\mathbf{a}_1)(\mathbf{a}_1\mathbf{a}_2\ldots\mathbf{a}_k) = \mathbf{I}$, and

$$\begin{aligned}\mathbf{I} &= \left(\mathbf{I}\alpha - \mathbf{e}_{23}\beta^1 - \mathbf{e}_{31}\beta^2 - \mathbf{e}_{12}\beta^3\right)\left(\mathbf{I}\alpha + \mathbf{e}_{23}\beta^1 + \mathbf{e}_{31}\beta^2 + \mathbf{e}_{12}\beta^3\right) \\ &= \mathbf{I}\left((\alpha)^2 + (\beta^1)^2 + (\beta^2)^2 + (\beta^3)^2\right).\end{aligned}$$

Since $(\alpha)^2 + (\beta^1)^2 + (\beta^2)^2 + (\beta^3)^2 = 1$, there exists an angle ψ such that

$$\cos\psi = \alpha \quad \text{and} \quad \sin\psi = \sqrt{(\beta^1)^2 + (\beta^2)^2 + (\beta^3)^2}.$$

Furthermore, if at least one of the β^k's is not zero, we can define the direction cosines for the axis of rotation by

$$n^k = \beta^k / \sqrt{(\beta^1)^2 + (\beta^2)^2 + (\beta^3)^2} = \beta^k / \sin\psi \quad \text{for } k = 1, 2, \text{and } 3.$$

(Note! this definition guarantees that $(n^1)^2 + (n^2)^2 + (n^3)^2 = 1$.) We now have shown:

$$\mathbf{a}_1\mathbf{a}_2\ldots\mathbf{a}_k = \mathbf{I}\cos\psi + (n^1\mathbf{e}_{23} + n^2\mathbf{e}_{31} + n^3\mathbf{e}_{12})\sin\psi.$$

If the $\sin\psi = 0$, $\mathbf{a}_1\mathbf{a}_2\ldots\mathbf{a}_k = \pm\mathbf{I}$. Otherwise, we have a nontrivial rotation operator. From Fig. 2.2, it is clear that this rotation operator can be replaced by a product of two reflections.

Case 2. The number of reflections k is odd.

In this case, we can multiply out the first k-1 reflections to get a rotation operator and we then have:

$$\begin{aligned}\mathbf{a}_1\mathbf{a}_2\ldots\mathbf{a}_k &= \left[\mathbf{I}\cos\psi + (n^1\mathbf{e}_{23} + n^2\mathbf{e}_{31} + n^3\mathbf{e}_{12})\sin\psi\right]\mathbf{a}_k \\ &= \left[\mathbf{I}\cos\psi + (n^1\mathbf{e}_{23} + n^2\mathbf{e}_{31} + n^3\mathbf{e}_{12})\sin\psi\right](k^1\mathbf{e}_1 + k^2\mathbf{e}_2 + k^3\mathbf{e}_3).\end{aligned}$$

If $\sin\psi = 0$ or $k^1n^1 + k^2n^2 + k^3n^3 = 0$, our product $\mathbf{a}_1\mathbf{a}_2\ldots\mathbf{a}_k$ reduces to a 1-vector. Otherwise after factoring the rotation into two reflections, we have the product of three reflections. □

Now we are in a position to have a reasonably intelligent discussion of symmetry groups. Generally, the set of multiple reflections that bring a particular finite body into its original position in space is called a *point group* for two reasons. One is due to the fact that at least one point remains fixed under all symmetry transformations associated with a particular body. The second is due to the fact that the set of the symmetry transformations identified with a particular body forms a mathematical structure known as a *group*.

Definition 9. A group is a set of elements with a binary operation $\circ$ having the following properties:

(1) Closure: $g_1 \in G, g_2 \in G \Rightarrow g_1 \circ g_2 \in G$.
(2) Associativity: $(g_1 \circ g_2) \circ g_3 = g_1 \circ (g_2 \circ g_3)$.

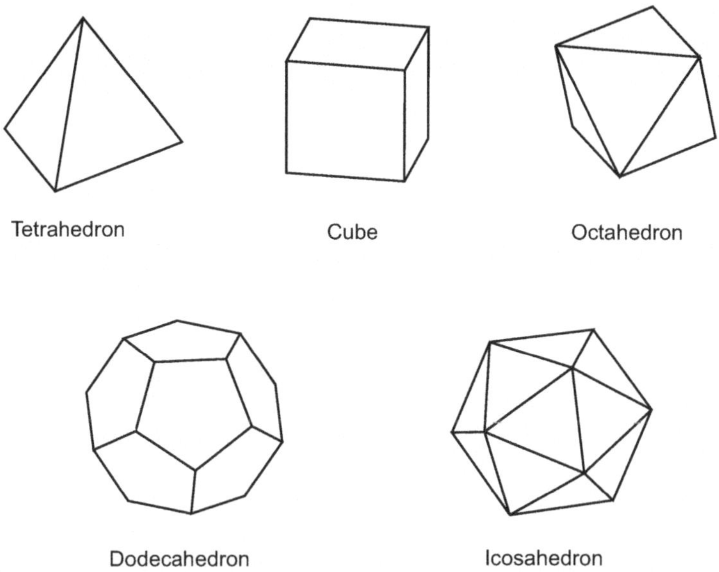

Fig. 2.4 The five regular polyhedrons

(3) Identity element: $\exists$ an element $e \in G$ such that $\forall\, g \in G,\, e \circ g = g \circ e = g$.
(4) Inverse: $\forall\, g \in G, \exists\, g^{-1} \in G$ such that $g \circ g^{-1} = g^{-1} \circ g = e$.

Examples of groups include the integers under addition, the positive rational numbers under multiplication, and nonsingular n×n matrices under matrix multiplication.

We will only give a short description of a few point groups – in particular the five point groups associated with the five regular polyhedrons. (See Fig. 2.4.) For each of the polyhedrons, we have a finite symmetry group. One way to verify we have a group is to run through the check list in the definition above.

The elements of a symmetry group for a finite solid are finite products of reflections. It is clear that the multiplication of two finite products results in a finite product, which preserves the original position of the relevant solid. Thus, the set of symmetry transformations satisfy the property of closure.

The identity element corresponds to the transformation that does nothing or rotates the solid some integral multiple of 360^0.

To obtain the inverse of a product of reflections, one simply constructs the product of the same reflections in the reverse order.

To show that the symmetry groups for the regular polyhedrons have only a finite number of members, let us consider the example of the cube. (See Fig. 2.5.) Applying Cartan's theorem, we know that an even number of reflections (a proper orthogonal transformation) can be reduced to either the identity element or a rotation. The possible symmetry rotations are not difficult to count.

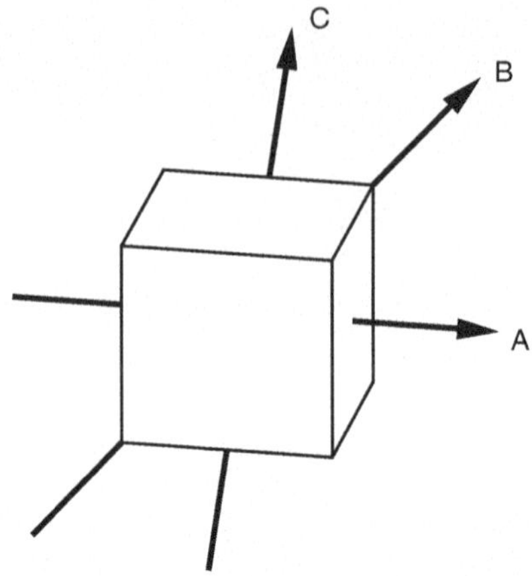

Fig. 2.5 Some symmetry axes of rotation for the cube

Perhaps, the most obvious symmetry rotations are those that correspond to the fourfold axes that pass through the centers of opposite faces. Not counting the 360^0 identity rotation, we have symmetry rotations of 90^0, 180^0, and 270^0. Since there are three such axes, this gives us $3 \times 3 = 9$ elements.

We also have some twofold axes that pass through the midpoints of opposite edges. Since there are 12 edges, there are six such axes and corresponding to each of these axes is a symmetry rotation of 180^0. This accounts for six more elements in the group. Then there are four threefold axes that pass through opposite vertices. This adds another eight members to the group.

Finally, there is the identity transformation. Thus, the total number of proper orthogonal members for the point group associated with the cube is $9 + 6 + 8 + 1 = 24$. (Because any product of reflections has two representations in the Clifford formalism ($\pm$), there are 48 Clifford numbers in the Clifford version of the proper orthogonal group for the cube.)

To obtain the number of improper orthogonal transformations by simply counting them is difficult because some members of this set are not simple reflections but products of three reflections. To complete our counting problem, we wish to apply the following theorem:

Theorem 10. *For a finite point group, the number of improper orthogonal transformations (products of an odd number of reflections) is equal to the number of proper transformations (products of an even number of reflections). Note! For those familiar with group theory, what is proven below is that the set of improper orthogonal transformations is a coset of the subgroup of proper orthogonal transformations.*

Proof. To establish the truth of this theorem, we choose a unit vector **a** corresponding to a simple reflection in the group and then show that any improper orthogonal transformation can be represented uniquely (aside from the sign ambiguity) in the form **Ra** where **R** is a rotation or $\pm$ the identity element **I**.

Consider a product of an odd number of reflections $\mathbf{a}_1\mathbf{a}_2\ldots\mathbf{a}_k$. If k is odd, we can multiply out the first k-1 reflections to get a rotation operator $\mathbf{R}$. So we have

$$\mathbf{a}_1\mathbf{a}_2\ldots\mathbf{a}_k = \mathbf{R}\mathbf{a}_k.$$

If $\mathbf{a}_k = \mathbf{a}$, we are incredibly lucky. Otherwise,

$$\mathbf{R}\mathbf{a}_k = \mathbf{R}\mathbf{a}_k(\mathbf{a})^2 = \mathbf{R}(\mathbf{a}_k\mathbf{a})\mathbf{a} = \mathbf{R}\acute{\mathbf{R}}\mathbf{a} = \acute{\acute{\mathrm{R}}}\mathbf{a}, \text{ where}$$

$$\acute{\acute{\mathrm{R}}} = \mathbf{R}\acute{\mathbf{R}}. \text{ Thus, we have}$$

$$\mathbf{a}_1\mathbf{a}_2\ldots\mathbf{a}_k = \acute{\acute{\mathrm{R}}}\mathbf{a}.$$

To show that this representation is unique aside from the sign ambiguity, suppose

$$\mathbf{R}\mathbf{a} = \pm\acute{\mathbf{R}}\mathbf{a}. \text{ Multiply both sides by } \mathbf{a} \text{ to get}$$

$$\mathbf{R}(\mathbf{a})^2 = \pm\acute{\mathbf{R}}(\mathbf{a})^2 \text{ or } \mathbf{R} = \pm\acute{\mathbf{R}}.$$

□

Applying this theorem to the cube, we see that the point group for the cube has 48 members (96 for the double valued Clifford version).

Using the terminology of group theory, we say the *order* of the point group for the cube is 48.

To get the orders for the point groups of the other polyhedrons, the chief problem is counting the edges and vertices. For example, the dodecahedron is constructed by assembling 12 regular pentagons. Before assembly, the 12 pentagons have a total of $12 \times 5 = 60$ edges. When assembled, one edge from one pentagon and one edge from a second pentagon align to become a single edge of the dodecahedron. Thus, the dodecahedron has $60/2 = 30$ edges, which correspond to $30/2 = 15$ twofold axes. Similarly, the 60 vertices of the 12 pentagons become $60/3 = 20$ vertices for the dodecahedron. In turn, this corresponds to ten threefold axes.

For four of the five regular polyhedrons, the axes of symmetry pass through pairs of faces, pairs of edges, or pairs of vertices. The one exception is the tetrahedron. For the tetrahedron, the twofold axes do indeed correspond to pairs of edges. However for the threefold axes, the situation is different. For the tetrahedron, each threefold axis passes through one vertex and one face.

When you determine the orders of the point groups (See Prob. 12.), you will see that the order of the point group for the cube is identical to the order of the point group for the octahedron. This raises the possibility that the two groups are *isomorphic*. Two groups are said to be isomorphic if one can set up a one-to-one correspondence between the groups is such a way that if $\mathbf{x}$ in one group corresponds to $\acute{\mathbf{x}}$ in the second group and $\mathbf{y}$ corresponds to $\acute{\mathbf{y}}$ then $\mathbf{x} \circ \mathbf{y}$ corresponds to $\acute{\mathbf{x}} \circ \acute{\mathbf{y}}$. For the cube and the octahedron, this is plausible because the numbers of fourfold, threefold, and twofold axes match up in the two groups. Nonetheless, it would

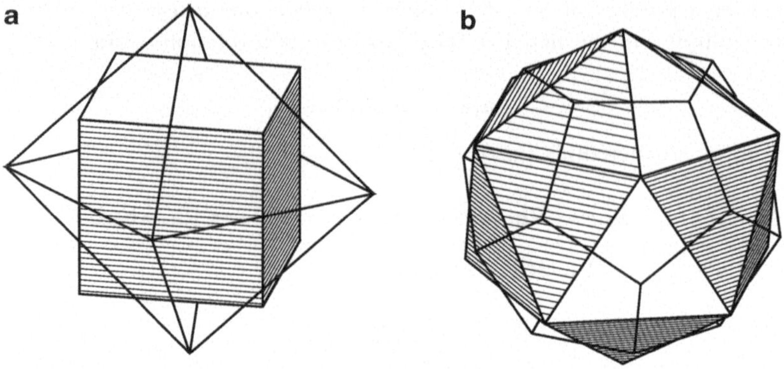

Fig. 2.6 (**a**) A cube aligned with a skeleton frame of an octahedron. (**b**) An icosahedron aligned with a skeleton frame of a dodecahedron

be very difficult to determine an isomorphic correspondence without resorting to geometry. However using geometry, it becomes a trivial exercise to establish the isomorphism. One merely matches the vertices of one with the face centers of the other. In Fig. 2.6a, we have aligned a cube with the skeleton frame of an octahedron is such a way that the symmetry axes of rotation for the two polyhedrons coincide. Thus we see that a proper symmetry transformation for one of the polyhedrons is a proper symmetry transformation for the other. The two point groups also contain the same improper symmetry transformations. (See Prob. 15.) Thus, the two point groups are isomorphic.

In Fig. 2.6b, we have aligned an icosahedron with the skeleton frame of a dodecahedron with similar consequences.

One can also demonstrate geometrically that the point group for the tetrahedron is a subgroup of the point groups for the other polyhedrons so that any symmetry transformation of the tetrahedron is also a symmetry transformation of the other polyhedrons.

One can imbed a tetrahedron inside a cube so that the threefold axes for the two polyhedrons coincide. (See Fig. 2.7a.) The twofold axes of the tetrahedron do not coincide with the twofold axes of the cube. However, the twofold axes of the tetrahedron do coincide with the fourfold axes of the cube. Thus, it becomes clear that any proper orthogonal transformation in the point group for the tetrahedron belongs to the point group for the cube. It can also be said that any improper transformation belonging to the point group for the tetrahedron is also an improper transformation belonging to the point group for the cube. (See Prob. 16.) Thus, it is clear that the point group for the tetrahedron is a subgroup of the point group for the cube.

It is more difficult to visualize but the point group for the tetrahedron is also a subgroup of the dodecahedron (or icosahedron). (See Fig. 2.7b.)

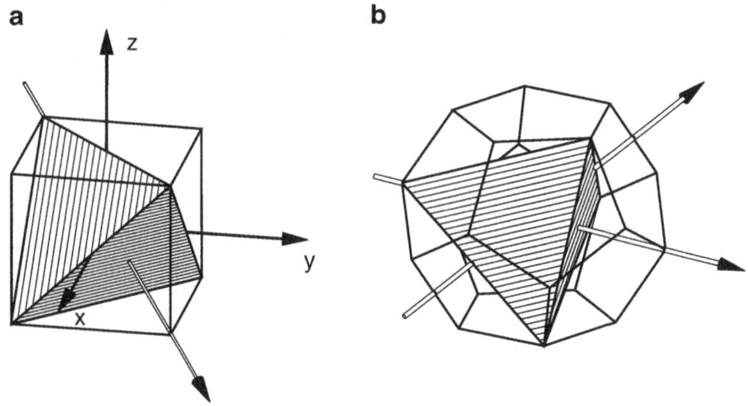

Fig. 2.7 (**a**) A tetrahedron aligned with the skeleton frame of a cube. (**b**) A tetrahedron aligned with the skeleton frame of a dodecahedron

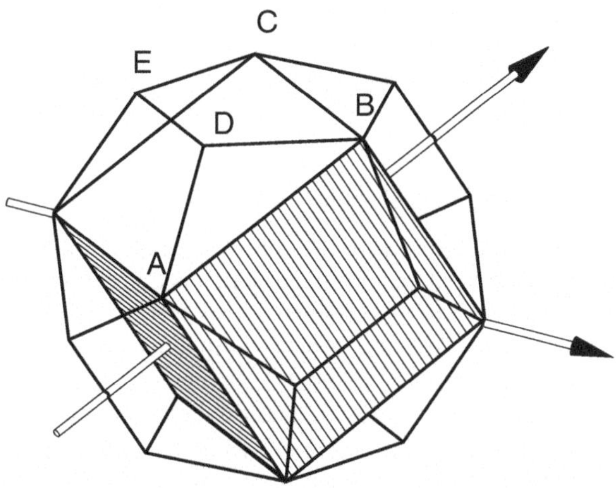

Fig. 2.8 A cube aligned with the skeleton frame of a dodecahedron.

It is also enlightening to examine Fig. 2.8. You may not be convinced that connecting some of the vertices of the dodecahedron as shown in Fig. 2.8 results in the edges of a cube. However, it should be clear that the direction of line segment AB is perpendicular to the direction of line segment DE. Furthermore, line segment DE is parallel to line segment BC. Thus, the edges of our suspect cube do indeed meet at right angles at each vertex. By studying the alignment of the various symmetry axes of rotation in Fig. 2.8, we reach the conclusion that the intersection of the point group for the cube (or octahedron) and the point group for the dodecahedron (or icosahedron) is the point group for the tetrahedron.

Problem 11. Prove that there are no more than five regular polyhedrons. Hint: What is the maximum number of equilateral triangles that can share a single vertex?

Problem 12. Determine the orders of the point groups for the tetrahedron, octahedron, dodecahedron, and icosahedron. Are your results consistent with Figs. 2.6a and 2.6b?

Problem 13. How does the result of Prob. 1 relate to the point group for the cube? What is the consequence of two successive 90^0 rotations about two non-aligned fourfold axes?

Problem 14. In view of Fig. 2.7a, the three twofold axes of the tetrahedron can be aligned with the **x**, **y**, and **z** axes. Suppose we designate a 180^0 rotation about the x-axis by $\mathbf{R}_x = \pm\mathbf{e}_{23}$. Suppose we also define $\mathbf{R}_y$ and $\mathbf{R}_z$ in a similar manner. Complete the following table:

$\circ$	$\mathbf{I}$	$\mathbf{R}_x$	$\mathbf{R}_y$	$\mathbf{R}_z$
$\mathbf{I}$				
$\mathbf{R}_x$	$\mathbf{R}_x$			
$\mathbf{R}_y$				
$\mathbf{R}_z$				

You will find that the 180^0 rotations commute, although the Clifford representations do not.

Problem 15. Consider Fig. 2.6a.

(a) Draw the figure with the cube and octahedron aligned with the **x**, **y**, and **z** axes.
(b) Describe a plane of reflection that is common to both the cube and octahedron.
(c) It has already been pointed out that if the cube and the octahedron are aligned as in Fig. 2.6a, the proper orthogonal symmetry transformations for the two point groups are identical. Use your result in part b) to show that the improper symmetry transformations for the two point groups are identical.
(d) Explain why the improper symmetry transformations for the icosahedron are the same as the improper symmetry transformations for the dodecahedron.

Problem 16. (a) Prove that any improper orthogonal symmetry transformation for the tetrahedron is also an improper orthogonal symmetry transformation for the cube. (If you get stuck, review the approach used in the proof of Theorem 10.)
(b) Explain why any improper orthogonal symmetry for the tetrahedron is also an improper orthogonal symmetry transformation for the dodecahedron (or icosahedron).

Problem 17. If a tetrahedron is aligned with the **x**, **y**, and **z** axes as shown in Fig. 2.7a, then the rotations about the threefold axis shown are

$$\pm\left[\mathbf{I}\cos 60^0 + \sin 60^0\left(\frac{1}{\sqrt{3}}\mathbf{e}_{23} + \frac{1}{\sqrt{3}}\mathbf{e}_{31} - \frac{1}{\sqrt{3}}\mathbf{e}_{12}\right)\right] = \pm\frac{1}{2}\left[\mathbf{I} + \mathbf{e}_{23} + \mathbf{e}_{31} - \mathbf{e}_{12})\right]$$

and

$$\pm\left[\mathbf{I}\cos 120^0 + \sin 120^0\left(\frac{1}{\sqrt{3}}\mathbf{e}_{23} + \frac{1}{\sqrt{3}}\mathbf{e}_{31} - \frac{1}{\sqrt{3}}\mathbf{e}_{12}\right)\right] = \mp\frac{1}{2}\left[\mathbf{I} - \mathbf{e}_{23} - \mathbf{e}_{31} + \mathbf{e}_{12})\right]$$
$$= \pm\frac{1}{2}\left[\mathbf{I} - \mathbf{e}_{23} - \mathbf{e}_{31} + \mathbf{e}_{12})\right].$$

(a) List all of the rotations for both the twofold and threefold axes. (Don't compute them all – after computing a few, you should see patterns.)
(b) Write down the Clifford representation of a reflection and use this to construct a list of the improper orthogonal symmetry for the tetrahedron.
(c) In the list constructed in part b), which are simple reflections and which cannot be achieved by fewer than three reflections?

Problem 18. Euler's Formula

In 1750, Leonard Euler made the conjecture that for any convex polyhedron, $F - E + V = 2$, where F equals the number of faces, E equals the number of edges, and V equals the number of vertices (James 2002, p. 5). Determine whether this formula is valid for the five regular polyhedrons. Suppose you slice off a corner of a cube. Does the resulting solid satisfy Euler's formula?

2.4 *Élie Cartan 1869–1951

The way mathematicians deal with differential geometry was significantly altered by the work of Élie Cartan. In 1993, the American Mathematical Society published a 301-page translation from Russian of a summary of his work. This short biography is extracted from that source.

The authors of that summary are two Russian mathematicians: M.A. Akivis and B.A. Rosenfeld (1993). Élie Cartan's contributions to mathematics are so deep and broad that these two accomplished geometers felt compelled to include a virtual apology in their preface: "Of course the authors are only able to describe in detail Cartan's results connected with those branches of geometry in which the authors are experts." (Akivis and Rosenfeld 1993, p. xi).

Élie Joseph Cartan was born on April 9, 1869 in Dolomieu, a small village in southeastern France of less than 2,000 people. At the time of his birth, no one would have predicted that Élie Cartan would become a world renowned mathematician. His father was a blacksmith. His older sister, Jeanne-Marie, became a dressmaker, and his younger brother, Leon, would eventually join the family business as another blacksmith.

Élie seemed destined for a similar career in rural France until a fateful visit to Élie's elementary school by the up and coming politician, Antonin Dubost (1844–1921). This event would change Élie's direction in life.

When Élie's teachers described their very remarkable student to Dubost, Dubost encouraged the young Cartan to compete for a scholarship at a more competitive lycée. Antonin Dubost eventually became the Minister of Justice under one administration and later became President of the French Senate for what was essentially the last 14 years of his life. Throughout his life, Antonin Dubost maintained a fatherly interest in Cartan's career.

To help Élie obtain the desired scholarship, one of his teachers, M. Dupuis, supervised his preparation for the required exam. Cartan scored well on the exam, received the scholarship, and left home at the age of 10.

At the age of 17, Cartan decided to become a mathematician and enrolled at l'École Normale Supérieure in Paris. During the next three years, Cartan not only attended lectures at l'École Normale Supérieure but also at the Sorbonne. In this way, he became exposed to many outstanding mathematicians including Henri Poincaré. After graduation, he was drafted into the French army for one year. He then returned to Paris and received his doctorate at the Sorbonne two years later in 1894 while attracting the attention of prominent mathematicians including Sophus Lie at Leipzig University in Germany.

Early in his career, Cartan developed aspects of Lie groups and Lie algebras that could be applied to differential geometry. Later, his work on differential forms led him to develop methods that are now commonly used to deal with differential equations. In 1910, Cartan began to perfect the method of moving frames to deal with problems in differential geometry (Cartan 1910a, 1910b). (You will encounter this method in later chapters of this book.)

In 1915, when Cartan was 46, he was again drafted into the French army soon after World War I broke out. However, he was not sent to the front. Instead, he was assigned to a hospital set up in the building of l'École Normale Supérieure. This situation allowed him to continue his mathematical research during the war years.

During these same war years, Einstein living in Berlin, discovered that a slight variation of Riemannian geometry was necessary to express his general theory of relativity. After the war, Einstein and others sought out mathematical structures that could be used to construct a unified field theory. With this motivation, Cartan turned his attention to extracting properties of more general geometric spaces that might be useful. (His correspondence with Einstein was edited by Robert Debever and published by Princeton University Press in 1979 under the title *Élie Cartan and Albert Einstein: Letters on Absolute Parallelism, 1929–1932.*)

To summarize, Cartan was prolific. Akivis and Rosenfeld attribute over 200 publications to Cartan, and this includes several books that have been republished in recent years.

Cartan was also successful as a family person. In 1903, he married Marie-Louise Bianconi (1880–1950) and soon became the father of three sons: Henri (1904–2008), Jean (1906–1932), and Louis (1909–1943). Later Élie and Marie-Louise had a daughter Hélene (1917–1952). His first son, Henri, became a world renowned mathematician in his own right. (Henri Cartan died on August 13, 2008 at the age of 104!) His second son, Jean, seemed headed for a promising career as a music composer but he died of tuberculosis at the age of 25. The third son, Louis, was a talented physicist, but during World War II, he was arrested by Vichy government police for his activities in the French resistance. He was then turned over to the Germans who held him in captivity for 15 months before executing him by decapitation. The daughter Hélene taught mathematics at several lycées and authored several math papers before she died at the age of 34.

During most of his adult life, Élie Cartan made his home in Paris or within commuting distance of Paris. He had spent much of his boyhood away from his hometown but he always maintained his ties there. He encouraged his younger sister Anna to pursue a career in math education. She taught at several secondary schools for girls and authored two textbooks, which were reprinted many times.

In 1909, Cartan built a vacation home in Dolomieu and sometimes he could be seen at the family blacksmith shop helping his father and brother to blow the blacksmith bellows.

Cartan's sister Anna and daughter Hélene were not the only women to receive Cartan's encouragement to study mathematics. After he retired from his professorial position at the Sorbonne in 1940, he devoted the last years of his life in his 70s to teaching mathematics at the École Normale Supérieure for girls.

After a long illness, he died in Paris on May 6, 1951.

2.5 *Suggested Reading

Milton Hamermesh 1962. *Group Theory.* Reading, Massachusetts, U.S.A: Addison-Wesley Publishing Company, Inc. Also reprint edition 1990. New York. Dover Publications, Inc.

The second chapter is devoted to the point groups.

Leo Dorst, Chris Doran, and Joan Lasenby (Editors) 2002. *Applications of Geometric Algebra in Computer Science and Engineering.* Boston: Birkhäuser. Chapter I entitled "Point Groups and Space Groups in Geometric Algebra"by David Hestenes is devoted to the application of Geometric Algebra (Clifford Algebra) to the classification of symmetry groups.

D.M.Y. Sommerville 1958. *An Introduction to the Geometry of N Dimensions.* New York: Dover Publications, Inc.

This book includes a discussion of regular polyhedrons in higher dimensions.

Chapter 3
Clifford Algebra in Minkowski 4-Space

3.1 A Small Dose of Special Relativity

When the speed of sound is measured, it is found that the speed is independent of direction only if it is measured with respect to the air. If the air is moving at a rate of 20 km per hour, an observer on the ground will discover that sound moving in the direction of the wind will move 20 km per hour faster than it would when the air is still. Similarly, sound moving against the wind will be slowed down.

During the nineteenth century, it was generally believed that light traveled through some kind of "ether" in much the same way as sound travels through air. In 1881, in an effort to measure the velocity of this ether with respect to earth, Albert Michelson designed an experiment that would compare the speed of light in different directions (1881). The result of the experiment was that the speed of light was the same in all directions. The experiment was refined and repeated 6 years later by Albert Michelson and Edward Morley, but the result was the same (1887).

It might be argued that at the time the measurements were taken, the earth was moving downstream in the same direction and with the same speed as the ether. However, in the course of a year as the earth moves around the sun, the earth hurtling through space in different directions and Michelson and Morley got the same null result at different times of the year.

To this day, there are some holdouts who argue that the ether is dragged along by the earth. However, the overwhelming majority of physicists have abandoned the ether concept.

Aside from the result of Michelson and Morley, there were also some anomalies that appeared in the study of Maxwell's equations for electromagnetic fields. According to Maxwell's equations, a magnetic field is generated by a moving charge. But what about an observer who moves with the same velocity as the charge? For such an observer, the charge is stationary. However, if the magnetic field exists, such an observer should be able to measure it.

J. Snygg, *A New Approach to Differential Geometry using Clifford's Geometric Algebra*, DOI 10.1007/978-0-8176-8283-5_3,

To resolve this kind of paradox, some mathematically inclined physicists investigated the mathematical symmetries of Maxwell's equations. Hendrik A. Lorentz (1904), Henri Poincaré (1905), and Albert Einstein (1905) published separate papers that presented a set of equations, which have since become known as the Lorentz transformation.

It was Einstein who saw that these equations made sense without the concept of ether. For this reason, he is generally credited with the introduction of the special theory of relativity.

The derivation shown below was lifted from *Modern University Physics* by Richards et al. (1960).

The Lorentz transformation can be derived from the assumption that the speed of light in a vacuum is a constant independent of direction for any observer who is moving at constant speed with respect to the source. Suppose we station a team of physicists on top of a speeding freight train and another team of physicists on the ground alongside the railroad track. Both teams adjust their watches so that when the exact middle of the train passes a designated point next to the track, all watches will agree that the time is zero. At the moment, the middle of the train is aligned with the designated point, a flash is set off either on board the train at the exact middle or at the designated point. (It does not matter.)

If the team on board monitors the position of the front edge of the expanding light wave, they will find that they are dealing with an expanding sphere whose radius is increasing with the constant speed of light c. Thus, if $(\bar{t}, \bar{x}^1, \bar{x}^2, \bar{x}^3)$ represents the coordinates of a point on the expanding sphere in a 4-dimensional space-time frame observed by the physicists on board, then

$$(\bar{x}^1)^2 + (\bar{x}^2)^2 + (\bar{x}^3)^2 = (c\bar{t})^2. \tag{3.1}$$

The team of physicists on board would discover that the wave front would arrive at the front end of the train and the rear end simultaneously.

For the team on the ground the result would be quite similar. That is, they also would observe that the wave front is an expanding sphere. Thus if (t, x^1, x^2, x^3) represents the coordinates of a point on the wavefront observed by the team on the ground, then

$$(x^1)^2 + (x^2)^2 + (x^3)^2 = (ct)^2. \tag{3.2}$$

However in this frame, the wavefront would arrive at the rear end of the train before it arrived at the front end. What are simultaneous events in one frame are not necessarily simultaneous in an alternate frame. This is a little baffling but it is a consequence of what is meant by the speed of light being constant.

If the train is speeding in the x^1 direction, then according to Newtonian mechanics, the relationship between the two coordinate systems is known as the *Galilean transformation*

$$t = \bar{t}, \tag{3.3}$$

$$x^1 = \bar{x}^1 + v\bar{t}, \tag{3.4}$$

$$\bar{x}^1 = x^1 - vt, \tag{3.5}$$

$$x^2 = \bar{x}^2, \text{ and} \tag{3.6}$$

$$x^3 = \bar{x}^3. \tag{3.7}$$

However, this system of equations is inconsistent with (3.1) and (3.2). (For one thing, (3.3) implies that events that are simultaneous in one frame are simultaneous in all other frames.) To adjust the Galilean transformation to make it consistent with the constant speed of light, one can make the following modification to (3.4) and (3.5):

$$x^1 = \bar{k}\left(\bar{x}^1 + v\bar{t}\right) \text{ and} \tag{3.8}$$

$$\bar{x}^1 = k\left(x^1 - vt\right). \tag{3.9}$$

If neither coordinate system is preferred, then we must require that

$$\bar{k} = k.$$

These equations imply that (3.3) must be adjusted. If we substitute the formula for $\bar{x}^1$ from (3.9) into (3.8), we have

$$x^1 = k\left[k\left(x^1 - vt\right) + v\bar{t}\right] = (k)^2 x^1 - (k)^2 vt + kv\bar{t}.$$

Solving this for $\bar{t}$, we get

$$\bar{t} = kt - \frac{(k)^2 - 1}{kv}x^1 = k\left(t - \frac{(k)^2 - 1}{(k)^2 v}x^1\right). \tag{3.10}$$

Substituting the values of $\bar{t}$ from (3.10), $\bar{x}^1$ from (3.9), $\bar{x}^2$ from (3.6), and $\bar{x}^3$ from (3.7) into (3.1), we get

$$(k)^2(x^1 - vt)^2 + (x^2)^2 + (x^3)^2 = (k)^2 c^2\left(t - \frac{(k)^2 - 1}{(k)^2 v}x^1\right)^2. \tag{3.11}$$

This should match with (3.2). To determine the formula for k, we will consider the special case for which $x^1 = 0$. In that case,

$$(k)^2(vt)^2 + (x^2)^2 + (x^3)^2 = (k)^2 c^2 (t)^2 \text{ or}$$

$$(x^2)^2 + (x^3)^2 = (k)^2 c^2 (t)^2 - (k)^2 (vt)^2 \text{ should match with}$$

$$(x^2)^2 + (x^3)^2 = c^2 (t)^2.$$

This implies that

$$k^2c^2 - k^2v^2 = c^2 \text{ or}$$

$$k^2 = \frac{c^2}{c^2 - v^2} = \frac{1}{1 - (v^2/c^2)} \text{ or}$$

$$k = \frac{1}{\sqrt{1 - (v^2/c^2)}}$$

When this value of k is substituted back into (3.11), it will indeed become (3.2). (See Problem 19.) Summarizing, we now have

$$c\bar{t} = k\left(ct - \frac{v}{c}x^1\right) \qquad ct = k\left(c\bar{t} + \frac{v}{c}\bar{x}^2\right) \tag{3.12}$$

$$\bar{x}^1 = k\left(x^1 - \frac{v}{c}ct\right) \qquad x^1 = k\left(\bar{x}^1 + \frac{v}{c}c\bar{t}\right) \tag{3.13}$$

$$\bar{x}^2 = x^2 \qquad x^2 = \bar{x}^2 \tag{3.14}$$

$$\bar{x}^3 = x^3 \qquad x^3 = \bar{x}^3, \tag{3.15}$$

where

$$k = \frac{1}{\sqrt{1 - (v^2/c^2)}}. \tag{3.16}$$

You should note that if $v \ll c$, then $k \approx 1$ and Einstein's equations are nearly identical to the Galilean transformation that we use for the physics of Newton.

The equations above are said to represent a *boost* in the x^1 direction. These equations can be reformulated to take on an appearance similar to that of a rotation in the $x^1 - t$ plane.

If we define

$$\cosh\phi = k = \frac{1}{\sqrt{1 - (v^2/c^2)}}, \tag{3.17}$$

then

$$\sinh^2\phi = \cosh^2\phi - 1 = \frac{1}{1 - (v^2/c^2)} - \frac{1 - (v^2/c^2)}{1 - (v^2/c^2)} = \frac{(v^2/c^2)}{1 - (v^2/c^2)}.$$

Thus, we can define

$$\sinh\phi = \frac{v/c}{\sqrt{1 - (v^2/c^2)}} = k\frac{v}{c}. \tag{3.18}$$

Now Einstein's equations for a boost in the x^1direction become

$$c\bar{t} = ct\cosh\phi - x^1\sinh\phi \qquad ct = c\bar{t}\cosh\phi + \bar{x}^1\sinh\phi \tag{3.19}$$

$$\bar{x}^1 = x^1\cosh\phi - ct\sinh\phi \qquad x^1 = \bar{x}^1\cosh\phi + c\bar{t}\sinh\phi \tag{3.20}$$

$$\bar{x}^2 = x^2 \qquad x^2 = \bar{x}^2 \tag{3.21}$$

$$\bar{x}^3 = x^3 \qquad x^3 = \bar{x}^3. \tag{3.22}$$

To accommodate these equations to the formalism of Clifford algebra, we represent a four-dimensional vector in space–time in the form:

$$\mathbf{s} = \mathbf{e}_0 ct + \mathbf{e}_1 x^1 + \mathbf{e}_2 x^2 + \mathbf{e}_3 x^3. \tag{3.23}$$

However for special relativity, we must use a non-Euclidean metric! In particular

$$(\mathbf{e}_0)^2 = -(\mathbf{e}_1)^2 = -(\mathbf{e}_2)^2 = -(\mathbf{e}_3)^2 = \mathbf{I} \text{ and} \tag{3.24}$$

$$\mathbf{e}_{ij} = \mathbf{e}_i\mathbf{e}_j = -\mathbf{e}_j\mathbf{e}_i = -\mathbf{e}_{ji} \text{ for } i \neq j. \tag{3.25}$$

With these definitions, a boost in the x^1 direction can be written as

$$\bar{\mathbf{s}} = \mathbf{B}^{-1}\mathbf{s}\mathbf{B}, \text{ where} \tag{3.26}$$

$$\bar{\mathbf{s}} = \mathbf{e}_0 c\bar{t} + \mathbf{e}_1\bar{x}^1 + \mathbf{e}_2\bar{x}^2 + \mathbf{e}_3\bar{x}^3, \tag{3.27}$$

$$\mathbf{B} = \mathbf{I}\cosh\frac{\phi}{2} + \mathbf{e}_{10}\sinh\frac{\phi}{2} = \exp\left(\mathbf{e}_{10}\frac{\phi}{2}\right), \text{ and} \tag{3.28}$$

$$\mathbf{B}^{-1} = \mathbf{I}\cosh\frac{\phi}{2} - \mathbf{e}_{10}\sinh\frac{\phi}{2} = \exp\left(-\mathbf{e}_{10}\frac{\phi}{2}\right). \tag{3.29}$$

Note!

$$\begin{aligned}
\bar{\mathbf{s}} &= \mathbf{B}^{-1}\mathbf{s}\mathbf{B} \\
&= \left(\mathbf{I}\cosh\frac{\phi}{2} - \mathbf{e}_{10}\sinh\frac{\phi}{2}\right)\left(\mathbf{e}_0 ct + \mathbf{e}_1 x^1 + \mathbf{e}_2 x^2 + \mathbf{e}_3 x^3\right)\left(\mathbf{I}\cosh\frac{\phi}{2} + \mathbf{e}_{10}\sinh\frac{\phi}{2}\right) \\
&= \left(\mathbf{I}\cosh\frac{\phi}{2} - \mathbf{e}_{10}\sinh\frac{\phi}{2}\right)\left(\mathbf{e}_0 ct + \mathbf{e}_1 x^1\right)\left(\mathbf{I}\cosh\frac{\phi}{2} + \mathbf{e}_{10}\sinh\frac{\phi}{2}\right) \\
&\quad + \left(\mathbf{I}\cosh\frac{\phi}{2} - \mathbf{e}_{10}\sinh\frac{\phi}{2}\right)\left(\mathbf{e}_2 x^2 + \mathbf{e}_3 x^3\right)\left(\mathbf{I}\cosh\frac{\phi}{2} + \mathbf{e}_{10}\sinh\frac{\phi}{2}\right).
\end{aligned}$$

Since $\mathbf{e}_{10}$ commutes with $\mathbf{e}_2$ and $\mathbf{e}_3$ and anticommutes with $\mathbf{e}_0$ and $\mathbf{e}_1$, we have

$$\begin{aligned}
\bar{\mathbf{s}} &= \mathbf{e}_0 c\bar{t} + \mathbf{e}_1 \bar{x}^1 + \mathbf{e}_2 \bar{x}^2 + \mathbf{e}_3 \bar{x}^3 \\
&= \left(\mathbf{e}_0 ct + \mathbf{e}_1 x^1\right)\left(\mathbf{I}\cosh\frac{\phi}{2} + \mathbf{e}_{10}\sinh\frac{\phi}{2}\right)\left(\mathbf{I}\cosh\frac{\phi}{2} + \mathbf{e}_{10}\sinh\frac{\phi}{2}\right) \\
&\quad + \left(\mathbf{e}_2 x^2 + \mathbf{e}_3 x^3\right)\left(\mathbf{I}\cosh\frac{\phi}{2} - \mathbf{e}_{10}\sinh\frac{\phi}{2}\right)\left(\mathbf{I}\cosh\frac{\phi}{2} + \mathbf{e}_{10}\sinh\frac{\phi}{2}\right).
\end{aligned}$$

Since

$$\begin{aligned}
&\left(\mathbf{I}\cosh\frac{\phi}{2} + \mathbf{e}_{10}\sinh\frac{\phi}{2}\right)\left(\mathbf{I}\cosh\frac{\phi}{2} + \mathbf{e}_{10}\sinh\frac{\phi}{2}\right) \\
&\quad = \mathbf{I}\left(\cosh^2\frac{\phi}{2} + \sinh^2\frac{\phi}{2}\right) + \mathbf{e}_{10} 2\sinh\frac{\phi}{2}\cosh\frac{\phi}{2} \\
&\quad = \mathbf{I}\cosh\phi + \mathbf{e}_{10}\sinh\phi,
\end{aligned}$$

and

$$\begin{aligned}
&\left(\mathbf{I}\cosh\frac{\phi}{2} - \mathbf{e}_{10}\sinh\frac{\phi}{2}\right)\left(\mathbf{I}\cosh\frac{\phi}{2} + \mathbf{e}_{10}\sinh\frac{\phi}{2}\right) \\
&\quad = \mathbf{I}\left(\cosh^2\frac{\phi}{2} - \sinh^2\frac{\phi}{2}\right) = \mathbf{I},
\end{aligned}$$

we get

$$\begin{aligned}
\bar{\mathbf{s}} &= \mathbf{e}_0 c\bar{t} + \mathbf{e}_1 \bar{x}^1 + \mathbf{e}_2 \bar{x}^2 + \mathbf{e}_3 \bar{x}^3 \\
&= \left(\mathbf{e}_0 ct + \mathbf{e}_1 x^1\right)\left(\mathbf{I}\cosh\phi + \mathbf{e}_{10}\sinh\phi\right) + \left(\mathbf{e}_2 x^2 + \mathbf{e}_3 x^3\right) \\
&= \mathbf{e}_0(ct\cosh\phi - x^1\sinh\phi) + \mathbf{e}_1(x^1\cosh\phi - ct\sinh\phi) + \mathbf{e}_2 x^2 + \mathbf{e}_3 x^3.
\end{aligned}$$

We see that this system is an alternate form of Einstein's equations.

Note! It would have been more expeditious but less obvious to the beginner to write:

$$\begin{aligned}
\bar{\mathbf{s}} &= \mathbf{B}^{-1}\mathbf{s}\mathbf{B} \\
&= \exp\left(-\mathbf{e}_{10}\frac{\phi}{2}\right)\left(\mathbf{e}_0 ct + \mathbf{e}_1 x^1 + \mathbf{e}_2 x^2 + \mathbf{e}_3 x^3\right)\exp\left(\mathbf{e}_{10}\frac{\phi}{2}\right) \\
&= \left(\mathbf{e}_0 ct + \mathbf{e}_1 x^1\right)\exp\left(\mathbf{e}_{10}\frac{\phi}{2}\right)\exp\left(\mathbf{e}_{10}\frac{\phi}{2}\right) + \left(\mathbf{e}_2 x^2 + \mathbf{e}_3 x^3\right)\mathbf{I} \\
&= \left(\mathbf{e}_0 ct + \mathbf{e}_1 x^1\right)\exp(\mathbf{e}_{10}\phi) + \mathbf{e}_2 x^2 + \mathbf{e}_3 x^3 \\
&= \left(\mathbf{e}_0 ct + \mathbf{e}_1 x^1\right)\left(\mathbf{I}\cosh\phi + \mathbf{e}_{10}\sinh\phi\right) + \mathbf{e}_2 x^2 + \mathbf{e}_3 x^3) \\
&= \mathbf{e}_0(ct\cosh\phi - x^1\sinh\phi) + \mathbf{e}_1(x^1\cosh\phi - ct\sinh\phi) + \mathbf{e}_2 x^2 + \mathbf{e}_3 x^3.
\end{aligned}$$

For two successive boosts in Newtonian physics, we would have

$$\begin{aligned}
\acute{x} &= x - v_1 t \\
\acute{\acute{x}} &= \acute{x} - v_2 t = (x - v_1 t) - v_2 t \quad \text{or} \\
\acute{\acute{x}} &= x - vt, \quad \text{where} \\
v &= v_1 + v_2.
\end{aligned}$$

By contrast, for two successive boosts in special relativity, we have

$$\mathbf{B} = \mathbf{B}_1 \mathbf{B}_2 \quad \text{or}$$

$$\exp\left(\mathbf{e}_{10}\frac{\phi}{2}\right) = \exp\left(\mathbf{e}_{10}\frac{\phi_1}{2}\right)\exp\left(\mathbf{e}_{10}\frac{\phi_2}{2}\right), \quad \text{so}$$

$$\phi = \phi_1 + \phi_2.$$

From (3.17) and (3.18),

$$\begin{aligned}
\frac{v}{c} &= \frac{\sinh\phi}{\cosh\phi} = \frac{\sinh(\phi_1 + \phi_2)}{\cosh(\phi_1 + \phi_2)} \\
&= \frac{\sinh\phi_1 \cosh\phi_2 + \cosh\phi_1 \sinh\phi_2}{\cosh\phi_1 \cosh\phi_2 + \sinh\phi_1 \sinh\phi_2}.
\end{aligned} \tag{3.30}$$

Dividing both the numerator and denominator by $\cosh\phi_1 \cosh\phi_2$, (3.30) becomes

$$\frac{v}{c} = \frac{\frac{\sinh\phi_1}{\cosh\phi_1} + \frac{\sinh\phi_2}{\cosh\phi_2}}{1 + \frac{\sinh\phi_1}{\cosh\phi_1}\frac{\sinh\phi_2}{\cosh\phi_2}} = \frac{(v_1/c) + (v_2/c)}{1 + (v_1 v_2/c^2)}.$$

or

$$v = \frac{v_1 + v_2}{1 + (v_1 v_2/c^2)} \tag{3.31}$$

As you might expect, for low velocities $v_1 v_2/c^2 \approx 0$ so $v \approx v_1 + v_2$. You may have heard that a physical object cannot go faster than the speed of light. Suppose we return to the speeding train with the team of physicists on board. Suppose the train is moving at two thirds the speed of light and an exceptionally strong physicist hurls a spear in the forward direction also at two thirds the speed of light with respect to the train. According to Newtonian physics, the observers on the ground should see the spear moving at four thirds the speed of light. However according to the theory of special relativity, we must use (3.31). If $v_1 = \frac{2}{3}c$ and $v_2 = \frac{2}{3}c$, then

$$v = \frac{\frac{2}{3}c + \frac{2}{3}c}{1 + \frac{4}{9}} = \frac{\frac{4}{3}c}{\frac{13}{9}} = \frac{12}{13}c.$$

In general if $v_1 < c$ and $v_2 < c$, then $v < c$. (See Problem 22.)

You may have heard of shrinking rods at high velocities. This is known as *Lorentz contraction*. Why does this occur? Suppose you consider a rod that has length L_0, when it is observed in a frame in which it is stationary. For such a frame, we can place one end at $\bar{x}^1_A = 0$ and the other end at $\bar{x}^1_B = L_0$. In this frame, you do not care at what times you locate the two ends when you set about determining the length of the rod. Thus, we may represent one end in space–time by the vector

$$\mathbf{e}_0 c\bar{t}_A,$$

and the other end by the vector

$$\mathbf{e}_0 c\bar{t}_B + \mathbf{e}_1 L_0.$$

From (3.19) and (3.20), we can see that for the frame in which the rod is moving with speed v in the x^1 direction, these two points would become, respectively:

$$\mathbf{e}_0 ct_A + \mathbf{e}_1 x^1_A = \mathbf{e}_0 c\bar{t}_A \cosh\phi + \mathbf{e}_1 c\bar{t}_A \sinh\phi$$

and

$$\mathbf{e}_0 ct_B + \mathbf{e}_1 x^1_B = \mathbf{e}_0 (c\bar{t}_B \cosh\phi + L_0 \sinh\phi) + \mathbf{e}_1 (L_0 \cosh\phi + c\bar{t}_B \sinh\phi).$$

We see that in this frame, the difference of the spatial coordinates for the two points is

$$L = x^1_B - x^1_A = L_0 \cosh\phi + c(\bar{t}_B - \bar{t}_A)\sinh\phi. \tag{3.32}$$

However in this frame, the rod is moving so the notion of length is meaningful only if we locate the two ends simultaneously. Thus, we require that

$$\begin{aligned} ct_A &= ct_B. \text{ That is} \\ c\bar{t}_A \cosh\phi &= c\bar{t}_B \cosh\phi + L_0 \sinh\phi, \text{ or} \\ c(\bar{t}_B - \bar{t}_A) &= -L_0 \frac{\sinh\phi}{\cosh\phi}. \end{aligned}$$

Thus, (3.32) becomes

$$L = L_0 \cosh\phi - L_0 \frac{\sinh^2\phi}{\cosh\phi} = L_0 \frac{\cosh^2\phi - \sinh^2\phi}{\cosh\phi} = \frac{L_0}{\cosh\phi},$$

or

$$L = L_0 \sqrt{1 - (v^2/c^2)}. \tag{3.33}$$

Equation (3.33) is the formula for Lorentz contraction.

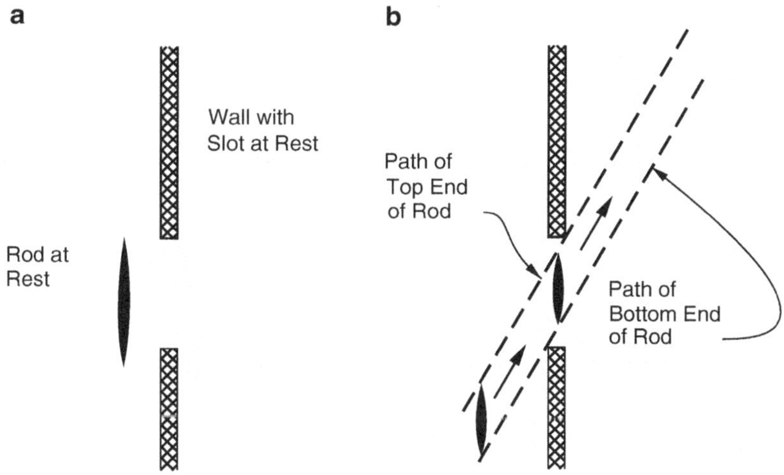

Fig. 3.1 (**a**) When the rod and the wall are both at rest, the rod is longer than the slot. (**b**) When the rod is given a sufficiently high velocity in a direction parallel to the wall with a slight sideways drift, the rod shrinks and passes through the slot in the stationary wall. What happens in the rest frame of the rod in which the length of the rod remains the same but the length of the slot shrinks?

One aspect of special relativity, which makes it fun, is the many "paradoxes" that seem to arise. One paradox that arises from this Lorentz contraction is as follows: Suppose we consider a rod that is somewhat longer than a vertical slot in a wall. According to the special theory of relativity, we should be able to give the rod sufficient speed in the vertical direction so as to shorten the rod down to that of the slot. If we then also added a small horizontal component to the velocity, the rod would pass through the slot. (See Fig. 3.1a, b.)

So far so good. But what happens in the rest frame of the rod? In the rest frame of the rod, the rod has its full rest frame length and the length of the slot is shorter. How would it then be possible for the rod to pass through the slot?

To understand the situation in the rest frame of the rod, compare Fig. 3.2a, b. For purposes of illustration, I have constructed the picture for the case for which the relative velocity of the rod and the wall is such that the Lorentz contraction factor is 1/2. I have identified the direction of motion with the x^1-axis. In this case, the distance from any point on the wall to the x^2-axis in Fig. 3.2b is half the corresponding distance in Fig. 3.2a. Furthermore, the distance of any point on the rod to the x^2-axis in Fig. 3.2b is twice the corresponding distance in Fig. 3.2a. What becomes clear when the two figures are compared is that what is parallel in one frame may not be parallel in another frame and events that are simultaneous in one frame may not be simultaneous in another frame. In the rest frame of the wall, the event of the top end of the rod passing the top end of the slot is simultaneous with event of the bottom end of the rod passing the bottom end of the slot. These two events are not simultaneous in the rest frame of the rod.

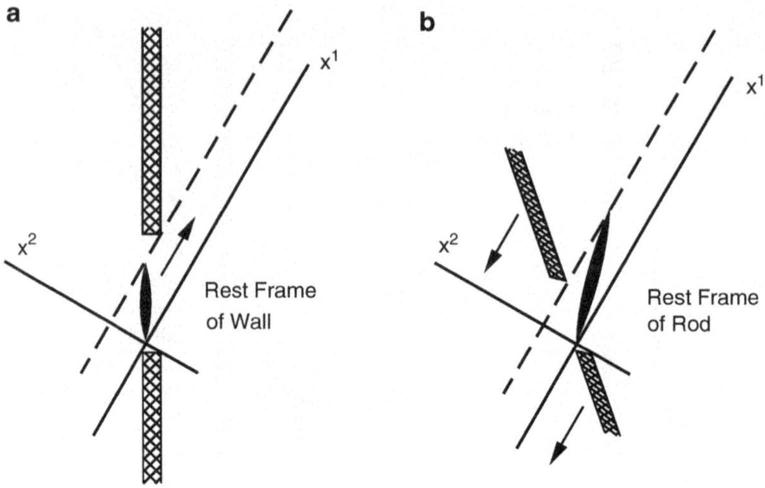

Fig. 3.2 (**a**) The change of lengths occurs in the direction of motion, that is in the direction of the x^1-axis. In the rest frame of the wall, the event of the top tip of the rod passing through the plane of the wall is simultaneous with the event of the bottom tip of the rod passing through the same plane. (**b**) Events which are simultaneous in the rest frame of the wall are not simultaneous in the rest frame of the rod

Problem 19. Show that if $1/\sqrt{1-(v^2/c^2)}$ is substituted for k in (3.11), the result is (3.2).

Problem 20. Use a Taylor's series expansion to show that $\exp(\mathbf{e}_{10}u) = \mathbf{I}\cosh u + \mathbf{e}_{10}\sinh u$.

Problem 21. Use the result of Problem 20 to show that $\cosh(\phi_1+\phi_2) = \cosh\phi_1\cosh\phi_2 + \sinh\phi_1\sinh\phi_2$ and $\sinh(\phi_1+\phi_2) = \sinh\phi_1\cosh\phi_2 + \cosh\phi_1\sinh\phi_2$.

Problem 22. Suppose $0 < v_1 < c$ and $0 < v_2 < c$, show $(v_1+v_2)/(1+(v_1v_2/c^2)) < c$. Hint: $(c-v_1)(c-v_2) > 0$.

Problem 23. Suppose we revisit the speeding train with the team of physicists on board. Suppose the velocity of the train is v and a physicist on board sets off a flash of light. What does (3.31) tell us about the velocity of the flash observed by the observers on the ground?

Problem 24. a. In this chapter, a boost in the direction of x^1 was represented by the operator $\exp(\mathbf{e}_{10}\phi/2)$. How would you represent a boost if the direction cosines for the space direction were (k_1, k_2, k_3)?

b. Show that in general the product of two boosts is not a boost. (To show this, choose two boosts with different directions.) Note! It can be shown that the product of any number of boosts and rotations can be represented as a product of a single boost and a single rotation.

Problem 25. Revisiting the speeding train once again, suppose the speed of the train is $100\,\text{km}\,\text{hr}^{-1}$ and the speed of the spear as observed by those on the train is also $100\,\text{km}\,\text{hr}^{-1}$. The speed of the spear observed by those on the ground is less than $200\,\text{km}\,\text{hr}^{-1}$. How much less? (The speed of light is $1.08 \times 10^9\,\text{km}\,\text{hr}^{-1}$.)

Problem 26. Suppose the team on the train and the team on the ground each have their own clocks. The moment that the two clocks are along side one another both clocks are set to zero. Suppose some time later the reading of the clock on the train is $\bar{t}$. What is the corresponding time and position for the observers on the ground? Note! If the corresponding time for the ground observers is t, then the corresponding position for the ground observers should be vt. Does this check out?

3.2 *Albert Einstein 1879–1955

Albert Einstein was born in Ulm, Germany on March 14, 1879. Shortly thereafter, the Einstein family decided to relocate to Munich, where Albert spent most of his youth.

From an early start, the task of formally educating Albert proved to be difficult. Inordinately bright, the young Einstein routinely questioned authority, possessed a fiercely independent streak, and rebelled at many turns. To complicate matters, financial setbacks for Einstein's father became an additional challenge to educating Albert. At the time the Einstein family arrived in Munich, parents were expected to send their children to schools operated by their respective religions. Although the Einsteins were born to the Jewish faith, they were not rigorous in their beliefs, and felt it made much more sense to send young Albert to a Catholic elementary school within walking distance, rather than a distant and more expensive Yeshiva.

German law did require religious training, so a relative provided a Judaic education during Albert's elementary school years. At age 11, Albert rebelled against his parents by scolding them for eating pork and not observing the Sabbath (Parker 2003, p. 26). Two years later, rabbis became the receiving end of Albert's rebellious spirit when he accused them of teaching falsehoods as their teachings could be disproved by scientific observation (Clark 1972, p. 36).

In the summer of 1894, the Einstein family decided to move to Italy. Albert's father Hermann had a kind nature but was not a natural business man. The move from Ulm to Munich had been prompted by Hermann's failed electrical and engineering workshop. In Munich, Hermann Einstein established an electric equipment business with Albert's uncle Jakob. However in time, this second business failed too. Extended family offered financial assistance if the Einstein family moved to Italy, but Albert's parents did not want to disrupt his education. Therefore, arrangements were made to leave Albert with a relative in Munich, while his parents departed for Italy, confident that he would continue his education at the Luitpold Gymnasium.

For its time, the Luitpold Gymnasium was considered a forward looking institution of the highest caliber. However, Albert found it to be repressive. Within months of his parents departure, Albert had left the gymnasium. It is unclear from the records whether Albert Einstein was allowed to quit of his own volition or if he was thrown out for disciplinary reasons. In either case, the departure was orchestrated by Einstein. Always resentful of authority, Albert Einstein formally renounced his German citizenship a year later at the age of 16 to avoid being drafted into the German army (Parker 2003, p. 38). He was willing to serve in the more democratic Swiss army and eventually became a Swiss citizen – a status he treasured throughout his adult life. Although he changed nationality two more times, he always maintained his Swiss citizenship. When he became a Swiss citizen in 1901, he happily presented himself for military duty but was disappointed when he was turned down because of flat feet and varicose veins (Clark 1972, p. 64).

Much of Einstein's disrespect for authority stemmed from his strong skepticism. This even applied to his own contributions to the advancement of physics. Three years after Einstein's General Theory of Relativity had received some observational confirmation and world acclaim, Cornelius Lanczos, a young physicist at the time, approached Einstein between two lectures. With great pride, Lanczos told him he had developed a method of successive approximations to deal with Einstein's gravitational equations. Einstein responded, "But why, should anybody be interested in getting precise solutions of such an ephemeral set of equations?" Lanczos would remember the word "ephemeral" many years later (Whitrow 1967, p. 49).

Without a diploma from a gymnasium, there were no universities that would consider Einstein for admission. The Swiss Federal Polytechnic School in Zurich provided Einstein an alternative solution. After some self-study and a year in the Swiss school system, Einstein was accepted for admission. Hermann Einstein was pleased with this turn of events as the Polytechnic school could well prepare his son for a career in electrical engineering. Albert, however, had different ideas and registered himself as a student of physics.

When Einstein graduated from the Swiss Federal Polytechnic School, he had extreme difficulty getting a job because he could not get a positive recommendation from any of his professors. This should not have surprised Einstein. He had alienated most of his professors and even antagonized some. Jean Pernet, his instructor in a laboratory course, was disturbed by the fact that more than once after giving Einstein written instructions for an experiment he would see Einstein wad up the instruction sheet, toss it in the wastebasket and then carry out the experiment in his own way (Parker 2003, p. 69). Einstein had a higher regard for the chairman of the department, Heinrich Weber, but Albert was prone to address him as "Herr Weber" instead of the more respectful and expected "Herr Professor."

During his 4 years at Polytechnic, Einstein became disappointed by the fact that none of his professors was inclined to discuss consequences of Maxwell's equations that Einstein thought were being overlooked by most of the physics community. In his disappointment, Einstein turned to self-study. In particular, he examined the works of Hermann Helmholtz, Ernst Mach, Rudolf Kirchoff, Heinrich Hertz, James Clerk Maxwell, and Paul Drude. Several of these authors expressed scepticism of various aspects of Newton's theories (Parker 2003, pp. 72–73).

A consequence of this self-study was a high rate of absenteeism in class. This could not have gone unnoticed. Polytechnic had a large number of engineering students but in Einstein's class there were only two physics majors and three math majors.

To graduate, Einstein had to pass two sets of exams – the intermediate exams normally taken at the end of 2 years and the final exams taken at the end of 4 years. Due to his poor attendance, in normal circumstances it would have been impossible for him to pass either set. Fortunately for Einstein, one of the math majors, Marcel Grossmann, came to his rescue. Marcel Grossmann had taken careful notes and helped Einstein cram for the exams. On the intermediate exams, Einstein got the highest score – even out scoring his friend Marcel Grossmann (Parker 2003, p. 68). On the final exams, he did not do so well. All three math majors out-scored him. The other physics major failed but Einstein passed (Parker 2003, p. 86).

Unfortunately for Einstein, without a good recommendation from any of his professors, he could not get the kind of academic position he had hoped for. For 2 years, he scraped by with a sequence of poorly paid part-time or temporary positions. Once again, Marcel Grossmann came to his rescue. Einstein's predicament was explained to Marcel's father, who in turn spoke to Friedrich Haller, the Director of the Patent Office in Bern. When the next job opened up in the patent office, the job description would fit Einstein's resume. As a result, Einstein eventually got a job in the patent office and he started work on June 23, 1902 at the age of 23.

Approximately 6 months later, Einstein married the other physics major in his class at Polytechnic. She was Mileva Marić. Six years earlier, in the fall of 1896, when Albert and Mileva began their studies at Polytechnic, Mileva was one of only twenty women in all of the Prussian and Swiss universities to study a natural science or mathematics (Zackheim 1991, pp. 14–15). However, Mileva failed her final exams twice so she never got her degree. (She was pregnant the second time she took the exams and may have been suffering from morning sickness.) Although her grades on the physics components of the exam were almost as high as Einstein's, her grades on the math components were far lower.

Albert and Mileva had become intimate in defiance of both sets of their parents. About a year before their marriage, on January 27, 1902, Mileva had given birth to a girl. The birth of Einstein's daughter, Lieserl, took place in Mileva's home in Serbia but Einstein never saw her. The existence of Einstein's daughter was generally unknown until some love letters from Albert to Mileva were discovered about 30 years after Einstein's death. What became of Lieserl is not certain. Michele Zackheim who has investigated this matter concludes that Lieserl was a victim of Down's Syndrome who died of scarlet fever at the age of 21 months on September 15, 1903. Some of the Zackheim's conclusions are speculative, but the most crucial ones are based on the contents of a letter Einstein wrote to console his wife on September 19, 1903 (Zackheim, p. 47 and p. 252).

Einstein's 7 years at the patent office in Bern proved to be fortuitous and productive. Not only did the job pay well, but there was also sufficient flexibility to allow him to pursue his passion for physics and obtain a Ph.D.

During this phase of Einstein's life, in 1905 to be precise, Einstein had his *annus mirabilis* (miracle year). He was only 26 at the time and yet in one volume of the prestigious journal *Annalen der Physik* he published three papers of historical importance on three diverse topics. The first paper demonstrated that the photoelectric effect could be nicely explained if one hypothesized the existence of quanta (photons). The second described how Brownian motion could be ascribed to molecular motion and also could be used to determine Avogadro's number. The third was the Theory of Special Relativity. Each of these papers received nominations for the Nobel Prize for physics at one time or another between 1910 and 1922 (Pais 1982, pp. 502–511). In the subsequent volume of *Annalen der Physik*, he published his famous equation, $E = mc^2$. In between all this, in the same year, Einstein also obtained his Ph.D. from the University of Zurich. (The Ph.D. thesis was dedicated to Marcel Grossman.)

Before this time, Einstein had published only a handful of papers. However, he had been thinking, reading, and discussing these topics for a long time. Ten years earlier at the age of 16, before he started his formal studies at Polytechnic, he wrote to his Uncle Cäsar a long letter outlining his plans to tackle the relationship between electricity, magnetism, and the hypothesized ether (Clark 1972, p. 41) and (Pyenson 1985, pp. 8–9). While cutting classes at Polytechnic he had been devouring all relevant material he could lay his hands on.

Many would have thought that Einstein was disadvantaged during his years working at the patent office as it did not allow him to exchange ideas with other academics at a university. Einstein took a different view. On his seventieth birthday, he wrote, "It gave me the opportunity to think about physics." Without pressure to publish he could delve into questions more deeply. He wrote, "– an academic career compels a young man to scientific production, and only strong characters can resist the temptation of superficial analysis." (Clark 1972, p. 75)

Author Peter Galison suggests that Einstein's job at the patent office may have given him an edge for another reason. During his employment, patent applications were being submitted to deal with the problem of synchronizing clocks. It was difficult to publish meaningful train schedules without synchronizing clocks of towns that were miles apart (Galison 2003). Galison's speculation is supported by the fact that when Einstein set out to explain the Special Theory of Relativity to a lay readership, he described the problem of synchronizing two clocks at opposite ends of a moving train (Einstein 2005, p. 13–55).

It should also be noted that Einstein was not intellectually isolated in Bern. Soon after moving to Bern, Einstein was meeting on a fairly regular basis with two friends. The three labeled themselves the "Olympia Academy." (Clark 1972, pp. 78–81) In 1903, this group would break up when the two friends Solovine and Habicht moved away. However with Einstein's encouragement, Michelangelo Besso, a friend from his days in Zurich, applied for and was accepted for a job in the Bern patent office (Parker 2003, p. 138). In his paper on Special Relativity, Einstein would later thank Besso for "a number of valuable suggestions." (Clark 1972, p. 101).

Despite the fact that his 1905 papers would revolutionize physics, it took a while before Einstein could move out from the patent office into the academic world.

A year after his "annus mirabilis", Einstein was turned down for a low paying part-time position at the University of Bern (Parker 2003, p. 155). In 1909 after being in the patent office for 7 years, he was finally given serious consideration by the University of Zurich. There was some reluctance at the University to hire a Jew and Einstein's teaching abilities were unknown. His thesis advisor, Alfred Kleiner, argued with his colleagues and overcame these reservations. Einstein was offered the position of associate professor but initially Einstein turned down the offer because he would have to take a substantial pay cut. Eventually, the University of Zurich agreed to match his salary at the patent office on July 6, 1909 (Parker 2003, pp. 164–165).

Years prior to his move to the University of Zurich, Einstein had been in steady communication with Nobel Prize winners and future Nobel Prize winners. These associations served Einstein well. His academic career would soon rise at a meteoric pace. After only one academic year in Zurich, Einstein accepted a position as full professor at the German University in Prague, Czechoslovakia. In August 1912 (Clark 1972, p. 194), he was back in Zurich – this time, ironically, at the Swiss Federal Polytechnic School. On April 6, 1914 not long before the outbreak of World War I, which would begin on August 1, Einstein moved again. This time he moved to Berlin to accept a post, which had no teaching obligations.

By now Einstein had two sons, but his relationship with Mileva had deteriorated. Judging from the love letters that Mileva had received from Albert before they were married, Mileva had reason to believe that her relation with Albert would be one of sharing. Instead Albert decided he could not use her as a sounding board for his ideas in physics and aside from his interest in music he did not have much else to share. During their married years in Bern and Zurich, Mileva was relegated to the task of doing cooking and laundry for student boarders living in the Einstein home to supplement his income. Shortly before their marriage, Einstein's father died in debt and presumably Mileva's efforts were devoted to helping Albert support his mother. Even though Einstein's mother never lived in the same household as Mileva, this burden imposed on her must have been particularly onerous. Einstein's mother, Pauline, believed that Mileva was never good enough for her Albert and she was outspoken about it. Nevertheless, it was Albert not Mileva who wanted to end the marriage. Einstein enjoyed being amused and Mileva had become joyless and less attractive to Albert (Zackheim 1999, p. 66).

In 1913, Albert wrote to Elsa Lowenthal, who would eventually become his second wife, "I treat my wife as an employee I cannot fire." (Levenson 2003, p. 28).

Mileva and Albert moved to Berlin together but within 3 months Mileva was on her way back to Zurich with Albert's two sons. She would live in Zurich for the rest of her life. Einstein would develop his General Theory of Relativity during the war years in Berlin without family distractions.

For several years, Einstein had recognized that the theory of Special Relativity dealt nicely with coordinate frames that moved at constant speed with respect to one another – but not with frames involving acceleration. To attack this question, he realized that he must also construct a new theory of gravity. (A passenger in a windowless spaceship with a quiet nonvibrating rocket engine would have difficulty knowing whether he or she was feeling the effect of gravity while the spaceship was parked on some planet or was feeling the effect of acceleration out in open space.)

To construct a successful theory, he would need to know more mathematics than he had previously thought necessary to do physics. Again his Polytechnic classmate, Marcel Grossmann, came to his rescue. Marcel Grossmann directed his attention to Riemannian geometry, which was then a fairly obscure topic in mathematics. After many failed efforts, it was late in 1915 that Einstein published his General Theory of Relativity in its completed form.

One virtue of the General Theory of Relativity was the fact that Einstein could use it to explain the deviation of the orbit of Mercury from the laws of Newton. This was a surprise. Although it was recognized that the orbit of Mercury had an odd behavior, it was thought that eventually the anomaly could be explained in Newtonian terms.

Earlier in the nineteenth century, astronomers had encountered a similar peculiarity related to the orbit of Uranus. What at first appeared to be a counter example of the laws of Newton became a dramatic confirmation. According to the laws of Newton, if there was only one planet in the solar system, its path would be that of an ellipse. The presence of other planets would alter this path. Generally, Newton's laws could consistently describe the actual orbit of any given planet when the influence of the other known planets were taken into account. Unfortunately for the disciples of Newton, the orbit of Uranus seemed to be an exception. The Newtonian calculations could not explain the orbit of Uranus.

However, two astronomers conjectured that Uranus' orbit could be explained by the existence of an undiscovered planet. Lengthy calculations were carried out independently by the British astronomer John Couch Adams and the French astronomer Urbain Jean Joseph Le Verrier to predict the position of the then unknown planet. Le Verrier sent his results to the German astronomer Johann Gottfried Galle of the Berlin Observatory. With this information, Galle was able to locate the planet, that is now known as Neptune on September 23, 1846.

This was a dramatic success for Newton's theory. Almost a decade later in 1855, Le Verrier discovered a discrepancy in the orbit of Mercury. He attributed this to another unknown planet, which he called "Vulcan." His calculations indicated that it had to be so close to the sun that it would be very difficult to observe. Nevertheless, over the next 60 years, many astronomers, professional and amateur alike, announced that they had observed this hypothetical planet. However, none of these observations could be confirmed with confidence. This search finally ended when Einstein's General Theory of Relativity provided an explanation for the behavior of Mercury's orbit without necessitating the presence of an additional planet.

It is nice to construct a theory that explains things previously observed and measured. However to really impress physicists, the proposer of a theory should successfully predict something previously unobserved. Einstein predicted that a light ray from a distant star passing near the sun would be deflected slightly by the local curvature of space–time created by the gravitational field of the sun. This would be observable only during a solar eclipse.

The opportunity to test Einstein's prediction came on May 29, 1919 – less than 6 months after the end of World War I. A team of British scientists under the leadership of Arthur Eddington organized two expeditions: one to Sobral in northern Brazil and the other to Principe Island off the coast of Africa. Fortunately, it was not too cloudy on the appointed day and enough data was collected to confirm Einstein's prediction.

The results were made public on November 6, 1919 in London at a joint meeting of the Royal Society and the Royal Astronomical Society. Members of the press were invited to attend and were present when Einstein's theory was described as "one of the greatest achievements in the history of human thought –." (Clark 1971, pp. 289–290) Neither the reporters nor most of their readers would ever be able to understand all the details of Einstein's theory but they could understand the significance of the fact that Einstein had shown that we do not live in a world ruled by the laws of Newton nor the axioms of Euclid.

In the following days as the reports of this meeting spread around the world, Einstein attained an international celebrity status that he retained to the end of his life.

For many he symbolized a hope for mankind. The effects of the "war to end all wars" had been horrific. For 4 years, the most technologically advanced nations of the world had exhausted their economic resources to achieve the senseless slaughter of young men in muddy trenches. Einstein's achievement showed that an enlightened civilization was capable of something better. The fact that a British team had verified a prediction of a German meant to some that nations could work together in peace.

However, many people in Germany took a quite different view. The German army had lost the war without losing a battle on German soil. In this circumstance, ultra nationalists were able to promote the idea that Germany had been "stabbed in the back" by disloyal people on the home front who were not true Germans (i.e., Jews).

Einstein could be hated for simply being Jewish, but there were other reasons why he drew negative attention. For most of his life, Einstein was strongly anti-German. While in Switzerland, Einstein had renounced his German citizenship and had become a citizen of Switzerland long before the war started. As a pacifist, and one not to shy from expressing his opinion, Einstein believed Germany was guilty of war crimes in Belgium and France (Clark 1971, p. 277). Moreover, Einstein had avoided making any contribution to Germany's war effort. (Most of Germany's scientists, Jews and non-Jews alike, had enthusiastically worked on such projects as the development of better aircraft and the perfection of gas warfare.) That said, when the war ended, Einstein went through the most pro-German phase in his life. He had hopes that the victorious allies would establish conditions that would allow the new democratic Weimar Republic to thrive. In December 1918, Einstein was one of the 100 intellectuals in Europe and the United States who signed a petition addressed to the heads of states about to meet in Versailles. The petition asked those heads of state to avoid creating conditions for a future war (Clark 1971, p. 274). In 1920, Einstein even reestablished his German citizenship (Clark 1971, p. 315). Nonetheless, when Germans began to suffer from the severe terms of the Versailles Treaty, Einstein became a special target for German nationalists.

For those who wished to equate Jews with vermin, it was particularly distressful to see the name Einstein become synonymous with "genius." As a result, the Nazis organized public meetings to belittle Einstein's theories of special and general relativity. Einstein found himself being portrayed as a member of some international Jewish effort to contaminate German science with false theories. For the unsophisticated, these criticisms had some credibility because a leader of those attempting to discredit Einstein was Philipp Lenard who had won the Nobel Prize for physics in 1905 for his revealing investigations of the photoelectric effect.

Ironically, Einstein was awarded the 1921 Nobel Prize in physics for developing a theoretical explanation of Lenard's experimental results. (Einstein never personally enjoyed the financial rewards of this prize, as he had agreed to give the money to Mileva as part of their divorce settlement in 1919.)

At the end of 1932, Hitler came to power and Einstein left Germany, never to return. In October of 1933, Einstein arrived in Princeton, New Jersey with his second wife Elsa, where they lived for the rest of their lives. Albert and Elsa had developed a relationship in Berlin, and were married in 1919, shortly after his divorce to Mileva was finalized.

During the 1930s, Einstein became convinced that pacifism would not thwart the aims of Hitler. Extreme force would be necessary. In 1939, he joined other scientists in urging President Roosevelt to establish an organized effort to develop the atomic bomb.

Due to his bitter experiences in Germany, Einstein felt that it was important to stand up against those who would suppress freedom to advance their own concept of an ideal society. He participated in fund-raising efforts to aid the refugees of Franco's Spain. He supported anti-lynching legislation and helped organize a chapter of the NAACP in Princeton. Princeton University was all white when Einstein first came to the Institute for Advanced Study and the public schools were segregated until 1948.

In 1937, Einstein became the focus of a notorious incident. On April 16, 1937, Marian Anderson gave a concert to a standing-room audience at Princeton's McCarter Theatre. The concert received rave reviews but Princeton's Nassau Inn refused to provide a room for the black contralto. In response, Einstein invited her to stay with him and his stepdaughter Margot. Marian Anderson accepted the invitation and she stayed with them thereafter whenever she came to Princeton (Jerome 2002, pp. 77–78). In earlier years, Einstein had urged the youth of the world to resist military service. Each of these political activities incurred the wrath of the director of the F.B.I., J. Edgar Hoover.

J. Edgar Hoover persuaded the U.S. Army to deny Einstein security clearance, so Einstein never participated in the Manhattan Project (the development of the atomic bomb). However, the U.S. Navy ignored Hoover and hired Einstein as a consultant to carry out various computations on the propagation of shock waves.

Shortly after World War II ended, the cold war with Russia began when Russia installed puppet governments in Eastern Europe. Passions were inflamed in 1949 when Mao Tse Tung took power in China and Russia exploded its first atomic bomb.

Under Truman, J. Edgar Hoover was directed to set up a loyalty-security program of all government employees. As a result, several thousand people lost their jobs. It did not matter that in normal times, very few of those jobs would have required any kind of security clearance. Others were called upon to testify at congressional hearings on their political affiliations and those of their acquaintances.

Einstein recognized that Russia had an undesirable government but he viewed these congressional committees along with J. Edgar Hoover to be much greater threats to freedom in America than the American Communist Party.

In 1953, Einstein urged intellectuals to insist on their First Amendment right of free speech and risk imprisonment and financial ruin rather than rely on the Fifth Amendment. Using the Fifth Amendment to avoid testifying was usually interpreted as admission of some kind of disloyalty. On June 12, The New York Times published a letter from Einstein expressing this view but denounced him in an editorial on the following day for urging civil-disobedience (Jerome 2002, p. 239).

Always the maverick, Einstein was never the blind adherent of any one group. In January 1953, he wrote a letter to Truman urging the commutation of the death sentence given to the Rosenbergs for soviet espionage (Jerome 2002, p. 140). Less than 2 weeks later, the Newark Star Ledger carried a United Press story stating: "Scientist Albert Einstein yesterday condemned the wave of anti-Semite purges behind the iron curtain" (Jerome 2002, p. 147). Einstein was an avid Zionist and in 1952, the Israeli government offered him the presidency. However, he often clashed with Zionist leaders for not pursuing better relations with the Arabs.

Despite his political activism, Einstein's life in Princeton was sheltered as much as possible but his world status could have consequences even for his neighbors.

My father knew a member of Princeton's psychology department who had a recurring nightmare involving Einstein. Einstein never learned to drive a car and each morning he walked from his home to his office at the Institute for Advanced Study on a route that took him past the home of the psychology professor. Often Einstein seemed to be engrossed in thought, totally oblivious to things around him. The psychology professor feared that on some morning when he was late for class and a little inattentive himself, he would suddenly become world famous for being the person that backed his car out of his driveway and ran over Einstein.

Fortunately for both Einstein and the psychology professor, this never happened. Because of an aneurysm on his aorta, Einstein died a peaceful death on April 18, 1955 at the age of 76.

3.3 *Suggested Reading

James A. Richards; Francis Weston Sears; M. Russell Wehr; Mark W. Zemansky 1960. *Modern University Physics.* Reading, Massachusetts: Addison-Wesley Publishing Company, Inc.

A nice elementary derivation of Einstein's $E = mc^2$ appears on pages 773-779.

M. Russell Wehr, James A. Richards, Jr., and Thomas W. Adair III, 1984. *Physics of the Atom,* 4th Ed. Reading, Massachusetts: Addison-Wesley Publishing Company, Inc.

The same derivation of $E = mc^2$, mentioned in the book above, appears in this book on pages 180–184. It also appears in the 3rd Ed. on pages 161–165.

John Snygg 1997. *Clifford Algebra – A Computational Tool for Physicists.* New York: Oxford University Press.

If you cannot get your hands on any of the books mentioned above, buy my book and read an outline of the same derivation of $E = mc^2$ on pages 38-40.

A. Shadowitz 1968. *Special Relativity.* Philadelphia: W. B. Saunders Company.

This book has a lot of very enlightening diagrams using the space-time geometry of Loedel. The book also contains an excellent discussion of the twin paradox.

In 1953, Albert Shadowitz attracted the attention of Senator Joseph McCarthy for his activity as a union organizer against a defense contractor. Having received a subpoena to appear before McCarthy's Senate committee, Shadowitz figured he was headed for jail for contempt of Congress. This was because he was determined not to name others who would then be subjected to vituperation and possible loss of employment. When Shadowitz sought out guidance from Einstein, Einstein's long time secretary, Helen Dukas, gave him access (Jerome 2002, p. 245). Shadowitz has been ever grateful for the resultant public support he received from Einstein and presumably for this reason, *Special Relativity* is dedicated to Helen Dukas.

Albert Einstein, Robert W. Lawson (translator) 2001, Relativity: *The Special and General Theory.* Dover Publications, Inc.

This book, originally published in 1920, is directed to the reader with no knowledge of physics or mathematics beyond the high school level.

Ronald W. Clark 1971. *Einstein: The Life and Times.* New York: Avon Books.

This is an outstanding biography of Einstein for a general audience.

Michele Zackheim 1999. *Einstein's Daughter – The Search for Lieserl.* New York: Riverhead Books a member of Penguin Putnam Inc.

In 1902 approximately 1 year before their marriage, Mileva gave birth to Albert's daughter Lieserl. This book describes an odyssey resulting in a partially successful effort to determine the fate of that daughter.

Chapter 4
Clifford Algebra in Flat n-Space

4.1 Clifford Algebra

The word "geometry" is derived from a greek word meaning "to measure land." The starting point for differential geometry is the definition of an infinitesimal distance. Generally, such an infinitesimal distance ds is defined in terms of a coordinate system. For the Cartesian coordinate system applied to an *n-dimensional Euclidean space*, we have

$$(\mathrm{d}s)^2 = \sum_{j=1}^{n} (\mathrm{d}x^j)^2. \tag{4.1}$$

For some purposes, Euclidean spaces are not sufficiently general as we saw in the last chapter. For an *n-dimensional pseudo-Euclidean space*, we have

$$(\mathrm{d}s)^2 = \sum_{j=1}^{p} (\mathrm{d}x^j)^2 - \sum_{j=p+1}^{n} (\mathrm{d}x^j)^2. \tag{4.2}$$

For such a space (p, q) is said to be the *signature* of the metric, where $q = n - p$.

Frequently, alternate coordinate systems are useful. For example, in three dimensions, it is often useful to use the spherical coordinate system. (See Fig. 4.1.) In that case

$$(\mathrm{d}s)^2 = (\mathrm{d}r)^2 + r^2(\mathrm{d}\theta)^2 + r^2 \sin^2\theta(\mathrm{d}\phi)^2. \tag{4.3}$$

To study the two-dimensional saddle surface for which $z = xy$, we may wish to use a coordinate system that results in the equation

$$(\mathrm{d}s)^2 = (1 + (u^2)^2)(\mathrm{d}u^1)^2 + 2u^1u^2\mathrm{d}u^1\mathrm{d}u^2 + (1 + (u^1)^2(\mathrm{d}u^2). \tag{4.4}$$

J. Snygg, *A New Approach to Differential Geometry using Clifford's Geometric Algebra*, DOI 10.1007/978-0-8176-8283-5_4, © Springer Science+Business Media, LLC 2012

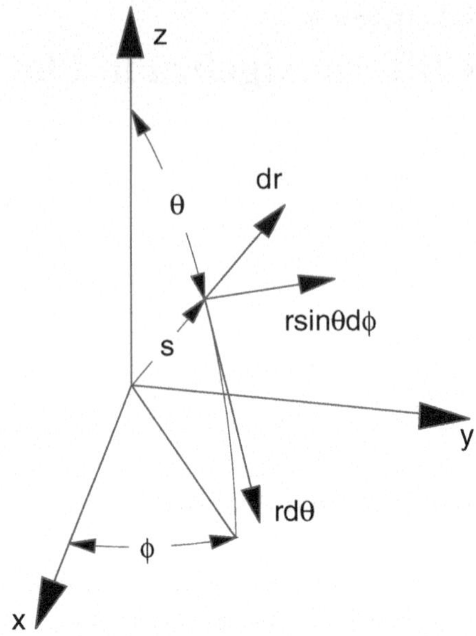

Fig. 4.1 The directions associated with $\mathrm{d}r$, $r\,\mathrm{d}\theta$, and $r \sin\theta\,\mathrm{d}\phi$ at the point $\mathbf{s}$ for spherical coordinates are mutually perpendicular

In general, we have

$$(\mathrm{d}s)^2 = \sum_{i=1}^{n}\sum_{j=1}^{n} g_{ij}(u^1, u^2, \ldots, u^n)\mathrm{d}u^i\,\mathrm{d}u^j. \tag{4.5}$$

(It is understood that $g_{ij} = g_{ji}$.)

Equation (4.5) can be abbreviated by using the *Einstein summation convention.* In particular,

$$(\mathrm{d}s)^2 = g_{ij}\,\mathrm{d}u^i\,\mathrm{d}u^j. \tag{4.6}$$

The "Einstein summation convention" is the convention that an index repeated as both a superscript and a subscript is a *dummy index* that is summed over even though no summation sign is written down. This saves a lot of writing. The range of the summation depends on the context and if the range is not obvious, the convention should not be used. In general, a dummy index appearing with or without a summation sign can be replaced in a mathematical expression without changing the meaning of that expression. For example, the i and the j that appear in (4.6) can be replaced by p and q. That is

$$(\mathrm{d}s)^2 = g_{ij}\,\mathrm{d}u^i\,\mathrm{d}u^j = g_{pq}\,\mathrm{d}u^p\,\mathrm{d}u^q.$$

For so-called "flat" Euclidean or pseudo-Euclidean spaces, we can start with a position vector

$$\mathbf{s} = \mathbf{e}_j x^j.$$

We then have

$$\mathbf{ds} = \mathbf{e}_j \, dx^j .$$

If we then write

$$(ds)^2 = (\mathbf{ds})^2 = \mathbf{e}_i \mathbf{e}_j \, dx^i \, dx^j = \frac{1}{2}(\mathbf{e}_i \mathbf{e}_j + \mathbf{e}_j \mathbf{e}_i) \, dx^i \, dx^j ,$$

it becomes natural to define

$$\mathbf{e}_i \mathbf{e}_j + \mathbf{e}_j \mathbf{e}_i = 2 n_{ij} \mathbf{I}, \tag{4.7}$$

where $\mathbf{I}$ is the identity matrix, $n_{kk} = 1$ if $1 \le k \le p$, $n_{kk} = -1$ if $p + 1 \le k \le p + q = n$, and $n_{ij} = 0, i \ne j$.

If we use some alternate coordinate system, we have

$$\mathbf{s} = \mathbf{e}_j x^j (u^1, u^2, \ldots, u^n) \tag{4.8}$$

and

$$\mathbf{ds} = \mathbf{e}_j \, dx^j = \mathbf{e}_j \frac{\partial x^j}{\partial u^k} \, du^k = \boldsymbol{\gamma}_k \, du^k , \tag{4.9}$$

where it is obvious that

$$\boldsymbol{\gamma}_k = \mathbf{e}_j \frac{\partial x^j}{\partial u^k} . \tag{4.10}$$

In this situation

$$(\mathbf{ds})^2 = \frac{1}{2}(\boldsymbol{\gamma}_j \boldsymbol{\gamma}_k + \boldsymbol{\gamma}_k \boldsymbol{\gamma}_j) du^j \, du^k . \tag{4.11}$$

Thus, we have

$$\boldsymbol{\gamma}_j \boldsymbol{\gamma}_k + \boldsymbol{\gamma}_k \boldsymbol{\gamma}_j = 2 g_{jk} \mathbf{I}. \tag{4.12}$$

(We will use $\mathbf{e}_1, \mathbf{e}_2, \ldots, \mathbf{e}_n$ to indicate a Cartesian frame for a Euclidean or pseudo-Euclidean space; $\boldsymbol{\gamma}_1, \boldsymbol{\gamma}_2, \ldots, \boldsymbol{\gamma}_n$ to indicate some alternative coordinate frame which we call *Dirac vectors*; and $\mathbf{E}_1, \mathbf{E}_2, \ldots, \mathbf{E}_n$ to indicate an orthonormal noncoordinate frame.)

With these thoughts in mind, we can present some appropriate definitions. To begin with:

Definition 27. We designate a set of n vectors $\{\mathbf{e}_1, \mathbf{e}_2, \ldots, \mathbf{e}_n\}$ as an n-dimensional Cartesian frame for a Euclidean or pseudo-Euclidean space if the vectors have the following properties.

(1)

$$\mathbf{e}_j \mathbf{e}_k + \mathbf{e}_k \mathbf{e}_j = 2 n_{jk} \mathbf{I}, \tag{4.13}$$

where $\mathbf{I}$ is the identity matrix if we use a matrix representation, $n_{kk} = 1$ if $1 \leq k \leq p$, $n_{kk} = -1$ if $p+1 \leq k \leq p+q = n$, and $n_{ij} = 0$ if $i \neq j$. (As we note below, a matrix representation is unnecessary. Nonetheless, $\mathbf{I}$ should be manipulated as if it were an identity matrix.)

(2) By taking all possible products of the members of the frame, one can form a set of 2^n linearly independent vectors. (These products may be written in the form $\mathbf{M}_1\mathbf{M}_2 \ldots \mathbf{M}_n$, where $\mathbf{M}_k = \mathbf{e}_k$ or $\mathbf{I}$.)

It should be noted that various mathematicians and physicists have different agendas. As a result, many authors use a different sign convention for condition 1. Namely, they require that $\mathbf{e}_j\mathbf{e}_k + \mathbf{e}_k\mathbf{e}_j = -2n_{jk}\mathbf{I}$, where n_{jk} is defined in the same manner I have.

A distinguishing feature of pseudo-Euclidean spaces is the *signature matrix* n_{jk}. From the discussion following (4.13), it is clear that n_{jk} is a diagonal matrix whose diagonal elements are ± 1.

From a theorem proved by Sylvester (Cartan 1966, pp. 5–6), it is known that regardless of what orthonormal basis is used to span a given pseudo-Euclidean space, the number of positive entries and the number of negative entries on the diagonal of the signature matrix remain invariant. For example, regardless of the orthonormal basis chosen to span Minkowski 4-space, one diagonal element of the signature matrix will be $+1$ and the other three will be -1. (This signature matrix is frequently designated by the notation $(+, -, -, -)$ or $(1, 3)$.)

It is not difficult to show that one cannot obtain more than 2^n linearly independent products by using the same $\mathbf{e}_k$ more than once in a single product of Dirac vectors. Since $\mathbf{e}_j\mathbf{e}_k = -\mathbf{e}_k\mathbf{e}_j$ for $j \neq k$, it is clear that any finite product of members of a Cartesian frame can be rewritten in the form:

$$\pm(\mathbf{e}_1)^{k_1}(\mathbf{e}_2)^{k_2} \ldots (\mathbf{e}_n)^{k_n}.$$

Furthermore, since $(\mathbf{e}_m)^{2j} = \pm\mathbf{I}$, we can replace $(\mathbf{e}_m)^{k_m}$ by $\pm\mathbf{I}$, if k_m is even or by $\pm\mathbf{e}_m$ if k_m is odd.

One might ask, "Is it always possible to construct a set of matrices with properties 1 and 2 in Def. 27?" As you might suspect, it is indeed always possible. (This is done in Appendix A.)

In a set of lecture notes published in 1958 and since republished in 1993, Marcel Riesz (Riesz 1993, pp. 10–12), "proved" that condition 2 follows from condition 1. This is true most of the time but Marcel Riesz overlooked some exceptional cases. Approximately ten years later, Ian Porteus introduced the necessary correction to the Riesz proof. The corrected proof appears in Appendix A of this book.

As we shall see shortly, Def. 27 in its generalized form is essentially a definition of a Clifford algebra. What is an algebra? What is a Clifford algebra? Before answering these questions, I will burden you with some additional terminology.

A 0-*vector* is a scalar multiple of $\mathbf{I}$.

If $\mathbf{a} = A^j\mathbf{e}_j$, where the A^j's are real numbers, $\mathbf{a}$ is said to be a 1-*vector*.

If

$$\mathbf{A} = \sum_{i_1 < i_2 < \cdots < i_p} A^{i_1 i_2 \ldots i_p} \mathbf{e}_{i_1} \mathbf{e}_{i_2} \ldots \mathbf{e}_{i_p},$$

A is said to be a *p-vector*.

Any linear combination of a 0-vector, a 1-vector, a 2-vector, etc. is said to be a *Clifford number*. Sometimes the terms *scalar, vector, bivector,* and *trivector* are used in place of the terms 0-vector, 1-vector, 2-vector, and 3-vector.

In Chap. 3, I used Clifford numbers to discuss special relativity without reference to any matrix representations. This raises the question: Can we define Clifford numbers without reference to matrix representations? The answer is yes. However, the mathematical machinery required to do that is beyond the stated prerequisites for this text. My reliance on matrix representations allow us to be certain we are not defining entities with inconsistent properties.

If you think of Clifford numbers as matrices, you will avoid making illegal manipulations. However, for actual computations, the use of any matrix representation would be a mechanical impediment. Furthermore, it would be an obstacle to any geometric insight. Therefore, you will not see any matrix representations after this section except in Appendices A and B. Rather than mention the matrix product for a matrix representation of two Clifford numbers, I will denote the corresponding product for the two Clifford numbers as the *Clifford product*. In view of these comments, you should feel free to eliminate **I** from (4.7), (4.12), and (4.13). (Most authors designate the identity element of a Clifford algebra by 1.)

Perhaps, I should add a cautionary note. The theory of matrices is well developed so those who wish to investigate the algebraic aspects of Clifford algebras may wish to use matrix representations.

Any Clifford number **A** can be decomposed into the sum of p-vectors. If we follow the lead of David Hestenes and Garret Sobczyk (1984, p. 3), we designate the p-vector component of **A** by $\langle \mathbf{A} \rangle_p$. We can then write

$$\mathbf{A} = \sum_{p=0}^{n} \langle \mathbf{A} \rangle_p \,. \tag{4.14}$$

Definition 28. An *algebra* consists of a vector space **V** over a field **F** (usually the set of real numbers **R** or the complex numbers **C**) with a binary operation of multiplication such that $\forall \alpha \in \mathbf{F}$ and $\forall \mathbf{A}, \mathbf{B}, \mathbf{C} \in \mathbf{V}$:

(1) $(\alpha \mathbf{A})\mathbf{B} = \mathbf{A}(\alpha \mathbf{B}) = \alpha(\mathbf{AB})$;
(2) $(\mathbf{A} + \mathbf{B})\mathbf{C} = \mathbf{AC} + \mathbf{BC}$; and
(3) $\mathbf{C}(\mathbf{A} + \mathbf{B}) = \mathbf{CA} + \mathbf{CB}$.
If in addition, we have
(4) $(\mathbf{AB})\mathbf{C} = \mathbf{A}(\mathbf{BC})$, the algebra is said to be an *associative algebra*.

Clearly, the set of Clifford numbers, identified with an n-dimensional Euclidean space or an n-dimensional pseudo-Euclidean space, forms a vector space of dimension 2^n that is closed under matrix multiplication. Thus, the set of Clifford numbers forms an associative algebra. If you are genuinely surprised to learn that this algebra is called a *Clifford algebra*, your ability to anticipate the obvious must be suspect.

If the field $\mathbf{F}$ is the set of real numbers, the corresponding Clifford algebra is said to be a *real Clifford algebra*. In this text, we will restrict ourselves to real Clifford algebras. The real Clifford algebra associated with the n-dimensional Euclidean space $\mathbf{R}^n$ is designated by $\mathbf{R}_n$ or $\mathbf{R}_{n,0}$. For a pseudo-Euclidean space with p $(+1)$'s and q (-1)'s in the signature matrix, the vector space is denoted by $\mathbf{R}^{p,q}$ and the corresponding real Clifford algebra is denoted by $\mathbf{R}_{p,q}$. (Here, again conventions vary among authors.)

It should be noted that requiring the field of scalars $\mathbf{F}$ to be the field of real numbers $\mathbf{R}$ does not imply that the components of a matrix representation for the $\mathbf{e}_j$'s have to be real.

Problem 29. Show that we can use the Pauli matrices $\boldsymbol{\sigma}_1$, $\boldsymbol{\sigma}_2$, and $\boldsymbol{\sigma}_3$ to generate the Clifford algebra $\mathbf{R}_3$. That is show that if

$$\mathbf{e}_1 = \boldsymbol{\sigma}_1 = \begin{bmatrix} 0 & 1 \\ 1 & 0 \end{bmatrix}, \quad \mathbf{e}_2 = \boldsymbol{\sigma}_2 = \begin{bmatrix} 0 & -i \\ i & 0 \end{bmatrix}, \quad \text{and} \quad \mathbf{e}_3 = \boldsymbol{\sigma}_3 = \begin{bmatrix} 1 & 0 \\ 0 & -1 \end{bmatrix},$$

then

(a) $\mathbf{e}_j\mathbf{e}_k + \mathbf{e}_k\mathbf{e}_j = 2n_{jk}\mathbf{I}$ and

(b) $\mathbf{I}$, $\mathbf{e}_1$, $\mathbf{e}_2$, $\mathbf{e}_3$, $\mathbf{e}_2\mathbf{e}_3$, $\mathbf{e}_3\mathbf{e}_1$, $\mathbf{e}_1\mathbf{e}_2$, and $\mathbf{e}_1\mathbf{e}_2\mathbf{e}_3$ are linearly independent. (Observe that $\mathbf{e}_1$ and $\mathbf{e}_2\mathbf{e}_3$ would not be linearly independent if we were using the field of complex numbers for our scalars instead of the field of real numbers.)

Problem 30. (a) Using (4.8)–(4.10), show that for spherical coordinates (r, θ, ϕ), where $x^1 = r\cos\phi\sin\theta$, $x^2 = r\sin\phi\sin\theta$, and $x^3 = r\cos\theta$, we have

$$\begin{aligned}
\boldsymbol{\gamma}_1 = \boldsymbol{\gamma}_r &= \mathbf{e}_1\cos\phi\sin\theta + \mathbf{e}_2\sin\phi\sin\theta + \mathbf{e}_3\cos\theta, \\
\boldsymbol{\gamma}_2 = \boldsymbol{\gamma}_\theta &= \mathbf{e}_1 r\cos\phi\cos\theta + \mathbf{e}_2 r\sin\phi\cos\theta - \mathbf{e}_3 r\sin\theta, \text{ and} \\
\boldsymbol{\gamma}_3 = \boldsymbol{\gamma}_\phi &= -\mathbf{e}_1 r\sin\phi\sin\theta + \mathbf{e}_2 r\cos\phi\sin\theta.
\end{aligned}$$

(b) Use (4.6) and (4.12) to compute g_{jk} for spherical coordinates. (Your result should be consistent with (4.3) and (4.5).

Problem 31. Repeat the computations of Prob. (30) for cylindrical coordinates (ρ, ϕ, z), where $x^1 = \rho\cos\phi$, $x^2 = \rho\sin\phi$, and $x^3 = z$.

Problem 32. (You will be referred to the results of this problem later in the text.) Consider the saddle surface defined by $z = xy$ or $x^3 = x^1x^2$. If we let $x^1 = u^1$, $x^2 = u^2$, and $x^3 = u^1u^2$, then the position vector corresponding to a point on the surface may be written in the form

$$\mathbf{s} = u^1\mathbf{e}_1 + u^2\mathbf{e}_2 + u^1u^2\mathbf{e}_3.$$

(a) Determine $\boldsymbol{\gamma}_{u^1}$ and $\boldsymbol{\gamma}_{u^2}$. (Later in the text, I will relabel these Dirac vectors by $\boldsymbol{\gamma}_1$ and $\boldsymbol{\gamma}_2$.)

(b) Use your result from part a) to determine g_{11}, g_{12}, g_{21}, and g_{22}. Then compare your result with (4.4).

Problem 33. Suppose $\mathbf{a} = a^j \mathbf{e}_j$ and $\mathbf{b} = b^k \mathbf{e}_k$. Show that

$$\mathbf{ab} + \mathbf{ba} = (2n_{jk} a^j b^k)\mathbf{I}.$$

(In a Euclidean space, this becomes

$$\mathbf{ab} + \mathbf{ba} = 2\left(\sum_{j=1}^{n} a^j b^j\right)\mathbf{I}.)$$

4.2 The Scalar Product and Metric Tensor

An arbitrary vector $\mathbf{a}$ can be written in the form $\mathbf{a} = a^j \mathbf{e}_j$. The result of Prob. 33 makes it possible to define a scalar product $\langle \mathbf{a}, \mathbf{b} \rangle$ of two vectors $\mathbf{a}$ and $\mathbf{b}$. Since $\mathbf{ab} + \mathbf{ba}$ is a 0-vector, we can define $\langle \mathbf{a}, \mathbf{b} \rangle$ as the coefficient of $\mathbf{I}$ that occurs in the following equation:

$$\frac{1}{2}(\mathbf{ab} + \mathbf{ba}) = \langle \mathbf{a}, \mathbf{b} \rangle \, \mathbf{I} = \langle \mathbf{b}, \mathbf{a} \rangle \, \mathbf{I}. \tag{4.15}$$

This definition of a scalar product generates a symmetric matrix known as the *metric tensor* g_{ij}. The components of the metric tensor are related to whatever coordinate system is being used. In particular,

$$g_{ij} = \left\langle \boldsymbol{\gamma}_i, \boldsymbol{\gamma}_j \right\rangle. \tag{4.16}$$

When the functional relationship between the Cartesian coordinates and the alternate coordinate system under consideration is known, we can determine the components of the metric tensor. In particular, since

$$\boldsymbol{\gamma}_\alpha = \frac{\partial x^j}{\partial u^\alpha} \mathbf{e}_j,$$

(4.16) can be rewritten as

$$g_{\alpha\beta} = \frac{\partial x^j}{\partial u^\alpha} \frac{\partial x^k}{\partial u^\beta} \left\langle \mathbf{e}_j, \mathbf{e}_k \right\rangle = \frac{\partial x^j}{\partial u^\alpha} \frac{\partial x^k}{\partial u^\beta} n_{jk}. \tag{4.17}$$

The inverse of the matrix $g_{\alpha\beta}$ is designated by $g^{\alpha\beta}$. Since $g_{\alpha\beta}$ is symmetric with respect to its two lower indices, the inverse matrix is also symmetric with respect to its two upper indices. That is $g^{\alpha\beta} = g^{\beta\alpha}$.

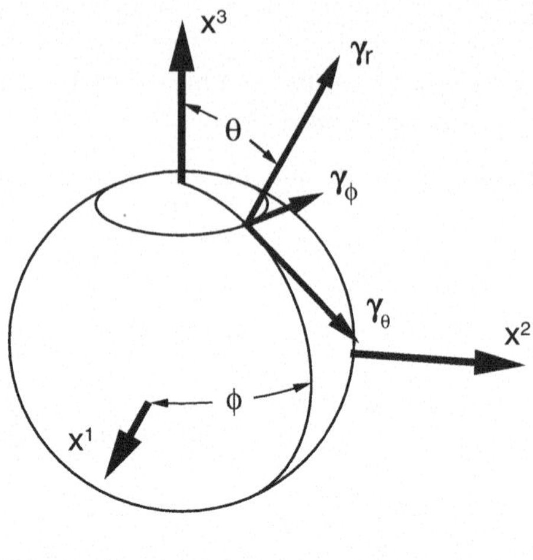

Fig. 4.2 For spherical coordinates or any other coordinate system, the lower index coordinate Dirac vectors are tangent to coordinate curves. What coordinate curve is left out in this diagram?

Assuming the inverse matrix exists, we can use it to construct an alternate basis for the n-dimensional vector space, which is said to be *biorthogonal* w.r.t. (with respect to) the coordinate basis $\{\boldsymbol{\gamma}_1, \boldsymbol{\gamma}_2, \ldots, \boldsymbol{\gamma}_n\}$. This alternate *biorthogonal basis* is designated by indices in the upper position and is defined by the relation:

$$\boldsymbol{\gamma}^\alpha = g^{\alpha\beta}\boldsymbol{\gamma}_\beta \quad \text{for } \alpha = 1, 2, \ldots, n. \tag{4.18}$$

We note that

$$\left\langle \boldsymbol{\gamma}^\alpha, \boldsymbol{\gamma}_\beta \right\rangle = g^{\alpha\eta}\left\langle \boldsymbol{\gamma}_\eta, \boldsymbol{\gamma}_\beta \right\rangle = g^{\alpha\eta} g_{\eta\beta} = \delta^\alpha_\beta.$$

That is

$$\left\langle \boldsymbol{\gamma}^\alpha, \boldsymbol{\gamma}_\beta \right\rangle = \delta^\alpha_\beta, \tag{4.19}$$

where δ^α_β is the *Kronecker delta symbol* defined by the relation:

$$\delta^\alpha_\beta = \begin{cases} 1 & \text{if } \alpha = \beta \\ 0 & \text{if } \alpha \neq \beta \end{cases} \tag{4.20}$$

Equation (4.18) may be reversed. That is

$$g_{n\alpha}\boldsymbol{\gamma}^\alpha = g_{\eta\alpha} g^{\alpha\beta}\boldsymbol{\gamma}_\beta = \delta^\beta_\eta \boldsymbol{\gamma}_\beta = \boldsymbol{\gamma}_\eta,$$

and thus

$$\boldsymbol{\gamma}_\eta = g_{n\alpha}\boldsymbol{\gamma}^\alpha. \tag{4.21}$$

It should be observed that the $\boldsymbol{\gamma}_\alpha$'s are tangent to coordinate curves. (See Fig. 4.2 for the example of spherical coordinates.) That is, if we consider the functions

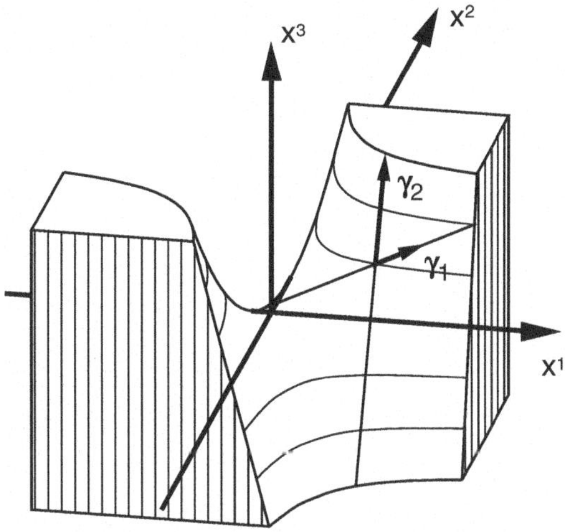

Fig. 4.3 For the two-dimensional saddle surface, the space spanned by $\boldsymbol{\gamma}_1$ and $\boldsymbol{\gamma}_2$ at one point is not the same as that spanned by $\boldsymbol{\gamma}_1$ and $\boldsymbol{\gamma}_2$ at another point. In this example, the coordinate curves are straight lines

$x^k(u^1, u^2, \ldots, u^n)$ and set each of u^j's except u^α equal to its individual constant c^j, then we have

$$x^k(u^\alpha) = x^k(c^1, \ldots, c^{\alpha-1}, u^\alpha, c^{\alpha+1}, \ldots, c^n) \quad \text{for } k = 1, 2, \ldots, n. \tag{4.22}$$

This equation defines a coordinate curve:

$$\mathbf{x}(u^\alpha) = x^j(c^1, \ldots, c^{\alpha-1}, u^\alpha, c^{\alpha+1}, \ldots, c^n)\mathbf{e}_j.$$

Since $\boldsymbol{\gamma}_\alpha = \frac{\partial}{\partial u^\alpha}\mathbf{x}(u^\alpha)$, we may interpret $\boldsymbol{\gamma}_\alpha$ to be a vector tangent to the given curve. We should also note that each $\boldsymbol{\gamma}_\alpha$ is a function of its location and thus both its magnitude and direction may vary from point to point. Thus each $\boldsymbol{\gamma}_\alpha$ and the space spanned by the $\boldsymbol{\gamma}_\alpha$'s should be identified with a particular point. If the $\boldsymbol{\gamma}_\alpha$'s span an n-dimensional space, then the spaces spanned by the $\boldsymbol{\gamma}_\alpha$'s at each point will be essentially the same except for the choice of origin.

However, in the example of the saddle surface described in Prob. 32, the situation is a little different. (See Fig. 4.3.) We have a two-dimensional curved surface imbedded in a flat three-dimensional space. The two-dimensional plane spanned by $\boldsymbol{\gamma}_1$ and $\boldsymbol{\gamma}_2$ is tangent to the saddle surface, where the point of tangency is the origin for the vectors $\boldsymbol{\gamma}_1$ and $\boldsymbol{\gamma}_2$.

Clearly, the plane tangent to the saddle surface at one point will have a different orientation and location than a plane tangent to the surface at another point. In this context, it makes sense to speak of any linear combination of the $\boldsymbol{\gamma}_\alpha$'s as a *tangent vector* and the space spanned by the $\boldsymbol{\gamma}_\alpha$'s at a given point as the *tangent space* for that given point. When we carry out any of the usual vector operations (for example, addition or scalar multiplication), the result is identified at that same given point.

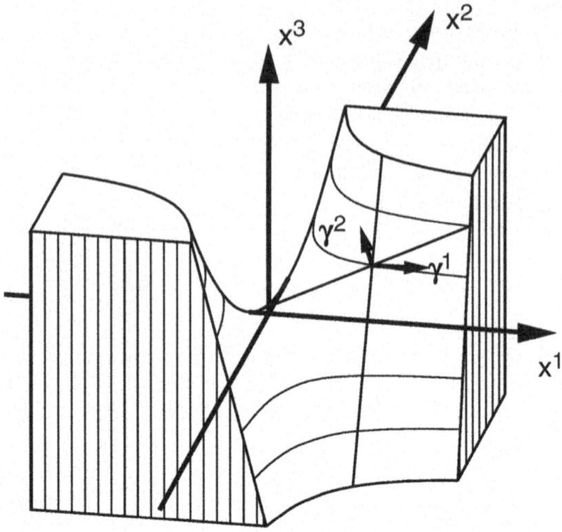

Fig. 4.4 On a two-dimensional curved surface, the upper index coordinate Dirac vectors $\boldsymbol{\gamma}^1$ and $\boldsymbol{\gamma}^2$ at a given point are each perpendicular to a coordinate curve at that point. However, $\boldsymbol{\gamma}^1$ and $\boldsymbol{\gamma}^2$ lie in the same plane as that spanned by $\boldsymbol{\gamma}_1$ and $\boldsymbol{\gamma}_2$. How does this notion generalize to higher dimensions?

If you have encountered or will encounter the formalism of tangent vectors and differential forms, you should note that the analog of $\boldsymbol{\gamma}_\alpha$ is the tangent vector $\partial/\partial u^\alpha$ and the analog of $\boldsymbol{\gamma}^\alpha$ is $\mathbf{d}u^\alpha$. Linear combinations of the $\partial/\partial u^\alpha$'s are said to be tangent vectors and linear combinations of the $\mathrm{d}u^\alpha$'s are said to be 1-forms.

The 1-forms form a vector space that is considered distinct from that spanned by the tangent vectors. Each 1-form is defined to be a linear function that maps tangent vectors onto the set of real numbers. In particular,

$$\mathrm{d}u^\alpha\left(\frac{\partial}{\partial u^\beta}\right) = \langle \boldsymbol{\gamma}^\alpha, \boldsymbol{\gamma}_\beta \rangle = \delta^\alpha_\beta.$$

The formalism of differential forms is both natural and necessary for the study of spaces without a metric tensor. However when a metric is added to the structure, the formalism of differential forms becomes both conceptually and computationally unwieldy when compared with the formalism of Clifford algebra.

Although it is appropriate to refer to the $\boldsymbol{\gamma}_\alpha$'s as tangent vectors, this is not a good way to distinguish them from the $\boldsymbol{\gamma}^\alpha$'s. It is true that the $\boldsymbol{\gamma}_\alpha$'s are tangent to coordinate curves while the $\boldsymbol{\gamma}^\alpha$'s may not be. (See Fig. 4.4.) However, the $\boldsymbol{\gamma}^\alpha$'s span the same vector space as the $\boldsymbol{\gamma}_\alpha$'s. Thus, the $\boldsymbol{\gamma}^\alpha$'s could also be referred to as "tangent vectors."

With these thoughts in mind, we will refer to the $\boldsymbol{\gamma}_\alpha$'s as *lower index coordinate Dirac vectors* and the $\boldsymbol{\gamma}^\alpha$'s as *upper index coordinate Dirac vectors.*

Using these upper and lower index coordinate Dirac vectors gives us two ways of representing the same tangent vector:

$$\mathbf{a} = A^j \boldsymbol{\gamma}_j = A_k \boldsymbol{\gamma}^k. \tag{4.23}$$

From (4.21)

$$\mathbf{a} = A^j \boldsymbol{\gamma}_j = A^j g_{jk} \boldsymbol{\gamma}^k. \tag{4.24}$$

Thus, it is clear that

$$A_k = A^j g_{jk}. \tag{4.25}$$

Similarly, it is easy to show that a reciprocal relation holds. Namely

$$A^i = g^{ij} A_j. \tag{4.26}$$

The A^i's are referred to as the *contravariant components of the vector* **a** while the A_j's are referred to as the *covariant components of the vector* **a**. (Some authors in other formalisms speak of "contravariant and covariant vectors." This terminology makes sense if one applies "contravariant vector" to the n-tuple $\left(A^1, A^2, \ldots, A^n\right)$ as Penrose and Rindler do in their *Spinors & Space-Time* (Penrose and Rindler 1984, p. 72). But it makes less sense when it is applied to $\mathbf{a} = A^j \boldsymbol{\gamma}_j$ since it is also true that the same vector $\mathbf{a} = A_j \boldsymbol{\gamma}^j$.

The terms *contravariant* and *covariant* refer to the way the entities behave under a coordinate transformation. Consider two systems of coordinates $\{u^1, u^2, \ldots, u^n\}$ and $\{\bar{u}^1, \bar{u}^2, \ldots, \bar{u}^n\}$. The corresponding lower index coordinate Dirac vectors are defined by the equations:

$$\boldsymbol{\gamma}_k = \frac{\partial x^j}{\partial u^k} \mathbf{e}_j \tag{4.27}$$

and

$$\bar{\boldsymbol{\gamma}}_p = \frac{\partial x^q}{\partial \bar{u}^p} \mathbf{e}_q. \tag{4.28}$$

From (4.27)

$$\frac{\partial u^k}{\partial x^r} \boldsymbol{\gamma}_k = \frac{\partial u^k}{\partial x^r} \frac{\partial x^j}{\partial u^k} \mathbf{e}_j = \frac{\partial x^j}{\partial x^r} \mathbf{e}_j = \delta^j_r \mathbf{e}_j = \mathbf{e}_r. \tag{4.29}$$

Thus, we have from (4.28) and (4.29)

$$\bar{\boldsymbol{\gamma}}_p = \frac{\partial x^q}{\partial \bar{u}^p} \mathbf{e}_q = \frac{\partial x^q}{\partial \bar{u}^p} \frac{\partial u^k}{\partial x^q} \boldsymbol{\gamma}_k = \frac{\partial u^k}{\partial \bar{u}^p} \boldsymbol{\gamma}_k.$$

That is to say

$$\bar{\boldsymbol{\gamma}}_p = \frac{\partial u^k}{\partial \bar{u}^p} \boldsymbol{\gamma}_k. \tag{4.30}$$

Furthermore, since

$$\mathbf{a} = \bar{A}^p \bar{\boldsymbol{\gamma}}_p = \bar{A}^p \frac{\partial u^k}{\partial \bar{u}^p} \boldsymbol{\gamma}_k = A^k \boldsymbol{\gamma}_k,$$

it follows that:

$$A^k = \bar{A}^p \frac{\partial u^k}{\partial \bar{u}^p}. \tag{4.31}$$

Similarly, it follows that

$$\bar{A}^r = A^k \frac{\partial \bar{u}^r}{\partial u^k}. \tag{4.32}$$

To figure out how the upper index coordinate Dirac vectors transform under a coordinate transformation, we observe that since the $\bar{\gamma}^j$'s and the γ^j's span the same space:

$$\bar{\boldsymbol{\gamma}}^j = \alpha_p^j \boldsymbol{\gamma}^p, \tag{4.33}$$

where the α_p^j's are yet to be determined. Also from (4.30), we have

$$\delta_k^j = \langle \bar{\boldsymbol{\gamma}}^j, \bar{\boldsymbol{\gamma}}_k \rangle = \left\langle \alpha_p^j \boldsymbol{\gamma}^p, \frac{\partial u^m}{\partial \bar{u}^k} \boldsymbol{\gamma}_m \right\rangle = \alpha_p^j \frac{\partial u^m}{\partial \bar{u}^k} \delta_m^p = \alpha_p^j \frac{\partial u^p}{\partial \bar{u}^k}$$

or restated

$$\delta_k^j = \alpha_p^j \frac{\partial u^p}{\partial \bar{u}^k}.$$

Multiplying this last equation by $\partial \bar{u}^k / \partial u^i$ and summing over k, gives us

$$\delta_k^j \frac{\partial \bar{u}^k}{\partial u^i} = \alpha_p^j \frac{\partial u^p}{\partial \bar{u}^k} \frac{\partial \bar{u}^k}{\partial u^i} = \alpha_p^j \frac{\partial u^p}{\partial u^i} = \alpha_p^j \delta_i^p.$$

That is

$$\alpha_i^j = \frac{\partial \bar{u}^j}{\partial u^i}.$$

Combining this equation with (4.33), we finally have

$$\bar{\boldsymbol{\gamma}}^j = \frac{\partial \bar{u}^j}{\partial u^i} \boldsymbol{\gamma}^i. \tag{4.34}$$

From this result, we can use essentially the same trick we used to prove (4.31) and (4.32) to get

$$A_k = \bar{A}_p \frac{\partial \bar{u}^p}{\partial u^k} \tag{4.35}$$

and

$$\bar{A}_p = A_j \frac{\partial u^j}{\partial \bar{u}^p}. \tag{4.36}$$

You should compare (4.32) and (4.34) to see that the upper index Dirac coordinate vectors transform in the same way that the contravariant components of a vector do. A similar observation follows from a comparison of (4.30) and (4.36).

The notion of covariant and contravariant components of a vector can be generalized. A real valued array of functions of space points $\{A_{j_1 j_2 \dots j_q}\}$ with q lower indices is said to be a *covariant tensor of rank q* if

$$\bar{A}_{j_1 j_2 \dots j_q} = A_{k_1 k_2 \dots k_q} \frac{\partial u^{k_1}}{\partial \bar{u}^{j_1}} \frac{\partial u^{k_2}}{\partial \bar{u}^{j_2}} \cdots \frac{\partial u^{k_q}}{\partial \bar{u}^{j_q}}. \tag{4.37}$$

Similarly, a real valued array of functions of space points $\{A^{k_1 k_2 \dots k_p}\}$ is said to be a *contravariant tensor of rank p* if

$$\bar{A}^{j_1 j_2 \dots j_p} = A^{k_1 k_2 \dots k_p} \frac{\partial \bar{u}^{j_1}}{\partial u^{k_1}} \frac{\partial \bar{u}^{j_2}}{\partial u^{k_2}} \cdots \frac{\partial \bar{u}^{j_p}}{\partial u^{k_p}}. \tag{4.38}$$

It should not be too surprising to be told that one can have *mixed tensors of rank* $p + q$. Frequently, the position of the indices can be significant. For example, it is quite possible that

$$A_{ij}{}^{k} \neq A_{i}{}^{k}{}_{j}.$$

At this point, the nature of coordinate transformations on a mixed tensor should be obvious. For example

$$\bar{A}^{k}{}_{ij} = A^{r}{}_{pq} \frac{\partial u^p}{\partial \bar{u}^i} \frac{\partial u^q}{\partial \bar{u}^j} \frac{\partial \bar{u}^k}{\partial u^r}.$$

Sometimes it is useful to refer to the *valence of a tensor*. A tensor with p upper indices and q lower indices is said to have *valence* $\begin{bmatrix} p \\ q \end{bmatrix}$.

It should be noted that just because an entity appears in this book with some indices does not mean that it is a tensor. The entities g_{jk}, g^{jk}, and δ^k_j are tensors but n_{jk} is not. (See Prob. 45.)

Problem 34. Use the results of Prob. 32 to compute the components of $g^{\alpha\beta}$ for the saddle surface. Then use (4.18) to compute $\boldsymbol{\gamma}^1$ and $\boldsymbol{\gamma}^2$.

Problem 35. Demonstrate that the Kronecker delta symbol δ^k_j that was defined by (4.20) is a tensor.

Problem 36. Using the relation

$$g_{jk} = \frac{1}{2}(\boldsymbol{\gamma}_j \boldsymbol{\gamma}_k + \boldsymbol{\gamma}_k \boldsymbol{\gamma}_j),$$

show g_{jk} is a covariant tensor of rank 2.

Problem 37. Show $\langle \boldsymbol{\gamma}^p, \boldsymbol{\gamma}^q \rangle = g^{pq}$.
Hint: $\boldsymbol{\gamma}^p = g^{p\alpha} \boldsymbol{\gamma}_\alpha$ and $\boldsymbol{\gamma}^q = g^{q\beta} \boldsymbol{\gamma}_\beta$.

(Using this result along with (4.34), it becomes a trivial task to show that g^{pq} is a contravariant tensor of rank 2.)

Problem 38. Using the fact that $\boldsymbol{\gamma}^{\alpha} = g^{\alpha\beta}\boldsymbol{\gamma}_{\beta}$ and $g^{jk}g_{km} = \delta^{j}_{m}$, demonstrate the fact that $\boldsymbol{\gamma}_j = g_{jk}\boldsymbol{\gamma}^k$.

Problem 39. Suppose $\mathbf{a} = A^{\alpha}\boldsymbol{\gamma}_{\alpha} = A_{\beta}\boldsymbol{\gamma}^{\beta}$. Use the relation $\boldsymbol{\gamma}^{\beta} = g^{\alpha\beta}\boldsymbol{\gamma}_{\alpha}$ to show that $A^{\alpha} = g^{\alpha\beta}A_{\beta}$. Also show $A_{\beta} = g_{\alpha\beta}A^{\alpha}$.

Problem 40. Suppose $A^{i_1 i_2 \ldots i_p}_{j_1 j_2 \ldots j_q}$ is a tensor with valence $\begin{bmatrix} p \\ q \end{bmatrix}$. Furthermore, suppose $A^{i_1 i_2 \ldots i_{p-1}}_{j_1 j_2 \ldots j_{q+1}} = A^{i_1 i_2 \ldots i_{p-1}\alpha}_{j_1 j_2 \ldots j_q} g_{\alpha\, j_{q+1}}$. Show $A^{i_1 i_2 \ldots i_{p-1}}_{j_1 j_2 \ldots j_{q+1}}$ is a tensor of valence $\begin{bmatrix} p-1 \\ q+1 \end{bmatrix}$.

(It should also be noted that if $A^{i_1 i_2 \ldots i_p}_{j_1 j_2 \ldots j_q}$ is a tensor with valence $\begin{bmatrix} p \\ q \end{bmatrix}$, then $A^{i_1 i_2 \ldots i_{p+1}}_{j_1 j_2 \ldots j_{q-1}}$ is a tensor with valence $\begin{bmatrix} p+1 \\ q-1 \end{bmatrix}$ if $A^{i_1 i_2 \cdots i_{p+1}}_{j_1 j_2 \ldots j_{q-1}} = A^{i_1 i_2 \ldots i_p}_{j_1 j_2 \ldots j_{q-1}\alpha} g^{\alpha\, i_{p+1}}$.)

Problem 41. Suppose $A^{i_1 i_2 \ldots i_{p-1}}_{j_1 j_2 \cdots j_{q-1}} = A^{i_1 i_2 \ldots i_{p-1}\alpha}_{j_1 j_2 \ldots j_{q-1}\alpha}$ and $A^{i_1 i_2 \ldots i_p}_{j_1 j_2 \cdots j_q}$ is a tensor of valence $\begin{bmatrix} p \\ q \end{bmatrix}$. Show $A^{i_1 i_2 \ldots i_{p-1}}_{j_1 j_2 \ldots j_{q-1}}$ is a tensor of valence $\begin{bmatrix} p-1 \\ q-1 \end{bmatrix}$.

Problem 42. Return to the saddle surface mentioned in Prob. 32.

(a) Determine the components of g^{jk}.
(b) Show

$$\boldsymbol{\gamma}^1 = \frac{\left[1 + (u^1)^2\right]\mathbf{e}_1 - u^1 u^2 \mathbf{e}_2 + u^2 \mathbf{e}_3}{1 + (u^1)^2 + (u^2)^2} \quad \text{and}$$

$$\boldsymbol{\gamma}^2 = \frac{-u^1 u^2 \mathbf{e}_1 + \left[1 + (u^2)^2\right]\mathbf{e}_2 + u^1 \mathbf{e}_3}{1 + (u^1)^2 + (u^2)^2}.$$

Problem 43. An *index-free Clifford number* $\mathbf{A}$ is a Clifford number that contains no unsummed indices. An example of such a Clifford number is the 3-vector $\mathbf{A} = A^{\alpha\beta}{}_{\nu}\boldsymbol{\gamma}_{\alpha}\boldsymbol{\gamma}_{\beta}\boldsymbol{\gamma}^{\nu}$ where the $A^{\alpha\beta}{}_{\nu}$'s are the components of a tensor. Demonstrate that an index-free Clifford number transforms under a change of coordinates as a scalar. That is $\bar{\mathbf{A}} = \bar{A}^{\alpha\beta}{}_{\nu}\bar{\boldsymbol{\gamma}}_{\alpha}\bar{\boldsymbol{\gamma}}_{\beta}\bar{\boldsymbol{\gamma}}^{\nu} = A^{\alpha\beta}{}_{\nu}\boldsymbol{\gamma}_{\alpha}\boldsymbol{\gamma}_{\beta}\boldsymbol{\gamma}^{\nu} = \mathbf{A}$.

Problem 44. Consider the two-dimensional surface of a sphere of radius R defined by the equation:

$$\mathbf{s} = \mathbf{e}_1 R\cos\phi\sin\theta + \mathbf{e}_2 R\sin\phi\sin\theta + \mathbf{e}_3 R\cos\theta.$$

Compute $\boldsymbol{\gamma}_{\theta}$, $\boldsymbol{\gamma}_{\phi}$, $g_{\theta\theta}$, $g_{\theta\phi}$, and $g_{\phi\phi}$. Then compute $g^{\theta\theta}$, $g^{\theta\phi}$, and $g^{\phi\phi}$. Finally, compute $\boldsymbol{\gamma}^{\theta}$, and $\boldsymbol{\gamma}^{\phi}$.

Problem 45. Suppose n_{ij} is the entity defined by (4.7). Also suppose that

$$\bar{n}_{rs} = n_{ij} \frac{\partial u^i}{\partial \bar{u}^r} \frac{\partial u^j}{\partial \bar{u}^s}.$$

Show that in general it cannot be said that $\bar{n}_{kk} = 1$ if $1 \leq k \leq p$, $\bar{n}_{kk} = -1$ if $p + 1 \leq k \leq n$, and $\bar{n}_{rs} = 0$ if $r \neq s$. (It is for this reason that n_{ij} is not a tensor.)

4.3 The Exterior Product for p-Vectors

As already indicated in Sect. 4.1, the Clifford product of p distinct $\mathbf{e}_j$'s is necessarily a p-vector. However, this does not mean that the Clifford product of p distinct 1-vectors is necessarily a p-vector. For example, suppose

$$\mathbf{a} = a^1\mathbf{e}_1 + a^2\mathbf{e}_2 + a^3\mathbf{e}_3 \quad \text{and}$$

$$\mathbf{b} = b^1\mathbf{e}_1 + b^2\mathbf{e}_2 + b^3\mathbf{e}_3.$$

Then

$$\mathbf{ab} = \left(a^1b^1 + a^2b^2 + a^3b^3\right)\mathbf{I} + \left(a^2b^3 - a^3b^2\right)\mathbf{e}_2\mathbf{e}_3 + \left(a^3b^1 - a^1b^3\right)\mathbf{e}_3\mathbf{e}_1 + \left(a^1b^2 - a^2b^1\right)\mathbf{e}_1\mathbf{e}_2.$$

Thus, it is not difficult to see that in general the Clifford product of two 1-vectors is a linear combination of a 0-vector and a 2-vector. However, a 2-vector can be constructed by taking a linear combination of **ab** and **ba**. In particular,

$$\frac{1}{2}(\mathbf{ab} - \mathbf{ba}) = \left(a^2b^3 - a^3b^2\right)\mathbf{e}_2\mathbf{e}_3 + \left(a^3b^1 - a^1b^3\right)\mathbf{e}_3\mathbf{e}_1 + \left(a^1b^2 - a^2b^1\right)\mathbf{e}_1\mathbf{e}_2.$$

More generally, suppose we consider p distinct Dirac vectors each one of which is expanded in terms of a Euclidean basis. That is $\boldsymbol{\gamma}_\upsilon = (\partial x^j/\partial u^\nu)\mathbf{e}_j$ for $\upsilon = 1, 2, \ldots, p$. Then

$$\boldsymbol{\gamma}_1\boldsymbol{\gamma}_2 \ldots \boldsymbol{\gamma}_p = \frac{\partial x^{j_1}}{\partial u^1}\frac{\partial x^{j_2}}{\partial u^2} \cdots \frac{\partial x^{j_p}}{\partial u^p}\mathbf{e}_{j_1}\mathbf{e}_{j_2} \ldots \mathbf{e}_{j_p}. \tag{4.39}$$

The terms on the R.H.S. (right-hand side) of (4.39) are of several types. If $\mathbf{e}_{j_1}$, $\mathbf{e}_{j_2}, \ldots, \mathbf{e}_{j_p}$ are all distinct, then their product is a p-vector. On the other hand, if some of the $\mathbf{e}_{j_k}$'s in a given product are identical, then they may be grouped

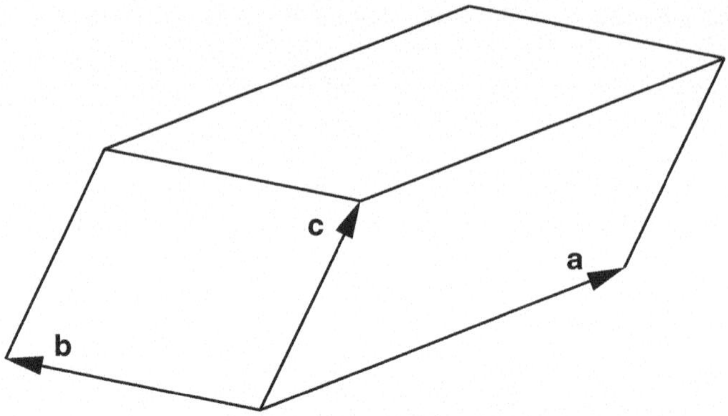

Fig. 4.5 $\mathbf{a} \wedge \mathbf{b} \wedge \mathbf{c} = |\mathbf{a} \cdot (\mathbf{b} \times \mathbf{c})|\, \mathbf{e}_1\mathbf{e}_2\mathbf{e}_3$. Volume of parallelepiped = $|\mathbf{a} \cdot (\mathbf{b} \times \mathbf{c})|$

together and multiplied out until the only remaining $\mathbf{e}_{j_k}$'s are distinct. Thus we see that the R.H.S. of (4.39) consists of a linear combination of p-vectors, $(p-2)$-vectors, $(p-4)$-vectors, and so forth on down to 1-vectors or 0-vectors.

One can project out the p-vector in the R.H.S. (right-hand side) of (4.39) by antisymmetrizing the L.H.S. To do this explicitly, it is useful to introduce the *generalized Kronecker delta symbol* $\delta^{i_1 i_2 \ldots i_p}_{j_1 j_2 \ldots j_p}$.

$$\delta^{i_1 i_2 \ldots i_p}_{j_1 j_2 \ldots j_p} = \begin{cases} 1 & \text{if the } j_k\text{'s are distinct and the sequence of } i_k\text{'s is} \\ & \quad \text{an even permutation of the } j_k\text{'s} \\ -1 & \text{if the } j_k\text{'s are distinct and the sequence of } i_k\text{'s is} \\ & \quad \text{an odd permutation of the } j_k\text{'s} \\ 0 & \quad \text{in all other cases} \end{cases}$$

With this generalized Kronecker delta function, we can define $\boldsymbol{\gamma}_{\nu_1\nu_2\ldots\nu_p}$ as the antisymmetric product of $\boldsymbol{\gamma}_{\nu_1}, \boldsymbol{\gamma}_{\nu_2}, \ldots, \boldsymbol{\gamma}_{\nu_p}$. That is

$$\begin{aligned} \boldsymbol{\gamma}_{\nu_1\nu_2\ldots\nu_p} &= \frac{1}{p!}\delta^{\eta_1\eta_2\ldots\eta_p}_{\nu_1\nu_2\ldots\nu_p}\boldsymbol{\gamma}_{\eta_1}\boldsymbol{\gamma}_{\eta_2}\cdots\boldsymbol{\gamma}_{\eta_p} \\ &= \frac{1}{p!}\delta^{\eta_1\eta_2\ldots\eta_p}_{\nu_1\nu_2\ldots\nu_p}\frac{\partial x^{j_1}}{\partial u^{\eta_1}}\frac{\partial x^{j_2}}{\partial u^{\eta_2}}\cdots\frac{\partial x^{j_p}}{\partial u^{\eta_p}}\mathbf{e}_{j_1}\mathbf{e}_{j_2}\ldots\mathbf{e}_{j_p} \\ &= \frac{1}{p!}\det\begin{bmatrix} \frac{\partial x^{j_1}}{\partial u^{\nu_1}} & \frac{\partial x^{j_2}}{\partial u^{\nu_1}} & \cdot & \cdot & \cdot & \frac{\partial x^{j_p}}{\partial u^{\nu_1}} \\ \frac{\partial x^{j_1}}{\partial u^{\nu_2}} & \frac{\partial x^{j_2}}{\partial u^{\nu_2}} & \cdot & \cdot & \cdot & \frac{\partial x^{j_p}}{\partial u^{\nu_2}} \\ \cdot & \cdot & \cdot & \cdot & \cdot & \cdot \\ \cdot & \cdot & \cdot & \cdot & \cdot & \cdot \\ \cdot & \cdot & \cdot & \cdot & \cdot & \cdot \\ \frac{\partial x^{j_1}}{\partial u^{\nu_p}} & \frac{\partial x^{j_2}}{\partial u^{\nu_p}} & \cdot & \cdot & \cdot & \frac{\partial x^{j_p}}{\partial u^{\nu_p}} \end{bmatrix}\mathbf{e}_{j_1}\mathbf{e}_{j_2}\ldots\mathbf{e}_{j_p}. \end{aligned} \tag{4.40}$$

It should be noted that if the $\boldsymbol{\gamma}_\nu$'s are linearly dependent, then each determinant on the R.H.S. of (4.40) will be zero and thus the antisymmetric product $\boldsymbol{\gamma}_{\nu_1\nu_2\ldots\nu_p}$ will be zero.

For the special case for which $p = n$, the determinants that appear on the R.H.S. of (4.40) are all identical so we can write

$$\boldsymbol{\gamma}_{12\cdots n} = \det \begin{bmatrix} \frac{\partial x^1}{\partial u^1} & \frac{\partial x^2}{\partial u^1} & \cdots & \frac{\partial x^n}{\partial u^1} \\ \frac{\partial x^1}{\partial u^2} & \frac{\partial x^2}{\partial u^2} & \cdots & \frac{\partial x^n}{\partial u^2} \\ \cdot & \cdot & \cdots & \cdot \\ \cdot & \cdot & \cdots & \cdot \\ \cdot & \cdot & \cdots & \cdot \\ \frac{\partial x^1}{\partial u^n} & \frac{\partial x^2}{\partial u^n} & \cdots & \frac{\partial x^n}{\partial u^n} \end{bmatrix} \mathbf{e}_{12\cdots n}, \quad \text{where } \mathbf{e}_{12\cdots n} = \mathbf{e}_1\mathbf{e}_2\ldots\mathbf{e}_n. \tag{4.41}$$

The determinant on the R.H.S. of (4.41) may be considered the volume of the n-dimensional parallelepiped spanned by the vectors $\boldsymbol{\gamma}_1, \boldsymbol{\gamma}_2, \ldots, \boldsymbol{\gamma}_n$.

The Clifford product of a p-vector and a q-vector results in a linear combination of vectors of the orders $p+q$, $p+q-2$, $p+q-4$, …, on down to $|p-q|$. However, for many purposes, it is useful to drop the lower order forms. With this in mind, one defines the *exterior product* of a p-vector and a q-vector as the projection of the Clifford product onto the space of vectors of order $p+q$. In particular,

$$\begin{aligned}\boldsymbol{\gamma}_{\nu_1\nu_2\ldots\nu_p} \wedge \boldsymbol{\gamma}_{\eta_1\eta_2\ldots\eta_q} &= \frac{1}{(p+q)!}\delta^{\beta_1\beta_2\beta_3\ldots\beta_{p+q}}_{\nu_1\nu_2\ldots\nu_p\eta_1\eta_2\ldots\eta_q}\boldsymbol{\gamma}_{\beta_1}\boldsymbol{\gamma}_{\beta_2}\cdots\boldsymbol{\gamma}_{\beta_{p+q}} \\ &= \boldsymbol{\gamma}_{\nu_1\nu_2\ldots\nu_p\eta_1\eta_2\ldots\eta_q},\end{aligned} \tag{4.42}$$

where it is understood that $\boldsymbol{\gamma}_{\nu_1\nu_2\ldots\nu_r} = 0$, if any two of the indices are identical.

This can be generalized further. Suppose

$$\mathbf{F} = \frac{1}{p!}F^{i_1i_2\ldots i_p}\boldsymbol{\gamma}_{i_1i_2\ldots i_p} \quad \text{and}$$

$$\mathbf{G} = \frac{1}{q!}G^{j_1j_2\ldots j_q}\boldsymbol{\gamma}_{j_1j_2\ldots j_q}.$$

Then

$$\mathbf{F}\wedge\mathbf{G} = \frac{1}{p!q!}F^{i_1i_2\ldots i_p}G^{j_1j_2\ldots j_q}\boldsymbol{\gamma}_{i_1i_2\ldots i_pj_1j_2\ldots j_q}. \tag{4.43}$$

It should be obvious to you that (4.41)–(4.43) can be adjusted for upper index coordinate Dirac vectors or for orthonormal noncoordinate Dirac vectors.

To summarize and review some of the consequences of our notation, we recall that any Clifford number can be written as a sum of p-vectors:

$$\mathbf{A} = \sum_{p=0}^{n} \langle \mathbf{A} \rangle_p = \sum_{p=0}^{n} \frac{1}{p!} A^{i_1 i_2 \ldots i_p} \boldsymbol{\gamma}_{i_1 i_2 \ldots i_p}. \tag{4.44}$$

For the example of spherical coordinates:

$$\mathbf{A} = A\mathbf{I} + A^r \boldsymbol{\gamma}_r + A^\theta \boldsymbol{\gamma}_\theta + A^\phi \boldsymbol{\gamma}_\phi + A^{\theta\phi} \boldsymbol{\gamma}_{\theta\phi} + A^{\phi r} \boldsymbol{\gamma}_{\phi r} + A^{r\theta} \boldsymbol{\gamma}_{r\theta} + A^{r\theta\phi} \boldsymbol{\gamma}_{r\theta\phi}. \tag{4.45}$$

(It is understood that $A^{\theta\phi} = -A^{\phi\theta}$ and $A^{r\theta\phi} = A^{\theta\phi r} = A^{\phi r\theta} = -A^{r\phi\theta} = -A^{\phi\theta r} = -A^{\theta r\phi}$.)

The set of pure p-vectors forms a vector subspace of Clifford numbers of dimension $\binom{n}{p}$. The dimension is $\binom{n}{p}$ since that is the number of ways you can choose a set of p $\mathbf{e}_k$'s from a set of n. For example, for a three-dimensional Euclidean space, the number of 2-vectors is $\binom{3}{2} = 3$ and is spanned by $\boldsymbol{\gamma}_{\theta\phi}$, $\boldsymbol{\gamma}_{\phi r}$, and $\boldsymbol{\gamma}_{r\theta}$. The same space is spanned by $\mathbf{e}_{23}$, $\mathbf{e}_{31}$, and $\mathbf{e}_{12}$ or $\boldsymbol{\gamma}^{\theta\phi}$, $\boldsymbol{\gamma}^{\phi r}$, and $\boldsymbol{\gamma}^{r\theta}$.

By taking the direct sum of the different p-vector spaces, we arrive at the vector space of all Clifford numbers. As previously indicated, the dimension of this larger space is 2^n. This follows from the binomial theorem:

$$\sum_{p=0}^{n} \binom{n}{p} = \sum_{p=0}^{n} \binom{n}{p} 1^p 1^{n-p} = (1+1)^n = 2^n.$$

In Sect. 4.2, we defined the scalar product for a pair of 1-vectors. We now turn to the problem of defining the scalar product for any two Clifford numbers.

To deal with this problem, we need to introduce the concept of the reverse of a Clifford number. Following the notation of Hestenes and Sobczyk (1984, p. 3) and the terminology of Marcel Riesz (1958, p.13), we define the *reverse* $\mathbf{A}^\dagger$ of a Clifford number $\mathbf{A}$ to be that Clifford number obtained by reversing the order of all products of Dirac vectors in the linear expansion of $\mathbf{A}$. For example, if $\mathbf{A}$ is the Clifford number that appears in (4.45), then

$$\begin{aligned}\mathbf{A}^\dagger &= A\mathbf{I} + A^r \boldsymbol{\gamma}_r + A^\theta \boldsymbol{\gamma}_\theta + A^\phi \boldsymbol{\gamma}_\phi + A^{\theta\phi} \boldsymbol{\gamma}_{\phi\theta} + A^{\phi r} \boldsymbol{\gamma}_{r\phi} + A^{r\theta} \boldsymbol{\gamma}_{\theta r} + A^{r\theta\phi} \boldsymbol{\gamma}_{\phi\theta r} \\ &= A\mathbf{I} + A^r \boldsymbol{\gamma}_r + A^\theta \boldsymbol{\gamma}_\theta + A^\phi \boldsymbol{\gamma}_\phi - A^{\theta\phi} \boldsymbol{\gamma}_{\theta\phi} - A^{\phi r} \boldsymbol{\gamma}_{\phi r} - A^{r\theta} \boldsymbol{\gamma}_{r\theta} - A^{r\theta\phi} \boldsymbol{\gamma}_{r\theta\phi}.\end{aligned} \tag{4.46}$$

We can now define the scalar product of two real Clifford numbers to be

$$\langle \mathbf{A}, \mathbf{B} \rangle = \left\langle \mathbf{A}^\dagger \mathbf{B} \right\rangle_0. \tag{4.47}$$

Problem 46. Harley Flanders (Flanders (1963), p.14) defined the scalar product for differential forms in a somewhat different manner. The Clifford algebra analogue would be

$$\left\langle \boldsymbol{\gamma}^{i_1 i_2 \ldots i_p}, \boldsymbol{\gamma}^{j_1 j_2 \ldots j_p} \right\rangle = \det \begin{bmatrix} \left\langle \boldsymbol{\gamma}^{i_1}, \boldsymbol{\gamma}^{j_1} \right\rangle & \left\langle \boldsymbol{\gamma}^{i_1}, \boldsymbol{\gamma}^{j_2} \right\rangle & \cdot \cdot \cdot & \left\langle \boldsymbol{\gamma}^{i_1}, \boldsymbol{\gamma}^{j_p} \right\rangle \\ \left\langle \boldsymbol{\gamma}^{i_2}, \boldsymbol{\gamma}^{j_1} \right\rangle & \left\langle \boldsymbol{\gamma}^{i_2}, \boldsymbol{\gamma}^{j_2} \right\rangle & \cdot \cdot \cdot & \left\langle \boldsymbol{\gamma}^{i_2}, \boldsymbol{\gamma}^{j_p} \right\rangle \\ \cdot & \cdot & \cdot \cdot \cdot & \cdot \\ \cdot & \cdot & \cdot \cdot \cdot & \cdot \\ \cdot & \cdot & \cdot \cdot \cdot & \cdot \\ \left\langle \boldsymbol{\gamma}^{i_p}, \boldsymbol{\gamma}^{j_1} \right\rangle & \left\langle \boldsymbol{\gamma}^{i_p}, \boldsymbol{\gamma}^{j_2} \right\rangle & \cdot \cdot \cdot & \left\langle \boldsymbol{\gamma}^{i_p}, \boldsymbol{\gamma}^{j_p} \right\rangle \end{bmatrix}. \tag{4.48}$$

For a slightly more general case, suppose

$$\mathbf{A} = \frac{1}{p!} A_{i_1 i_2 \ldots i_p} \boldsymbol{\gamma}^{i_1 i_2 \ldots i_p} \quad \text{and} \quad \mathbf{B} = \frac{1}{p!} B_{j_1 j_2 \ldots j_p} \boldsymbol{\gamma}^{j_1 j_2 \ldots j_p}.$$

Then

$$\langle \mathbf{A}, \mathbf{B} \rangle = \left(\frac{1}{p!} \right)^2 A_{i_1 i_2 \ldots i_p} B_{j_1 j_2 \ldots j_p} \left\langle \boldsymbol{\gamma}^{i_1 i_2 \ldots i_p}, \boldsymbol{\gamma}^{j_1 j_2 \ldots j_p} \right\rangle. \tag{4.49}$$

Show that for p-vectors, (4.48) and (4.49) agree with (4.47).

Suggestion: Try using some orthonormal basis.

Problem 47. Show that $\left\langle \mathbf{A}^{\dagger} \mathbf{B} \right\rangle_0 = \left\langle \mathbf{B}^{\dagger} \mathbf{A} \right\rangle_0$ where $\mathbf{A}$ and $\mathbf{B}$ are arbitrary Clifford numbers that are not necessarily index free. (This shows that $\langle \mathbf{A}, \mathbf{B} \rangle = \langle \mathbf{B}, \mathbf{A} \rangle$.)

Problem 48. Show that $\langle \mathbf{AB} \rangle_0 = \langle \mathbf{BA} \rangle_0$, where $\mathbf{A}$ and $\mathbf{B}$ are arbitrary Clifford numbers that are not necessarily index free.

Problem 49. Generalize the result of Prob. 48 to show that the scalar component of a product of several Clifford numbers is invariant under cyclic permutations. That is

$$\begin{aligned} \langle \mathbf{A}_1 \mathbf{A}_2 \ldots \mathbf{A}_n \rangle_0 &= \langle \mathbf{A}_2 \mathbf{A}_3 \ldots \mathbf{A}_n \mathbf{A}_1 \rangle_0 \\ &= \ldots = \langle \mathbf{A}_{k+!} \mathbf{A}_{k+2} \ldots \mathbf{A}_n \mathbf{A}_1 \mathbf{A}_2 \ldots \mathbf{A}_k \rangle_0 . \end{aligned} \tag{4.50}$$

Problem 50. Suppose in three dimensions $\mathbf{a} = a^k \mathbf{e}_k$, $\mathbf{b} = b \mathbf{e}_k$, and $\mathbf{c} = c^k \mathbf{e}_k$. Show that

$$\mathbf{a} \wedge \mathbf{b} \wedge \mathbf{c} = \det \begin{bmatrix} a^1 & a^2 & a^3 \\ b^1 & b^2 & b^3 \\ c^1 & c^2 & c^3 \end{bmatrix} \mathbf{e}_1 \mathbf{e}_2 \mathbf{e}_3 .$$

4.4 Some Useful Formulas

In this section, I will present some useful formulas for products of Dirac vectors. You should first note that

$$\begin{aligned}\mathbf{e}_j\mathbf{e}_k &= \frac{1}{2}\left(\mathbf{e}_j\mathbf{e}_k - \mathbf{e}_k\mathbf{e}_j\right) + \frac{1}{2}\left(\mathbf{e}_j\mathbf{e}_k + \mathbf{e}_k\mathbf{e}_j\right)\\ &= \mathbf{e}_{jk} + n_{jk}\mathbf{I}.\end{aligned}$$

For the product $\mathbf{e}_i\mathbf{e}_j\mathbf{e}_k$, there are five cases:

Case 1: All three indices are distinct. For this situation

$$\mathbf{e}_i\mathbf{e}_j\mathbf{e}_k = \mathbf{e}_{ijk}.$$

Case 2: $i = j \neq k$. Then

$$\mathbf{e}_i\mathbf{e}_j\mathbf{e}_k = n_{ij}\mathbf{e}_k.$$

Case 3: $i = k \neq j$. Then

$$\mathbf{e}_i\mathbf{e}_j\mathbf{e}_k = -\mathbf{e}_j\mathbf{e}_i\mathbf{e}_k = -n_{ik}\mathbf{e}_j.$$

Case 4: $i \neq j = k$. Then

$$\mathbf{e}_i\mathbf{e}_j\mathbf{e}_k = n_{jk}\mathbf{e}_i.$$

Case 5: $i = j = k$. Then

$$\mathbf{e}_i\mathbf{e}_j\mathbf{e}_k = n_{ij}\mathbf{e}_k = n_{ik}\mathbf{e}_j = n_{jk}\mathbf{e}_i.$$

Because of the fact that the expressions for most of the cases are zero in any given situation, we can summarize all five cases with a single equation. Namely

$$\mathbf{e}_i\mathbf{e}_j\mathbf{e}_k = \mathbf{e}_{ijk} + n_{ij}\mathbf{e}_k - n_{ik}\mathbf{e}_j + n_{jk}\mathbf{e}_i.$$

Using similar arguments, we can show that

$$\begin{aligned}\mathbf{e}_i\mathbf{e}_j\mathbf{e}_k\mathbf{e}_m &= \mathbf{e}_{ijkm} + n_{ij}\mathbf{e}_{km} - n_{ik}\mathbf{e}_{jm} + n_{im}\mathbf{e}_{jk} + n_{jk}\mathbf{e}_{im} - n_{jm}\mathbf{e}_{ik} + n_{km}\mathbf{e}_{ij}\\ &\quad + n_{ij}n_{km}\mathbf{I} - n_{ik}n_{jm}\mathbf{I} + n_{im}n_{jk}\,\mathbf{I}.\end{aligned}$$

One can easily use these formulas to obtain others. For example if we replace $\mathbf{e}_i\mathbf{e}_j$ by $\mathbf{e}_{ij}$ and $\mathbf{e}_k\mathbf{e}_m$ by $\mathbf{e}_{km}$ in the left hand side of the previous equation then the terms on the right hand side involving n_{ij} or n_{km} become zero and we get

$$\mathbf{e}_{ij}\mathbf{e}_{km} = \mathbf{e}_{ijkm} - n_{ik}\mathbf{e}_{jm} + n_{im}\mathbf{e}_{jk} + n_{jk}\mathbf{e}_{im} - n_{jm}\mathbf{e}_{ik} - n_{ik}n_{jm}\mathbf{I} + n_{im}n_{jk}\,\mathbf{I}.$$

Summarizing, we have

$$\mathbf{e}_j\mathbf{e}_k = \mathbf{e}_{jk} + n_{jk}\mathbf{I}. \tag{4.51}$$

$$\mathbf{e}_i\mathbf{e}_j\mathbf{e}_k = \mathbf{e}_{ijk} + n_{ij}\mathbf{e}_k - n_{ik}\mathbf{e}_j + n_{jk}\mathbf{e}_i. \tag{4.52}$$

$$\begin{aligned}\mathbf{e}_i\mathbf{e}_j\mathbf{e}_k\mathbf{e}_m &= \mathbf{e}_{ijkm} + n_{ij}\mathbf{e}_{km} - n_{ik}\mathbf{e}_{jm} + n_{im}\mathbf{e}_{jk} + n_{jk}\mathbf{e}_{im} - n_{jm}\mathbf{e}_{ik} + n_{km}\mathbf{e}_{ij}\\ &\quad + n_{ij}n_{km}\mathbf{I} - n_{ik}n_{jm}\mathbf{I} + n_{im}n_{jk}\,\mathbf{I}.\end{aligned} \tag{4.53}$$

$$\mathbf{e}_{ij}\mathbf{e}_{km} = \mathbf{e}_{ijkm} - n_{ik}\mathbf{e}_{jm} + n_{im}\mathbf{e}_{jk} + n_{jk}\mathbf{e}_{im} - n_{jm}\mathbf{e}_{ik} - n_{ik}n_{jm}\mathbf{I} + n_{im}n_{jk}\,\mathbf{I}. \tag{4.54}$$

These formulas can easily be modified for coordinate vectors. For example, if

$$\boldsymbol{\gamma}_i = \frac{\partial x^\alpha}{\partial u^i}\mathbf{e}_\alpha,$$

then

$$\begin{aligned}\boldsymbol{\gamma}_\alpha\boldsymbol{\gamma}_\beta\boldsymbol{\gamma}_\lambda &= \frac{\partial x^i}{\partial u^\alpha}\frac{\partial x^j}{\partial u^\beta}\frac{\partial x^k}{\partial u^\lambda}\mathbf{e}_i\mathbf{e}_j\mathbf{e}_k\\ &= \frac{\partial x^i}{\partial u^\alpha}\frac{\partial x^j}{\partial u^\beta}\frac{\partial x^k}{\partial u^\lambda}\left[\mathbf{e}_{ijk} + n_{ij}\mathbf{e}_k - n_{ik}\mathbf{e}_j + n_{jk}\mathbf{e}_i\right]\\ &= \boldsymbol{\gamma}_{\alpha\beta\lambda} + g_{\alpha\beta}\boldsymbol{\gamma}_\lambda - g_{\alpha\lambda}\boldsymbol{\gamma}_\beta + g_{\beta\lambda}\boldsymbol{\gamma}_\alpha.\end{aligned}$$

Organizing this along with several other equations, we have

$$\boldsymbol{\gamma}_\alpha\boldsymbol{\gamma}_\beta = \boldsymbol{\gamma}_{\alpha\beta} + g_{\alpha\beta}\mathbf{I}. \tag{4.55}$$

$$\boldsymbol{\gamma}_\alpha\boldsymbol{\gamma}_\beta\boldsymbol{\gamma}_\lambda = \boldsymbol{\gamma}_{\alpha\beta\lambda} + g_{\alpha\beta}\boldsymbol{\gamma}_\lambda - g_{\alpha\lambda}\boldsymbol{\gamma}_\beta + g_{\beta\lambda}\boldsymbol{\gamma}_\alpha. \tag{4.56}$$

$$\begin{aligned}\boldsymbol{\gamma}_\alpha\boldsymbol{\gamma}_\beta\boldsymbol{\gamma}_\lambda\boldsymbol{\gamma}_\nu &= \boldsymbol{\gamma}_{\alpha\beta\lambda\nu} + g_{\alpha\beta}\boldsymbol{\gamma}_{\lambda\nu} - g_{\alpha\lambda}\boldsymbol{\gamma}_{\beta\nu} + g_{\alpha\nu}\boldsymbol{\gamma}_{\beta\lambda} + g_{\beta\lambda}\boldsymbol{\gamma}_{\alpha\nu} - g_{\beta\nu}\boldsymbol{\gamma}_{\alpha\lambda} + g_{\lambda\nu}\boldsymbol{\gamma}_{\alpha\beta}\\ &\quad + g_{\alpha\beta}g_{\lambda\nu}\mathbf{I} - g_{\alpha\lambda}g_{\beta\nu}\mathbf{I} + g_{\alpha\nu}g_{\beta\lambda}\mathbf{I}.\end{aligned} \tag{4.57}$$

$$\boldsymbol{\gamma}_{\alpha\beta}\boldsymbol{\gamma}_{\lambda\nu} = \boldsymbol{\gamma}_{\alpha\beta\lambda\nu} - g_{\alpha\lambda}\boldsymbol{\gamma}_{\beta\nu} + g_{\alpha\nu}\boldsymbol{\gamma}_{\beta\lambda} + g_{\beta\lambda}\boldsymbol{\gamma}_{\alpha\nu} - g_{\beta\nu}\boldsymbol{\gamma}_{\alpha\lambda} - g_{\alpha\lambda}g_{\beta\nu}\mathbf{I} + g_{\alpha\nu}g_{\beta\lambda}\mathbf{I}. \tag{4.58}$$

Problem 51. Obtain an appropriate formula for $\boldsymbol{\gamma}_\alpha\boldsymbol{\gamma}_{\beta\lambda}$.

Problem 52. Obtain an appropriate formula for $\boldsymbol{\gamma}^\alpha\boldsymbol{\gamma}^\beta\boldsymbol{\gamma}^\lambda$.

Problem 53. (a) Use (4.47) and (4.58) to obtain a formula for $\langle\boldsymbol{\gamma}_{\alpha\beta}, \boldsymbol{\gamma}_{\lambda\nu}\rangle$.
(b) $\langle\boldsymbol{\gamma}_{\alpha\beta}, \boldsymbol{\gamma}_{\alpha\beta}\rangle = ?$

4.5 Gram–Schmidt Formulas

Sometimes it is useful to construct an orthonormal frame other than the Euclidean system. This is particularly true in the study of n-dimensional surfaces imbedded in an $(n+1)$-dimensional Euclidean space. (Such surfaces are called *hypersurfaces*.)

In this case, you may have a coordinate system of Dirac vectors $\{\boldsymbol{\gamma}_1, \boldsymbol{\gamma}_2, \ldots, \boldsymbol{\gamma}_n\}$ that are all tangent to the surface at a given point. You would like to have a system of orthonormal vectors $\{\mathbf{E}_1, \mathbf{E}_2, \ldots, \mathbf{E}_n\}$ that span the same space as $\{\boldsymbol{\gamma}_1, \boldsymbol{\gamma}_2, \ldots, \boldsymbol{\gamma}_n\}$ plus one additional unit vector $\mathbf{N}$ that is perpendicular to the surface at the given point.

Using the usual vector techniques, this can be a formidable computation. However, Hestenes and Sobczyk (1984, pp. 27–28) have worked out a variant of the Gram–Schmidt process that seems somewhat easier. For Euclidean spaces, you begin as usual by letting

$$\mathbf{E}_1 = \boldsymbol{\gamma}_1 / |\boldsymbol{\gamma}_1| \,. \tag{4.59}$$

To compute $\mathbf{E}_k$ for higher values of k, you take advantage of the fact that

$$\mathbf{E}_k = (\mathbf{E}_{k-1}\mathbf{E}_{k-2} \cdots \mathbf{E}_2\mathbf{E}_1)(\mathbf{E}_1\mathbf{E}_2 \cdots \mathbf{E}_{k-1}\mathbf{E}_k) \,. \tag{4.60}$$

On first impression, this formula may seem useless or even less than useless. However, the 1-vectors $\{\boldsymbol{\gamma}_1, \boldsymbol{\gamma}_2, \ldots, \boldsymbol{\gamma}_r\}$ span the same space as $\{\mathbf{E}_1, \mathbf{E}_2, \ldots, \mathbf{E}_r\}$. As a consequence

$$\mathbf{E}_1\mathbf{E}_2 \ldots \mathbf{E}_r = \boldsymbol{\gamma}_{12\cdots r} / |\boldsymbol{\gamma}_{12\cdots r}|$$

and (4.60) can be converted to a more useful form:

$$\mathbf{E}_k = \frac{\boldsymbol{\gamma}_{k-1\cdots 21}}{|\boldsymbol{\gamma}_{k-1\cdots 21}|} \frac{\boldsymbol{\gamma}_{12\cdots k}}{|\boldsymbol{\gamma}_{12\cdots k}|} \quad \text{for} \quad k = 2, 3, \ldots, n. \tag{4.61}$$

Finally to obtain a normalized version of a vector perpendicular to the surface, we take advantage of the fact that

$$\mathbf{E}_1\mathbf{E}_2 \cdots \mathbf{E}_n\mathbf{N} = \mathbf{e}_1\mathbf{e}_2 \cdots \mathbf{e}_{n+1} = \mathbf{e}_{12\cdots n+1}.$$

Using this relation and the idea used in the derivation of (4.61) from (4.60), we have

$$\mathbf{N} = \frac{\boldsymbol{\gamma}_{n\cdots 21}}{|\boldsymbol{\gamma}_{n\cdots 21}|} \mathbf{e}_{12\cdots n+1}. \tag{4.62}$$

It should be emphasized that the equations in this section apply only to Euclidean spaces. It is not clear to me whether or not these formulas can be generalized to pseudo-Euclidean spaces. In pseudo-Euclidean spaces or in spaces embedded in pseudo-Euclidean spaces, there are vectors with length zero. If any such vectors occur in the original basis, then the method outlined above will not work. However, using a completely different approach, one can obtain an orthonormal frame from a coordinate frame whenever the metric tensor is known. (See Appendix B.)

For some computations on hypersurfaces, it may be useful to retain the coordinate system of Dirac vectors. However, it is still useful to construct $\mathbf{N}$. In that case, you only need to use (4.62) without having to deal with (4.61).

You may also wish to deal with an n-dimensional curved surface embedded in an m-dimensional Euclidean space where $m - n > 1$. For this situation, it is not clear how useful it is to know that the n-dimensional surface is embedded in a flat m-dimensional Euclidean space. A lot of information can be obtained simply by taking measurements restricted to the n-dimensional surface. If the eigenvalues of the metric tensor are all positive, this approach is known as *Riemannian geometry*. If some of the eigenvalues of the metric tensor are negative, this approach is known as *non-Riemannian geometry*. The whole theory of general relativity is based on this intrinsic approach. The next chapter is also devoted to this angle of attack and you will see some significant results for two-dimensional surfaces. Later you will get an idea of what additional information can be obtained about two-dimensional surfaces by a three-dimensional observer. (For two-dimensional surfaces that can be embedded in a flat three-dimensional Euclidean space, a three-dimensional observer does have some advantages over the two-dimensional observer.) We will not attempt to reveal what advantage a 4 or five-dimensional observer might have over a two-dimensional observer for two-dimensional surfaces that cannot be embedded in a flat three-dimensional space. That is beyond the scope of this author and therefore beyond the scope of this book. On the other hand, we will devote some optional sections to n-dimensional surfaces viewed by an $(n + 1)$-dimensional observer.

Problem 54. In Prob. 32, you hopefully showed that for the saddle surface $\boldsymbol{\gamma}_1 = \mathbf{e}_1 + u^2\mathbf{e}_3$ and $\boldsymbol{\gamma}_2 = \mathbf{e}_2 + u^1\mathbf{e}_3$.

(a) Show $\boldsymbol{\gamma}_{12} = -u^2\mathbf{e}_{23} - u^1\mathbf{e}_{31} + \mathbf{e}_{12}$.
(b) Use the result from part a and whatever formulas you find necessary from this last section to compute $\mathbf{N}$, $\mathbf{E}_1$, and $\mathbf{E}_2$.

4.6 *The Qibla (Kibla) Problem

The Gram–Schmidt process can be used to dispose of a problem encountered and solved by the Islamic mathematicians. Muslims are expected to pray five times a day facing Mecca. However, even if you know your location and that of Mecca, the determination of qibla (the direction to Mecca) is not a trivial problem. Al-Khwarizmi (790–850AD) was one of the several Islamic mathematicians who developed an approximate method for dealing with this problem. Al-Biruni (973–1055AD) was one of the several mathematicians who achieved an exact solution using the knowledge of spherical trigonometry developed by Islamic mathematicians.

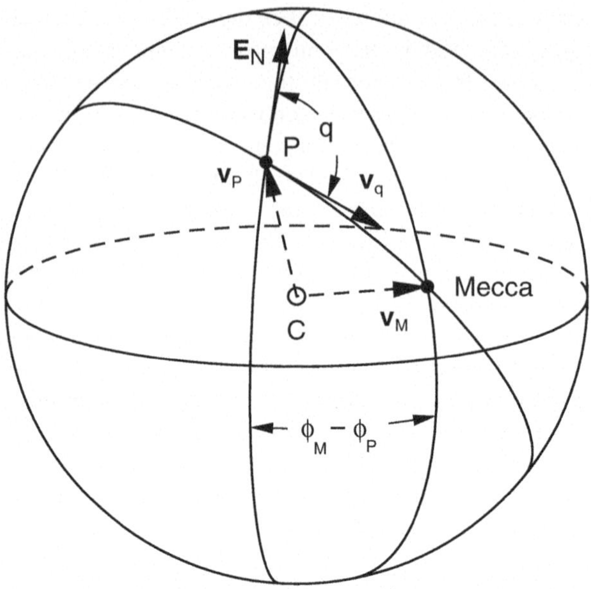

Fig. 4.6 The point P represents the location of the observant Moslem who wishes to face Mecca at time of prayer

The qibla problem can be solved using either vector algebra or Clifford algebra without a sophisticated knowledge of spherical trigonometry. In this section, I will show how the problem can be solved using Clifford algebra.

Using the center of the earth as our origin, we designate the position of our Moslem by $\mathbf{v}_P$ and the position of Mecca by $\mathbf{v}_M$. (See Fig. 4.6.) We designate the direction to Mecca by $\mathbf{v}_q$. If we treat the radius of the earth as our unit length, then

$$\mathbf{v}_q = (\mathbf{v}_P \mathbf{v}_P)\,\mathbf{v}_q = \mathbf{v}_P \mathbf{v}_P \mathbf{v}_q. \tag{4.63}$$

Since $\mathbf{v}_P$ and $\mathbf{v}_q$ are perpendicular, it follows that the product $\mathbf{v}_P \mathbf{v}_q = \mathbf{v}_P \wedge \mathbf{v}_q$. Furthermore, since $\mathbf{v}_q$ lies in the plane spanned by $\mathbf{v}_P$ and $\mathbf{v}_M$, the exterior product $\mathbf{v}_P \wedge \mathbf{v}_q$ is equal to some scalar multiple of $\mathbf{v}_P \wedge \mathbf{v}_M$. If we do not normalize $\mathbf{v}_q$, we can write

$$\mathbf{v}_q = \mathbf{v}_P(\mathbf{v}_P \wedge \mathbf{v}_M). \tag{4.64}$$

Now if θ is used to designate the latitude (the angle above the equator) and ϕ is used the designate the longitude (the number of degrees east of Greenwich, England), then

$$\mathbf{v}_P = \mathbf{e}_1 \cos\theta_P \cos\phi_P + \mathbf{e}_2 \cos\theta_P \sin\phi_P + \mathbf{e}_3 \sin\theta_P.$$

When our computation is complete, we want the qibla expressed in terms of cardinal directions (north, south, east, west) at the point P. To obtain a vector pointing east from point P, we compute

$$\frac{\partial \mathbf{v}_P}{\partial \phi_P} = -\mathbf{e}_1 \cos\theta_P \sin\phi_P + \mathbf{e}_2 \cos\theta_P \cos\phi_P.$$

Factoring out $\cos\theta_P$, we get a unit vector pointing east. Namely

$$\mathbf{E}_E = -\mathbf{e}_1 \sin\phi_P + \mathbf{e}_2 \cos\phi_P .$$

The unit vector pointing north is

$$\mathbf{E}_N = \frac{\partial \mathbf{v}_P}{\partial \theta_P} = -\mathbf{e}_1 \sin\theta_P \cos\phi_P - \mathbf{e}_2 \sin\theta_P \sin\phi_P + \mathbf{e}_3 \cos\theta_P .$$

Summarizing, we have

$$\begin{bmatrix} \mathbf{v}_P \\ \mathbf{E}_E \\ \mathbf{E}_N \end{bmatrix} = \begin{bmatrix} \cos\theta_P \cos\phi_P & \cos\theta_P \sin\phi_P & \sin\theta_P \\ -\sin\phi_P & \cos\phi_P & 0 \\ -\sin\theta_P \cos\phi_P & -\sin\theta_P \sin\phi_P & \cos\theta_P \end{bmatrix} \begin{bmatrix} \mathbf{e}_1 \\ \mathbf{e}_2 \\ \mathbf{e}_3 \end{bmatrix} . \tag{4.65}$$

The 3×3 matrix in (4.65) is orthogonal, so the inverse is simply the transpose. Thus,

$$\begin{bmatrix} \mathbf{e}_1 \\ \mathbf{e}_2 \\ \mathbf{e}_3 \end{bmatrix} = \begin{bmatrix} \cos\theta_P \cos\phi_P & -\sin\phi_P & -\sin\theta_P \cos\phi_P \\ \cos\theta_P \sin\phi_P & \cos\phi_P & -\sin\theta_P \sin\phi_P \\ \sin\theta_P & 0 & \cos\theta_P \end{bmatrix} \begin{bmatrix} \mathbf{v}_P \\ \mathbf{E}_E \\ \mathbf{E}_N \end{bmatrix} . \tag{4.66}$$

Now

$$\mathbf{v}_M = \begin{bmatrix} \cos\theta_M \cos\phi_M, & \cos\theta_M \sin\phi_M, & \sin\theta_M \end{bmatrix} \begin{bmatrix} \mathbf{e}_1 \\ \mathbf{e}_2 \\ \mathbf{e}_3 \end{bmatrix} . \tag{4.67}$$

Combining (4.67) and (4.66), we get

$$\begin{aligned} \mathbf{v}_M = {} & [\cos\theta_M \cos\theta_P (\cos\phi_M \cos\phi_P + \sin\phi_M \sin\phi_P) + \sin\theta_M \sin\theta_P]\, \mathbf{v}_P \\ & + [\cos\theta_M (\sin\phi_M \cos\phi_P - \cos\phi_M \sin\phi_P)]\, \mathbf{E}_E \\ & + [-\cos\theta_M \sin\theta_P (\cos\phi_M \cos\phi_P + \sin\phi_M \sin\phi_P) + \sin\theta_M \cos\theta_P]\, \mathbf{E}_N \end{aligned}$$

or

$$\begin{aligned} \mathbf{v}_M = {} & [\cos\theta_M \cos\theta_P \cos(\phi_M - \phi_P) + \sin\theta_M \sin\theta_P]\, \mathbf{v}_P \\ & + [\cos\theta_M \sin(\phi_M - \phi_P)]\, \mathbf{E}_E \\ & + [-\cos\theta_M \sin\theta_P \cos(\phi_M - \phi_P) + \sin\theta_M \cos\theta_P]\, \mathbf{E}_N . \end{aligned}$$

It then follows that

$$\begin{aligned} \mathbf{v}_P \wedge \mathbf{v}_M = {} & [\cos\theta_M \sin(\phi_M - \phi_P)]\, \mathbf{v}_P \mathbf{E}_E \\ & + [-\cos\theta_M \sin\theta_P \cos(\phi_M - \phi_P) + \sin\theta_M \cos\theta_P]\, \mathbf{v}_P \mathbf{E}_N . \end{aligned}$$

And from (4.64),

$$\begin{aligned}\mathbf{v}_q &= \mathbf{v}_P(\mathbf{v}_P \wedge \mathbf{v}_M) \\ &= [\cos\theta_M \sin(\phi_M - \phi_P)]\,\mathbf{E}_E + [-\cos\theta_M \sin\theta_P \cos(\phi_M - \phi_P) \\ &\quad + \sin\theta_M \cos\theta_P]\,\mathbf{E}_N .\end{aligned}$$

If we change the magnitude of $\mathbf{v}_q$ by dividing by $\cos\theta_M$, we have

$$\mathbf{v}_q = \sin(\phi_M - \phi_P)\mathbf{E}_E + [\tan\theta_M \cos\theta_P - \sin\theta_P \cos(\phi_M - \phi_P)]\,\mathbf{E}_N . \qquad (4.68)$$

Using this last equation and referring to Fig. 4.6, we see that for $\frac{-\pi}{2} < q < \frac{\pi}{2}$:

$$\tan q = \frac{\sin(\phi_M - \phi_P)}{\tan\theta_M \cos\theta_P - \sin\theta_P \cos(\phi_M - \phi_P)} .$$

So finally

$$q = \arctan\left(\frac{\sin(\phi_M - \phi_P)}{\tan\theta_M \cos\theta_P - \sin\theta_P \cos(\phi_M - \phi_P)}\right) . \qquad (4.69)$$

For $q > \frac{\pi}{2}$ or $q < -\frac{\pi}{2}$, these last two formulas have to be adjusted but (4.68) remains valid providing that you are intelligent about the sign conventions for θ_P and ϕ_P.

Problem 55. Kamal Abdali (Abdali 1997, p. 10) notes that in the classical Muslim scientific works on qibla determination, the prime meridian was usually taken to be located at the western coast of Africa or the Canary Islands because that was considered to be the edge of civilization. How would the choice of the prime meridian affect the formula of (4.68)?

Problem 56. According to Abdali (Abdali 1997, p. 2), the Mosque on Massachusetts Avenue in Washington D.C. is aligned so as to be consistent with q=56°, 33 min, and 15 s. This means that the qibla is roughly in the northeast direction. However, Washington D.C. is further north than Mecca so some observers have suggested that the qibla for that Mosque should be in a southeasterly direction. Make some intelligent comment.

Problem 57. The longitude of Washington D.C. is approximately 77° west of Greenwich. For Washington D.C., what value should be assigned to ϕ_P in (4.68)? The latitude for Capetown, Africa is approximately 34° south of the equator. For Capetown, what value should be assigned to θ_P in (4.68).

If you wish to learn some spherical trigonometry, consider the following two problems.

Problem 58. Suppose the sphere in Fig. 4.7 has unit radius. Also suppose the spherical triangle is oriented so that $\mathbf{v}_A$ is located at the northpole $(0, 0, 1)$ and

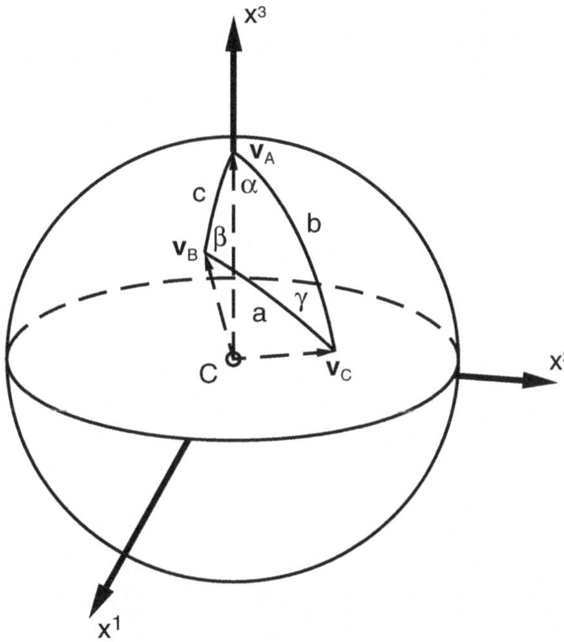

Fig. 4.7 Spherical triangle with angles α, β, and γ along with great circle arcs of lengths a, b, and c

$\mathbf{v}_B$ is in the x^1–x^3 plane. This implies that $\mathbf{v}_B = (\sin c, 0, \cos c)$ and $\mathbf{v}_C = (\cos\alpha \sin b, \sin\alpha \sin b, \cos b)$.

(a) Compute $\langle \mathbf{v}_B, \mathbf{v}_C \rangle$ to obtain a formula for $\cos a$ in terms of $\cos\alpha$, $\sin b$, $\cos b$, $\sin c$, and $\cos c$. Using symmetry without further computation, write down similar formulas for $\cos\beta$ and $\cos\gamma$.
(b) Suppose the sphere has radius R. Then the formulas in Part a) remain valid if a is replaced by a/R, b is replaced by b/R, and c is replaced by c/R. Show that if you then consider the limit when $R \to \infty$, you obtain the usual Law of Cosines.

Problem 59. Consider Figs. 4.5 and 4.7.

(a) Using the assumptions of Part a of Prob. 58, compute $\mathbf{v}_A \wedge \mathbf{v}_B \wedge \mathbf{v}_C$ to obtain a formula for the volume of the parallelepiped determined by $\mathbf{v}_A$, $\mathbf{v}_B$, and $\mathbf{v}_C$. Use symmetry to obtain two more formulas for the same volume.
(b) Equate the three formulas and obtain the equation,

$$\frac{\sin\alpha}{\sin a} = \frac{\sin\beta}{\sin b} = \frac{\sin\gamma}{\sin c}.$$

(c) Use the same approach used in Part (b) of Prob. 58 to obtain the usual Law of Sines.

Problem 60. Use the spherical trigonometry developed in Probs. 58 and 59 to obtain (4.69) without resorting to Clifford algebra. (According to Victor Katz, Al-Bīrūnī solved the qibla problem without using the results of Prob. 58. See (Katz 1998, pp. 277–281)).

4.7 *Mathematics of Arab Speaking Muslims

*4.7.1 *Greek Science and Mathematics in Alexandria*

In 332BC, shortly after he conquered Egypt, Alexander the Great founded the city of Alexandria at the mouth of the Nile. When Alexander died nine years later in 323BC, his empire was divided into three parts by warring factions of Macedonians who had served as his top-ranking officers. Ptolemy, son of Lagus, emerged as self-proclaimed king of Egypt. As Ptolemy I, he established a dynasty of Ptolemies that would last almost 300 years until the death of the famous Cleopatra in 30BC.

At least some of the Macedonians had a high regard for learning. Alexander the Great's father, Philip of Macedonia, had employed Aristotle to tutor the future world conquerer. Ptolemy I shared this high regard for learning. As an historian, he was the author of an authoritative account of Alexander's campaigns. However, it is possible he did not have a gift for mathematics. According to a popular story, he once asked Euclid if he could learn geometry without struggling through Euclid's *Elements*. Noting that there might be limits to the powers of a seemingly omnipotent king, Euclid was said to respond, "There is no royal road to geometry" (Casson 2001, pp. 32–33).

Like Philip of Macedonia, Ptolemy I wanted his offspring to get a good education. Demetrius of Phalerum, a disciple of Aristotle and exiled governor of Athens, was recruited to tutor Ptolemy's family. Demetrius would soon become the first librarian in Alexandria. Associated with the library was the "Museum" so named because it was considered to be a "house of Muses." It was this institution that was designed to attract the top intellects of the Mediterranean world.

The library may have lasted for more than 500 years, but the impact of the Museum as a dominant center of cutting edge research was relatively short. At the web site of John J. O'Connor and Edmund F. Robertson at the University of St. Andrews, Scotland (http:www-history.mcs.st-andrews.ac.uk/), there is a list of outstanding mathematicians (broadly defined). On the list I was able to identify 15 who were associated with the Museum or at least the city of Alexandria. For one of them (Hypsicles), there seems to be a disagreement about whether he lived during the second century BC or the second century AD. (O'Connor and Roberson list his life span as 190–120BC.) Of the remaining 14, seven lived during a time span of two or three generations. The difference in time between the death of the first of the seven (Euclid) and the last of the seven (Apollonius of Perga) was only 75 years! The next mathematician on the list of 14 was Heron of Alexandria who was born about 200 years after the death of Apollonius. What was the cause of this time gap?

Historians point to the events of 144BC. Other things may have contributed to the decline of mathematics at the Museum. The date of 144BC was roughly 46 years after the death of Apollonius but what happened around 144BC explains a lot. For several years prior to 144BC, two brother Ptolemys were co-rulers of Egypt. They ruled as mutual enemies with Rome acting as a reluctant referee.

When the older brother died in battle in 145BC, his 16-year-old son was in line to become the sole ruler of Egypt. To prevent his loss of power, the younger brother, Physcon, known as (Potbelly), took preemptive action. He persuaded his brother's widow to marry him and then had his nephew assassinated on the day of the wedding. As you might imagine, this put a strain on the marriage. Physcon's bride became enraged further, when Physcon took her daughter as a second wife. Physcon now had two wives, who both had the name "Cleopatra." You might think that this would lead to confusion in conversations. However, Physcon's first wife was also his sister. In this situation, the first wife became known as "Cleopatra the Sister" while the second wife became known as "Cleopatra the Wife" (Green 1990, pp. 537–538).

Physcon's controversial political power plays caused a split in public opinion and it was not long before civil war broke out. The residents of Alexandria – particularly the intellectuals and the Jewish community rallied behind the banner of Cleopatra the Sister while the outlying towns rallied behind the banner of Physcon. When Physcon prevailed, he carried out mass purges and expulsions of those who had opposed him in Alexandria.

Some historians suggest that this dispersal of intellectuals was a benefit to the rest of the Hellenic world. However, if there existed mathematicians with the ability of Euclid, Aristarchus, Archimedes, Eratosthenes, or Apollonius, they died as relative unknowns and their contributions to the advancement of mathematics did not survive.

The role of the Alexandrian Museum as a truly dominant center of cutting edge science ended in 144BC. But what of the fabled library? The library continued during the reign of Physcon, but the job of head librarian became a patronage position – assigned to mediocre toadies.

Under earlier Ptolemys, vast resources were allocated to establish the library. Michael Harris writes, "Thousands upon thousands of rolls were bought, copied, stolen, and compiled for its shelves until it contained, according to some estimates 600,000 rolls"% (Harris 1995, p. 45).

There was a competition between libraries of the ancient world for prominence and even dominance. According to one version of history, Ptolemy V tried to restrict the growth of rival libraries by banning the export of papyrus. The chief rival library to the one in Alexandria was that in Pergamum located in modern day Turkey. In response to the export ban, an alternative writing material, parchment, was developed in Pergamum. This was a fortunate event since ancient works that would not have survived written on papyrus have survived on the more durable parchment. Parchment is the result of processing animal skin so that it can be written on both sides.

Writing on leather had occurred earlier and it is not clear how much innovation can be attributed to the residents of Pergamum. However, it is clear that Pergamum became a manufacturing center for parchment and the label "parchment" is a corruption of the name "Pergamum."

The ultimate fate of the library in Alexandria has been the subject of much speculation. A thoughtful evaluation of ancient historical accounts is contained in an essay by Robert Barnes (MacLeod 2000, pp. 61–77). An early account attributed the destruction of the library to Julius Caesar in 48BC. According to this version, Caesar set fire to the ships in the harbor of Alexandria to prevent the enemies of Cleopatra (the Cleopatra of 'Anthony and Cleopatra' fame) from taking the city by sea. Using current military terminology, the library suffered substantial "collateral damage" when the fire spread to the wharves. However, the main library was located a substantial distance from the harbor inside the palace grounds where Caesar was residing. No mention of fire encroaching palace grounds appears in Caesar's chronicles. In a continuation of the Caesar destruction narrative, Marc Anthony was said to have offered Cleopatra 200,000 books looted from the library in Pergamum as compensation. This would have been the entire contents of the Pergamum library but the account does not indicate whether or not the promised gift was delivered.

Confusion existed amongst the few ancient historians who discussed the fate of the Alexandrian library because as a general rule, they were unaware of the fact that there were two significant libraries in Alexandria. The main library was located on the palace grounds and a smaller satellite library was located in a pagan temple dedicated to the Egyptian god Serapis.

At the time of Caesar's visit to Alexandria, there may have been some scrolls destroyed in a warehouse and there was possibly some looting by Romans eager to establish personal libraries. (The first significant Roman library was formed from the spoils of war obtained by Paulus Aemilius, a Roman general, after his victory over King Perseus of Macedonia in 165BC. Later Roman generals would become a little competitive in the formation of personal libraries "collected" from the fringes of the Empire. Shortly before his death, Julius Caesar set up plans to create a public library that would equal or surpass that of Alexandria. His assassination preempted these plans and Caesar's library never came to fruition (Harris 1995, pp. 56–57).)

There are few contemporary references or even near contemporary references to the libraries of Alexandria. One such reference presented in Barnes' essay relates to Domitian who was emperor of Rome from 81AD to 96AD. Referring to Domitian, Sustonius, and early second century biographer wrote, "At the beginning of his reign he neglected liberal studies, although he had arranged for the libraries [in the Porticus Octaviae in Rome], which were destroyed by fire, to be replaced at great expense, seeking everywhere for copies of books, and sending scholars to Alexandria to transcribe and correct them." Thus, we can infer that there was a significant library in Alexandria at that time (MacLeod 2000, pp. 72–73).

If the main library continued to exist, it was certainly destroyed in 272AD when the area of Alexandria that included the palace was obliterated as the result of civil strife (MacLeod 2000, p. 73). At some time, the actual site disappeared. This may have occurred in 365AD. On July 21, 365AD, an earthquake, centered in Crete, created a tsunami that had devastating effects at Alexandria. In later years, significant earthquakes also had impacts on Alexandria. Whatever caused the ultimate disappearance of the library, underwater archaeologists are now convinced that they have discovered the site of the royal palace and thus the site of the ancient library on the floor of the Alexandrian harbor (Goddio 1998, 2008).

The arrival of Christianity put an end to the satellite library. When Christianity was on the rise, Romans burned whatever Christian literature they could seize. When the Christians gained ascendancy, they behaved in a similar manner. In 391 AD, the Temple of Serapis was destroyed on order of Bishop Theophilus (Harris 1995, p. 47). (The Bishop may have been more interested in destroying a pagan temple than a pagan library.)

4.7.2 *Hypatia

One cannot end a discussion of mathematics in Alexandria without mention of Hypatia. Hypatia has been labeled the first woman to make a substantial contribution to the development of mathematics. Her father, Theon of Alexandria who died about 405 AD, is believed to be one of the last and perhaps the last member of the Museum. For many centuries, the only Greek versions of Euclid's *Elements* were attributed to Theon. It was only in the late nineteenth century that an earlier version was discovered in the Vatican (Heath 1921, vol. I, p. 360).

By some accounts, Hypatia outshone her father as a mathematician. It is believed that she was responsible for publishing the results of Diophantus and Apollonius in a form that would be accessible to beginning students (Heath 1921, vol. II, p. 528). The *Almagest,* the 13 volume work by Claudius Ptolemy (ca. 90–ca. 168 AD), was the mathematical foundation for the geocentric theory of the solar system using deferents and epicycles which was to be accepted for roughly 1,400 years. Some scholars now believe that at least some Greek versions of this work that have survived antiquity were edited by Hypatia (Dzielska 1995, pp. 71–72).

However, Hypatia is best known as a Neoplatonic philosopher who suffered a dramatic death at the hands of a vicious mob. In March 415 AD, she was dragged from her chariot into a church where she was stripped naked, dismembered and later burned. Since her death was the result of machinations of Cyril, the church patriarch in Alexandria, many authors of plays, novels, and histories have treated this event as a defeat for Greek pagan reason, logic, and science and a victory for mindless superstition that ushered in the dark ages. Historian Maria Dzielska has studied the events surrounding Hypatia's death along with the correspondence of one of her students, Synesius of Cyrene. She reaches the conclusion that Hypatia's death was the result of a conflict between Christians. I am not convinced that this fact makes the historic implications of Hypatia's death any less horrific.

On October 15, 412 AD, Theophilus died. Theophilus was the church patriarch who had been responsible for the destruction of the Temple of Serapis. Some of the ruling class had tried to defend the pagans (Dzielska 1995, pp. 79–80). However, the members of the political elite were predominately Christians and they did not feel directly threatened by the campaigns of Theophilus against pagans. However, when Cyril was named to succeed Theophilus, it was recognized that he would not only try to suppress the pagan community in Alexandria but he would also be heavy handed toward many of his fellow Christians. As a result, three days of fighting broke out in Alexandria. After attaining victory, Cyril, as expected, set out to suppress Jews, Nestorian Christians, and other heretics. He also set out to gain power at the expense of civil authority represented by Orestes, prefect of Alexandria. Orestes had been sent by Rome to Alexandria only a short time before and probably seemed vulnerable.

Cyril recruited 500 monks from outside Alexandria who confronted Orestes and accused him of paganism. Despite his denials, one of the monks, Ammonius, severely injured the prefect by hitting him on the head with a stone (Dzielska 1995, pp. 86–87). Fortunately for Orestes, a crowd of Alexandrians came to his rescue. When Ammonius was eventually captured, Orestes ordered him to be tortured. When the torture resulted in the death of Ammonius, Cyril tried to gain advantage by portraying Ammonius as a martyr. This ploy did not influence moderate Christians who were well aware of what Ammonius had done.

Both Orestes and Cyril appealed to Rome for support without much response. Cyril then turned to another tactic which would prove much more effective. He suggested that Orestes and he would be like-minded if it were not for the fact that Orestes was a gullible dimwit under the control of a satanic witch – Hypatia. Early in his term in office, Orestes had turned to Hypatia for advice and Hypatia had devoted herself to bolstering Orestes' efforts to maintain secular authority. Paganism was not a primary issue. Hypatia may have been a Christian herself. Certainly, many of her students were Christians who saw no conflict with her teachings and their faith. Two of her students became bishops (Dzielska 1995, p. 105).

Hypatia is believed to be a virgin at the time of her death. Usually "virgin" is associated with youth and many accounts of Hypatia describe her as a young Aphrodite at the time of her death. However, Dzielska has concluded that she was probably about sixty years old at the time of her death. She had many years to attain respect and esteem amongst the elite of Alexandria. From the correspondence of Synesius of Cyrene (one of her students), we can infer that indeed, Hypatia had many devoted followers amongst the social elite of Alexandria. Hypatia and her students were inclined to use their influence to benefit one another but not for members of lower classes (Dzielska 1995, p. 41 and p. 61). Thus, it is plausible that her esteem did not extend to the lower classes. It is also possible that her influence in the upper class may have given credence to Cyril's charge that she could cast spells on individuals to attain her evil ends.

Another fact that gave credibility to Cyril's accusations was that Hypatia and her students formed what was essentially a secret society. They believed that there were some aspects of higher learning that would be dangerous to share with naive members of the lower classes. In one of his letters, Synesius quoted Lysis, a Pythagorean, "To explain philosophy to the mob is only to awaken among men a great contempt for things divine" (Dzielska 1995, p. 60). This left outsiders free to speculate on all sorts of possible satanic rituals carried out by her followers. Hypatia was an accomplished astronomer but Greek astronomy was synonymous with astrology. It is also known that her students were engaged in the interpretation of dreams (Dzielska 1995, p. 63 and p. 79). All these facts made Hypatia vulnerable to Cyril's charges.

It is surmised that the murder of Hypatia was carried out by a society of men known as parbolans. They were assigned the task of rounding up sick, disabled and homeless people and transporting them to hospitals or church houses for the poor. They also served as a militia at the disposal of the Alexandrian patriarch. They had been employed by the previous patriarch to attack paganism and they were used to spread the charge that Hypatia was an evil witch.

Cyril's machinations were quite successful. No one was prosecuted for Hypatia's murder and Orestes disappeared from Alexandria. Presumably, he either resigned his post or was recalled by Rome. Some efforts were made to reduce Cyril's control of the parabolans but the effect of these efforts was short lived (Dzielska 1995, p. 96). Cyril's dogged pursuit of his goals was deeply appreciated by some. After his death in 444AD, the mother church conferred sainthood on him.

Maria Dzielska points out that the works of Plato and Aristotle continued to be taught in the Roman Empire after the death of Hypatia (Dzielska 1995, p. 105). However, it is clear that Alexandria was no longer the city that could attract scholars dedicated to open inquiry from the four corners of the world.

*4.7.3 *The Rise of Islam and the House of Wisdom*

The Romans never did establish a library of the same scale as that of Alexandria or Pergamum. Generally, the libraries that were established disappeared during the 5th and 6th centuries when the Roman Empire crumbled (Harris 1995, pp. 66–67).

If it had been left up to the Romans, many of the Greek accomplishments would have been lost and forgotten. The Roman impact on the progress of mathematics was not positive. In 212BC, after a three-year siege, the Romans conquered the city of Syracuse on the island of Sicily. During the subsequent pillaging, a Roman soldier ignoring a standing order, killed Archimedes. (The siege would not have been so long had it not been for the effective war machines designed by Archimedes to defend his home city.) Some Romans were great engineers but no evidence survives that demonstrates they had any interest in pure mathematics or science.

What saved many mathematical and other works of the Greeks from eventual destruction was the rise of Islam. Because of his religious convictions, Mohammad had to flee Mecca in fear of his life in 622AD, but he was able to return in triumph eight years later. Within 100 years, Islam was spread by force of arms from Spain to Persia. This expansion ushered in a golden age of science in the Arabic world. According to Islamic historian C.A. Qadir,

> "In the eyes of the Prophet, knowledge ranked higher than worship, for he said, 'Man's glance at knowledge for an hour is better for him than prayer for sixty years.' He therefore commanded all believers to seek knowledge and to go to China in search of knowledge, if required." (Qadir 1988, pp. 15–16)

In this spirit, Muslims would eventually establish pre-eminent learning centers in many places including Damascus and Baghdad in the East and Toledo and Cordova in the West. Some of the problems that became the focus of investigation arose from Islam. I have already discussed the Qibla problem. It was also considered important to calculate the exact times of prayer and the dates for religious ceremonies along with the month of Ramadan (Ifrah 2000, p. 514). The inheritance laws in the Mideast had been complicated before the rise of Islam and Muslims did nothing to simplify them.

It is interesting to note that Islamic inheritance problems had a prominent place in the earliest text books on algebra. This suggests to some that algebra was invented by the Arabs to deal with their inheritance law.

George Saliba, author of *Islamic Science and the Making of the European Renaissance*, suggests that at least some of the surge of scientific progress that occurred under Arab rule can be attributed to baser motivations. Abd al-Malik (646–705 AD) ruled the Arab empire as caliph (685–705 AD). During his reign, he mandated that Arabic become the official language of government throughout the empire. This action threatened the livelihood of several communities of non-Arabs – in particular Zoroastrians in Persia along with Jews and Nestorian Christians who had found refuge in Syria and Persia from the Church of Rome.

For generations, many bureaucratic positions had been reserved for members of families in these communities. These non-Arab families had protected their status by maintaining a monopoly on the knowledge necessary for administration tasks such as taxation, inheritance law, and surveying. (Before the introduction of arabic numerals, these tasks required computations that were quite difficult even with the use of an abacus.) Abd al-Malik demanded more than having government employees talk to one another in Arabic. He demanded that whatever learning materials necessary for the education of an effective bureaucrat be translated into Arabic.

It is the contention of George Saliba (2007, pp. 60–72) that this edict set off a competition between members of several ethnic groups who wished to obtain or maintain positions of influence in the court of the caliph. This competition resulted in a rapid expanse of scientific knowledge. Soon Muslim scholars would be familiarizing themselves with the mathematical achievements of India. In India, progress had been made in trigonometry (Katz 1998, pp. 212–218) and algebra (Stillwell 2001, p. 82). Although it is now recognized that the Babylonians invented zero as a place keeper, it is also believed that the concept was reinvented in India (Ifrah 2000, p. 146 and pp. 399–421). The Persian astronomer and mathematician, al-Khwarizmi ca. (780–850 AD, is credited with popularizing the Indian place-value system for Arab speakers (Ifrah 2000, p. 521). Using zero along with the decimal system enabled scientists in the Islamic world to carry out lengthy computations that were heretofore nearly impossible.

Later, scientists philosophers and mathematicians, in the courts of the caliphs, would turn to Greek texts for further learning (Saliba 2007, p. 75). Assimilating the achievements of Greece and India, the Arabs not only introduced decimal fractions but made significant advances in algebra, trigonometry, spherical trigonometry, and number theory. In 830AD, al-Khwarizmi published a book on the solution of equations entitled *Hasib a-jabr w'al mûqabala* (Science of Restoring and Opposition). The words "al-jabr" and "al-mûqabala" refer to steps used in the solution of an algebraic equation. The word "al-jabr" evolved into the word "algebra." From the author's name, al-Khwarizmi, we get the word "algorithm." The Arabs also made significant advances in chemistry, medicine, and observational astronomy. Other scientific words derived from Arabic are *azimuth (al-sumūt), nadir (na zīr),* and *zenith (al-samt).*

For 350–400 years,[1] mathematics flourished in the Islamic world while virtually nothing of mathematical importance was happening in Europe. As the boundaries of the Islamic world expanded, Nestorian Christians in Syria, Zorastrians in Persia, along with Jews and pagans from all over were absorbed into an empire. According to some western historians, pagans and unbelievers were given a stark choice: convert to Islam or die. George Saliba tells me that reality was much more nuanced. First of all the Quran states that religion should not be a matter of compulsion. Second, the treatment of pagans depended on the whims of widely varying governmental authorities. In practice, most pagans converted after several generations, often for personal advancement. On the other hand, Christians, Jews, and even Zoroastrians were regarded as "peoples of the book." They were not allowed to become full citizens but were allowed to pursue their religions as long as they paid a special tax.

For its time, this was an extremely liberal approach. In the ninth century, the Muslim governor of Antioch had to appoint a guard to keep Christian sects from massacring one another at church (Durant 1950, p. 218). A caliph of particular note, who promoted science, was al-Mamun who ruled from 813 to 833. In his state council, he included Christians, Jews, Sabians, and Zoroastrians. He gave strong support to arts, sciences, letters, and philosophy. Members of his court went to Constantinople, Alexandria, Antioch, and elsewhere for the writings of the Greek masters, which were then translated into Arabic. He established an academy of

[1]For a number of years, John J. O'Connor and Edmund F. Robertson have posted profiles at their web site (www-history.mcs.st-and.ac.uk) or google "MacTutor History of Mathematics" on people who have made significant contributions to mathematics. In November 2011, the list of those born between 700 and 1050AD was dominated by members of the Islamic world. Out of a total of 44, 31 or 70% were Arab speaking Muslims. The remaining 13, consists of nine Indians, two Chinese, one Englishman, and one German. Neither the Englishman nor the German is included for making original contributions. The Englishman, Alcuin, is included for preserving the contents of some ancient Greek documents. The German, Hermann of Reichenau, is included for describing some Arabic science in the Latin language.

During the following 50 years (1050–1100), six more people were listed. The six consisted of two Indians, two Spanish Jews, and one English Christian, and only one Arab speaking Muslim. However, it should be noted that the two Spaniards and the Englishman were on the list for their roles in transferring Arab mathematics to the European community.

science at Baghdad that became known as "The House of Wisdom." He funded salaries for physicists, jurists, musicians, poets, mathematicians, and astronomers. The greatest of the mathematicians in the court of al-Mamun was al-Khwarizmi but there were others, including al-Kindi, better known as a philosopher.

Al-Mamun set an example that set the stage for an Islamic "golden age." During this period many rulers of various realms in the Arab world provided patronage for various intellectual pursuits. However, eventually this patronage would disappear. Why?

Much of the early Arab success in science stemmed from the willingness of important rulers to tolerate non-Muslims. However, Muslims like Christians of the same period found it difficult to settle differences without resorting to arms. Centers of scholarly research did not remain islands of serenity.

One of many divisions that would be settled by force was focused on the methodology used to interpret the Quran. In the early days of Islam, religious scholars would argue over interpretation of the Quran but eventually something like a body of legal precedents was established. In this circumstance, a rigid orthodox theology emerged for some which was immune from new knowledge gained from the sciences.

By contrast, others had a more flexible and allegorical interpretation of scripture. These people felt that the study of the universe created by God would lead to a better understanding of God's nature and thereby a better understanding of God's intentions as described in the Quran.

If the proponents of these opposing views had agreed to disagree, the golden age of science might have continued much longer. However, the inclination to rely on violence to settle disputes was deeply embedded in Arab society at the time. Much of this stemmed from clan rivalries that existed before the rise of Islam.

At the moment of Mohammad's death in 632, there was division over who would inherit the role of Mohammad as leader of what soon became an Islamic empire. Some thought Mohammad's son-in-law, Ali, should become caliph while others thought that Mohammad's closest advisor, Abu Bakr, should be awarded the position. This split was essentially a split between rival clans. Since the caliph would be both spiritual and civil ruler, this division was worth a war in the seventh century. This division continues to have lethal consequences in the twenty-first century. Those who think Ali was the rightful heir are known as Shiites and those who think Abu Bakr was deserving are known as Sunnis.

The Umayyad clan supporting Abu Bakr initially took control. However, after 24 years in 656, there was a revolt and Ali became caliph. Then in 661, Ali was killed by a member of a group that had seceded from Ali's army. Soon after, the Umayyad clan regained control and Mu'awiya became caliph. Things did not end there. Ali's son, Hasan, became leader of the Shiites but he was forced to submit to the Umayyad clan and in 669 he died. Shiites believe that he died because Mu'awiya persuaded one of Hasan's wives to poison him.

When Mu'awiya died in 680, another war of succession broke out and Husayn (another son of Ali) was killed in battle with many members of his immediate family. In Karbala, where Husayn was killed, the Shiites built a shrine. This shrine was destroyed by a Sunni caliph in 850 (Durant 1950, p. 252). Nevertheless the Shiites enact an annual passion play at the site.

The Umayyads ruled from Damascus until 750, when they were overturned by the rival Abbasid clan. The Abbasid clan was a Sunni clan that was able to exploit the grievances of Persian Shiites to obtain supremacy.

The Abbasids were not kind to the defeated Umayyads. The head of the previous caliph was delivered to Abu al-Abbas – the new caliph. Even then, the Abbasids were not satisfied. An uncle of the new caliph announced an amnesty to the Umayyads and eighty of their leaders were lured to a dinner. When seated, the eighty were put to the sword by soldiers who had been hidden for the occasion (Durant 1950, p. 196). A few members of the Umayyad clan were able to escape and a rival caliphate was established on the Iberian peninsula.

Politics could be rough even between members of the same family. Al-Mamun, an Abbasid caliph, was discussed earlier in this section as the enlightened patron of arts and science in his court in Baghdad. However, he had to compete with a brother to become caliph. He became informed that he would become caliph in 813 when the severed head of his older brother was delivered to him.

The liberal attitudes that al-Mamun had toward Christians, Jews, Sabians, and Zoroastrians did not extend to many of his fellow Muslims. This was particularly true near the end of his reign. A version of Islam, influenced by Greek philosophy was imposed on the general public. According to Durant (1950, pp. 250–251):

> – al-Mamun in 832 issued a decree requiring all Moslems to admit that the Koran had been created in time; a later decree ruled that no one could be a witness in law, or a judge, unless he declared his acceptance of the new dogma; further decrees extended this obligatory acceptance to the doctrines of free will, and the impossibility of the soul ever seeing God with a physical eye; at last, refusal to take these tests and oaths was made a capital crime. Al-Mamun died in 833, but his successors al-Mutassim and al-Wathiq continued his campaign. The theologian Ibn Hanbal denounced this inquisition; summoned to take the tests, he answered all questions by quoting the Koran in favor of the orthodox view. He was scourged (flogged) to unconsciousness and cast into jail; but his sufferings made him, in the eyes of the people, a martyr and a saint, and prepared for the reaction that overwhelmed Moslem philosophy.

Al-Mamun's efforts to impose his views on the populace were counterproductive. To enforce their will on the general population, al-Mamun, along with his successors, al-Mutassim and al-Wathiq became more and more dependent on an army of Turkish mercenaries who became more and more powerful. When al-Wathiq died, it was these Turkish mercenaries who chose the following caliph. The Turks observed a very orthodox version of Islam and when their choice for caliph, al-Mutawakkil assumed the reins of power in 847, he immediately released Iman Ahmad ibn Hanbal from prison and reversed the policies of the previous caliphs. The version of Islam that al-Mamun had tried to eradicate now became the official religion. When ibn Hanbal died in 855, virtually the entire population of Baghdad turned out for his funeral.

During the Medieval Age, a Muslim mathematician or scientist was usually involved in such a broad range of investigations that he could not avoid religious controversy. An extreme example was al-Kindi (801–873). For what its worth, according to Wikipedia (wikipedia.org/wiki/Al-kindi), "He wrote at least 250 books, contributing heavily to geometry (32 books), medicine and philosophy (22 each), logic (nine books), and physics (12 books)." In geometry, he gave a lemma that considered the possibility of having lines in the plane that are simultaneously nonparallel and nonintersecting (Bulgakov and Ahmedov 1977, pp. 30–36). (Such a lemma would have made al-Kindi a focus of controversy in Europe during the first half of the nineteenth century.)

Al-Kindi flourished in the court of al-Mamun. He also thrived in the court of his successors, al-Mutassim and then al-Wathig. However, the situation for al-Kindi changed dramatically when al-Mutawakkil became caliph in 847.

During his reign, al-Mutawakkil confiscated al-Kindi's library and then had him subjected to 50 strokes of a whip (Hayes 1992, p. 99). It appears that this flogging stemmed from the machinations of Jafar Muhammad Banu Musa, another court mathematician. Indeed, in this incident, Jafar Muhammad was able to gain possession of al-Kindi's library. However, these machinations would not have been possible without the bitter religious divisions of the time. (The flogging of al-Kindi occurred about six years after the funeral of Iman Amad ibn Hanbal that I mentioned above. Ibn Hanbal had opposed the kind of allegorical interpretation of the Quran advocated by al-Kindi and had won the adulation of the populace when he refused to recant his views in the face of harsh retribution, which had also included a public flogging.)

Later, a friend of al-Kindi, skilled in court intrigue, was able to get the library returned to al-Kindi. Al-Kindi died in 873 at the age of 72. In some sense, the fate of al-Kindi was symptomatic of the state of affairs in the court of al-Mutawakkil.

As I mentioned above, al-Mutawakkil was quite dogmatic. It was al-Mutawakkil that ordered the destruction of the Shiite shrine dedicated to Husayn ibn Ali. (This shrine has been damaged or destroyed many times over the centuries. Each time, it has been reconstructed with greater grandeur. Under Saddam Hussein, pilgrimages were forbidden. Thousands were arrested and hundreds were killed to enforce his edict. Today, Shiites protect the site with extreme vigilance.)

Under al-Mutawakkil, Jews and Christians were ousted from government posts and required to wear a distinctive color of dress, put colored patches on the garments of their slaves, ride only on mules and asses, and attach wooden devils to their doors (Durant 1950, p. 252). It should be noted that al-Mutawakkil may have been as capricious as he was dogmatic. We know with certainty that at least one member of his court, Hunayn ibn Ishaq (808–873), was a Nestorian Christian. Hunayn was not only the most prolific translator in the court of al-Mutawakkil but also one of the most prolific translators of Greek science in the history of the Islamic empire. Hunayn also held the title of chief physician in the caliph's court.

According to at least one report, Hunayn was one of the 57 physicians, most of whom were also Christians, who were sustained in the caliph's court (Saliba 2000, p. 92). For Hunayn, this support may have come at a high price. According to the report just mentioned, Hunayn life in the court was so miserable that he considered suicide at times because of the intrigues of his fellow Christians. When they falsely accused him of being an atheist, Hunayn was flogged, imprisoned, and nearly executed (Saliba 2000, pp. 95–96). George Saliba interprets this episode as evidence that al-Mutawakkil was willing to permit the Christian authorities to police their own heretics. Indeed, according to the story, the caliph did consult with a Christian cleric before deciding on the appropriate punishment.

In this same story, Hunayn was saved from execution when the caliph had a vision in which Christ appeared before him to intercede on behalf of Hunayn. Hunayn was then released after six months in prison to treat an illness that had painfully incapacitated the caliph for almost two months. The ministrations of the other court physicians had been ineffective so the caliph was very grateful to Hunayn when he regained his health soon after consuming the potion prescribed by Hunayn (Saliba 2000, pp. 95–98).

Due to the oppressive and divisive policies of the Abbasid caliphs, their authority soon deteriorated after al-Mutawakkil and independent centers of power cropped up all over the Islamic world. Mathematicians and scientists could find patronage at some of these centers but frequently when one ruler died there would be a profound shift in policy. Philosophers in particular often had to flee to avoid imprisonment or death when a new ruler assumed the reins of power.

This was the fate of al-Biruni (973–1048) who was mentioned in the last section for his contribution to the qibla problem. In 973, he was born in present day Uzbekistan. As he matured, he became closely identified with the Banu Iraq family. When he reached the age of 22, in 995, the Banu Iraq were overthrown in a coup and al-Biruni fled. During the next few years, he was able to collaborate with the astronomer al-Khufandi at Rayy near the present city of Tehran, Iran. However, without a patron he was living in poverty. Sometime before the middle of 1004, he returned to his homeland where he now had a patron. In 1017, his patron was killed and the region fell under the control of Mahmud. Al-Biruni now became a virtual prisoner of Mahmud and was taken to Ghazna in present day Afghanistan. To please his new patron, he determined the latitude and longitude of Ghazna and then computed the qibla.

Around 1022, Mahmud invaded India and al-Biruni accompanied him as a member of his entourage. During his time in India, al-Biruni not only learned Sanskrit but he was able to familiarize himself with many aspects on Indian culture. In his voluminous work, *India*, he not only described the Indian system of writing and numbers, but he also discussed Indian religion, philosophy, medicine, geography, astronomy, astrology, and the calender.

More is known about al-Biruni than most Islamic mathematicians because he included bits of autobiographical writings in some of his academic publications. In one of these, *Shadows*, he relates an encounter with a hard-line orthodox cleric. The cleric admonished al-Biruni because he had used an astronomical instrument with Byzantine months engraved on it to determine the time of prayers. Al-Biruni replied:

> "The Byzantines also eat food. Then do not imitate them in this!"

4.7.4 *The Impact of Al-Ghazālī

Like al-Kindi, most other mathematicians and scientists in the Arab world accepted the Quran. On the other hand, as a rule, they felt that logic and knowledge gained by scientific investigation would lead them to a deeper and more meaningful understanding of the Quran than otherwise possible. This approach might require an allegorical interpretation of some parts of the Quran but they were convinced that the Quran was not in conflict with rational thought. Nonetheless, most if not all of them thought that members of the general public should not be encouraged to apply their logic to any serious question. Using his (or her) inferior skills of logic, a member of the general public might persuade himself (or herself) that some heretical idea might be valid.

The orthodox clerics had a different position. They were fearful that *anyone* who studied the works of Aristotle, Plato, and others could be led astray. Furthermore, a philosopher, who became skilled with the philosophical arguments of the Greeks or the logic of mathematics could become very persuasive with the youth and was therefore a threat to the community. Al-Ghazālī (1058–1111) was a prominent theologian who advanced this position. Al-Ghazālī once wrote:

> Even if geometry and arithmetic do not contain notions that are harmful to religious belief, we nevertheless fear that one might be attracted through them to doctrines that are dangerous (Swartz 1981, p. 195).

Al-Ghazālī had more credibility than most Islamic jurists because of the fact that he studied Greek philosophy and was able to use philosophical arguments to make his points. In one of his best known works, entitled, *The Incoherence of Philosophers*, he indicated that a heretical rationalist philosopher would contend that fire burns cotton. Al-Ghazālī then continued:

> This we deny. The agent of the burning is God, through His creating the black in the cotton and the disconnection of its parts, and it is God who made the cotton burn and made it ashes through the intermediation of the angels or without mediation. For fire is a dead body which has no action, and what is the proof that it is the agent? Indeed, the philosophers have no other proof than the observation of the occurrence of the burning, when there is contact with fire, but observation proves only a simultaneity, not a causation, and, in reality, there is no cause but God (Hoodbhoy 1991, p. 105).

Al-Ghazālī may have been instrumental in the decline of patronage for scientific endeavors. It is plausible that some elements of the Islamic world were particularly receptive to al-Ghazālī's teachings because of Arab reversals on the battlefield during his lifetime. One way to pursue an argument is to marginalize your opponents by characterizing them as unpatriotic, heretical, or advocates of some foreign ideology. This tactic is particularly effective when there is a foreign threat (real or imagined). In 1085 Toledo, at the west end of the Islamic world, fell into Christian hands. In 1099, the army of the First Crusade took possession of Jerusalem. In this context, we can more easily understand the popularity of *The Incoherence of Philosophers,* which was written sometime between 1095 and 1111.

The teachings of al-Ghazālī provided ammunition to those inclined to denounce those who wished to build on the achievements of the Greeks. Much of the luster of the golden age was lost during the lifetime of al-Ghazālī. The impact of al-Ghazālī and his adherents is illustrated by the fate of a contemporary, namely Omar Khayyam (1048–1131). Omar Khayyam is best known in the West for his poetry ("– A jug of wine, a loaf of bread, and thou –"). However, he was primarily a mathematician – perhaps the foremost mathematician of his time. He was able to show that the solutions of cubic equations were equivalent to determining the points of intersections of conic sections (circles, ellipses, parabolas, and hyperbolas). He was employed to set up an observatory in Esfahan, which lies in present day Iran. He accomplished much at this observatory but his funding came to an end when the ruler of the city died in 1092. Generally, Omar Khayyam was able to find patrons for his endeavors throughout his life but like others I have mentioned above he came under attack from orthodox Muslims who criticized his tendencies toward free inquiry.

Al-Ghazālī spent a substantial part of his life in Baghdad, but he had an impact on the entire Islamic world that continues to this day. A futile counterattack to the teachings of al-Ghazālī was launched by Ibn Rushd (1126–1198) from the opposite end of the Islamic world. Ibn Rushd was born about 15 years after the death of al-Ghazālī in Cordova, Spain.. Like al-Ghazālī, ibn Rushd would have a significant impact as a philosopher. In Europe, he would become known as Averroës. And it was in Europe that he would have his impact – not in the world of Islam.

Averroës wrote a rebuttal to al-Ghazālī entitled, *The Incoherence of the Incoherence*. In this work, he quoted al-Ghazālī's discussion of burning cotton and then wrote:

> To deny the existence of efficient causes which are observed in sensible things is sophistry, and he who defends this doctrine either denies with his tongue what is present in his mind or is carried away by a sophistical doubt which occurs to him concerning this question. – Denial of cause implies the denial of knowledge, and denial of knowledge implies that nothing in this world can be really known, and that what is supposed to be known is nothing but opinion, that neither proof nor definition exist, and that the essential attributes which compose definitions are void (Averroës 1969, pp. 318–319).

During the prime of his life, Averroës received the patronage of Abu Yaqub Yusuf, the caliph of Morocco. It was this caliph who urged Averroës to write the commentaries on Aristotle that made Averroës famous. In 1184, the caliph was fatally wounded in battle while trying to suppress some dissident Muslims in Spain. Abu Yaqub Yusuf was then succeeded by his son, Abu Yusuf Yaqub. The new caliph continued to support Averroës for several years. But soon, the caliphate became threatened by an energized coalition of Christian powers in Spain (Hayes 1992, p. 100). To unite the Muslim factions, Averroës became a sacrificial lamb. Due to his commentaries on Aristotle and his refutation of al-Ghazālī, he had aroused the anger of orthodox clerics. Around 1194, the caliph accused Averroës of heresy or near heresy and ordered the burning of his books. In addition, Averroës was ousted from the caliph's court in Marrakesh, Morocco and banished to the present day, Lucena, Spain, which is near Cordova. This was probably a political success.

At any rate, the caliph obtained enough unity among the fractious Muslims so that he won a decisive battle over the Christians at Alarcos in 1195. Thereafter, the caliph was known as "al-Mansur" (The Invincible).

When things settled down and al-Mansur was able to return to Morocco, he lifted the edicts on Averroës. And soon after, Averroës rejoined the caliph at his court in Marrakesh. Averroës would die roughly one year later on December 10, 1198 at the age of 72.

In any society, there is tension between those who have questions they wish to be answered and those who have answers they wish not to be questioned. The kind of anger and rage directed toward Averroës by dogmatic clerics made it difficult for science to progress in the Arab world.

It is interesting to note that by this time, Christian scholars were translating the works of Aristotle, Euclid and others from Arabic into Latin. In 1085, Toledo with its great Muslim library had fallen into the hands of Christian Spaniards. Around 1135, the Archbishop of Toledo, Don Raimundo, established a translation center in Toledo. The same Greek works that had ushered in a golden age of math and science in the world of Islam along with the advances of the Muslims would now usher in the Renaissance and the Age of Reason in Europe.

The works of Averroës that became marginalized in the Arab world would become influential in Europe. His arguments that the works of Aristotle were compatible with the Quran could also be used to persuade religious authorities in Europe that the works of Aristotle were compatible with the Christian bible. In Europe, the battle for and against the idea that people should or could be allowed to think for themselves would be long and difficult.

Over 350 years after Archbishop Don Raimundo established his translation center, another Archbishop of Toledo, Francisco Jiménez de Cisneros, (confessor for Queen Isabella) organized a book burning. On January 2, 1492, King Ferdinand of Spain had obtained the surrender of the last Muslim stronghold in Spain. The terms of surrender stipulated that the resident Muslims would continue the freedom to practice their faith. However, Cisneros was unhappy with this provision and he had enough influence to overrule the local archbishop who was observing the stipulated deference to the Moslem population. He initiated forceful conversions from Islam to Christianity and in 1499 he presided over the burning of more than 5,000 Moslem books with ornamental bindings, even of gold and silver with exceptional artistry (Harvey 1992, pp. 328–333).

For reasons stated above, Western historians assumed for many years that the "golden age" of Islamic science came to an end a few generations after the fall of Jerusalem in the First Crusade. The medieval Arab world would never create universities with the long-term permanence that were later created in Europe. It was thought that if the golden age did not end before the 13th century, it certainly ended with the destruction of Baghdad with its great library in 1258 by Hulagu Khan, grandson of Genghis Khan. This point of view was radically changed by an event that occurred in 1957. In that year, the science historian, Edward S. Kennedy, discovered an Arab document written by Ibn al-Shātir circa (1304–1375AD) in the Bodleian Library of Oxford University. He presented the document to Otto Neugebauer, an authority on the details of the work of Copernicus. It was immediately recognized that a mathematical advance, that is central to the work of Copernicus, appeared over 100 years earlier in this work by Ibn al-Shātir. See (Roberts 1957; Kennedy and Victor 1959).

This discovery led to a serious reevaluation of Islamic science. Western historians were well aware of the advances of Islamic science before the Toledo library fell into Christian hands in 1085. This was simply because the contents of the Toledo library were familiar to these scholars. However, as historian George Saliba has suggested, you do not find what you do not look for.

When science historians investigated further, they discovered that Islamic scientists had made significant scientific progress after the destruction of Baghdad. Indeed, meaningful patronage of scientific endeavor had come to an end before the fall of Baghdad. However, there remained some members of the Arab world who were able to continue the advance of science. In particular, Islamic clerics, like Copernicus (a Catholic cleric) at a later time, had enough leisure time to make important scientific contributions – at least in astronomy.

An obvious case in point is Nasīr al-Din al-Tūsī (1201–1274AD) who was an astronomer and Shiite scholar. As he approached Baghdad, Hulagu Khan recruited segments of the local population who were either disenchanted with the Sunni caliphate or eager to please the conquering Mongol. As a respected member of this entourage, al-Tūsī was able to rescue a large number of library manuscripts before the sack of Baghdad. He is now best known for the "Tūsī Couple" which was used by other Islamic astronomers and eventually by Copernicus to deal with latitudinal motion in modified versions of Ptolemy's models.

Not long after the destruction of Baghdad, al-Tūsī was able to persuade the son of Hulagu Khan to finance the construction of an observatory in Marāgha. This would soon become one of the most and perhaps the most significant observatory in the Arab speaking world.

To understand the contributions of the post Baghdad Islamic astronomers, it is useful to survey the achievements of Ptolemy and Copernicus.

4.7.5 *Claudius Ptolemy, Al-Tūsī, Al-'Urdī, Ibn al-Shātir, Nicholas Copernicus, Tycho Brahe, Johannes Kepler, and Isaac Newton

To understand the achievement of Nicholas Copernicus (1473–1543), it is important to understand the achievements of Claudius Ptolemy and the Islamic astronomers. The most casual observer will see that, in the course of a night, the stars appear to move with a uniform circular motion from east to west as if they were attached to some giant rotating sphere. Even in ancient times, more careful observers noted that the five known planets along with the Sun and the Moon were exceptions. Each planet in its own way would spend most of the time moving in an easterly direction with respect to the stars. But then from time to time a planet would assume a *retrograde motion*. That is, it would reverse direction with respect to the stars for a number of weeks before resuming its easterly motion. (The word "planet" is derived from a Greek word for "wanderer.")

From esthetical considerations and crude observations, Aristotle (384–322BC) concluded that, for the heavenly realm, uniform rotation of spheres was essentially the law of physics. It was not until Newton came along that it became understood that the laws of physics in the heavenly realm were the same as the laws of physics for the earthly realm. A heavenly body moving in a force free environment would move not in a circle but in a straight line. Later Einstein showed that the apparent gravitational forces that appear to deflect planets from straight line paths could be better explained by the geometry of the world we live in. In Einstein's version of the cosmos, planets move along geodesics in a four-dimensional curved time-space world.

Aristotle was confident that the Earth was stationary at the center of the universe and the stars were attached to a celestial sphere. It would be left to others to fill in the details. Around 280BC, Aristarchus would challenge this vision with a heliocentric theory. He hypothesized that the Earth rotated about its own axis in a day and revolved about the Sun in the course of a year. It was easy to dismiss such a theory. First of all, one does not feel any motion. If the Earth rotated about an axis, a citizen of Athens would be moving at a rate of 1330 km/h (820 mph)! A person moving that fast would surely feel a strong wind.

A more critical defect of the heliocentric theory was a consequence of the revolution of the Earth about a fixed Sun. In the course of a night, the stars appear to rotate about a fixed point in the sky. (Currently, the "north star" is located near this point. Because of a wobble in the earth's axis that has a 26,000-year period, this point changes over a period of centuries but this change was difficult to detect in the lifetime of an ancient astronomer.) In Aristarchus' heliocentric theory, this point in the sky would lie along an extended version of the Earth's axis of rotation. In the course of the year as the Earth revolved about the Sun, the axis of rotation would move with it and the "fixed point" in the sky would have to change. If the axis of the Earth was aligned with the North star at one time of year, one should reasonably expect it to be aligned with some other point in the sky six months later when the Earth was on the opposite side of its orbit about the Sun.

An advocate of a heliocentric theory would be forced to conclude that the nearest star was many times as far as the distance of the Earth to the Sun. To investigate the consequence of this conclusion, Aristarchus devised a method for measuring the distance of the Earth to the Sun. Unfortunately, his method relied on a measurement that is theoretically possible but nearly impossible in practice – at least with any useful precision. Some speculate that Aristarchus was satisfied with his method but took a wild guess at the quantity that required measurement (Evans 1998, pp. 68–72). The result was an estimate for the radius of the Earth's orbit about the Sun that was too small by a factor of 20. Even with this low estimate, the heliocentric theory implied a distance to the stars that seemed inconceivable.

(Astronomers failed to improve on this estimate until the seventeenth century, sometime after the deaths of Nicholas Copernicus, Tycho Brahe, and Johannes Kepler. (O'Connor and Robertson: *The Size of the Universe* and Giovanni *Domenico Cassini*)).

Thus, it was easy to dismiss the heliocentric theory of Aristarchus. On the other hand, it was very difficult to construct a model, which would match reality with the constraints of Aristotle. It was probably Claudius Ptolemy circa (90–168AD) building on the results of Hipparchus circa (190–120BC) who was the first to construct a predictive model for each of the planets. To get a small idea of the task faced by Ptolemy, let us consider an easier problem. Knowing the exact solution, how could we construct an approximate solution which would conform to the dictates of Aristotle? Johannes Kepler (1571–1630), using the data of Tycho Brahe (1546–1601), was able to infer that the path of any planet is an ellipse with the Sun at a focal point. This is known as Kepler's first law. According to Newtonian physics, these elliptical orbits would be exact in the limiting approximation that the ratio of the mass of the planets to that of the Sun be zero. As any mathematician knows there is nothing inherently wrong with a geocentric coordinate system. The relation between the heliocentric and geocentric coordinate systems is neatly illustrated in Fig. 4.8, which I have stolen from James Evans with a slight upgrade (Evans 1998, p. 411). In the diagram, the orbits of Earth and Mars are not accurately scaled and the eccentricities are greatly exaggerated. I have used Mars as an illustration but the same diagram would apply to the two other "superior" planets known to the ancient Greeks (Jupiter and Saturn). I will leave it to you to figure out the corresponding diagram for the "inferior" planets (Venus and Mercury). (See Prob. 61.)

The geocentric model in Fig. 4.8 represents Ptolemy's model as "it should have been." In reality, following the dictates of Aristotle, Ptolemy had to make adjustments so all motions were restricted to combinations of circles.

In the geocentric coordinate system, the motion of Mars with respect to Earth can be visualized as a point moving counter clockwise about an elliptical *epicycle* one of whose focal points orbits an elliptical *deferent*. Over a period of centuries, the axis of both the Earth orbit and the Mars orbit would rotate with respect to the "fixed stars." This is mainly due to the gravitational forces from the other planets. However, for a short-term approximation, the angle between the major axis of the epicycle and the major axis of the deferent would remain constant. It should also be noted that the epicycle and the deferent do not lie in the same plane.

In the geocentric model, retrograde motion for a superior planet occurs when such a planet is on a portion of the epicycle which is closest to Earth. In the heliocentric model, retrograde motion is an illusion, which occurs when the Earth with its faster angular velocity crosses an imaginary line joining the Sun with the superior planet. In this circumstance, the superior planet appears to be moving backward in a westerly direction with respect to the background stars. However, most of the time, an outer planet appears to be moving easterly. (Actually because the planes of the orbits are different, Earth will generally miss the imaginary line but I hope you get the idea.)

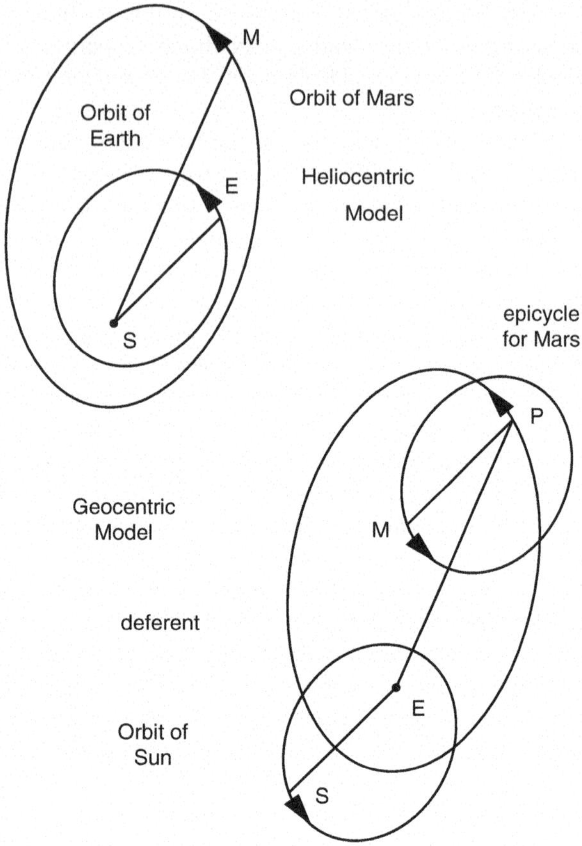

Fig. 4.8 Heliocentric and geocentric coordinate systems

An obvious advantage for the heliocentric point of view is that it can explain retrograde motion without hypothesizing a weird motion to the outer planets.

Another triumph for the heliocentric point of view was a meaningful interpretation of the relative sizes of the deferents and epicycles in the Ptolemaic model. From the geocentric point of view, the epicycle for a superior planet has the same size and shape as the orbit of the Sun around Earth. However, this was not realized before Copernicus. The ancient astronomers recognized that there was a correlation between the motion of the Sun about its orbit around Earth and the motion of Mars, Jupiter, and Saturn around their respective epicycles, but they did not realize that the size of the Sun's orbit was identical to the size of the epicycles for the superior planets. This was because they could not measure interplanetary distances. They could measure the relative size of a deferent and an epicycle, but they had no way of relating either to the size of the Sun's orbit about the Earth. (There was a similar but different problem for the inferior planets.)

Before Copernicus, astronomers assumed that with the Earth at the center of the universe, all heavenly bodies could be ranked by their distance from Earth.

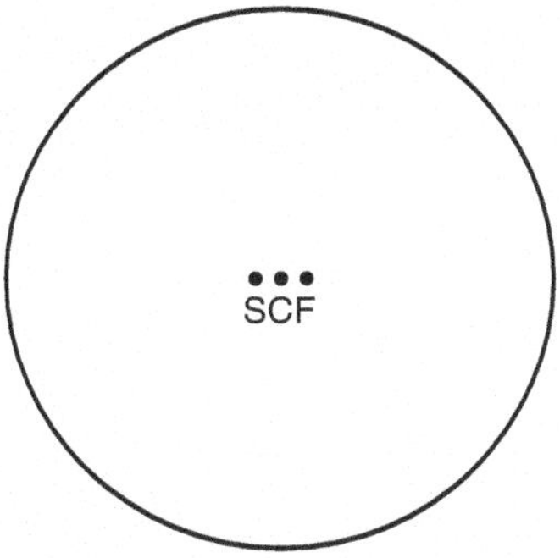

Fig. 4.9 A circle with a radius equal to the length of the semimajor axis is superimposed on the ellipse corresponding to the orbit of Mars in the heliocentric theory or the elliptical deferent in the geocentric theory. Because of the thickness of the lines and the scale of the page, the perimeter of the two curves is indistinguishable

They generally agreed that Saturn was the furthest of the planets because it was the slowest moving with respect to the "fixed stars." However, on average, Venus, Mercury, and the Sun each take one Earth year to complete a circuit through the zodiac. (Sometimes an inferior planet will be ahead of the Sun and other times it will appear to lag behind the Sun.)

At times Venus or Mercury can be seen shortly before the Sun rises in the morning or shortly after the Sun sets in the evening. Other times one or both of the planets disappears. With naked eye observation, it is impossible to determine whether they disappear behind the Sun or in front of the Sun. Thus, the ranking of these bodies was subject to debate. Ptolemy chose the sequence in order of increasing distance from Earth to be: Moon, Mercury, Venus, Sun, Mars, Jupiter, and finally Saturn. Other astronomers had placed the Sun closer to Earth than Mercury and Venus and still others had reversed the order of Mercury and Venus (Evans 1998, p. 348). According to Copernicus' heliocentric model, these speculations were answers to a wrong question.

Now let us examine some problems faced by Copernicus and Ptolemy.

For an ellipse with the major axis aligned with the x-axis, the equation is

$$\frac{x^2}{a^2} + \frac{y^2}{b^2} = 1, \text{ where}$$

"a" is the length of the semimajor axis, "b" is the length of the semiminor axis. The distance of either focal point from the center is $c = ae$, where $c^2 = a^2 - b^2$, and e is said to be the *eccentricity*. The eccentricity e for the orbit of Mars is now known to be 0.0934.

For the Mars deferent (or orbit) drawn to scale see Fig. 4.9. The perimeter of this ellipse is very close to that of a circle. Actually, a circle with radius equal to the semimajor axis of the ellipse is superimposed on the ellipse in my drawing. However, because of the thickness of the lines and the scale of the page, the circle and the ellipse are indistinguishable. If my calculations are correct, the minor axis

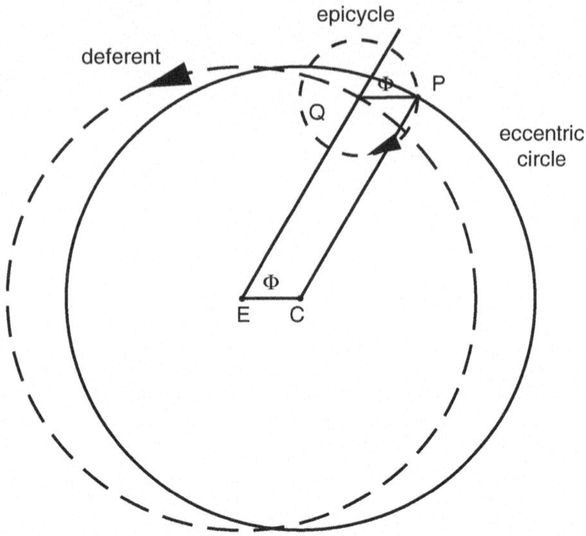

Fig. 4.10 An eccentric circle centered at C can be reproduced by a deferent centered at E and an epicycle centered at Q. Depending on your point of view the direction fo QP remains constant with respect to the fixed stars or rotates with respect to EQ. In either case, EQPC is always a parallelogram while EQ rotates counterclockwise and the path P is the eccentric circle

is less than $1/2$ of one percent shorter than the major axis. Except for Mercury, Mars has the greatest eccentricity of all the planets known to Ptolemy. Because of the proximity of Mercury to the Sun, the determination of the orbit of Mercury was difficult to obtain. Thus, the use of circles to represent the orbits of planets was within the experimental error for the instruments available until the time of Tycho Brahe (1546–1601).

On the other hand, for Mars, the separation between the center of the ellipse and a focal point is over 9% of the semimajor axis. Thus the approximating circle centered at the center of the ellipse is quite distinct from a circle of the same size centered at the focal point where the Earth, is located. In this context Ptolemy's approximating circle became known as an *eccentric*. Since the center of the eccentric did not coincide with the location of Earth it seemed at odds with the physics of Aristotle. In his work, *Almagest*, Ptolemy noted that Apollonius of Perga (ca. 262BC–ca. 190BC) had devised an alternate method for dealing with this problem over 300 years earlier. Apollonius demonstrated that an eccentric circle could be replaced by combining a *deferent* circle of equal size centered at the Earth and an epicycle with a radius equal to the focal distance. (See Fig. 4.10.)

Guided by their personal philosophical outlook, some later astronomers would use it and others would not. Copernicus used the eccentric in some of his writings and the epicycle in other writings. Which representation of planetary motion was more Aristotelian was subject to debate. Either of these two representations could be accepted as satisfying Aristotle's axioms.

However, Ptolemy encountered another feature of planetary motion that did not seem to lend itself to an Aristotelian representation. The motion of a planet along its elliptical orbit is not uniform. The closer the planet is to the Sun, the faster it moves. According to Kepler's second law, the radial vector from the Sun to the

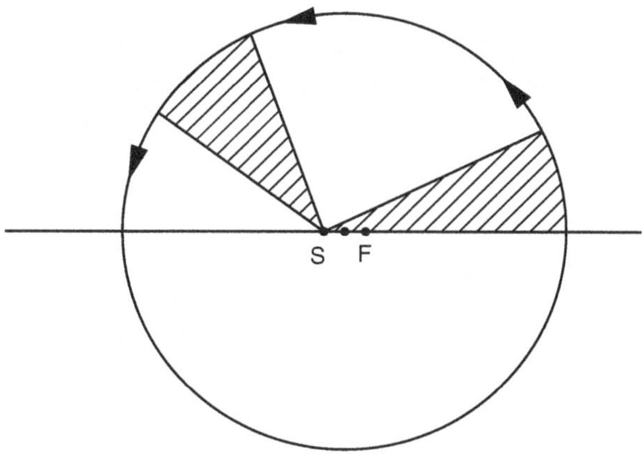

Fig. 4.11 According to Kepler's second law, the radial line joining the Sun and a planet sweeps out equal areas for equal time intervals as the planet moves along its orbit

planet sweeps out an area at a constant rate. (It is said that the *areal velocity* is constant.) (See Fig. 4.11.) In Newton's theory of physics, this is a consequence of the conservation of angular momentum. By contrast, according to the axioms of Aristotle, a planet moves on the orbit of a circle with uniform angular velocity. Ptolemy devised a method for dealing with this problem but it did not have an Aristotelian flavor. If Ptolemy applied the method he devised for a geocentric model to a heliocentric model, Mars would follow a circular path with a center at the center of the ellipse but it would do so with a constant angular speed about what we now recognize as the empty focal point. The empty focal point became known as the *equant*. To make sure you understand this, refer to Fig. 4.9. For Copernicus, the Sun would be located at point S, the approximating circle would be the orbit of Mars, and point F would be the equant point. For Ptolemy, Earth would be located at point S, the approximating circle would be the eccentric, which would carry an epicycle for Mars, and point F would be again be the equant point.

The predictive value of Ptolemy's model is illustrated in Fig. 4.12 for the planet Mars. Dividing the year into 12 equal periods, using the laws of Kepler, the location of Mars at the beginning of each period is determined by the intersection of the ellipse with a solid line passing through the Sun. According to the approximation of Ptolemy, the location of the same point is determined by the intersection of a circle virtually indistinguishable from the ellipse (the eccentric) with a dotted line passing through the empty focal point. As you see, the difference is essentially indistinguishable for the human eye for the scale of the figure. The difference was also within the experimental error from the time of Claudius Ptolemy circa (90–168 AD) to the time of Tycho Brahe (1546–1601).

The notion of the equant was never satisfactory to the esthetic outlook of ancient and medieval astronomers. Even if you were not a slavish adherent of Aristotle, it was clear that Ptolemy who accepted Aristotle's axioms was being logically inconsistent. For Aristotle, the planets were embedded in a nested set of spherical shells of a transparent, weightless, and crystalline material known as *ether*. Deferents and epicycles could be translated into models of spherical

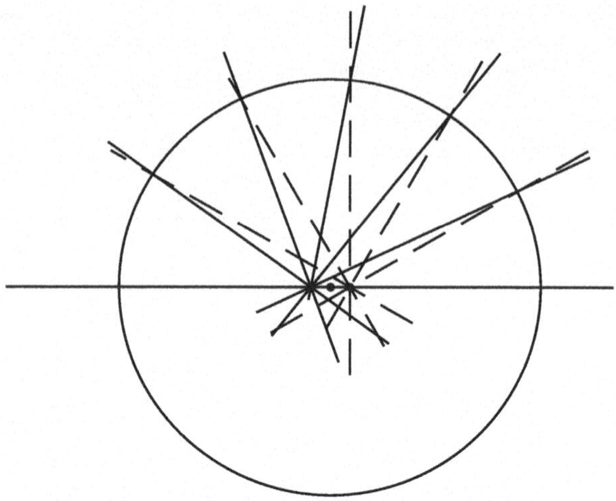

Fig. 4.12 For Mars, the elliptical orbit is closely approximated by a circle. The points defined by the intersection of the solid lines passing through the left focal point and the orbit correspond to the location of Mars spaced 1/12 of a year apart according to Kepler's second law. The points defined by the dotted lines passing through the right focal point (equant point) are the corresponding points defined by Ptolemy's rule

shells rotating at uniform speeds. Equants could not. Equants seemed to imply spherical shells rotating at uniform rates about points other than their centers. Since this mathematical machinery of Ptolemy could not be translated into a plausible physical model, ancient astronomers who used this machinery were dismissed by the Greek intellectual elite for being, "less than truth seekers." Eventually, some Islamic astronomers found a way of dealing with the equant problem, which was much closer to meeting the demands of Aristotle's dictates.

The notion of the equant also disturbed Copernicus. A few decades before he published his classic *De revolutionibus orbium caelestium*, Copernicus distributed a hand written outline of his heliocentric theory that is known as *The Commentariolus*. Starting with the third paragraph in that manuscript, Copernicus wrote:

> Yet the planetary theories of Ptolemy and most other astronomers although consistent with the numerical data, seemed likewise to present no small difficulty. For these theories were not adequate unless certain equants were also conceived; it then appeared that a planet moved with uniform velocity neither on its deferent nor about the center of its epicycle. Hence a system of this sort seemed neither sufficiently absolute nor sufficiently pleasing to the mind.
>
> Having become aware of these defects, I often considered whether there could perhaps be found a more reasonable arrangement of circles, from which every apparent inequality would be derived and in which everything would move uniformly about its proper center, as the role of absolute motion requires. After I had addressed myself to this very difficult and almost insoluble problem, the suggestion at length came to me how it could be solved with fewer and much simpler constructions than were formerly used, if some assumptions (which are called axioms) were granted me. They follow in this order (Copernicus 1959, pp. 57–58).

Copernicus then set forth seven axioms, the third being "All the spheres revolve about the sun as their mid-point, and therefore the sun is the center of the universe." Later in the same document, he writes in reference to the superior planets,

> Each deferent has two epicycles, one of which carries the other, in much the same way as was explained in the case of the moon, but with a different arrangement. For the first epicycle revolves in the direction opposite to that of the deferent, the periods of both being equal. The second epicycle, carrying the planet, revolves in the direction opposite to that of the first with twice the velocity (Copernicus 1959, pp. 74–75).

After describing the relevant distances for each of the three superior planets, he comments, "Thus the radius of the first epicycle is three times as great as that of the second" (Copernicus 1959, p. 77).

This combination of a deferent and two epicycles appears to be quite complicated. There is only one exact solution to a problem but there are many possible approximations. Thus, when a manuscript by Ibn al-Shātir(d. 1375) was discovered in 1957 in the Bodleian Library with the same complicated approximation to the Kepler orbit, it seemed plausible that Copernicus had adopted a device developed by Ibn al-Shātir for a geocentric theory and integrated it into his heliocentric theory without acknowledgement of his source.

An aspect of this alternative to the equant, which may have added fuel to suspicions, is the fact that it results in an approximate circle with a bulge, but with a bulge in the wrong direction! Students who present the same bad answers for an exam or homework assignment become much more suspect than students who present the same good answers. It should be noted that Copernicus took the trouble to show that this alternate orbit was close to that of a circle. In any case, it has become conventional wisdom amongst historians that Copernicus adopted the Islamic solution for the equant problem and adjusted it for his heliocentric theory (Swerdlow and Neugebauer 1984).

I am inclined to take issue with this point of view. The quote from *The Commentariolus* cited above represents Copernicus' first known talking point for his heliocentric theory. I find it difficult to believe that he would present a solution of the equant problem as a talking point for a heliocentric theory if he had lifted it from a geocentric theory. This approximation of the Kepler orbit is not as idiosyncratic as it first appears. The larger epicycle is that of Apollonius. This epicycle can be removed to get an equivalent version of the same approximation. In this equivalent version, the approximation is achieved by an eccentric circle and a single epicycle. This second version appears much easier to discover. Indeed, the flurry of activity stimulated by the discovery of the Ibn al-Shātir manuscript turned up this second version in the Arab literature. In particular, about 100 years before Ibn al-Shātir, a colleague of al-Tūsī by the name of Mu'ayyad al-Dīn al-'Urdī (d. 1266AD) is now credited with this alternate version. It is important to note that this "easier to discover" version was also known to Copernicus.

To understand the level of difficulty faced by 'Urdī, consider Fig. 4.13. In a geocentric theory there, was always the possibility that one could find a combination of epicycles that did not involve the point P in Fig. 4.8. With his heliocentric viewpoint, Copernicus could ignore this possibility. Thus, it would have been easier for Copernicus to rediscover this solution of the equant problem than it was for 'Urdī to discover the solution originally. For Copernicus, the possibilities of what can be

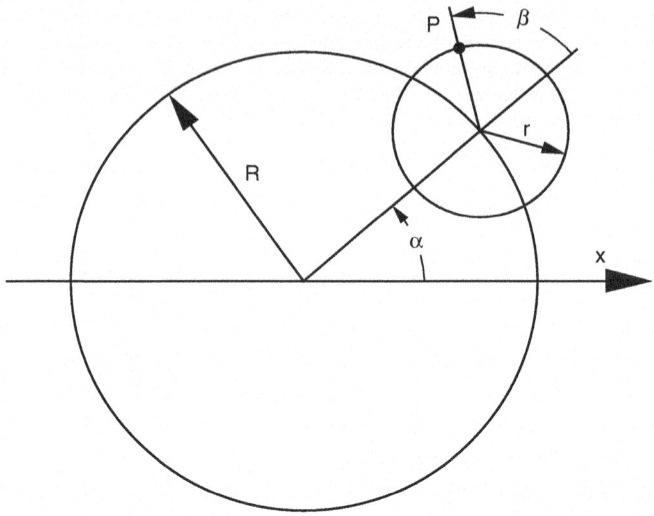

Fig. 4.13 The position of point P on the epicycle is determined by the angles α and β

done with an eccentric circle and a single epicycle to approximate the Kepler orbit are surprisingly limited:

(1) If α and β change at uniform rates, then $\alpha = k\beta + \beta_0$, where k and β_0 are constants.
(2) If the path of P is a closed orbit, we must have $\beta = n\alpha + \beta_0$, where n is an integer. (More precisely, I am making a stronger condition. That is when the center of the epicycle makes one complete circuit, point P returns to its initial position.)
(3) In order for the orbit of P to be symmetric with respect to the x-axis, we must have $\beta_0 = 0$ or π.
(4) In order for the orbit of P to be reasonably smooth, the value of n must be quite small.
(5) If n is an even integer, the orbit and its associated velocities will be symmetric with respect to a 180° rotation. This results in equal speeds at both ends of the x-axis. This is unacceptable since it is contrary to Kepler's second law.
(6) The value of -1 cannot be accepted for n because it corresponds to the epicycle of Apollonius, which results in a copy of the eccentric circle shifted either to the right or left depending on the value chosen for β_0.

What's left? $n = 1$! Except for the adjustment of some parameter, this is the epicycle of al-'Urdī. Condition (5) is obvious to a modern mathematician trained to look for symmetries but it might not have been obvious to Copernicus (or al-'Urdī). Nonetheless, it would have been easy for Copernicus to eliminate the cases for which $n = -2, 0$, or 2 and after that condition (4) would have become a major consideration.

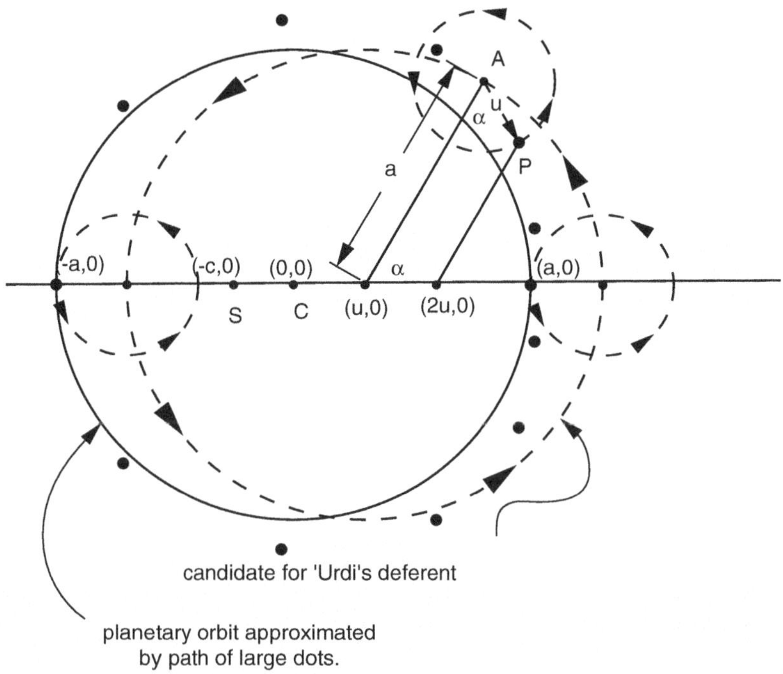

Fig. 4.14 The ’Urdī epicycle

What parameters are left to be determined? I will limit my discussion to Mars. (Refer to Fig. 4.14.) In Fig. 4.14, point S represents the location of the Sun at the left focal point $(-c, 0)$. The circle (or ellipse) centered at point C, with a solid line perimeter represents the orbit of Mars. The circle centered at $(2u, 0)$ with the dotted line perimeter is the eccentric circle, which carries the epicycle. To get the proposed orbit to agree with the Ptolemaic orbit at perihelion and aphelion, the radius of the epicycle must be u. (*Perihelion* is the point of the orbit at which the planet is closest to the Sun and has its greatest speed. *Aphelion* is the point of the orbit at which the planet is farthest from the Sun and has its least speed.) If I locate the center of the eccentric circle at $(u, 0)$, then the radius of the epicycle must be u and the value of β_0 must be π. The only thing left to consider is the value of u.

How can the radius of the epicycle be determined? ’Urdī proved that the figure that looks like a trapezoid is a trapezoid. This implies that point P moves at a uniform angular velocity about the point at $(2u, 0)$. Thus, to get the proposed orbit to be as close as possible to the Ptolemaic model, the point $(2u, 0)$ must correspond to the equant point and thus $(2u, 0) = (c, 0) = (ea, 0)$. Thus the radius of the epicycle would be $\frac{1}{2}ae$. Furthermore, if the epicycle of Apollonius was used to center the deferent at the Sun, it would have a radius of $\frac{3}{2}ae$ which is three times the radius of the ’Urdī epicycle. Copernicus did not indicate that he saw the relation between this innovation and the Ptolemaic model and at least one historian has cited this situation

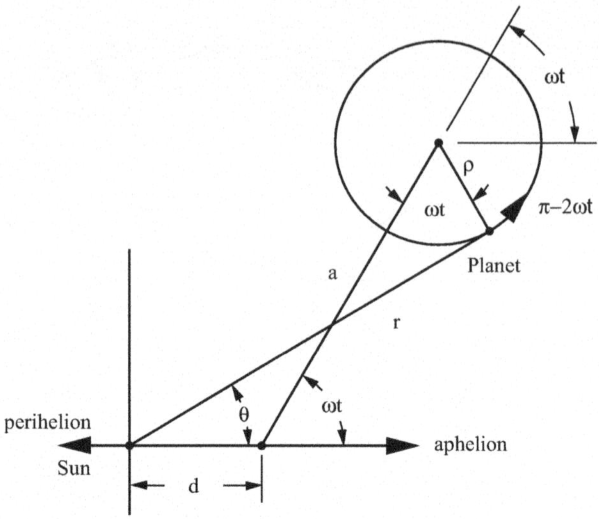

Fig. 4.15 Parameters for the Urdī model

as evidence that Copernicus copied the 'Urdī epicycle without fully understanding what he was copying:

> Copernicus, however, did not apparently realize the full significance of the two components of Ibn al-Shātīr's model (the Apollonius and the 'Urdī components), and simply used the model as a whole, by transposing it to heliocentrism as we just said (Saliba 2007, p. 205).

I speculate that the reason that Copernicus did not see how he could adjust the radius of the 'Urdī epicycle to fit closely with Ptolemy's model is that he never drew the line connecting P with $(2u, 0)$ in Fig. 4.14. What Copernicus did do is to fit the parameters of the model with data collected by Ptolemy and himself. The process of dealing with this data was far more complex that the task of understanding Urdī's argument for adjusting his parameters the way he did.

What happened when Copernicus set out to adjust the parameters of Urdī's model to fit real data?

Referring to Fig. 4.15 to understand the parameters involved, it is possible to determine the following relations for the Urdī–Copernicus model:

$$\sin\theta = \sin\omega t\left[1 - \frac{d+\rho}{a}\cos\omega t - \frac{(d+\rho)^2}{2a^2}\sin^2\omega t + \frac{d^2-\rho^2}{a^2}\cos^2\omega t\right] + \tag{4.70}$$

higher powers of d/a and ρ/a.

And

$$r = a\left[1 + \frac{d-\rho}{a}\cos\omega t + \frac{(d+\rho)^2\sin^2\omega t}{2a^2}\right] + \tag{4.71}$$

$$a\,[\text{higher powers of } d/a \text{ and } \rho/a]\,.$$

(See Appendix C for the relevant computations.)

From actual data computed from Kepler's Laws:

$$\sin\theta = \sin\omega t\left[1 - 2e\cos\omega t - 2e^2\sin^2\omega t + \frac{5}{2}e^2\cos^2\omega t\right] + \text{higher powers of } e. \tag{4.72}$$

And

$$r = a\left[1 + e\cos\omega t + e^2\sin^2\omega t\right] + \text{higher powers of } e. \tag{4.73}$$

(See Appendix D for the relevant computations.)

For the Urdī–Copernicus model, a is the radius of the deferent, d is the distance between the Sun and the center of the deferent, and ρ is the radius of the epicycle. For the Kepler version, a is the semimajor axis of the elliptical orbit and e is the eccentricity. In both cases, $\omega = 2\pi/T$, where T is the time for the given planet to complete a revolution.

For Mars, $e = 0.0934$ so e^2 is less than 1/100. Thus, for the astronomical instruments available to Ptolemy and Copernicus, we can ignore the second-order terms in (4.72) and (4.73). If we equate the first-order terms in (4.70) with the first-order terms in (4.72) and then equate the first-order terms in (4.71) with the first-order terms in (4.73), we get

$$\frac{d+\rho}{a} = 2e \quad \text{and} \quad \frac{d-\rho}{a} = e.$$

From this result, we get

$$d = \frac{3}{2}ae \quad \text{and} \rho = \frac{ae}{2}.$$

This is exactly what al-'Urdī got by matching his model with that of Ptolemy. (It should be noted that for Copernicus, e was not the eccentricity of an ellipse, but $2ae$ was the distance between the Sun and the equant point for whatever planet was in question.)

For Copernicus, there were some problems which I have hidden from you. First of all, he could not stand on the Sun to take his measurements. He could deal with that problem by taking measurements when Mars and the Sun were in opposition – that is lined up with Earth as shown in Fig. 4.16. (In Fig. 4.16, α is the angle between the reference star and aphelion. Copernicus was able to use methods described in Ptolemy's *Almagest* to compute this angle from data collected at three times when the Sun and an outer planet were in opposition as shown in Fig. 4.16.) A more serious problem was that there was no obvious way to measure r – at least directly. Copernicus did take measurements that gave him the relative size of the orbit of Earth with each of the planetary orbits. However, he did not take any measurement, which would give him r or some alternate entity as a function of time. Clearly, some measurement or measurements were necessary to conclude that $(d-\rho)/e = a$.

Ptolemy did not reveal how he concluded that the center of his deferent (Copernicus' orbit) was half-way between his basic point of reference and the

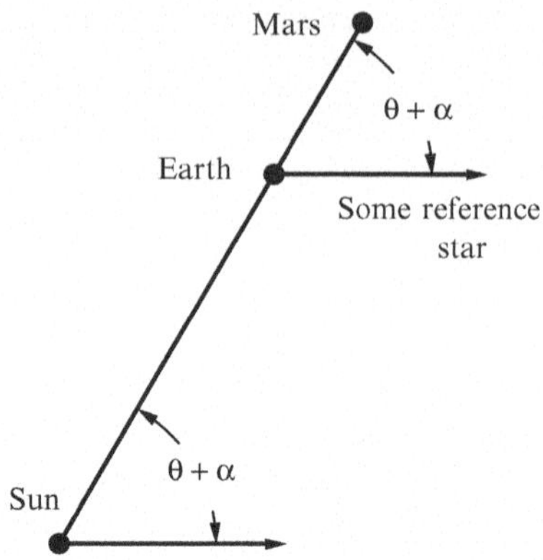

Fig. 4.16 Measurement of θ when the Sun and Mars are in opposition

equant point. Roughly 60 years after the death of Copernicus, Kepler was able to make this conclusion by studying the latitudinal motion of Mars. See Evans (1998), p. 1019 for the relevant diagram and explanation. Kepler concluded that Ptolemy must have reached this bisecting conclusion by making similar observations. However, historian James Evans has concluded that Ptolemy arrived at an equivalent result by studying the retrograde motion of Mars (Evans 1984, pp. 1080–1089) When Ptolemy's conclusion is translated into a heliocentric theory, the center of the planetary orbit must lie half-way between the Sun and the equant point. From Fig. 4.15, we see for the Urdī–Copernicus model, the aphelion occurs at $(d + a - \rho, 0)$ and the perihelion occurs at $(d - a - \rho, 0)$. The center or average of these two points is $(d - \rho, 0)$ and the midpoint for the Sun and the equant point is $(ae, 0)$. Thus, using Ptolemy's conclusion for retrograde motion, Copernicus was able to conclude that $d - \rho = ae$ and therefore $d = 3\rho$.

The equant problem posed an inconsistency with the concept of rotating spherical shells of ether. As we noted above, Copernicus testified that this was an important motivation for him to seek out an alternative to Ptolemy's geocentric theory. Ironically, if he had been born 50 years later, this inconsistency would not have confronted Copernicus. In 1577 AD (34 years after the death of Copernicus), Tycho Brahe carefully tracked the path of a comet in a way that clearly showed that it moved freely between the planets – in space supposedly filled by solid shells of ether. Those shells simply did not exist (Abers and Kennel 1977, pp. 109–110).

It is interesting to note that in view of this situation, Kepler seriously considered reintroducing the concept of the equant point in planetary theory before he arrived at his first two laws (Koyré 1973, pp. 176–177).

To repeat myself, I am inclined to believe that Copernicus rediscovered the epicycle of al-'Urdī. Western science historians have now uncovered the names of many other Islamic astronomers who made contributions to the progress of their science. Nonetheless, no document has been uncovered that would show a transfer of knowledge from al-'Urdī (or Ibn al-Shātir) to Copernicus. It has been established that some documents were translated from Arabic into Byzantine Greek and then transferred to Europe where people like Copernicus could read Greek. It has also been established that there were Arab speakers in Italy, where Copernicus received much of his education. On the other hand, if you assert that Copernicus copied the efforts of Ibn al-Shātir and his predecessors and then claimed that he originated these devices, you have another problem. To claim credit for the al-'Urdī–Ibn al Shātir devices for dealing with the equant problem, Copernicus would have had to be confident that any manuscript written by one of these Islamic astronomers would never see the light of day in Europe. It is one thing to steal an idea from your graduate student who has not had a chance to establish a reputation. You can always claim that your graduate student only filled in few trivial details of your brilliant idea. It is another thing to claim false credit for a result that has been published many years before. This form of plagiarism is sometimes attempted but usually in some work that is going to be read by very few people such as a term paper or possibly a Ph.D. thesis.

Moving on to another topic, Copernicus was not the first to propose a heliocentric theory. He was aware that Aristarchus had proposed such a theory many centuries before.

By carrying out the mathematical details necessary to compete with the Ptolemaic model, Copernicus was able to let scholars make a fair comparison. The model of Copernicus was not much simpler than that of Ptolemy. However, the Copernican model generated insights not possible with the Ptolemaic model. As I mentioned above, astronomers contemporary to Copernicus had great difficulty measuring distances but they could measure the ratio of the radius of an epicycle to the radius of the deferent in the Ptolemaic theory. This ratio was meaningless in the Ptolemaic theory but corresponds to the ratio of the radius of a planet's orbit to that of Earth's orbit in the Copernican model. (This is true for Mars, Jupiter, and Saturn, which lie outside the orbit of Earth. The ratio is reversed for Venus and Mercury, which lie inside the Earth's orbit. See Prob. 61.) This enabled Copernicus to compute the relative sizes of all planetary orbits.

As I also mentioned above, another insight from the Copernican theory was an explanation for retrograde motion. In the Ptolemaic theory, retrograde motion seemed to be an idiosyncrasy of each planet. In the Copernican theory, it could be seen that retrograde motion is an illusion that occurs when the Earth and a given planet are in certain relative positions in their respective orbits.

Later developments occurring within a century of the death of Copernicus would provide further evidence for his theory. With the invention of the telescope, Galileo could see that the planet Venus went through a sequence of phases just as the Moon does. This was consistent with the theory that Venus revolves around the Sun. In addition, Galileo could observe that Jupiter had at least four Moons. This tended to put Jupiter on an equal footing with Earth.

However, the most significant confirmation of a heliocentric theory arose from Tycho Brahe's efforts to prove it to be wrong. Tycho Brahe was convinced that the Earth was stationary. He proposed that yes the five known planets revolved around the Sun but the Sun revolved around the Earth. Tycho Brahe (1546–1601) constructed instruments far more accurate than those used before. He used these instruments to collect a wealth of data that he hoped would disprove the theory of Copernicus.

Copernicus had challenged the geocentrism of Aristotle. Johannes Kepler (1571–1630) approached the data collected by Tycho Brahe with the thought that another aspect of Aristotle's vision should be questioned. Namely that the heavens were ruled by uniform circular motion. His most important discovery was probably his second law that the areal velocity of a planet is constant. However, the discovery of his first law that the orbit of any planet is an ellipse with the Sun located at a focal point would not come easily. He did this by examination of the data Tycho Brahe had collected for the planet Mars. He had boasted to Tycho that he would determine the orbit of Mars in eight days. It took him eight years.

The process that Kepler went through is nicely outlined by Alexandre Koyré (Koyré 1973). Kepler could not determine the orbit of Mars without determining the orbit of Earth and he could not determine the orbit of Earth without determining the orbit of Mars. Furthermore, he could not determine the orbit of either without making some shrewd guesses. An additional complicating factor of the analysis is the fact that the planes of the two orbits do not coincide, although both planes pass through the Sun. (In case you are curious, the third law of Kepler is that the square of the period of revolution divided by the cube of the major axis is the same for each planet.)

Later Isaac Newton was able to show all three laws were a consequence of his law of gravity.

Problem 61. Consider Fig. 4.8.

(a) How could you construct a diagram to conform to the theory that the Earth is stationary, the Sun moves in an orbit about the Earth, and Mars moves in an orbit about the Sun.
(b) Suppose you are a Martian. Construct a diagram to represent the view that Mars is stationary, the Sun moves in an orbit about Mars, and the Earth moves in an orbit about the Sun.
(c) Figure 4.8 remains valid if Mars is replaced by Jupiter or Saturn which lie outside the Earth's orbit. Adjust the two diagrams so that they would be appropriate for Venus or Mercury which lie inside the Earth's orbit. (If you assume that the deferent is larger than any epicycle, then the deferent in the geocentric model will correspond to the orbit of the Earth about the Sun in the heliocentric model.)

Problem 62. Refer to Fig. 4.17. According to Kepler's second law, the areal velocity about the point E is constant. That is

$$\frac{1}{2}r^2\frac{\mathrm{d}\phi}{\mathrm{d}t} = h, \quad \text{where } h \text{ is a constant.}$$

According to Ptolemy's rule,

$$\frac{\mathrm{d}\theta}{\mathrm{d}t} = \omega, \quad \text{where } \omega \text{ is a constant.}$$

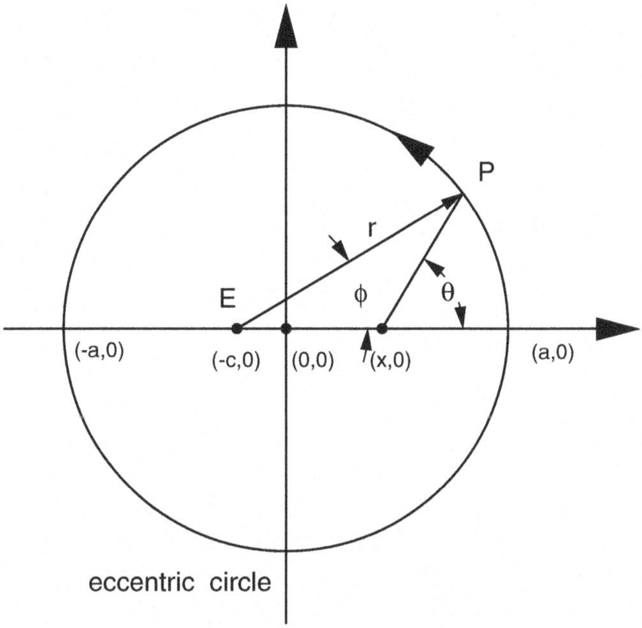

Fig. 4.17 The value of x can be chosen so that Ptolemy's model closely agrees with Kepler's second law at the ends of the diameter passing through the location of earth and the center of the eccentric circle

Determine the value of x which is used to locate Ptolemy's equant point, so that Ptolemy's rule agrees with Kepler's second law at the ends of the diameter passing through the location of Earth and the center of the eccentric circle. (You should get $x = c$.) (In reality, Ptolemy used information from other orbital points to locate his equant point. Nonetheless, he arrived at the value "c" for his equant point.)

Problem 63. TŪSĪ COUPLE I. Consider Fig. 4.13. Suppose $\beta = -2\alpha + \beta_0$. Show that for both $\beta_0 = 0$ and $\beta_0 = \pi$, you get an ellipse. Ellipses constructed in this manner could not be used for planetary orbits without modification because the rate of movement at each end of the major axis is the same. However, the limiting case when the radius of the epicycle equals the radius of the deferent results in an oscillatory straight line motion. It is believed that al-Tūsī was the first to discover this construction. Al-Tūsī, along with succeeding Islamic astronomers and Copernicus, used this device to deal with latitudinal motion and the orbital motion of Mercury. This device has become known as the "Tūsī Couple."

Clearly, Copernicus did not rediscover this device. He did not claim to do so. In a passage deleted from Book III, Chap. 5 of *De Revolutionibus,* Copernicus wrote that other persons before him knew and had used this theorem (Copernicus 1992, pp. 126, 384, and 385). It is also understood that in 1501, Copernicus became

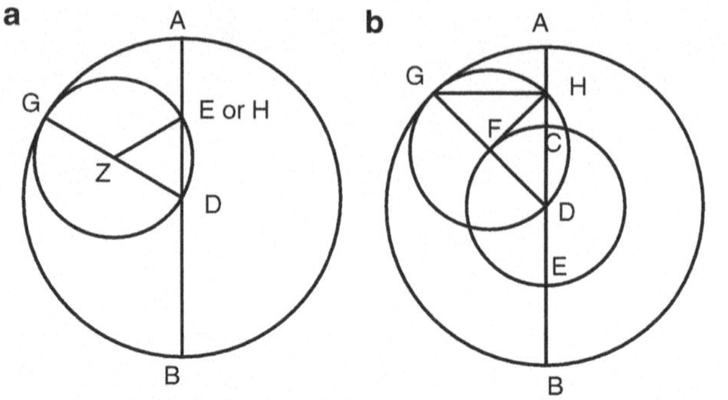

Fig. 4.18 (**a**) appears in Nasīr al-Din al-Tūsī's *Memoir on Astronomy.* (**b**) appears in Copernicus' *De revolutionibus orbium coelestium* (Book III, Chap. 4)

acquainted with Girolamo Fracastoro, a professor of medicine at the University of Padua. Copernicus was 28 at the time and seemed to have chosen "graduate student" as his career. His teacher, Fracastoro, was more than five years younger than he was having been appointed professor at the age of 19. Copernicus would eventually obtain a diploma of Doctor of Canon Law two years later at the University of Ferrara and then return to the University of Padua for several years (Armitage 1963, pp. 63–67). Astrology was considered a necessary part of a physician's education at the time and Fracastoro not only published works in medicine but he was also considered to be a scholar in mathematics, geography, and astronomy. It is known that Fracastoro along with others in Italy were familiar with the Tūsī device (Di Bono 1995, pp. 133–154).

Nonetheless, Copernicus has been accused of lifting the Tūsī couple directly from al-Tūsī himself. Fig. 4.18a appears in Nasīr al-Din al-Tūsī's *Memoir on Astronomy* (Ragep 1993, vol. I, p. 198). In 1973, science historian, Willy Hartner, noticed that the lettering of Fig. 4.18a was very similar to the corresponding diagram in *De revolutionibus*, namely Fig. 4.18b (Hartner 1973). Of course, in the original work Arabic letters were used in Fig. 4.18a. Those who see a strong resemblance between the two diagrams transcribe the Arabic letter "haa" into an "H." When F. Jamil Ragep translated al-Tūsī's *Memoir on Astronomy,* he transcribed the same letter into an "E." I contacted Prof. Ragep on this matter on the internet and he informed me that there are indeed two conventions for transcribing the letter "haa."

George Saliba, who feels that the two diagrams are essentially identical, has pointed out that, someone unfamiliar with Arabic transcript traditions could easily misread an Arabic "*zain*" in a medieval Arabic manuscript as a "*fā.*" The correct *zain* would be transcribed into a "Z", but if it were misread as a *fā*, it would be translated into an "F", in which case there would be total agreement in the lettering (Saliba 2007, p. 200).

It should be noted that a modern mathematician would draw the diagram and then sprinkle in the letters. In this context, the agreement of the lettering is extraordinary. However, Copernicus lived in a time when Euclidean traditions prevailed and discussion of the construction of the diagram was an important part of the presentation. When Copernicus presented his diagram, he wrote,

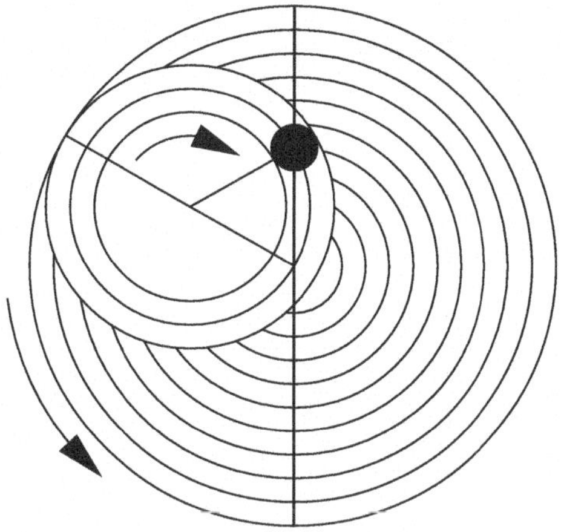

Fig. 4.19 Rotating spherical shells of ether resulting in a straight line oscillatory motion

> Let there be a straight line AB. Let it be divided into four equal parts at points C, D, and E. Around D, draw the circles ADB and CDE, with the same center and in the same plane. On the circumference of the inner circle, take any point F at random. With F as center, and with radius FD, draw the circle GHD. Let this intersect the straight line AB at the Point H. Draw the diameter DFG (Copernicus 1992, p. 125).

Notice that each letter is introduced in alphabetical order. In the Euclidean tradition, it becomes almost a little difficult to see how Copernicus could have lettered his diagram any other way.

For me any debate about the extent of agreement in the lettering is irrelevant because from a mathematical point of view, the diagrams are completely different. In the Copernicus version, no reference is made to the largest circle in Copernicus's proof that Copernicus used to prove the straight line oscillatory motion of point H. The outer most circle is used only to prove a follow-up corollary. Copernicus also mentions the fact that if the deferent and epicycle are unequal in size, one gets an ellipse. This observation does not readily follow from al-Tūsī's diagram. We should note that the diagram of al-Tūsī becomes meaningless if the outer circle is eliminated.

Nasīr al-Din al-Tūsī's *Memoir on Astronomy* has been translated by F. Jamil Ragep (Ragep 1993). As translated and interpreted by Ragep, the large circle rotates counterclockwise while the circle with half its radius rotates clockwise with twice the angular speed of the large circle. Although the two circles slide past one another, the location of the point of contact is determined by the motion of the large circle. Meanwhile, the straight line oscillatory motion is generated by point H (or E), which is attached to the smaller circle. This seems somewhat bizarre but it is consistent with the Greek concept that the system of deferent and epicycles associated with each planet was constrained by a spherical shell composed of a solid transparent substance known as "ether." That is between each planet, there was a spherical shell of ether. (See Fig. 4.19.)

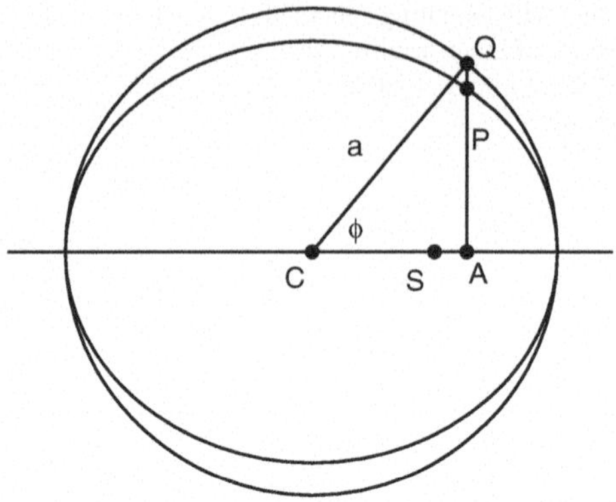

Fig. 4.20 Kepler discovered a remarkable relation

Problem 64. TŪSĪ COUPLE II. Show that a straight line can also be generated by a point attached to a circle rolling without slipping inside a stationary circle with a radius twice the size of the rolling circle.

Problem 65. According to Fred Hoyle, a critical step for Kepler's derivation of his first law was a remarkable observation (Hoyle 1962, p. 117). See Fig. 4.20. Kepler discovered that for a planet located at point P with the Sun located at point S:

$$SP = a - ae\cos\phi, \text{ where} \tag{4.74}$$

$a =$ the radius of the circle, e is a constant, and $\angle CAQ$ is a right angle.

(a) Show $CS = ae$.
(b) Show (4.74) implies that the orbit of the planet is and ellipse with eccentricity e. Furthermore, demonstrate that the Sun is located at a focal point.

For the instruments available to Ptolemy and Copernicus, measurement of θ in Fig. 4.16 could be used to determine the sum $d + \rho$ but neither d nor ρ separately. However, Tycho Brahe was able to design and build instruments with higher precision. Using data collected using these instruments, Kepler was able to determine values for both d and ρ from measurements of θ.

Problem 66. Comparing (4.70) and (4.72), determine the ratio d/ρ that Kepler should have gotten. How does this ratio compare with that Kepler actually obtained: $(14948/3616)$? (Koyré 1973, p. 176)

When Kepler later inferred the relevant distances from his latitudinal observations, he realized there was an inconsistency between reality and the model of Copernicus. (In this circumstance, the comparison of (4.71) and (4.73) come into play.) No matter how he adjusted the values of d and ρ, the discrepancy between the theoretical position and observed position of Mars in some parts of the orbit was at least $8'$ (8 min or $8/60$ of a degree). The precision of Tycho Brahe's instruments was too good to ignore this small discrepancy. The model of Copernicus had to be abandoned for something better.

4.8 *Christopher Columbus and Some Bad Geography

According to American mythology, Columbus had to argue with those that thought the earth was flat. Indeed, at one time the prevailing view in Europe was that the earth might not be entirely flat but it certainly was not a sphere where you would have people living thousands of kilometers apart standing on earth with their heads pointed in opposite directions. This was the view of Saint Augustine (354–450) who is one of the most respected theologians in the history of the Christian church. Saint Augustine was convinced that there was enough geographical information in biblical scripture to refute those that thought the earth was spherical. He even labeled those who dared suggest that the earth was spherical to be heretics (Boorstin 1983, pp. 107–111).

Nonetheless, at the time that Columbus was presenting his proposal, the scientific advisors in the courts of Spain and Portugal were much more sophisticated. They not only recognized that the earth is spherical but they also knew that Eratosthenes had measured the size of the earth and that Arab scientists had checked his measurements. The American myth, that Columbus had to argue that the earth is round, stems from a biography of Columbus written by Washington Irving. In a later biography of Columbus, this point is discussed by the author, Samuel Eliot Morison: He wrote,

> What then, becomes of the celebrated sessions of the University of Salamanca, before whose professors of mathematics, geography and astronomy Columbus argued his case, and was turned down because he could not convince them that the world was round? That is pure moonshine. Washington Irving, scenting his opportunity for a picturesque and moving scene, took a fictitious account of this nonexistent university council published 130 years after the event, elaborated on it, and let his imagination go completely (Morison 1942, p. 89).

Scientific advisors in both Spain and Portugal advised against financing Columbus' proposed voyage to Asia because the ships available at that time were not suitable for sailing the long distance from Europe to Asia. It was Columbus who had it all wrong. Columbus did not realize that Arabic miles were longer than Roman or Italian miles and concluded that the earth was 25% smaller than it actually is (Morison 1942, p. 65). Furthermore, he concluded that Marco Polo had traveled much further east than he actually did. Columbus estimated it would be about four time zones to Japan going west from the Canary Islands. With his 25% "correction", this would be equivalent to three time zones. In reality, going west, it is 12 time zones to Tokyo and 15 time zones to Shanghai. Fortunately for Columbus and his crew, he encountered the "New World" after sailing slightly less than four time zones.

Columbus was first turned down by Don João II of Portugal in 1484–1485. In 1485, he went to Spain to present his proposal to King Ferdinand and Queen Isabella. Between 1485 and 1492, Columbus was tentatively turned down at least twice in Spain and once more in Portugal. However, in Spain he maintained his hopes because Ferdinand and Isabella were focused on eliminating Islamic rule from its last foothold in Spain. After the Christians had gained control of Cordova in 1236, Islamic rule had been restricted to the Kingdom of Granada – a mountainous strip of territory along the southern coast of Spain roughly 100 km wide and 300 km long.

Christopher Columbus must have been elated to be present at the formal surrender of Granada on January 2, 1492. Now Ferdinand and Isabella could turn their attention to his proposal. However, a few days later Queen Isabella turned down his proposal with a finality he had not encountered before. Christopher's brother Bartholomew had presented his proposal to King Henry VII of England and King Charles VIII of France and both courts had turned Christopher down.

Always the optimist, Columbus almost immediately set out to France by mule to make a personal appeal to King Charles VIII of France. Fortunately for Columbus, his situation changed suddenly. The same day that Columbus set out for France, Luis de Santangel, keeper of the privy purse for Ferdinand, obtained an audience with the Queen. He argued that although it might be a long shot, Columbus might be right. (Even the most knowledgeable scholar of Spain could not say with certainty where the east coast of Japan was.) Furthermore, the payoff would be tremendous for a small investment. Thus persuaded, the Queen sent out a messenger to Columbus with a command for him to return to her court (Morison 1942, pp. 101–103).

Indeed, the royal investment in Columbus' voyage of discovery was relatively modest. Much more time and effort would be devoted to another pet project of Queen Isabella – the expulsion of all Jews from Spain. The same summer that Columbus set sail with 90 men and boys from the port of Palos, an estimated 8,000 Jewish families sailed out of the nearby port of Cadiz into exile (Morison 1942, p. 109).

By contrast, the terms of the 1492 surrender allowed Muslims to continue the privilege of being Muslims. However, in succeeding years, these terms would be ignored and Muslims would be given the same stark choice as the Jews: convert to Christianity or face expulsion.

Problem 67. Go to a library or search the internet to find out how Eratosthenes (born around 284 BC) was able to measure the size of the earth. Then write up an explanation with a suitable diagram in a form that can be understood by your slowest classmate.

4.9 *Clifford and Grassmann

*4.9.1 *William Kingdon Clifford 1845–1879*

William Clifford is currently recognized for two concise papers, published at the end of his short life, that laid the foundations of what is now known as Clifford algebra. The first was nine pages long with the title, "Applications of Grassmann's Extensive Algebra" (Clifford 1878). This was published in 1878 at a time when his health was steadily declining. The second paper, five pages in length, was entitled,

"On the Classification of Geometric Algebras." This was published in unfinished form in 1882 after his death (Clifford 1882).

The importance of these papers was not recognized for a long time. In a book published in 1919, the mathematics historian, Florian Cajori, described William Clifford as one of the four mathematicians (two English and two German) who dominated the study of geometry in the late nineteenth century (Cajori 1919, p. 278). Cajori listed several of Clifford's papers he thought to be significant but made no mention of Clifford's geometric algebra (Cajori 1919, p. 307). In a book first published in 1923, another historian of mathematics, David Eugene Smith like Cajori pointed out Clifford's contributions to Riemann surfaces, biquaternions, and other fields but made no mention of what we now label as Clifford algebra (Smith 1958, pp. 467–468).

William Kingdon Clifford was born on May 4, 1845 in Starcross, England not far from Exeter. His father (also William) made his living as owner of a bookstore. Although he was frequently in poor health, the older William served his city both as Alderman and Justice of the Peace. At age 15, the young William won a Mathematical and Classical Scholarship to King's College London. While at King's College, at age 18, he was able to get a paper published in *The Quarterly Journal of Pure and Applied Mathematics* in March 1863 (Chisholm 2002, pp. 16–17). He then won a Foundation Scholarship to Trinity College (part of Cambridge University).

The pathetic trajectory of Clifford algebra as an accepted framework for the study of geometry appears to be linked to the respect that Clifford had for fellow mathematician James Joseph Sylvester (1814–1897). J.J. Sylvester also attended Cambridge University – a generation before Clifford. In 1837, Sylvester scored second highest on the very challenging Mathematical Tripos exam. Because of this achievement he was given the prestigious title of "second wrangler." Nonetheless, he was not awarded a degree. At that time, a student could not obtain a degree unless he signed an oath swearing adherence to each of the Thirty-Nine Articles of the Church of England. As a Jew, J.J. Sylvester refused to take the oath.

Almost 30 years later in 1866, William Clifford encountered the same requirement. He too finished second on the Mathematical Tripos. Perhaps, he would have done better but he devoted some of his time studying the works of Sylvester and others – material he knew would not be on the exam (Chisholm 2002, p. 22). At the time he entered Cambridge, Clifford was a devout Christian. While at King's College he had even won a Divinity Prize (Chisholm 2002, p. 16). However, the dogmatic position of the Anglican Church against Darwin's theory of evolution induced him to reexamine his beliefs. He soon evolved into an agnostic and then an atheist. Clifford signed the required religious oath in 1863 and 1864 without reservation. In 1865, he did so reluctantly and in 1866 he refused (Chisholm 2002, pp. 21–22).

The religious requirement was still in place but authorities were more liberal then they had been for Sylvester. The regulation was ignored and Clifford got his degree. A few years later in 1871 while Gladstone was prime minister, the requirement was abolished (Chisholm 2002, p. 29).

In the Spring of 1871, Clifford became Professor of Applied Mathematics and Mechanics at University College, London. Relations with J.J. Sylvester and other mathematicians had been established during his years at Cambridge and now became closer (Chisholm 2002, p. 29).

Four years after he was denied a degree at Cambridge University, Sylvester was awarded a B.A. and M.A. by Trinity College, Dublin. He won recognition for his many publications and in 1866 he was elected President of the London Mathematical Society. Nonetheless, for most of his life, he was unable to find steady employment at an academic position. To make living he studied law and worked as an actuary.

Finally at age 63, he obtained a position that matched his abilities. He was given the responsibility to organize the first American Ph.D. program at John Hopkins University. One year later in 1878, he founded the *American Journal of Mathematics.* Clifford's paper on what he called "geometric algebra" appeared in the very first volume of that journal. Today, the *American Journal of Mathematics* is one of the most prestigious math journals in the world, but in 1878 it was less than obscure. Why did Clifford submit his paper to Sylvester? Presumably because of his political outlook, he wanted to endorse Sylvester's efforts.

University College where Clifford was employed was considered either "forward looking" or "godless" for admitting Dissenters, Jews, and Roman Catholics. Although University College was at the cutting edge of social reform, women were taught in separate classes. William Clifford was the first mathematician at University College to admit both sexes to his lectures.

In 1873, William met and fell in love with Sophia Lucy Jane Lane (better known as "Lucy"). They eventually got married on April 7, 1875 and their home soon became an intellectual salon. Visitors included Thomas Huxley, the principal exponent of Darwinism in England, and Robert Louis Stevenson. Lucy and William particularly valued their friendship with George Eliot.

Clifford had a broad intellect. He not only mastered several modern languages but he also studied Arabic, Greek, and Sanskrit. Due to his abilities in exposition, he was frequently asked to give lectures on a wide range of topics in science, philosophy, and even history. Despite the demands on his time, he was somehow able to publish a fairy tale entitled, *The Giant's Shoes.*

He would frequently work through a night after giving his lectures the previous day. To complete an 18-page essay, *The Unseen Universe*, Clifford took up his pen at nine forty-five one evening and worked straight through till nine the next morning (Chisholm 2002, p. 50). Eventually, these work habits took their toll. At Cambridge, William was known for his physical prowess. He could do one-armed chin ups on a bar with either arm. However, he had developed respiratory problems and soon after he arrived in London in 1876 he suffered a physical collapse.

It eventually became clear that he had tuberculosis. Typically, tuberculosis was contracted in childhood but would remain dormant for many years, often for a lifetime. However, once tuberculosis flared up a second time, the victim was in serious trouble. Furthermore, nineteenth century London did not enjoy clean air. During winter fogs, when people were burning coal to keep warm, the air quality was perhaps the worst in the world. In the opening paragraph of the first chapter of *Bleak House*, Charles Dickens gave a vivid description of how bad it could be,

> Implacable November weather. – Smoke lowering down from chimney-pots, making a soft black drizzle, with flakes of soot in it as big as full-grown snow-flakes – gone into mourning, one might imagine, for the death of the sun.

The air quality in London did not improve for many years. During the "Great Smog", which lasted for less than a week in early December of 1952, it is estimated that there were "at least 1,600 (perhaps 4,000) excess deaths in London, mainly from bronchitis, pneumonia and coronary disease among infants and the old" (Inwood 1998, p. 838).

At a time before antibiotics, Clifford's only hope of recovery was to get rest somewhere away from the filthy air of London. (In Germany, the medical profession were discovering that, without antibiotics, the best treatment for tuberculosis was bedrest in a hospital. In England, it was still thought that sea voyages were beneficial, although it was also recognized that rest was essential.) A change of climate did seem to have a beneficial effect on some victims.

In 1873, believing that he had only a few months to live, a young American dentist moved west from Atlanta, Georgia, in the hopes of extending his life. This action probably did extend his life, but it clearly shortened the lives of many others. His violent coughing spells forced him to give up dentistry Subsequently, he became a successful gambler. To survive in this second profession, he became very proficient at killing people. "Doc" Holliday eventually died from tuberculosis in 1887, but that was 14 years after he left Atlanta and six years after his participation in the famous gunfight near the OK Corral in Tombstone, Arizona.

By contrast, Clifford was not inclined to take the advice of his doctors. Clifford's friends had more concern for his health than he did. Thomas Huxley and Frederick Pollock raised funds so that he and Lucy could take a seven week cruise in the Mediterranean. However, when they returned to London, Clifford ignored the advice of friends and resumed his academic duties at University College.

Not only did he resume his lectures but he also poured himself into criticism of contemporary religious thought and practice. His efforts included two essays, *Ethics of Belief* and *Ethics of Religion.* (*Ethics of Belief* has recently been republished.)

His health continued to deteriorate. In February 1878, his father died at the age of 58 during a visit to southern France and William was too ill to attend the funeral. Nonetheless, he continued to ignore the consequences of his work habits. He worked through a night to complete the treatise, *Virchow on the Teaching of Science*. Two years before in 1876, he had presented an abstract on what is now known as "Clifford algebras" to the London Mathematical Society. Now he developed his preliminary results further and submitted it to the *American Journal of Mathematics* under the title, "Applications of Grassmann's extensive algebra." The status of his health became obvious when he collapsed while trying to finish a lecture (Chisholm 2002, p. 55).

His friends became alarmed and financed another cruise on the Mediterranean Sea and a placement in a Swiss sanatorium. However, in Switzerland, William and

Lucy encountered some brutal weather. Lucy wrote to a friend complaining that they might freeze to death in the "coldest, most shivery, chatter-your teeth, sort of place you could possibly imagine" (Chisholm 2002, p. 56).

The Cliffords now had two daughters who had been left behind in England. Since William's health was not improving, they returned to England. In a last desperate effort to save his life, William and Lucy embarked on a five-day voyage from England to Madeira. Madeira is a Portuguese island about 650 km (400 miles) west of Morocco. Lucy and William were accompanied by a daughter and son-in-law of Thomas Huxley. The climate in Madeira was more pleasant but William died less than two months later on March 3, 1879.

Lucy had established a career as a writer before she met William. Despite the burdens of being a mother of two daughters and a nurse to William, she had continued her writing during her short four-year marriage. After William's death, she had a long and illustrious career as a writer. She always retained fond memories of William and when she died in 1929, she was buried alongside him.

During his life, William Clifford became convinced that we live in a space that is curved. In 1870, he presented a paper entitled, *On the Space Theory of Matter* to the Cambridge Philosophical Society. In this paper, Clifford proposed the idea that matter and field energy are simply manifestations of curvature of space (Chisholm 2002, p. 39). It is thus fun to speculate on whether or not Clifford could have been a major contributor to the theory of relativity (either the special or general theory). Clearly, Clifford died too soon. He died 11 days before Einstein was born. But was he born too soon? Perhaps not. What were the key ideas and events that paved the way for Einstein's special theory of relativity? Most of the groundwork arose in the decade following Clifford's death.

First were the Michelson–Morley experiments showing that the speed of light was the same in frames moving at constant speed with respect to one another. The first experiment was carried out by Michelson alone in 1881 and the second was carried out with Morley in 1887. Second was the influence of Ernst Mach. Mach challenged Newton's concepts of "absolute space" and "absolute time" (Clark 1971, p. 61). The work by Mach that Einstein admired so much was *Die Mechanik in ihrer Entwicklung,* which was published in 1883. (Ten years later, an English version would be published under the title *The Science of Mechanics: A Critical and Historical Account of Its Development.*) The other important inputs came from Henri Poincaré and Hendrick Lorentz who were, respectively, eight and nine years younger than Clifford. They made their contributions somewhat later.

However, Clifford might have had an advantage that was not available to Lorentz, Poincaré, or Einstein. All three of those men spent considerable time trying to determine the nature of the world we live in by studying Maxwell's equations. However, even today virtually every text book on electro-magnetic fields presents Maxwell's equations in a form that is in some sense wrong. Certainly, they are written in a form that is a severe hindrance for someone trying to invent the special theory of relativity. In this standard formulation, the electric and magnetic fields are each represented by what we can now interpret as two separate1-vectors and the time variable does not appear to be on an equal footing with the space variables.

However, in reality, the electric and magnetic fields are both part of a single 2-vector. If you know the coordinates of the magnetic field in Cartesian coordinates and wish to determine the coordinates in polar coordinates, you will get the wrong answer if you think that the magnetic field is a 1-vector.[2]

Correct versions of Maxwell's equations, using the formalism of tensors, generally appear in text books on relativity and only sometimes in a closing chapter of a text on electro-magnetism. According to Wolfgang Pauli, Hermann Minkowski was the first one to arrive at a proper four-dimensional space–time version of Maxwell's equations (Pauli 1958, p. 78). Minkowski presented this formulation in a lecture entitled, "*Das Relativitätsprinzip*" to the Mathematische Gesellschaft Göttingen on November 7, 1907 – two years after Einstein published his special theory. Minkowski died about 14 months later on January 12, 1909 and the contents of his lecture were not published until 1915 (Minkowski 1915a, 1915b).

It is plausible that Clifford could have arrived at an equivalent formulation much sooner. Shortly after Clifford died, a battle developed between those who advocated quaternions and those who advocated vectors for the formulation of the equations of physics. On p. 154 of the book, *A History of Vector Analysis*, the author Michael J. Crowe writes, "In 1879 Gibbs gave a course in vector analysis with applications to electricity and magnetism, and in 1881 he arranged for the private printing of the first half of his *Elements of Vector Analysis*; the second half appeared in 1884. In an effort to make his system known, Gibbs sent out copies to more than 130 scientists and mathematicians." On the same page, Crowe indicates that Willard Gibbs cited Clifford in the introductory paragraph of his vector analysis book. Thus, we can be confident that Clifford would have received a copy of Gibb's work. In this circumstance, Clifford would have been eager to enter the quaternion-vector battle with his geometric algebra.

It is probably impossible to present a truly correct (covariant) version of Maxwell's equations using either quaternions or vectors. However, it is possible using Clifford algebra. Done properly the time variable appears on an equal footing with the space variables. Having translated Riemann's ground breaking work on geometry in 1873, Clifford would have been suitably equipped to deal with this proper formulation problem. Had Clifford succeeded, the development of special relativity could have been accelerated and math undergraduates would have been studying Clifford algebra 100 years ago. It is even plausible that Clifford could have spent the last thirty or forty years of his life fruitlessly seeking out the kind of unified field theory that eluded Einstein.

---

[2]I have discussed this problem in my book *Clifford Algebra* (Snygg 1997, pp. 137–144 and 154–161). However, the reader should be forewarned of two errors that I made. One, I made the claim that using Clifford algebra, one can write Maxwell's equations as a single equation. This is clearly not true in the presence of a dielectric. Second, Bernard Jancewicz has pointed out that the approximate magnetic field that I use for Maglich's Migma Chamber is not a solution of Maxwell's equation. Nonetheless, the discussion of the two-vector nature of the electro-magnetic field is correct.

4.9.2 *Hermann Günther Grassmann 1809–1877

> I propose to communicate in a brief form some applications of Grassmann's theory which it seems unlikely that I shall find time to set forth at proper length, though I have waited long for it. Until recently I was unacquainted with the *Ausdehnungslehre*, and knew only so much of it as is contained in the author's geometrical papers in *Crelle's Journal* and in *Hankel's Lectures on Complex Numbers*. I may, perhaps, therefore be permitted to express my profound admiration of that extraordinary work, and my conviction that its principles will exercise a vast influence upon the future of mathematical science. (The opening paragraph of Clifford's *Applications of Grassmann's Extensisve Algebra* (Clifford 1878, p. 350)).

The math historian Florian Cajori once attempted to summarize Grassmann's life with one sentence, "He started as a theologian, wrote on physics, composed texts for the study of German, Latin, and mathematics, edited a political paper and a missionary paper, investigated phonetic laws, wrote a dictionary to the Rig-Veda, translated the Rig-Veda in verse, harmonized folk songs in three voices, carried on successfully the regular work of a teacher and brought up nine of his eleven children – all this in addition to the great mathematical creations which we are about to describe" (Cajori 1919, pp. 335–336). Grassmann made substantial contributions to mathematics but during his lifetime he probably received more acclaim for his scholarship in Sanskrit.

Hermann Günther Grassmann was born on April 15, 1809 in Stettin, Prussia. (After World War II, Stettin became part of Poland and the name was changed to Szczecin.) He was the third of 12 children. Although his father, Justus Günther Grassmann, was trained mainly in theology, he taught mathematics and physical science at the Stettin Gymnasium. (Roughly equivalent to an elite American high school.)

In 1827, Hermann entered the University of Berlin where he spent six semesters studying philology and theology. Although he attended no mathematical lectures, he did read some mathematical texts written by his father (Crowe 1985, p. 55).

After leaving Berlin, Grassmann returned to Stettin and prepared himself for the state examinations required to become a teacher. In this process, he studied mathematics, physics, natural history, theology, and philology. In 1834, he accepted a position at a Berlin technical school but he returned to Stettin after a little over a year. He would remain in Stettin the rest of his life teaching at various schools. He frequently strove to obtain a position at a university but never succeeded (Crowe 1985, p. 55).

To prove his suitability for a university position, Grassmann wrote a paper on the tides with a length of more than 200 pages. The paper entitled, *Theorie der Ebbe und Flut (Theory of Low and High Tides)*, was significant because it contained the first presentation of a system of spatial analysis based on vectors. Grassmann was the first to introduce the notion of vectors as directed line segments that could be added in a geometrically meaningful way. Hermann's father Justus had introduced the idea of a geometric product in a couple of his books. However, Hermann generalized his father's idea and is thus credited with the invention the exterior product. (Grassmann referred to what is now known as the "exterior product" as the "geometric product.")

In 1840, the "Ebb and Flow" paper was submitted as part of a job application but it was ignored (Crowe 1985, pp. 55–57).

Undeterred, Grassmann generalized these concepts to n-dimensions and in 1844 he published his new ideas in a book entitled *Die lineale Ausdehnungslehre, ein neuer Zweig der Mathematik dargestellt and durch Anwendungen auf die übrigen Zweige der Mathematik, wie auch auf der Statik, Mechanik, die Lehre von Magnetismus und Krystallonomie erlaütert (Linear Extension Theory representing a new Branch of Mathematics illustrated by Applications to other Branches of Mathematics and also Statics, Mechanics, and the Theory of Magnetism and Crystallography).*

Generalizing a formalism to n-dimensions was a radical departure in an era in which geometry was considered a study of the space we live in. Indeed, he had opened up a new branch of mathematics with a new approach, which forced him to deal with a lot of new questions. For example, if we assign a direction to a finite line segment, should we or can we regard it as being equal to a parallel line segment with the same length if it lies on the same infinite straight line? – if it lies in the same plane? If we introduce a noncommutative multiplication, can we still have a multiplicative unit? Dealing with these questions, Grassmann developed an insight into the importance not only of the commutative law but also of the associative and distribution laws.

Grassmann sent out copies of his book to several mathematicians but got very little response – positive or negative. For example, Gauss sent him a letter of thanks on December 14, 1844. In the letter, Gauss stated that he had worked on similar ideas nearly half a century before and published his results in 1831. Michael Crowe has concluded that Gauss was probably referring to some work he had done on the graphical representation of complex numbers (Crowe 1985, p. 78). At any rate, it appears that Gauss did not read deeply enough into Grassmann's work to grasp the significance of Grassmann's achievement.

Why was Grassmann's work ignored? Grassmann's lack of credentials certainly contributed to the situation. But the American Josiah Willard Gibbs at Yale faced a similar problem 40 years later when he wanted to promote his *Elements of Vector Analysis*. In 1884, a Yale professor of mathematical physics with a Ph.D. in engineering did not have the prestige in Europe that he would have today. The only significant graduate mathematics department in America at that time was at John Hopkins University. John Hopkins University opened its doors in 1876 and James Joseph Sylvester was imported from Great Britain to organize the math department.

At age 61, Sylvester was willing to take on this task because for most of his life, being a Jew had made him ineligible for the most prestigious academic positions in England. Although he had been a very productive mathematician, he had been forced to turn to a career as a lawyer to make a living. Sylvester had satisfied the academic requirements for a mathematics degree at Cambridge in 1834. However, he did not actually obtain the degree until 1872 when the religious requirement was dropped.

Gibbs is now recognized as one of the giants of the nineteenth century in physics because of his contributions to thermodynamics. However, when Gibbs set out to promote his approach to vectors, his work in thermodynamics was largely unrecognized because it had been published in an obscure journal (*Transactions of the Connecticut Academy*). At Yale, Gibbs prestige was so low that for many years Yale did not see fit to pay him. He lived on inherited wealth. It was only when he got a job offer from John Hopkins that the authorities at Yale put Gibbs on the payroll.

Despite his lowly status, Gibbs was able to have a much greater impact on how mathematics and physics would be done during the twentieth century than Grassmann. Why? A clue appears in Gauss's thank you note. In that note, Gauss indicated he was very busy and to understand the real thrust of the work he would have to familiarize himself with the peculiar terminology used in the text (Crowe 1985, p. 78). Grassmann was presenting radically new material and even today he is difficult to read. At the time Grassmann was writing, the study of geometry was considered the study of the space we live in. Grassmann was taking a different approach. As a modern mathematician would do, Grassmann was defining his own mathematical structures and then investigating their properties. However, when Grassmann set out to explain his approach, his readers did not have an easy time. (The following passage is a translated version but judging from the way people responded to the *Ausdehnungslehre,* the exposition in the original German was not any plainer.)

> Thought exists only in reference to an existent that confronts it and is portrayed by the thought: but in the real sciences this existent is independent, existing for itself outside of thought, whereas in the formal it is established by thought itself, when a second thought process is confronted as an existent. – Thus proof in the formal sciences does not extend beyond the sphere of thought, but resides purely in the combination of different thought processes. Consequently the formal sciences cannot begin with postulates, as do the real; rather, definitions comprise their foundations (Grassmann 1995, p. 23).

When he got into the mathematics, the reading did not get easier:

> – The concept of continuous becoming is more easily grasped if one first treats it by analogy with the more familiar discrete mode of emergence. Thus since in continuous generation what has already become is always retained in that correlative thought together with the newly emerging at the moment of its emergence, so by analogy one discerns in the concept of the continuous form a twofold act of placement and conjunction, but in this case the two are unified in a single act, and thus proceed together as an indivisible unit. Thus, of the two parts of the conjunction (temporarily retaining this expression for the sake of the analogy), the one has already become, but the other – (Grassmann 1995, p. 25).

By contrast, when Gibbs set out to sell his vector formulation, he touched base with what was familiar to his audience. He adopted the **i**, **j**, **k** notation for unit vectors in the x, y, and z directions. This does not generalize to higher dimensions but readers acquainted with quaternions could feel that they were on familiar ground. Even among the few mathematicians that recognized the greatness of Grassmann's work, there were several who confessed in private correspondence to others that they did not have the fortitude to master much of Grassmann's *Ausdehnungslehre.* In a letter dated January 5, 1846, Möbius wrote to Ernst Friedrich Apelt: "You ask me whether, I have read Grassmann's *Ausdehnungslehre.* To this I answered that Grassmann himself presented me with a copy and that I have on numerous occasions attempted to study it but have never gone beyond the first sheets, since, as you mentioned, intuitiveness, the essential character of mathematical thought, is not to be found in the work. – " (Crowe 1985, p. 79)

Grassmann published a second edition of his *Ausdehnungslehre* in 1862. But that edition was not any easier to read. In 1888, the year that Grassmann died, Gibbs wrote to Victor Schlegel:

> – I procured the two Ed. of the Ausd. but I cannot say that I found them easy reading. In fact I have never had the perseverance to get through with either of them, & have perhaps got more ideas from his miscellaneous memoirs than from those works. – (Crowe 1985, p. 153)

Both Möbius and Gibbs recognized that results they had achieved in linear algebra had been developed earlier by Grassmann in the context of a more substantial work. However, neither Möbius nor Gibbs was able to absorb enough of Grassmann's work to carry his ideas forward. Indeed, very few mathematicians were able to use Grassmann's work to make new advancements. More frequently, they independently reinvented his work piecemeal in bits and pieces. Two outstanding exceptions were Clifford with his geometric algebra and Cartan with his differential forms.

Mathematics was not the only thing that attracted the attention of Hermann Grassman. The year 1848 was a year of upheaval in continental Europe. This was the year Karl Marx and Friedrich Engels published the *Communist Manifesto* and began publication of a newspaper in Cologne, Germany. In the midst of an economic depression that had been preceded by a couple of years of bad harvests, many laborers, out of work, faced starvation.

On February 22, mass demonstrations broke out in Paris and two days later King Louis Phillipe abdicated. News of this event spread across Europe and soon there were revolts in Italy and Austria. In Berlin, it appeared that the Prussian army was successfully suppressing a violent revolt with much loss of life. However, King Frederick William IV of Prussia, fearing massive bloodshed, ordered his troops back to their barracks on March 19. The King also authorized a constitutional convention and these gestures ended the violence.

Generally, members of the professional class were eager for the right to vote, a free press, and a unified Germany. However, they were appalled by the possibility of "mob rule" by those demanding such things as guaranteed work, a minimum wage, and a ten hour day. Hermann Grassmann was no exception.

Responding to these events, Hermann and his brother Robert like Marx and Engels started publishing a newspaper (Engel 1911, p. 140). The title of their paper was *Deutsche Wochenschrift für Staat, Kirche, und Volksleben* (*German Weekly for State, Church, and Home*). The first issue appeared on May 20. The first few editions advocated a unified Germany headed by a hereditary king with a Reichstag (elected parliament) in an advisory role.

In 1849, the members of the constitutional convention completed their work and offered King Frederick William IV the role of constitutional monarch of what they hoped would become a unified Germany. However by this time, the royal houses of Europe had regrouped and successfully put down the revolts against them. Even in France, democracy was short-lived. In the summer of 1848, a constitution was put in place that guaranteed universal suffrage (except for women). However, in the first national election, people outside Paris, alarmed by the disorder in Paris, overwhelmingly voted for Louis Napoleon Bonaparte, nephew of Napoleon I, to be their president. This occurred in December 1848 and about three years later this president became Emperor.

By 1849, the liberals in Germany, who had constructed their proposed constitution, had lost what little leverage they might have had by completely ignoring the aspirations of the workers and peasants who had initiated the demonstrations against the King of Prussia the year before. The King was not prepared to surrender the concept that he ruled by divine right so he turned down the role of constitutional monarch and hopes for a more democratic government was delayed for many years.

If Hermann Grassmann were alive today, he would be glad to see that Germany is now unified. He might be disappointed by the fact that Germany is not ruled by a hereditary king. He would certainly be disappointed by the fact that after World War II, the boundary between Germany and Poland was moved west and his home city of Stettin became part of Poland and is now better known by its Polish name *Szczecin.*

During his lifetime, Grassmann received little of the recognition he deserved for his accomplishments in mathematics. In fact his greatest recognition was received for his work in philology. In the early 1870s, he became discouraged by the lack of response he received in mathematics and turned to the study of Sanskrit. His study of Sanskrit begun in 1849 culminated with the publication of his *Wörterbuch zum Rig-Veda (Lexicon for the Rig-Veda)* (1784 pp) and his translation of the *Rig-Veda.* (The *Rig-Veda* contains over 1,000 hymns of praise to different Hindu gods.) For these achievements, he received an honorary doctorate from the University of Tübingen in 1876 (Crowe 1985, p. 93). He died the following year on September 26, 1877.

Chapter 5
Curved Spaces

5.1 Gaussian Curvature (Informal)

The simplest example of a curved surface is the ordinary 2-dimensional sphere in Euclidean 3-space. Furthermore the geometry of the sphere serves a major motivation for much of the mathematical work that has been done for more general surfaces. For this reason it is useful to examine some basic results associated with the geometry of a sphere.

Given two points on a 2-dimensional surface in ordinary 3-space , a *geodesic* is a path of minimum length passing through the two points.[1] For a sphere, a geodesic is a segment of a great circle, that is a circle formed by the intersection of a sphere with a plane passing through the center. A *lune* is the region bounded by two great circles on the surface of a sphere. (See Fig. 5.1a.)

The area of a sphere is $4\pi r^2$, where r is the radius. The area of a lune is the fraction $\theta/2\pi$ of the sphere. Thus the area of a lune is $(\theta/2\pi)(4\pi r^2)$ or $2r^2\theta$. From this result, it is not too difficult to determine the area of a figure formed by three great circles. Such a figure is known as a *spherical triangle*.

Theorem 68. *(Refer to Fig. 5.1b) The area of the spherical triangle* $\triangle ABC = [(A + B + C) - \pi]\, r^2$ *where* A, B, *and* C *are the vertex angles measured in radians. (One of the three great circles bounding* $\triangle ABC$ *is the silhouette* $CB\acute{C}\acute{B}C$*!)*

Proof. (For shorthand purposes, we will designate the area of a figure simply by the label for the figure.)

[1] Most mathematicians will cringe at this definition with good reason. When one travels by plane from New York to London by a great circle route over the Atlantic Ocean, one is following a geodesic. However if one continues over the same great circle, one is still following a geodesic even though it is not the shortest route from London to New York. Nevertheless, I will rely on your intuition and your common sense until I have developed enough mathematical machinery to give a more sophisticated definition in Sect. 5.4 of this chapter.

J. Snygg, *A New Approach to Differential Geometry using Clifford's Geometric Algebra*,
DOI 10.1007/978-0-8176-8283-5_5,

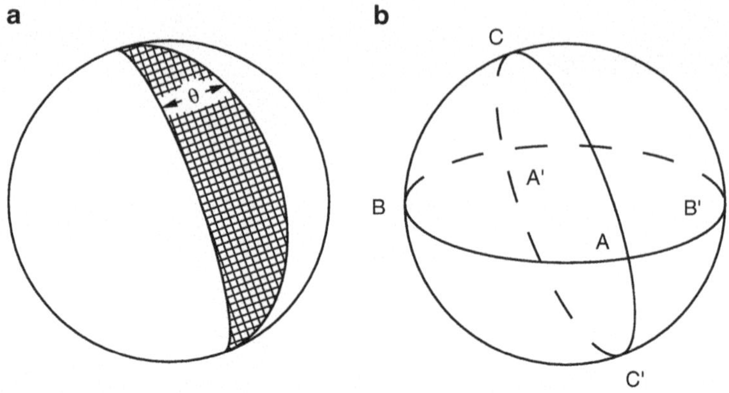

Fig. 5.1 (**a**) A lune is the region bounded by two great circles. (**b**) The area of $\triangle ABC = [(A + B + C) - \pi]\, r^2$, where A, B, and C are the vertex angles measured in radians

$$\triangle ABC + \triangle A'BC = 2r^2 A \text{ (This is the area of lune } ACA'BA.)$$
$$\triangle ABC + \triangle ABC' = 2r^2 C$$
$$\triangle ABC + \triangle AB'C = 2r^2 B$$
$$\triangle A'B'C - \triangle ABC' = 0 \text{ (By symmetry.)}$$

Adding up the two sides of the equations above, we get

$$2\triangle ABC + (\triangle ABC + \triangle A'BC + \triangle A'B'C + \triangle AB'C) = 2r^2(A + B + C).$$

Observing that the four spherical triangles in the parentheses on the L.H.S. of the equation above add up the area of the upper hemisphere in Fig. 5.1b, we now have

$$2\triangle ABC + 2\pi r^2 = 2r^2(A + B + C) \text{ or}$$
$$\triangle ABC = [(A + B + C) - \pi]\, r^2 \tag{5.1}$$

□

The theorem above is usually attributed to Adrien Marie Legendre (1752–1833). However, according to historian J.A. Lohne, Albert Girard published a proof in his *Invention nouvelle en algèbre* (1629). Furthermore a proof of the theorem dated September 18, 1603 was found among the personal effects of Thomas Harriot (Lohne 1979, pp. 300–301). In fact the proof that I have used above is identical to that of Thomas Harriot. This theorem was later generalized to regions bounded by geodesic curves on other surfaces by Karl Friedrich Gauss (1777–1855) and then again for non-geodesic boundaries by Pierre Ossian Bonnet (1819–1892) (Bonnet 1848, pp. 1–146). The results of Gauss and Bonnet will be discussed in Chap. 6.

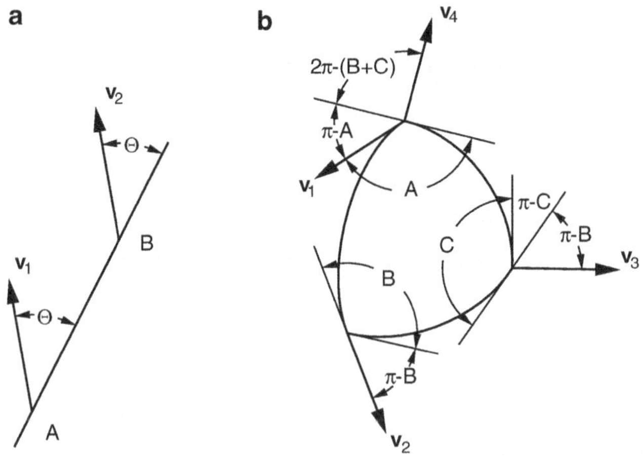

Fig. 5.2 (**a**) The vector $\mathbf{v}_2$ is the result of the parallel transport of $\mathbf{v}_1$ along path AB. (**b**) The vector $\mathbf{v}_1$, moved along AB by parallel transport, becomes $\mathbf{v}_2$ at vertex B. Continuing the process, the vector becomes $\mathbf{v}_3$ at C and $\mathbf{v}_4$ when it is returned to its original location. Careful computation reveals that $\mathbf{v}_1$, relabeled as $\mathbf{v}_4$, has undergone a rotation of $A + B + C - \pi$

The quantity $[(A + B + C) - \pi]$ is called the *spherical excess* of $\triangle ABC$. This concept can be generalized to polygons whose edges may or may not be geodesics. To do this, it is useful to obtain an alternate geometric interpretation of the spherical excess of a spherical triangle.

Suppose we consider the problem of transporting a vector tangent to the spherical surface around the perimeter of a spherical triangle without rotation. When I say "without rotation", I mean without rotation detectable to an intrinsic observer who is required to make all measurements in the surface of the sphere.

In a flat 2-dimensional plane, we can move a vector from point A to point B by sliding it along the straight line joining the two points while maintaining a constant angle between the vector and the straight line. Such a displacement is called *parallel transport*. (See Fig. 5.2a.)

On the surface of a sphere, the analog of a straight line is a geodesic or great circle. With that in mind, we now try to move a vector around the perimeter of a spherical triangle by parallel transport. (Refer to Fig. 5.2b.)

Suppose we consider the vector $\mathbf{v}_1$ tangent to geodesic AB at vertex A. Moving the vector by parallel transport along AB, it becomes vector $\mathbf{v}_2$ at vertex B. Continuing the process, the vector becomes $\mathbf{v}_3$ at vertex C and finally vector $\mathbf{v}_4$ when it is returned to its original location. This is perhaps most easily visualized for the example in which A is at the North Pole and BC is a segment of the equator. In that case, our parallel transported vector will be pointing South at all times during the motion.

Although the vector is moved by parallel transport along all points of the perimeter, we discover that when the vector is returned to its original location, it has become rotated counterclockwise from its original orientation. Referring again

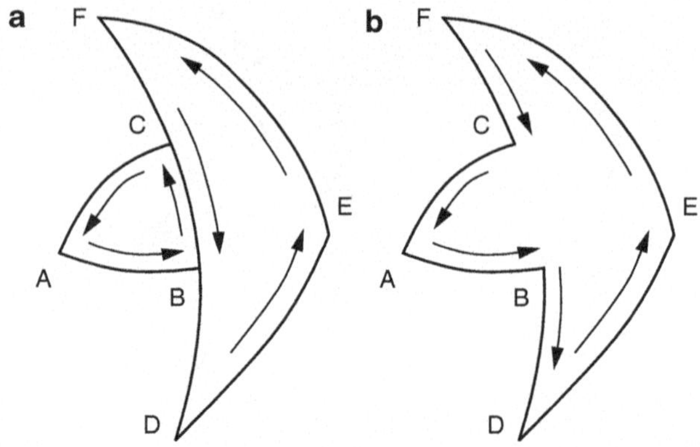

Fig. 5.3 The total rotation corresponding to the parallel transport of a vector around the boundary in figure B is equal to the sum of the rotations corresponding to each of the two geodesic triangles in figure A

to Fig. 5.2b, it is not too difficult to convince oneself that the angle of this rotation is $[(A + B + C) - \pi]$, which is exactly the spherical excess of the spherical triangle.

This last result can be generalized in several ways. First: One can patch spherical triangles together to obtain any spherical polygon with geodesic edges. (We assume that you do your patching in such a way that the resulting spherical polygon is simply connected – it has no holes in it.) (See Fig. 5.3.) Secondly: One can consider closed boundaries composed of connecting segments that are not necessarily geodesics. If a path segment is reasonably well behaved, one can determine the consequence of parallel transport along such a path segment by approximating the path segment by a string of short geodesics and then taking the appropriate limit. However, this approach is difficult and can be avoided. We will return to this later in the chapter. (See Prob. 80.)

It should now be obvious how to generalize the concept of spherical excess to 2-dimensional surfaces other than those of spheres. In general, the angle of rotation that results from the parallel transport of a vector around a closed path will depend on the shape of the surface. For a sphere or ellipsoid, the angle will be counterclockwise (positive). For a flat surface, the angle will be zero. And for a saddle surface, the angle will be clockwise (negative). (See Fig. 5.4b.)

This observation opens up the possibility of defining *curvature* in terms of the angle of rotation resulting from the parallel transport of a vector around a closed path. We would like to assign a number for the curvature at each point on the given surface. For a sphere, this number should be the same at all points but for an ellipsoid, the number should vary from point to point.

To assign a number for the curvature to a point, we would want to consider a sequence of ever shorter loops passing through the given point. The problem with this approach is that for small regions in the neighborhood of a point on a smooth surface, the region will be nearly flat and the corresponding angle of rotation will be close to zero. (Why else did many people in ancient cultures believe that the earth is flat?)

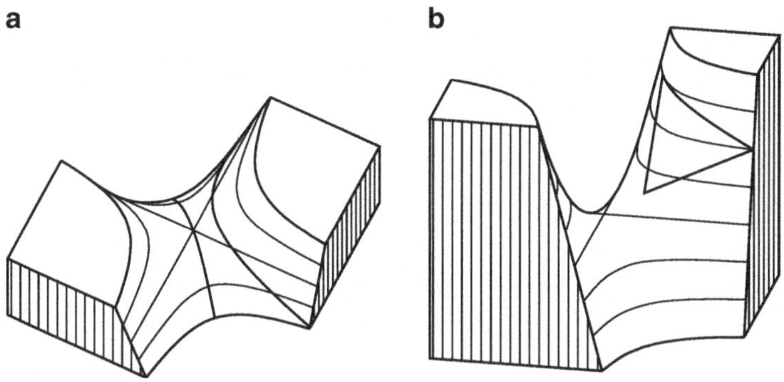

Fig. 5.4 (**a**) On a surface of negative curvature, geodesics that appear parallel at one location diverge elsewhere. (**b**) A triangle formed from geodesics on a surface of negative curvature has the property that the sum of its angles is less than π

Theorem 68 gives us a way of dealing with this problem. For a spherical triangle,

$$\text{Area of } \Delta ABC = r^2\theta, \tag{5.2}$$

where θ is the angle of rotation for a vector that is parallel transported around the perimeter of the spherical triangle. From the discussion relating to Fig. 5.3, we know that this can be generalized. That is

$$\text{Area of loop} = r^2\theta. \tag{5.3}$$

By "loop", I mean a closed path that does not cross itself. We now see that at least for the surface of a sphere, the following limit makes sense:

$$\lim_{d\to 0} \theta/(\text{Area of loop}),$$

where d is the maximum distance between two points on each closed loop in some appropriate sequence of loops. This suggests that we can define the curvature $K(\mathbf{x})$ at a point $\mathbf{x}$ by the limit:

$$K(\mathbf{x}) = \lim_{d\to 0} \theta/(\text{Area of loop}), \tag{5.4}$$

where each loop in the sequence passes through the point $\mathbf{x}$ and has a non-zero area. This limit is known as the *Gaussian curvature*. From (5.3), we see that the Gaussian curvature for any point on a sphere is $1/r^2$.

There are certain deficiencies in this definition of the Gaussian curvature. For one thing, it is not straight forward (at least to me) to show that the limit will exist for all points on any smooth surface. In addition, the definition is useless for computational purposes. This situation will be remedied later in this chapter. In

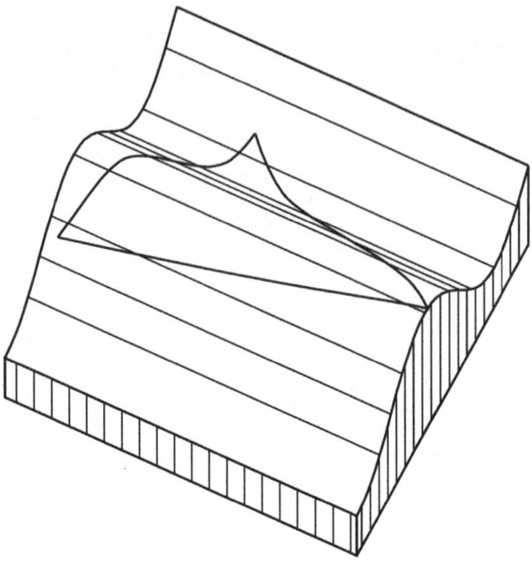

Fig. 5.5 An equilateral geodesic triangle on a sinusoidal surface. Each vertex angle is 60° and the Gaussian curvature of the surface is zero

Sect. 5.5 of this chapter where I present the Riemann curvature tensor, I will present an alternate definition that will be computationally useful. That definition will not be obviously equivalent to the one just given. However later in Chapt. 6, I will prove the Gauss–Bonnet formula that shows that the two definitions are equivalent. Still later in Chapt. 7, I will give an extrinsic definition for 2-dimensional surfaces embedded in a flat 3-dimensional Euclidean space.

For now we can apply our preliminary definition of Gaussian curvature to a few examples. The most trivial example is a flat surface. Consider a flat sheet of paper. On such a sheet of paper, geodesics are straight lines, which can be constructed with a pencil and ruler. In this case a "geodesic triangle" is an ordinary triangle. When we parallel transport a vector about such a triangle, the result is a zero rotation. Thus by our definition, the Gaussian curvature is zero. This is indeed no great surprise.

What is somewhat less obvious is the fact that if the paper is bent or even folded without stretching, the resulting surface still has a zero value for its Gaussian curvature at all points. If for example the paper is bent into a cylinder or a cone, the edges of a penciled triangle are no longer straight lines to the extrinsic observer in 3-space. Nevertheless they are still geodesics in the surface of the paper. An intrinsic observer confined to the 2-dimensional surface would not be able to distinguish the surface of the flat sheet of paper from the surface of the cylinder.

Perhaps we should insert a condition or reservation in this last statement. If the paper were infinite and it was bent into an infinite sine wave, then the statement would indeed be true. (See Fig. 5.5.) On the other hand, if the paper were rolled into a cylinder, there would be some geodesics along which the intrinsic observer could travel in one direction and eventually return to some point encountered previously on the trip. In this situation the observer could safely conclude that he or she was not dealing with a flat surface. However, if we require the observer to take all measurements not only in the surface but also in some sufficiently small region then it would indeed be impossible for the intrinsic observer to determine whether or not the local region was flat or bent in 3-space.

This illustrates the limitation of the intrinsic observer. A point on a sphere and a point on an ellipsoid may have the same Gaussian curvature but that does not mean that the two surfaces have the same shape even in the respective neighborhoods of the two points. Nonetheless, the sign and magnitude of the Gaussian curvature is informative.

On the spherical surface of the earth, two meridians that appear parallel at the equator converge and intersect at the North and South Poles. By contrast, on a surface of negative curvature, geodesics that appear parallel at one location diverge from one another. (See Fig. 5.4a.) As a result, triangles formed from geodesics on a surface of negative curvature have the property that the sum of their angles is less than π. (See Fig. 5.4b.) A vector that is parallel transported around the perimeter of a triangle formed by geodesics on a surface of negative curvature will be rotated in the negative or clockwise direction.

Problem 69. Consider Figs. 5.3a and 5.3b. Suppose ΔABC and ΔDEF are geodesic triangles. Convince yourself that the angle of rotation caused by the parallel transport of a vector about the contour of Fig. 5.3b is the same as the rotation caused by successive parallel transport of a vector around one triangle and then the other in Fig. 5.3a. (In the case of a spherical surface, the total rotation for both Fig. 5.3a and Fig. 5.3b is $(1/r^2) \times$ (area of figure bounded by the curve $ABCDEFCA$.)

5.2 n-Dimensional Curved Surfaces (Spaces)

If $n < m$, we can characterize an n-dimensional surface embedded in a flat m-dimensional Euclidean space E^m or a pseudo-Euclidean space by expressing the m components of the position vector $\mathbf{s}$ as functions of n parameters. That is

$$x^j = x^j(u^1, u^2, \ldots, u^n) \quad \text{for } j = 1, 2, \ldots, m \text{ and}$$

$$\mathbf{s} = x^j \mathbf{e}_j .$$

To avoid topological pathologies, we also require that each parameter is restricted to some interval that may be closed, open, or half open and either bounded or unbounded. To legitimately describe our set of parameterized points in E^m as being

"n-dimensional", we also need to require that the set of n coordinate Dirac vectors $\{\boldsymbol{\gamma}_1, \boldsymbol{\gamma}_2, \ldots, \boldsymbol{\gamma}_n\}$ to be linearly independent where

$$\boldsymbol{\gamma}_k = \frac{\partial \mathbf{s}}{\partial u^k} = \frac{\partial x^j}{\partial u^k}\mathbf{e}_j.$$

(This is of course equivalent to the requirement that $\boldsymbol{\gamma}_{12\ldots n} \neq 0$.)

For the study of extrinsic differential geometry, we usually require that $\mathbf{s}$ (and therefore each of the x^j's) has continuous partial derivatives at least up to third order. For the study of the intrinsic Riemannian geometry in which the function $\mathbf{s}$ is considered to be unknown, we require that the metric tensor $g_{jk} = \langle \boldsymbol{\gamma}_j, \boldsymbol{\gamma}_k \rangle$ has continuous partial derivatives at least up to second order.

Generally one refers to the set of parameters as "coordinates." Frequently, one cannot cover an entire surface with a single coordinate system. For example, consider the 2-dimensional surface of a sphere. For the usual spheroidal coordinate system:

$$\begin{aligned} x^1(\theta, \phi) &= R\cos\phi\sin\theta, \\ x^2(\theta, \phi) &= R\sin\phi\sin\theta, \text{ and} \\ x^3(\theta, \phi) &= R\cos\theta. \end{aligned}$$

This implies that

$$\begin{aligned} \boldsymbol{\gamma}_\theta &= R\cos\phi\cos\theta\mathbf{e}_1 + R\sin\phi\cos\theta\mathbf{e}_2 - R\sin\theta\mathbf{e}_3 \text{ and} \\ \boldsymbol{\gamma}_\phi &= -R\sin\phi\sin\theta\mathbf{e}_1 + R\cos\phi\sin\theta\mathbf{e}_2. \end{aligned}$$

We note that at the North and South Poles where $\sin\theta = 0$, $\boldsymbol{\gamma}_\phi = 0$. Furthermore $\boldsymbol{\gamma}_\theta$ is undefined at those same two points. (What is ϕ when $\theta = 0$ or $\pi/2$?) Thus we see that the usual spherical coordinate system breaks down at those two points. For this kind of situation, it sometimes becomes necessary to "patch" together different overlapping coordinate systems. In the region of overlap, we require that the functions

$$u^k = u^k(\bar{u}^1, \bar{u}^2, \ldots, \bar{u}^n)$$

and their inverses not only exist but have continuous partial derivatives up to at least second order.

Among other things this implies that the Jacobian

$$\det\left[\frac{\partial u^k}{\partial \bar{u}^j}\right] \neq 0.$$

In turn, this means that we can not only write

$$\bar{\boldsymbol{\gamma}}_k = \frac{\partial \mathbf{s}}{\partial \bar{u}^k} = \frac{\partial \mathbf{s}}{\partial u^j}\frac{\partial u^j}{\partial \bar{u}^k} = \boldsymbol{\gamma}_j \frac{\partial u^j}{\partial \bar{u}^k} \text{ but also the inverse equation;} \tag{5.5}$$

$$\boldsymbol{\gamma}_j = \bar{\boldsymbol{\gamma}}_k \frac{\partial \bar{u}^k}{\partial u^j}. \tag{5.6}$$

For example, in the case of the sphere, you could use an alternative coordinate system that works on the northern hemisphere. Namely:

$$\begin{aligned} x^1(\theta,\phi) &= R\cos\phi\sin\theta, \text{ and} \\ x^2(\theta,\phi) &= R\sin\phi\sin\theta. \end{aligned} \tag{5.7}$$

So we have

$$\begin{aligned} \det\begin{bmatrix} \dfrac{\partial x^1}{\partial\theta} & \dfrac{\partial x^1}{\partial\phi} \\ \dfrac{\partial x^2}{\partial\theta} & \dfrac{\partial x^2}{\partial\phi} \end{bmatrix} &= \det\begin{bmatrix} R\cos\phi\cos\theta & -R\sin\phi\sin\theta \\ R\sin\phi\cos\theta & R\cos\phi\sin\theta \end{bmatrix} \\ &= R^2\sin\theta\cos\theta. \end{aligned} \tag{5.8}$$

We observe that the Jacobian $R^2\sin\theta\cos\theta \neq 0$ for $0 < \theta < \frac{\pi}{2}$, which is the region of overlap where both coordinate systems are meaningful.

As we shall see in the next few sections of this book, a lot of information can be obtained about an n-dimensional surface (or "space") without knowing what m-dimensional space it might be embedded in. For some purposes it might be pointless to view a curved surface or space as being embedded in a higher dimensional space. The general theory of relativity is based on the geometry of a curved 4-dimensional surface or space. If we do not have any way of taking measurements in any additional dimension it would seem that we would not be able to observe any physically observable consequence of anything that occurred in those extra dimensions. In the past, it has been suggested that some of the stochastic behavior observed by elementary particles obeying the laws of quantum mechanics could be explained by "hidden variables." However those proposed hidden variables do not explain what is observed.

This does not deter theoretical physicists from proposing higher dimensional theories that might unify the theories of quantum mechanics and general relativity. It is argued that measurements in these higher dimensions are surely difficult but not necessarily impossible.

Another aspect of embedding has intrigued mathematicians for a long time. For many years it was speculated that there might be curved spaces characterized by a metric tensor that could not be embedded in a higher dimensional flat space. As a result, entire chapters of books have been written on how one can define and intelligently discuss an n-dimensional space (or "manifold") without embedding it in a flat space. (Perhaps that is part of the reason that a formalism was developed in which directional derivatives are labeled "tangent vectors." Users of directional derivatives as tangent vectors can avoid thinking of tangent vectors as straight arrows sticking out into a conceivably non-existent higher dimensional flat space.)

With the exception of applications to special and general relativity, this book is devoted to the study of spaces (or surfaces) for which the metric tensor is *positive definite*. For such surfaces, the length of any curve connecting two distinct points is positive definite. Alternatively, we can define a positive definite metric tensor as one for which the g_{jk} matrix has only positive definite eigenvalues. John Nash was able to show that any finite dimensional surface whose geometry is determined by a positive definite metric can be embedded in a higher (but finite) dimensional flat Euclidean space (Nash 1956, pp. 20–63).

This is the same John Nash that was the subject of book and movie both entitled *A Beautiful Mind*. Part of the *A Beautiful Mind* was filmed a few blocks from my home in East Orange, New Jersey. It was thought that parts of present day East Orange can be considered a credible facsimile of what residential Princeton was like in the early 1960s. One of my claims to fame is that Jim Gerulski (the same man that keeps my car working) supplied the filmmakers with a 1952 light gray Chevy Deluxe to give authenticity to a street scene.

For many years it was thought that in his 1956 paper, Nash had also shown that for sufficiently smooth surfaces, this embedding could be carried out without introducing self-intersections. However after emerging from a long mental illness, John Nash received an e-mail message in June 1998 from Professor R.M. Solovay pointing out a flaw in Nash's argument. Nash conceded the flaw (Kuhn and Nasar 2002, p. 209). Rather than repair the flaw, Nash states that his work has now been superseded by stronger results developed by Mikhail Gromov (Gromov 1986). The Nash embedding theorem has been generalized to spaces that do not have positive definite metrics (Clarke 1970; Greene 1970).

The Nash result is not usually as useful as one might think. If only one extra dimension is required to get a flat space then some mathematical machinery can be applied to obtain lots of useful results. As far as I know, there is currently little or no mathematical machinery available to take advantage of the situation when two or more extra dimensions are required to get a flat space. It is worthwhile to note that generally the 4-dimensional spaces of general relativity cannot be embedded in a flat 5-dimensional space. Nonetheless, the Nash result does allow us to think of tangent vectors as sticking out into empty space without being considered naive.

Meanwhile, substantial information can be obtained about a surface intrinsically from its metric tensor without knowing the dimension of whatever flat space in which the curved surface could be embedded. The extraction of geometric information from a surface using only the metric tensor is known as *Riemannian geometry* if the metric is positive definite. For the study of special and general relativity, we need to deal with metrics that are not positive definite. If the metric is not positive definite, the extraction of geometric information from the metric tensor is said to be *non-Riemannian geometry*.

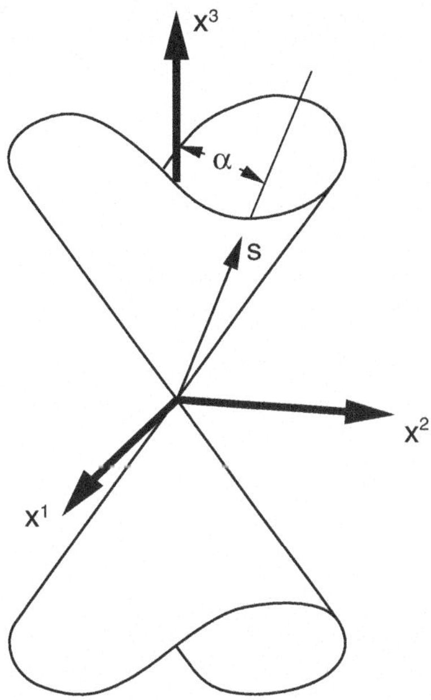

Fig. 5.6 A right circular cone

Riemannian and non-Riemannian geometry will be the focus of the remainder of this chapter and the next chapter.

Problem 70. Note that from (5.8), the Jacobian is zero when $\theta = \pi/2$. Since the spherical coordinate system does not exhibit any pathology for those points, (5.8) suggests that there is some pathology at the equator for the coordinate system defined by (5.7). What is it?

Problem 71. The pathological points already discussed arise because of a failure in the chosen coordinate system. Such points are described in Kreyszig's text, *Differential Geometry* (1991, p.73) as *singular points with respect to the representation.* Other pathological points are intrinsic to the geometric object. These he describes more simply as *singular points*. Consider a right circular cone (See Fig. 5.6.) One coordinate system is:

$$x^1(u,\phi) = u \sin\alpha \cos\phi,\ x^2(u,\phi) = u \sin\alpha \sin\phi, \text{ and}$$

$$x^3(u,\phi) = u \cos\alpha.$$

(Note! This is a slight variation of spherical coordinates where the cone is determined by the value of the acute angle α and u is allowed to assume negative values.)

Compute $\boldsymbol{\gamma}_u$ and $\boldsymbol{\gamma}_\phi$ and show that there is a pathology at $u = 0$. Discuss why this pathology is intrinsic – that is choosing an alternative coordinate system will not eliminate the singularity.

5.3 The Intrinsic Derivative ∇_k

For the study of a curved n-dimensional space (or "surface") embedded in some higher m-dimensional Euclidean space, some things can be measured by observers restricted to the n-dimensional space (or "surface") and some cannot. In this section, we will discuss a differential operator that would make sense to the n-dimensional observer who does not have access to any higher dimension.

For the purpose of illustration, consider the 2-dimensional saddle surface embedded in E^3. That is $u^3 = u^1 u^2$, or

$$\mathbf{s} = u^1\mathbf{e}_1 + u^2\mathbf{e}_2 + u^1u^2\mathbf{e}_3.$$

Thus

$$\boldsymbol{\gamma}_1 = \mathbf{e}_1 + u^2\mathbf{e}_3 \text{ and} \tag{5.9}$$

$$\boldsymbol{\gamma}_2 = \mathbf{e}_2 + u^1\mathbf{e}_3. \tag{5.10}$$

From Prob. 54,

$$\mathbf{N} = \frac{-u^2\mathbf{e}_1 - u^1\mathbf{e}_2 + \mathbf{e}_3}{\sqrt{1 + (u^1)^2 + (u^2)^2}}. \tag{5.11}$$

From Prob. 42,

$$\boldsymbol{\gamma}^1 = \frac{\left[1 + (u^1)^2\right]\mathbf{e}_1 - u^1u^2\mathbf{e}_2 + u^2\mathbf{e}_3}{1 + (u^1)^2 + (u^2)^2} \text{ and} \tag{5.12}$$

$$\boldsymbol{\gamma}^2 = \frac{-u^1u^2\mathbf{e}_1 + \left[1 + (u^2)^2\right]\mathbf{e}_2 + u^1\mathbf{e}_3}{1 + (u^1)^2 + (u^2)^2} \tag{5.13}$$

Now to investigate the nature of this surface, we would like to see how $\boldsymbol{\gamma}_1$ and $\boldsymbol{\gamma}_2$ change in magnitude and direction when one moves about the surface. Clearly $(\partial\boldsymbol{\gamma}_k/\partial u^j)$ must be a linear combination of $\boldsymbol{\gamma}_1$, $\boldsymbol{\gamma}_2$, and $\mathbf{N}$. Thus we can write

$$\frac{\partial\boldsymbol{\gamma}_k}{\partial u^j} = \Gamma^i_{kj}\boldsymbol{\gamma}_i + h_{jk}\mathbf{N}. \tag{5.14}$$

From (5.9) and (5.10),

$$\frac{\partial\boldsymbol{\gamma}_1}{\partial u^1} = \frac{\partial\boldsymbol{\gamma}_2}{\partial u^2} = 0, \text{ and} \tag{5.15}$$

$$\frac{\partial\boldsymbol{\gamma}_2}{\partial u^1} = \frac{\partial\boldsymbol{\gamma}_1}{\partial u^2} = \mathbf{e}_3. \tag{5.16}$$

From (5.14),

$$\frac{\partial\boldsymbol{\gamma}_2}{\partial u^1} = \Gamma^1_{21}\boldsymbol{\gamma}_1 + \Gamma^2_{21}\boldsymbol{\gamma}_2 + h_{12}\mathbf{N}. \tag{5.17}$$

This implies

$$\left\langle \boldsymbol{\gamma}^1, \frac{\partial \boldsymbol{\gamma}_2}{\partial u^1} \right\rangle = \Gamma^1_{21} \langle \boldsymbol{\gamma}^1, \boldsymbol{\gamma}_1 \rangle + \Gamma^2_{21} \langle \boldsymbol{\gamma}^1, \boldsymbol{\gamma}_2 \rangle + h_{12} \langle \boldsymbol{\gamma}^1, \mathbf{N} \rangle,$$

or using (5.12) and (5.16),

$$\Gamma^1_{21} = \left\langle \boldsymbol{\gamma}^1, \frac{\partial \boldsymbol{\gamma}_2}{\partial u^1} \right\rangle = \frac{u^2}{1 + (u^1)^2 + (u^2)^2}. \tag{5.18}$$

Similarly

$$\Gamma^2_{21} = \left\langle \boldsymbol{\gamma}^2, \frac{\partial \boldsymbol{\gamma}_2}{\partial u^1} \right\rangle = \frac{u^1}{1 + (u^1)^2 + (u^2)^2}, \tag{5.19}$$

and

$$h_{12} = \frac{1}{\sqrt{1 + (u^1)^2 + (u^2)^2}}. \tag{5.20}$$

Combining (5.16), (5.17), (5.18), (5.19), and (5.20), we have

$$\begin{aligned} \frac{\partial \boldsymbol{\gamma}_2}{\partial u^1} &= \frac{\partial \boldsymbol{\gamma}_1}{\partial u^2} \\ &= \frac{u^2}{1 + (u^1)^2 + (u^2)^2} \boldsymbol{\gamma}_1 + \frac{u^1}{1 + (u^1)^2 + (u^2)^2} \boldsymbol{\gamma}_2 + \frac{1}{\sqrt{1 + (u^1)^2 + (u^2)^2}} \mathbf{N}. \end{aligned} \tag{5.21}$$

Equations (5.15) and (5.21) indicate the change in $\boldsymbol{\gamma}_1$ and $\boldsymbol{\gamma}_2$ along the coordinate curves in the saddle surface as seen by an observer who has access to all three dimensions. However an intrinsic observer confined to the saddle surface would not be able to measure the change in $\boldsymbol{\gamma}_1$ and $\boldsymbol{\gamma}_2$ in the $\mathbf{N}$ direction. Therefore if we wish to compute the rate of change of $\boldsymbol{\gamma}_1$ and $\boldsymbol{\gamma}_2$ on the saddle surface as seen by our 2-dimensional observer then we are compelled to suppress the $\mathbf{N}$ component. Thus in place of ∂/u^1 and ∂/u^2, it becomes necessary to introduce two new operators ∇_1 and ∇_2. Suppressing the $\mathbf{N}$ component, (5.15) and (5.21) become

$$\nabla_1 \boldsymbol{\gamma}_1 = \nabla_2 \boldsymbol{\gamma}_2 = 0, \text{ and} \tag{5.22}$$

$$\nabla_1 \boldsymbol{\gamma}_2 = \nabla_2 \boldsymbol{\gamma}_1 = \frac{u^2}{1 + (u^1)^2 + (u^2)^2} \boldsymbol{\gamma}_1 + \frac{u^1}{1 + (u^1)^2 + (u^2)^2} \boldsymbol{\gamma}_2 \tag{5.23}$$

This situation is easy to generalize to other surfaces where possibly $(m - n) > 1$ but I will use a more intrinsic approach to define the *intrinsic derivative* ∇_k. (Also known as the *geodesic derivative*.)

In passing, it is important to note that it was no accident that $\nabla_1 \boldsymbol{\gamma}_2 = \nabla_2 \boldsymbol{\gamma}_1$. For any embedded surface,

$$\frac{\partial}{\partial u^j}\boldsymbol{\gamma}_k = \frac{\partial^2}{\partial u^j u^k}\mathbf{s} = \frac{\partial^2}{\partial u^k u^j}\mathbf{s} = \frac{\partial}{\partial u^k}\boldsymbol{\gamma}_j$$

and thus

$$\nabla_j \boldsymbol{\gamma}_k = \nabla_k \boldsymbol{\gamma}_j. \tag{5.24}$$

Using (5.24) as a motivating factor, we can now construct a definition.

Definition 72. An *intrinsic derivative* (or *geodesic derivative*) ∇_α is defined by the following five properties:

1) $\nabla_\alpha \boldsymbol{\gamma}_\beta$ is a linear combination of $\boldsymbol{\gamma}_\nu$'s. (We can write $\nabla_\alpha \boldsymbol{\gamma}_\beta = \Gamma^\nu_{\beta\alpha}\boldsymbol{\gamma}_\nu$ where the $\Gamma^\nu_{\beta\alpha}$'s are known as *Christoffel symbols.*)
2) $\nabla_\alpha \boldsymbol{\gamma}_\beta = \nabla_\beta \boldsymbol{\gamma}_\alpha$ (∇_α is said to be "torsion free" and $\Gamma^\nu_{\beta\alpha} = \Gamma^\nu_{\alpha\beta}$.)
3) ∇_α acting on a scalar or the tensor component of any p-vector coincides with ∂/u^α.
4) If **A** and **B** are any Clifford numbers (not necessarily index free) and both $\nabla_\alpha\mathbf{A}$ and $\nabla_\alpha\mathbf{B}$ are defined, then

$$\nabla_\alpha(\mathbf{A}+\mathbf{B}) = \nabla_\alpha\mathbf{A} + \nabla_\alpha\mathbf{B}.$$

5) ∇_α satisfies the Leibniz property. That is

$$\nabla_\alpha(\mathbf{AB}) = (\nabla_\alpha\mathbf{A})\mathbf{B} + \mathbf{A}(\nabla_\alpha\mathbf{B}),$$

where both $\nabla_\alpha\mathbf{A}$ and $\nabla_\alpha\mathbf{B}$ are defined and **A** and **B** are otherwise arbitrary Clifford numbers that are not necessarily index free.

To see that these five conditions do indeed give a precise definition of ∇_α, we need to only see how these five conditions determine a formula for the Christoffel symbols in terms of the metric tensor. We first note that it is not difficult to demonstrate that $\nabla_\alpha\mathbf{I} = \mathbf{0}$. Essentially, if follows from the Leibniz property:

$$\nabla_\alpha\mathbf{I} = \nabla_\alpha(\mathbf{I}\cdot\mathbf{I}) = (\nabla_\alpha\mathbf{I})\mathbf{I} + \mathbf{I}\nabla_\alpha\mathbf{I} = 2\nabla_\alpha\mathbf{I}\,. \text{ Thus}$$

$$\nabla_\alpha\mathbf{I} = \mathbf{0}. \tag{5.25}$$

To get a formula for the Christoffel symbols in terms of the metric tensor, we first note that

$$2g_{\alpha\beta}\mathbf{I} = \boldsymbol{\gamma}_\alpha\boldsymbol{\gamma}_\beta + \boldsymbol{\gamma}_\beta\boldsymbol{\gamma}_\alpha, \quad \text{so}$$

$$\nabla_\nu(2g_{\alpha\beta}\mathbf{I}) = \nabla_\nu(\boldsymbol{\gamma}_\alpha\boldsymbol{\gamma}_\beta + \boldsymbol{\gamma}_\beta\boldsymbol{\gamma}_\alpha). \tag{5.26}$$

Using condition 3 and (5.25) on the left hand side of (5.26) and conditions 4 and 5 on the right hand side, we get

$$2\frac{\partial g_{\alpha\beta}}{\partial u^\nu}\mathbf{I} = (\nabla_\nu \boldsymbol{\gamma}_\alpha)\boldsymbol{\gamma}_\beta + \boldsymbol{\gamma}_\alpha(\nabla_\nu \boldsymbol{\gamma}_\beta) + (\nabla_\nu \boldsymbol{\gamma}_\beta)\boldsymbol{\gamma}_\alpha + \gamma_\beta(\nabla_\nu \boldsymbol{\gamma}_\alpha). \tag{5.27}$$

Now applying condition 1 to the right hand side of (5.27) and regrouping terms, we have

$$2\frac{\partial g_{\alpha\beta}}{\partial u^\nu}\mathbf{I} = \Gamma^\eta_{\alpha\nu}(\boldsymbol{\gamma}_\eta \boldsymbol{\gamma}_\beta + \gamma_\beta \boldsymbol{\gamma}_\eta) + \Gamma^\eta_{\beta\nu}(\boldsymbol{\gamma}_\alpha \boldsymbol{\gamma}_\eta + \gamma_\eta \boldsymbol{\gamma}_\alpha)$$

or

$$\frac{\partial g_{\alpha\beta}}{\partial u^\nu} = \Gamma^\eta_{\alpha\nu} g_{\eta\beta} + \Gamma^\eta_{\beta\nu} g_{\alpha\eta}. \tag{5.28}$$

By cyclic permutation of the unsummed indices, we obtain two other equations:

$$\frac{\partial g_{\beta\nu}}{\partial u^\alpha} = \Gamma^\eta_{\beta\alpha} g_{\eta\nu} + \Gamma^\eta_{\nu\alpha} g_{\beta\eta} \text{ and} \tag{5.29}$$

$$-\frac{\partial g_{\nu\alpha}}{\partial u^\beta} = -\Gamma^\eta_{\nu\beta} g_{\eta\alpha} - \Gamma^\eta_{\alpha\beta} g_{\nu\eta}. \tag{5.30}$$

From the torsion free condition 2, $\boldsymbol{\Gamma}^\eta_{\lambda\mu} = \Gamma^\eta_{\mu\lambda}$. Thus if we add the last three equations, we get

$$2\Gamma^\eta_{\alpha\nu} g_{\eta\beta} = \frac{\partial g_{\alpha\beta}}{\partial u^\nu} + \frac{\partial g_{\beta\nu}}{\partial u^\alpha} - \frac{\partial g_{\nu\alpha}}{\partial u^\beta}.$$

Multiplying both sides of this last equation by $(g^{\lambda\beta}/2)$ and noting that $g_{\eta\beta}g^{\lambda\beta} = \delta^\lambda_\eta$, one immediately obtains the equation:

$$\Gamma^\lambda_{\alpha\nu} = \frac{g^{\lambda\beta}}{2}\left(\frac{\partial g_{\alpha\beta}}{\partial u^\nu} + \frac{\partial g_{\beta\nu}}{\partial u^\alpha} - \frac{\partial g_{\nu\alpha}}{\partial u^\beta}\right). \tag{5.31}$$

The operator ∇_α may be applied to any Clifford number with differentiable components including upper index Clifford numbers. It is not too difficult to show that

$$\nabla_\alpha \boldsymbol{\gamma}^\beta = -\Gamma^\beta_{\eta\alpha}\boldsymbol{\gamma}^\eta. \tag{5.32}$$

(See Prob. 73.)

It is important to note that Christoffel symbols do not transform as tensors. To see that, we first observe that from (5.26)

$$\nabla_\nu(2g_{\alpha\beta}\mathbf{I}) = \nabla_\nu(\boldsymbol{\gamma}_\alpha \boldsymbol{\gamma}_\beta + \gamma_\beta \boldsymbol{\gamma}_\alpha).$$

It was shown that this equation essentially defines ∇_ν. Therefore under a change of coordinates, ∇_ν behaves in the same manner as $\partial/\partial u^\nu$. That is

$$\bar{\nabla}_\alpha = \frac{\partial u^\eta}{\partial \bar{u}^\alpha}\nabla_\eta. \tag{5.33}$$

Thus

$$\bar{\nabla}_\alpha \bar{\boldsymbol{\gamma}}^\beta = \left(\frac{\partial u^\eta}{\partial \bar{u}^\alpha} \nabla_\eta\right) \left(\boldsymbol{\gamma}^\nu \frac{\partial \bar{u}^\beta}{\partial u^\nu}\right).$$

Using (5.32), this becomes

$$-\bar{\Gamma}^\beta_{\lambda\alpha} \bar{\boldsymbol{\gamma}}^\lambda = -\frac{\partial u^\eta}{\partial \bar{u}^\alpha} \frac{\partial \bar{u}^\beta}{\partial u^\nu} \Gamma^\nu_{\mu\nu} \boldsymbol{\gamma}^\mu + \frac{\partial u^\eta}{\partial \bar{u}^\alpha} \frac{\partial^2 \bar{u}^\beta}{\partial u^\eta \partial u^\nu} \boldsymbol{\gamma}^\nu.$$

It then follows that

$$\bar{\Gamma}^\beta_{\lambda\alpha} \bar{\boldsymbol{\gamma}}^\lambda = \frac{\partial u^\eta}{\partial \bar{u}^\alpha} \frac{\partial \bar{u}^\beta}{\partial u^\nu} \Gamma^\nu_{\mu\nu} \frac{\partial u^\mu}{\partial \bar{u}^\lambda} \bar{\boldsymbol{\gamma}}^\lambda - \frac{\partial u^\eta}{\partial \bar{u}^\alpha} \frac{\partial^2 \bar{u}^\beta}{\partial u^\eta \partial u^\nu} \frac{\partial u^\nu}{\partial \bar{u}^\lambda} \bar{\boldsymbol{\gamma}}^\lambda.$$

And finally

$$\bar{\Gamma}^\beta_{\lambda\alpha} = \frac{\partial u^\eta}{\partial \bar{u}^\alpha} \frac{\partial u^\mu}{\partial \bar{u}^\lambda} \frac{\partial \bar{u}^\beta}{\partial u^\nu} \Gamma^\nu_{\mu\nu} - \frac{\partial u^\eta}{\partial \bar{u}^\alpha} \frac{\partial u^\nu}{\partial \bar{u}^\lambda} \frac{\partial^2 \bar{u}^\beta}{\partial u^\eta \partial u^\nu}. \tag{5.34}$$

It is because of the second term on the right hand side of (5.34) that it cannot be said that Christoffel symbols are tensors. Equation (5.34) can be proven directly from (5.31) but with much more difficulty.

Although Christoffel symbols do not transform as tensors, we see from (5.33) that $\nabla_\alpha \mathbf{A}$ does transform as a tensor if $\mathbf{A}$ is index free. This enables us to introduce an alternate differential operator, which unlike the intrinsic operator ∇_k, always maps tensors into tensors. Consider the example $\mathbf{A} = A^\eta_{\ \beta\nu} \boldsymbol{\gamma}^\beta \boldsymbol{\gamma}^\nu \boldsymbol{\gamma}_\eta$. Then

$$\nabla_\alpha \mathbf{A} = \left(\frac{\partial A^\eta_{\ \beta\nu}}{\partial u^\alpha}\right) \boldsymbol{\gamma}^\beta \boldsymbol{\gamma}^\nu \boldsymbol{\gamma}_\eta - A^\eta_{\ \beta\nu} \left(\Gamma^\beta_{\lambda\alpha} \boldsymbol{\gamma}^\lambda\right) \boldsymbol{\gamma}^\nu \boldsymbol{\gamma}_\eta - A^\eta_{\ \beta\nu} \boldsymbol{\gamma}^\beta \left(\Gamma^\nu_{\lambda\alpha} \boldsymbol{\gamma}^\lambda\right) \boldsymbol{\gamma}_\eta$$

$$+ A^\eta_{\ \beta\nu} \boldsymbol{\gamma}^\beta \boldsymbol{\gamma}^\nu \left(\Gamma^\lambda_{\eta\alpha} \boldsymbol{\gamma}_\lambda\right).$$

Relabeling some of the dummy indices, this equation becomes

$$\nabla_\alpha \mathbf{A} = \left(\frac{\partial A^\eta_{\ \beta\nu}}{\partial u^\alpha} - A^\eta_{\ \lambda\nu} \Gamma^\lambda_{\beta\alpha} - A^\eta_{\ \beta\lambda} \Gamma^\lambda_{\nu\alpha} + A^\lambda_{\ \beta\nu} \Gamma^\eta_{\lambda\alpha}\right) \boldsymbol{\gamma}^\beta \boldsymbol{\gamma}^\nu \boldsymbol{\gamma}_\eta. \tag{5.35}$$

This motivates the introduction of the *semicolon notation* for the covariant derivative. In our example

$$A^\eta_{\ \beta\nu;\alpha} = \frac{\partial A^\eta_{\ \beta\nu}}{\partial u^\alpha} - A^\eta_{\ \lambda\nu} \Gamma^\lambda_{\beta\alpha} - A^\eta_{\ \beta\lambda} \Gamma^\lambda_{\nu\alpha} + A^\lambda_{\ \beta\nu} \boldsymbol{\Gamma}^\eta_{\lambda\alpha}. \tag{5.36}$$

With this notation, (5.35) becomes

$$\nabla_\alpha \mathbf{A} = A^\eta_{\ \beta\nu;\alpha} \boldsymbol{\gamma}^\beta \boldsymbol{\gamma}^\nu \boldsymbol{\gamma}_\eta. \tag{5.37}$$

Both the entities $\nabla_\alpha \mathbf{A}$ and $\boldsymbol{\gamma}^\beta \boldsymbol{\gamma}^\nu \boldsymbol{\gamma}_\eta$ transform as tensors and therefore $A^\eta_{\ \beta\nu;\alpha}$ also transforms as a tensor. (See Prob. 76.) That is

$$\bar{A}^\delta_{\ \mu\lambda;\rho} = \frac{\partial \bar{u}^\delta}{\partial u^\eta} \frac{\partial u^\beta}{\partial \bar{u}^\mu} \frac{\partial u^\nu}{\partial \bar{u}^\lambda} \frac{\partial u^\alpha}{\partial \bar{u}^\rho} A^\eta_{\ \beta\nu;\alpha}. \tag{5.38}$$

$A^\eta_{\ \beta\nu;\alpha}$ is the *covariant derivative* of the tensor $A^\eta_{\ \beta\nu}$ with respect to u^α. In general when one computes the covariant derivative of a tensor, one gets a Christoffel symbol with a negative sign for each lower index and a Christoffel symbol with a positive sign for each upper index.

The reader should be forewarned that many authors now use ∇_α to designate the covariant derivative. Furthermore to add to possible confusion I referred to the intrinsic derivative ∇_α as "a covariant derivative" in my previous book, *Clifford Algebra – A Computational Tool for Physicists*. That was a serious mistake.

I believe that it is easier to incorporate the formalism of differential forms into the formalism of Clifford algebra using the intrinsic derivative rather than the covariant derivative. Thus I only introduce the covariant derivative here so you will not be confused, troubled, or disturbed when you read the works of other authors.

Before going on to the next section, I will pause long enough to determine the covariant derivative of a lower index coordinate Dirac matrix $\boldsymbol{\gamma}_{\beta;\alpha}$. First, we require that for any index free Clifford number $\mathbf{A}$,

$$\nabla_\alpha \mathbf{A} = \mathbf{A}_{;\alpha}. \tag{5.39}$$

Secondly, we require that the covariant derivative satisfies the Leibniz property. As a consequence, $\mathbf{A} = A^\beta \boldsymbol{\gamma}_\beta$ implies

$$\mathbf{A}_{;\alpha} = A^\beta_{\ ;\alpha} \boldsymbol{\gamma}_\beta + A^\beta \boldsymbol{\gamma}_{\beta;\alpha} = \left(\frac{\partial A^\beta}{\partial u^\alpha} + A^\eta \Gamma^\beta_{\ \eta\alpha} \right) \boldsymbol{\gamma}_\beta + A^\beta \boldsymbol{\gamma}_{\beta;\alpha}, \text{ but}$$

$$\nabla_\alpha \mathbf{A} = \frac{\partial A^\beta}{\partial u^\alpha} \boldsymbol{\gamma}_\beta + A^\beta \Gamma^\eta_{\ \beta\alpha} \boldsymbol{\gamma}_\eta = \left(\frac{\partial A^\beta}{\partial u^\alpha} + A^\eta \Gamma^\beta_{\ \eta\alpha} \right) \boldsymbol{\gamma}_\beta.$$

From (5.39), it immediately follows that

$$\boldsymbol{\gamma}_{\beta;\alpha} = 0. \tag{5.40}$$

Problem 73. Since $\boldsymbol{\gamma}^\beta$ is a linear combination of $\boldsymbol{\gamma}_\mu$'s, it is clear that $\nabla_\alpha \boldsymbol{\gamma}^\beta$ is also a linear combination of $\boldsymbol{\gamma}_\mu$'s and therefore also a linear combination of $\boldsymbol{\gamma}^\eta$'s. Thus $\nabla_\alpha \boldsymbol{\gamma}^\beta$ can be written in the form $G^\beta_{\ \eta\alpha} \boldsymbol{\gamma}^\eta$. Applying the operator ∇_α to the equation $\boldsymbol{\gamma}^\beta \boldsymbol{\gamma}_\eta + \boldsymbol{\gamma}_\eta \boldsymbol{\gamma}^\beta = 2\delta^\beta_\eta \mathbf{I}$, show $G^\beta_{\ \eta\alpha} = -\Gamma^\beta_{\ \eta\alpha}$.

Problem 74. For the surface of a sphere, use either (5.31) or the results of Prob. 44 to compute the Christoffel symbols $\Gamma^\theta_{\ \theta\theta}, \Gamma^\theta_{\ \theta\phi}, \ldots$. Which way is easier?

Problem 75. Show $\boldsymbol{\gamma}^\beta_{\ ;\alpha} = 0$.

Problem 76. Suppose

$$\nabla_\alpha \mathbf{A} = A^{\eta}{}_{\beta\nu;\alpha}\boldsymbol{\gamma}^\beta\boldsymbol{\gamma}^\nu\boldsymbol{\gamma}_\eta$$

and both $\nabla_\alpha \mathbf{A}$ and $\boldsymbol{\gamma}^\beta\boldsymbol{\gamma}^\nu\boldsymbol{\gamma}_\eta$ are tensors. Show that $A^{\eta}{}_{\beta\nu;\alpha}$ is a tensor.

Problem 77. Use (5.40) to show that $g_{\alpha\beta;\eta} = 0$.

Problem 78. Difficult!

a) Suppose $g = \det\left[g_{\alpha\beta}\right]$, show $\partial g/\partial u^\nu = g g^{\alpha\beta}\left(\partial g_{\alpha\beta}/\partial u^\nu\right)$.
b) Use (5.31) to show $\Gamma^\eta_{\nu\eta} = (g^{\alpha\beta}/2)(\partial g_{\alpha\beta}/\partial u^\nu)$.
c) Combine results from a and b to show $\Gamma^\eta_{\nu\eta} = (1/\sqrt{g})(\partial\sqrt{g}/\partial u^\nu)$. (This last formula assumes g is positive. How can you adjust it if g is negative?)

5.4 Parallel Transport and Geodesics

Having defined the intrinsic operator ∇_α, it is now possible to define the notion of *parallel transport* of a Clifford number along a curve in an n-dimensional surface. If we embedded our surface in a higher dimensional Euclidean space of m dimensions, we could write

$$\mathbf{x}(t) = \sum_{i=1}^{m} x^i(u^1(t), u^2(t), \ldots, u^n(t))\mathbf{e}_i. \tag{5.41}$$

From this equation, we could compute a velocity vector $\mathbf{v}(t) = d\mathbf{x}(t)/dt$. That is

$$\mathbf{v}(t) = \frac{\partial x^i}{\partial u^\nu}\frac{du^\nu}{dt}\mathbf{e}_i = \frac{du^\nu}{dt}\boldsymbol{\gamma}_\nu. \tag{5.42}$$

For an arbitrary metric, $\langle \mathbf{v}(t), \mathbf{v}(t)\rangle$ may be positive, zero, or negative. If $\langle \mathbf{v}(t), \mathbf{v}(t)\rangle$ is non-zero, it is useful (at least for theoretical discussions) to change the parameter t so that $\langle \mathbf{v}, \mathbf{v}\rangle = \pm 1$. We can do this by introducing a strictly increasing function $t(s)$. This enables us to reparameterize our curve so that we have

$$\bar{\mathbf{x}}(s) = \mathbf{x}(t(s)).$$

This implies that

$$\frac{d\bar{\mathbf{x}}(s)}{ds} = \frac{d\mathbf{x}(t)}{dt}\frac{dt}{ds} = \mathbf{v}(t)\frac{dt}{ds}. \tag{5.43}$$

Thus

$$\left\langle \frac{d\bar{\mathbf{x}}}{ds}, \frac{d\bar{\mathbf{x}}}{ds}\right\rangle = \langle \mathbf{v}(t), \mathbf{v}(t)\rangle \left(\frac{dt}{ds}\right)^2.$$

Therefore $\left\langle \frac{d\bar{\mathbf{x}}}{ds}, \frac{d\bar{\mathbf{x}}}{ds}\right\rangle$ has the same sign as $\langle \mathbf{v}(t), \mathbf{v}(t)\rangle$. Furthermore

$$\left\langle \frac{d\bar{\mathbf{x}}}{ds}, \frac{d\bar{\mathbf{x}}}{ds}\right\rangle = \pm 1 \text{ if } \frac{ds}{dt} = |\mathbf{v}(t)|.$$

This implies that

$$s(t) = \int_a^t |\mathbf{v}(u)|\,du.$$

It should now be clear that when $\langle \mathbf{v}(t), \mathbf{v}(t)\rangle > 0$, we can interpret s as the *arc length* of the curve from some chosen point. In either the positive or the negative case, $s(t)$ is differentiable and strictly increasing. This implies that the inverse function $t(s)$ exists and is also differentiable and strictly increasing. For the remainder of the book, I will try to remember to use the parameter s when I am discussing a curve $\bar{\mathbf{x}}$ that has been parameterized so that

$$\left\langle \frac{d\bar{\mathbf{x}}}{ds}, \frac{d\bar{\mathbf{x}}}{ds} \right\rangle = \pm 1.$$

That will make it unnecessary to use a "bar" over the $\mathbf{x}$. In this context, it is useful to use a lower case bold face $\mathbf{t}$ to designate a normalized velocity or "unit tangent" vector. Dropping the bar, we now have

$$\mathbf{t}(s) = \frac{d\mathbf{x}(s)}{ds}. \tag{5.44}$$

Since

$$\mathbf{x}(s) = \sum_{i=1}^{m} x^i(u^1(s), u^2(s), \ \ldots, u^n(s))\mathbf{e}_i,$$

it follows that

$$\mathbf{t}(s) = \frac{d\mathbf{x}(s)}{ds} = \frac{\partial x^i}{\partial u^\nu}\frac{du^\nu}{ds}\mathbf{e}_i.$$

Or

$$\mathbf{t}(s) = \frac{du^\nu}{ds}\boldsymbol{\gamma}_\nu. \tag{5.45}$$

This last equation makes sense without reference to a higher dimensional flat space. Since the curve lies in the possibly curved n-dimensional surface, the unit tangent vector $\mathbf{t}(s)$ is tangent to that surface.

$\mathbf{t}(s)$ is not the only example of a function that assigns a Clifford number to each point of a given curve. Suppose $\mathbf{A}(s)$ represents a differentiable function that assigns an index free Clifford number to each point on a given curve. An observer in the large m-dimensional space who wished to compute the derivative of the Clifford number $\mathbf{A}(s)$ with respect to s would simply use the formula

$$\frac{d\mathbf{A}(s)}{ds} = \frac{\partial \mathbf{A}}{\partial u^\alpha}\frac{du^\alpha}{ds}.$$

However an observer constrained to take all measurements on the n-dimensional surface would detect only components of $d\mathbf{A}(s)/ds$ projected onto the space of

Clifford numbers identified with the tangent plane. Thus such an observer would compute

$$\nabla_s \mathbf{A}(s) = \nabla_\alpha \mathbf{A}(s)(du^\alpha/ds). \tag{5.46}$$

An index free Clifford number $\mathbf{A}(s)$ that is *parallel transported* along a curve (not necessarily a geodesic) is one for which

$$\nabla_s \mathbf{A}(s) = 0. \tag{5.47}$$

I can now give a more sophisticated definition of a geodesic than the informal one that I gave in Sect. 5.1. A *geodesic* is a curve such that the unit tangent vector $\mathbf{t}(s)$ is parallel transported along the curve. That is

$$\nabla_s \mathbf{t}(s) = 0. \tag{5.48}$$

In other words for the observer constrained to the n-dimensional surface, $\mathbf{t}(s)$ appears to be constant.

From (5.45), (5.46), and (5.48);

$$\begin{aligned}\nabla_s \mathbf{t}(s) = \nabla_s \left(\frac{du^\alpha}{ds}\boldsymbol{\gamma}_\alpha\right) &= \frac{d^2u^\alpha}{ds^2}\boldsymbol{\gamma}_\alpha + \frac{du^\alpha}{ds}\frac{du^\beta}{ds}\nabla_\beta \boldsymbol{\gamma}_\alpha \\ &= \frac{d^2u^\eta}{ds^2}\boldsymbol{\gamma}_\eta + \frac{du^\alpha}{ds}\frac{du^\beta}{ds}\Gamma^\eta_{\ \alpha\beta}\boldsymbol{\gamma}_\eta.\end{aligned}$$

Therefore the equations for the coordinates of a geodesic may be written as

$$\frac{d^2u^\eta}{ds^2} + \Gamma^\eta_{\ \alpha\beta}\frac{du^\alpha}{ds}\frac{du^\beta}{ds} = 0. \tag{5.49}$$

This last equation also serves as a definition of a geodesic when the tangent or velocity vector has zero length.

Of course if the u^α's are Euclidean coordinates in an Euclidean plane, then the Christoffel symbols are zero and the equations are easily solved. You should not be too surprised to observe that in this situation the general solution is an arbitrary straight line in the n-dimensional Euclidean plane. That is

$$u^\alpha = c^\alpha s + u_0^\alpha \text{ for } \alpha = 1, 2, \ldots, n,$$

where the c^α's and the u_0^α's are arbitrary constants. (If you require that $\langle \mathbf{t}(s), \mathbf{t}(s)\rangle = 1$, then you have the constraint that $\sum_{\alpha=1}^n (c^\alpha)^2 = 1$.)

Problem 79. From the theory of the calculus of variations, it is known that an alternate form of the equations that determine a geodesic is

$$\frac{d}{ds}\left(\frac{\partial F}{\partial \dot{u}^\nu}\right) - \frac{\partial F}{\partial u^\nu} = 0, \text{ where} \tag{5.50}$$

$$F(u^1, u^2, \cdots, u^n, \dot{u}^1, \dot{u}^2, \cdots, \dot{u}^n) = g_{\alpha\beta}(u^1, u^2, \cdots, u^n)\dot{u}^\alpha \dot{u}^\beta,$$

$$\dot{u}^\nu = \frac{du^\nu(s)}{ds}, \text{ and}$$

$$\nu = 1, 2, \ldots, n.$$

For an n-dimensional space there are $(n^3+n^2)/2$ distinct Christoffel symbols. Using (5.31) to compute them can be a lengthy process. On the other hand one can compute (5.50) for each of the n values of ν and then by comparing the results with the form of (5.49), one can determine the formula for each Christoffel symbol. From these remarks, it is also clear that (5.50) is also a quick way of determining the equations that determine a geodesic.

a) Use (5.50) and (5.49) to determine the Christoffel symbols for the surface of a sphere where

$$F(\theta, \phi, \dot{\theta}, \dot{\phi}) = R^2(\dot{\theta})^2 + R^2 \sin^2\theta(\dot{\phi})^2.$$

b) Use (5.31) to demonstrate that (5.50) is equivalent (although not identical) to (5.49).

Problem 80. From Prob. 74 or Prob. 79, you have the Christoffel symbols for the surface of a sphere. Use these formulas along with (5.46) and (5.47) to treat the problem of parallel transporting a vector around a four-sided figure on a sphere formed by two parallels and two meridians. That is, first parallel transport the vector $A^\theta \boldsymbol{\gamma}_\theta + A^\phi \boldsymbol{\gamma}_\phi$ "south" along the path $(\theta, \phi) = (s/R + \theta_0, \phi_0)$ from $s = 0$ to $s = R(\theta_1 - \theta_0)$. Then parallel transport the vector "east" along the path $(\theta, \phi) = (\theta_1, s/(R\sin\theta_1) + \phi_0)$ from $s = 0$ to $s = R\sin\theta_1(\phi_1 - \phi_0)$. Then parallel transport the vector "north" along the path $(\theta, \phi) = (-s/R + \theta_1, \phi_1)$ from $s = 0$ to $s = R(\theta_1 - \theta_0)$. Finally, parallel transport the vector back to its original position by moving it "west" along the path $(\theta, \phi) = (\theta_0, -s/(R\sin\theta_0) + \phi_1)$ from $s = 0$ to $s = R\sin\theta_0(\phi_1 - \phi_0)$.

Several check points for the computation are as follows: For all path segments, you should get

$$\frac{\partial A^\theta}{\partial \theta}\frac{d\theta}{ds} + \frac{\partial A^\theta}{\partial \phi}\frac{d\phi}{ds} - A^\phi \sin\theta\cos\theta\frac{d\phi}{ds} = 0 \text{ and}$$

$$\frac{\partial A^\phi}{\partial \theta}\frac{d\theta}{ds} + \frac{\partial A^\phi}{\partial \phi}\frac{d\phi}{ds} + A^\phi\frac{\cos\theta}{\sin\theta}\frac{d\theta}{ds} + A^\theta\frac{\cos\theta}{\sin\theta}\frac{d\phi}{ds} = 0.$$

Integrating these equations along the first leg of the rectangular loop, you should get $A^\theta = A_0^\theta$ and $A^\phi \sin\theta = A_0^\phi \sin\theta_0$ where (A_0^θ, A_0^ϕ) are the initial components of the vector in the northwest corner of the loop.

Integrating along the second segment, you should get

$$A^\theta = A_1^\theta \cos\left[(\phi - \phi_0)\cos\theta_1\right] + \left(A_1^\phi \sin\theta_1\right)\sin\left[(\phi - \phi_0)\cos\theta_1\right] \text{ and}$$

$$A^\phi \sin\theta_1 = -A_1^\theta \sin\left[(\phi - \phi_0)\cos\theta_1\right] + \left(A_1^\phi \sin\theta_1\right)\cos\left[(\phi - \phi_0)\cos\theta_1\right].$$

where (A_1^θ, A_1^ϕ) are the components of the vector in the southwest corner of the loop. Note! The parallel transport of the vector along the curve of constant latitude in the east direction results in a clockwise rotation with respect to the curve of constant latitude. This is the same direction of rotation that would result if you approximated the parallel by a sequence of geodesics.

When you have completed the parallel transport about the rectangular loop, you should be able to show that the resulting angle of rotation is $(\phi_1 - \phi_0)(\cos\theta_0 - \cos\theta_1)$. Using calculus, you should be able to compute the area of the rectangular loop and then see that the result of this problem is consistent with the discussion immediately before and after (5.3).

5.5 The Riemann Tensor and the Curvature 2-form

The operator ∇_α behaves very much like the operator $\partial/\partial u^\alpha$. However there is one major exception. Unless the n-dimensional surface is intrinsically indistinguishable from a flat subspace, $\nabla_\alpha\nabla_\beta \neq \nabla_\beta\nabla_\alpha$. In particular

$$\left(\nabla_\alpha\nabla_\beta - \nabla_\beta\nabla_\alpha\right)\boldsymbol{\gamma}_\eta = R^\lambda{}_{\eta\alpha\beta}\boldsymbol{\gamma}_\lambda. \tag{5.51}$$

where $R^\lambda{}_{\eta\alpha\beta}$ is known as the *Riemann curvature tensor*. (The definition of (5.51) is pretty standard but it is different than the one I used in my text *Clifford Algebra.*) As you might suspect from the name, this is a tensor. (See Prob. 83.) As you might also guess from the name, the Riemann curvature tensor is also a measure of the curvature of the given space. In particular, I can now give a formal definition of the Gaussian curvature:

Definition 81. For 2-dimensional spaces, the Gaussian curvature

$$K = \frac{1}{2}R^{\alpha\beta}{}_{\alpha\beta} = R^{12}{}_{12}.$$

In the derivation of (6.52), I will show that this formal definition of the Gaussian curvature is equivalent to the informal definition that I gave in (5.4).

It is useful to state and then prove some symmetries for the indices of R_{kmij} where

$$R_{kmij} = g_{k\lambda}R^\lambda{}_{mij}.$$

First R_{ijkm} is antisymmetric with respect to its second two indices. That is

$$R_{kmij} = -R_{kmji}. \tag{5.52}$$

Secondly, the tensor is antisymmetric with respect to the first two indices:

$$R_{kmij} = -R_{mkij}. \tag{5.53}$$

Furthermore, there is a cyclic symmetry in the last three indices. In particular

$$R_{mkij} + R_{mijk} + R_{mjki} = 0. \tag{5.54}$$

(This is generalized in Prob. 87.)

Finally the curvature tensor is symmetric with respect to an exchange of the first pair of indices with the second pair:

$$R_{kmij} = R_{ijkm}. \tag{5.55}$$

Equation (5.52) is an immediate consequence of (5.51), which was used to define the Riemann tensor.

To verify (5.53) is more difficult. A first step is to show that

$$(\nabla_j \nabla_k - \nabla_k \nabla_j)(\mathbf{AB}) = \left[(\nabla_j \nabla_k - \nabla_k \nabla_j)\,\mathbf{A}\right]\mathbf{B} + \mathbf{A}\,(\nabla_j \nabla_k - \nabla_k \nabla_j)\,\mathbf{B} \tag{5.56}$$

where **A** and **B** are any Clifford numbers, which are not necessarily index free. (See Prob. 84.) Using (5.56) and then (5.51), we have

$$(\nabla_i \nabla_j - \nabla_j \nabla_i)\,\boldsymbol{\gamma}_k \boldsymbol{\gamma}_m = R^{\lambda}{}_{kij}\boldsymbol{\gamma}_\lambda \boldsymbol{\gamma}_m + R^{\lambda}{}_{mij}\boldsymbol{\gamma}_k \boldsymbol{\gamma}_\lambda.$$

Switching the k and m indices, we have

$$(\nabla_i \nabla_j - \nabla_j \nabla_i)\,\boldsymbol{\gamma}_m \boldsymbol{\gamma}_k = R^{\lambda}{}_{mij}\boldsymbol{\gamma}_\lambda \boldsymbol{\gamma}_k + R^{\lambda}{}_{kij}\boldsymbol{\gamma}_m \boldsymbol{\gamma}_\lambda.$$

Adding these last two equations gives us

$$(\nabla_i \nabla_j - \nabla_j \nabla_i)\,2g_{mk}\mathbf{I} = R^{\lambda}{}_{kij}2g_{\lambda m}\mathbf{I} + R^{\lambda}{}_{mij}2g_{k\lambda}\mathbf{I}. \tag{5.57}$$

Since ∇_i and ∇_j act on g_{mk} like $\partial/\partial u^i$ and $\partial/\partial u^j$, the left hand side of (5.57) is zero. When we carry out the summations over the λ index, (5.57) becomes

$$0 = R_{mkij} + R_{kmij}$$

which verifies (5.53).

To obtain the cyclic symmetry of (5.54), we make use of the torsion free condition. That is $\nabla_\eta \boldsymbol{\gamma}_\mu = \nabla_\mu \boldsymbol{\gamma}_\eta$. This implies that

$$\nabla_i \nabla_j \boldsymbol{\gamma}_k - \nabla_i \nabla_k \boldsymbol{\gamma}_j = 0,$$

$$\nabla_j \nabla_k \boldsymbol{\gamma}_i - \nabla_j \nabla_i \boldsymbol{\gamma}_k = 0, \text{ and}$$

$$\nabla_k \nabla_i \boldsymbol{\gamma}_j - \nabla_k \nabla_j \boldsymbol{\gamma}_i = 0.$$

Adding these three equations and then regrouping the terms gives us the equation

$$\left(\nabla_i \nabla_j - \nabla_j \nabla_i\right) \boldsymbol{\gamma}_k + \left(\nabla_j \nabla_k - \nabla_k \nabla_j\right) \boldsymbol{\gamma}_i + \left(\nabla_k \nabla_i - \nabla_i \nabla_k\right) \boldsymbol{\gamma}_j = 0,$$

which is equivalent to saying,

$$\left(R^{\lambda}{}_{kij} + R^{\lambda}{}_{ijk} + R^{\lambda}{}_{jki}\right) \boldsymbol{\gamma}_\lambda = 0. \tag{5.58}$$

Since the $\boldsymbol{\gamma}_\lambda$'s are linearly independent, we now have

$$R^{\lambda}{}_{kij} + R^{\lambda}{}_{ijk} + R^{\lambda}{}_{jki} = 0.$$

Therefore

$$\left(R^{\lambda}{}_{kij} + R^{\lambda}{}_{ijk} + R^{\lambda}{}_{jki}\right) g_{\lambda m} = 0$$

and thus

$$R_{mkij} + R_{mijk} + R_{mjki} = 0,$$

which is what we set out to prove.

Actually the Riemann curvature tensor has a similar cyclic symmetry with respect to any three of the indices. (See Prob. 87.)

To get the last symmetry of pair exchange in (5.55), we add up four versions of the cyclic symmetry equation that was just proven. That is

$$R_{kmij} + R_{kijm} + R_{kjmi} = 0,$$

$$R_{mkji} + R_{mjik} + R_{mikj} = 0,$$

$$-R_{ijkm} - R_{ikmj} - R_{imjk} = 0, \text{ and}$$

$$-R_{jimk} - R_{jmki} - R_{jkim} = 0.$$

When we add these last four equations together, we make use of the fact that $R_{\alpha\beta\eta\lambda} = R_{\beta\alpha\lambda\eta}$. As a result some terms add and some cancel and we arrive at the equation

$$2R_{kmij} - 2R_{ijkm} = 0$$

which verifies (5.55).

For many computations, it is useful to use the curvature 2-form. The *curvature 2-form* $\mathbf{R}_{ij}$ is defined by the equation:

$$\mathbf{R}_{ij} = \frac{1}{2} R_{pqij} \boldsymbol{\gamma}^{pq} = \frac{1}{2} R^{pq}{}_{ij} \boldsymbol{\gamma}_{pq}. \tag{5.59}$$

Theorem 82. $(\nabla_i \nabla_j - \nabla_j \nabla_i) \boldsymbol{\gamma}_k = \frac{1}{2}\mathbf{R}_{ij}\boldsymbol{\gamma}_k - \boldsymbol{\gamma}_k \frac{1}{2}\mathbf{R}_{ij}.$

Proof. We first note that

$$\begin{aligned}
\boldsymbol{\gamma}^p \boldsymbol{\gamma}^q \boldsymbol{\gamma}_k - \boldsymbol{\gamma}_k \boldsymbol{\gamma}^p \boldsymbol{\gamma}^q &= \boldsymbol{\gamma}^p \boldsymbol{\gamma}^q \boldsymbol{\gamma}_k + (\boldsymbol{\gamma}^p \boldsymbol{\gamma}_k \boldsymbol{\gamma}^q - \boldsymbol{\gamma}^p \boldsymbol{\gamma}_k \boldsymbol{\gamma}^q) - \boldsymbol{\gamma}_k \boldsymbol{\gamma}^p \boldsymbol{\gamma}^q \\
&= \boldsymbol{\gamma}^p (\boldsymbol{\gamma}^q \boldsymbol{\gamma}_k + \boldsymbol{\gamma}_k \boldsymbol{\gamma}^q) - (\boldsymbol{\gamma}^p \boldsymbol{\gamma}_k + \boldsymbol{\gamma}_k \boldsymbol{\gamma}^p) \boldsymbol{\gamma}^q \\
&= \boldsymbol{\gamma}^p 2\delta_k^q - 2\delta_k^p \boldsymbol{\gamma}^q.
\end{aligned}$$

This implies that

$$\begin{aligned}
\frac{1}{2}\mathbf{R}_{ij}\boldsymbol{\gamma}_k - \boldsymbol{\gamma}_k \frac{1}{2}\mathbf{R}_{ij} &= \frac{1}{4} R_{pqij} (\boldsymbol{\gamma}^p \boldsymbol{\gamma}^q \boldsymbol{\gamma}_k - \boldsymbol{\gamma}_k \boldsymbol{\gamma}^p \boldsymbol{\gamma}^q) \\
&= \frac{1}{4} R_{pqij} (2\boldsymbol{\gamma}^p \delta_k^q - 2\delta_k^p \boldsymbol{\gamma}^q) \\
&= \frac{1}{2} R_{pkij} \boldsymbol{\gamma}^p - \frac{1}{2} R_{kqij} \boldsymbol{\gamma}^q \\
&= R_{pkij} \boldsymbol{\gamma}^p = R^{\lambda}{}_{kij} \boldsymbol{\gamma}_\lambda.
\end{aligned}$$

Combining this last equation with (5.51), we have our desired result:

$$(\nabla_i \nabla_j - \nabla_j \nabla_i) \boldsymbol{\gamma}_k = \frac{1}{2}\mathbf{R}_{ij}\boldsymbol{\gamma}_k - \boldsymbol{\gamma}_k \frac{1}{2}\mathbf{R}_{ij}. \tag{5.60}$$

□

It is not too difficult to generalize (5.60). To do that the first step is to extend (5.56) by induction to any finite product of Clifford numbers. (See Prob. 85.) This implies that

$$\begin{aligned}
(\nabla_i \nabla_j - \nabla_j \nabla_i) \boldsymbol{\gamma}_1 \boldsymbol{\gamma}_2 \cdots \boldsymbol{\gamma}_p = &\left[(\nabla_i \nabla_j - \nabla_j \nabla_i) \boldsymbol{\gamma}_1\right] \boldsymbol{\gamma}_2 \boldsymbol{\gamma}_3 \cdots \boldsymbol{\gamma}_p \\
&+ \boldsymbol{\gamma}_1 \left[(\nabla_i \nabla_j - \nabla_j \nabla_i) \boldsymbol{\gamma}_2\right] \boldsymbol{\gamma}_3 \boldsymbol{\gamma}_4 \cdots \boldsymbol{\gamma}_p \\
&+ \cdots + \boldsymbol{\gamma}_1 \boldsymbol{\gamma}_2 \cdots \boldsymbol{\gamma}_{p-1} (\nabla_i \nabla_j - \nabla_j \nabla_i) \boldsymbol{\gamma}_p.
\end{aligned}$$

Combining this last equation with (5.60) and then using the telescopic property, we get

$$\begin{aligned}(\nabla_i\nabla_j - \nabla_j\nabla_i)\,\gamma_1\gamma_2\cdots\gamma_p &= \left(\frac{1}{2}\mathbf{R}_{ij}\gamma_1 - \gamma_1\frac{1}{2}\mathbf{R}_{ij}\right)\gamma_2\gamma_3\cdots\gamma_p \\ &\quad + \gamma_1\left(\frac{1}{2}\mathbf{R}_{ij}\gamma_2 - \gamma_2\frac{1}{2}\mathbf{R}_{ij}\right)\gamma_3\gamma_4\cdots\gamma_p \\ &\quad + \cdots + \gamma_1\gamma_2\cdots\gamma_{p-1}\left(\frac{1}{2}\mathbf{R}_{ij}\gamma_p - \gamma_p\frac{1}{2}\mathbf{R}_{ij}\right) \\ &= \frac{1}{2}\mathbf{R}_{ij}\,(\gamma_1\gamma_2\cdots\gamma_p) - (\gamma_1\gamma_2\cdots\gamma_p)\,\frac{1}{2}\mathbf{R}_{ij}.\end{aligned}$$

You should now be able to convince yourself that this last equality is valid for any twice differentiable Clifford number. That is

$$(\nabla_i\nabla_j - \nabla_j\nabla_i)\,\mathbf{A} = \frac{1}{2}\mathbf{R}_{ij}\mathbf{A} - \mathbf{A}\frac{1}{2}\mathbf{R}_{ij}, \tag{5.61}$$

where $\mathbf{A}$ is any twice differentiable Clifford number that is not necessarily index free.

Problem 83. Note that

$$(\bar{\nabla}_\alpha\bar{\nabla}_\beta - \bar{\nabla}_\beta\bar{\nabla}_\alpha)\,\bar{\gamma}_\eta = \left[\left(\frac{\partial u^\delta}{\partial \bar{u}^\alpha}\nabla_\delta\right)\left(\frac{\partial u^\sigma}{\partial \bar{u}^\beta}\nabla_\sigma\right) - \left(\frac{\partial u^\sigma}{\partial \bar{u}^\beta}\nabla_\sigma\right)\left(\frac{\partial u^\delta}{\partial \bar{u}^\alpha}\nabla_\delta\right)\right]\left(\frac{\partial u^\mu}{\partial \bar{u}^\eta}\gamma_\mu\right).$$

a) Use this relation to show that

$$(\bar{\nabla}_\alpha\bar{\nabla}_\beta - \bar{\nabla}_\beta\bar{\nabla}_\alpha)\,\bar{\gamma}_\eta = \frac{\partial u^\delta}{\partial \bar{u}^\alpha}\frac{\partial u^\sigma}{\partial \bar{u}^\beta}\frac{\partial u^\mu}{\partial \bar{u}^\eta}\,(\nabla_\delta\nabla_\sigma - \nabla_\sigma\nabla_\delta)\,\gamma_\mu.$$

Hint! $\left(\frac{\partial u^\delta}{\partial \bar{u}^\alpha}\nabla_\delta\right)\left(\frac{\partial u^\sigma}{\partial \bar{u}^\beta}\right) = \frac{\partial}{\partial \bar{u}^\alpha}\left(\frac{\partial u^\sigma}{\partial \bar{u}^\beta}\right) = \frac{\partial^2 u^\sigma}{\partial \bar{u}^\alpha\partial \bar{u}^\beta}$

b) Use the result of part a) to show that $R^\lambda{}_{\mu\delta\sigma}$ transforms as a tensor under a change of coordinates.

Problem 84. Prove (5.56).

Problem 85. Use induction to extend (5.56) to products of p Clifford numbers. That is show

$$\begin{aligned}(\nabla_i\nabla_j - \nabla_j\nabla_i)\,(\mathbf{A}_1\mathbf{A}_2\cdots\mathbf{A}_p) &= \left[(\nabla_i\nabla_j - \nabla_j\nabla_i)\,\mathbf{A}_1\right]\mathbf{A}_2\mathbf{A}_3\ldots\mathbf{A}_p \\ &\quad + \mathbf{A}_1\left[(\nabla_i\nabla_j - \nabla_j\nabla_i)\,\mathbf{A}_2\right]\mathbf{A}_3\mathbf{A}_4\cdots\mathbf{A}_p \\ &\quad + \cdots + \mathbf{A}_1\mathbf{A}_2\cdots\mathbf{A}_{p-1}\,(\nabla_i\nabla_j - \nabla_j\nabla_i)\,\mathbf{A}_p,\end{aligned}$$

where the $\mathbf{A}_k$'s are twice differentiable not necessarily index free Clifford numbers.

Problem 86. Formally prove (5.61). That is

$$\left(\nabla_i \nabla_j - \nabla_j \nabla_i\right) \mathbf{A} = \frac{1}{2}\mathbf{R}_{ij}\mathbf{A} - \mathbf{A}\frac{1}{2}\mathbf{R}_{ij}.$$

Problem 87. Assuming (5.54) is true.

a) Use (5.52), (5.53), and (5.55) to show that

$$R_{kmij} + R_{mikj} + R_{ikmj} = 0.$$

b) Pose another equation like (5.54) and then prove it using the same equations that you used in part a.

Problem 88. Use (5.51) to show that

$$R^{\lambda}{}_{\eta\alpha\beta} = \frac{\partial}{\partial u^{\alpha}}\Gamma^{\lambda}{}_{\eta\beta} - \frac{\partial}{\partial u^{\beta}}\Gamma^{\lambda}{}_{\eta\alpha} + \Gamma^{\nu}{}_{\eta\beta}\Gamma^{\lambda}{}_{\nu\alpha} - \Gamma^{\nu}{}_{\eta\alpha}\Gamma^{\lambda}{}_{\nu\beta}.$$

Problem 89. Show that $R_{\alpha\beta}{}^{\lambda\nu} = R^{\lambda\nu}{}_{\alpha\beta}$. (This shows that for 2-dimensions, the Gaussian curvature $R^{12}{}_{12} = R_{12}{}^{12}$.)

5.6 Fock–Ivanenko Coefficients

5.6.1 Moving Frames

For many computations, it is easier to use orthonormal non-coordinate frames rather than coordinate Dirac vectors. Using orthonormal frames, one can exploit a certain symmetry that one cannot do using coordinate frames. (To avoid some possible confusion later on, I will place bars over the numeric indices that refer to orthonormal bases. For example: $R_{12}{}^{12}\boldsymbol{\gamma}^{12} = R_{\bar{1}\bar{2}}{}^{12}\mathbf{E}^{\bar{1}\bar{2}}$ so usually $R_{12}{}^{12} \neq R_{\bar{1}\bar{2}}{}^{12}$.) Visually, this convention does not work well if it is extended to indices that are members of the alphabet. For indices that are members of an alphabet that refer to an orthonormal basis, I will use upper case letters.

To construct an orthonormal frame $\{\mathbf{E}_{\bar{1}}, \mathbf{E}_{\bar{2}}, \ldots, \mathbf{E}_{\bar{n}}\}$ from a coordinate frame, one can use the Gram–Schmidt method outlined in Sect. 4.5 at least for Euclidean spaces or spaces embedded in Euclidean spaces. For intrinsic observers (observers who are restricted to making all measurements in the embedded space), this task is more difficult but still possible. For example, let us consider the saddle surface. The intrinsic observer would begin with a 2-dimensional metric. One possible metric that was encountered in Prob. 32 was

$$\begin{bmatrix} g_{11} & g_{12} \\ g_{21} & g_{22} \end{bmatrix} = \begin{bmatrix} 1 + (u^2)^2 & u^1 u^2 \\ u^1 u^2 & 1 + (u^1)^2 \end{bmatrix}. \tag{5.62}$$

Using the Gram–Schmidt procedure, we get

$$\mathbf{E}_{\bar{1}} = \frac{\boldsymbol{\gamma}_1}{|\boldsymbol{\gamma}_1|} = \frac{\boldsymbol{\gamma}_1}{\sqrt{\langle \boldsymbol{\gamma}_1, \boldsymbol{\gamma}_1 \rangle}} = \frac{\boldsymbol{\gamma}_1}{\sqrt{1 + (u^2)^2}}, \tag{5.63}$$

and using (4.56),

$$\begin{aligned} \mathbf{E}_{\bar{2}} &= \frac{\boldsymbol{\gamma}_1}{|\boldsymbol{\gamma}_1|} \frac{\boldsymbol{\gamma}_{12}}{|\boldsymbol{\gamma}_{12}|} = \frac{\langle \boldsymbol{\gamma}_1, \boldsymbol{\gamma}_1 \rangle \boldsymbol{\gamma}_2 - \langle \boldsymbol{\gamma}_1, \boldsymbol{\gamma}_2 \rangle \boldsymbol{\gamma}_1}{\sqrt{\langle \boldsymbol{\gamma}_1, \boldsymbol{\gamma}_1 \rangle} \sqrt{g_{11} g_{22} - g_{12} g_{21}}} \\ &= \frac{-u^1 u^2 \boldsymbol{\gamma}_1 + \left(1 + (u^2)^2\right) \boldsymbol{\gamma}_2}{\sqrt{1 + (u^2)^2} \sqrt{1 + (u^1)^2 + (u^2)^2}}. \end{aligned} \tag{5.64}$$

After a long calculation using (5.31), we get

$$\Gamma^1_{11} = \Gamma^2_{11} = 0, \tag{5.65}$$

$$\Gamma^1_{12} = \Gamma^1_{21} = \frac{u^2}{1 + (u^1)^2 + (u^2)^2}, \tag{5.66}$$

$$\Gamma^2_{12} = \Gamma^2_{21} = \frac{u^1}{1 + (u^1)^2 + (u^2)^2}, \text{ and} \tag{5.67}$$

$$\Gamma^1_{22} = \Gamma^2_{22} = 0. \tag{5.68}$$

Now applying these equations to (5.63), we discover that

$$\nabla_1 \mathbf{E}_{\bar{1}} = 0, \text{ and} \tag{5.69}$$

$$\nabla_2 \mathbf{E}_{\bar{1}} = \frac{u^1}{\left(1 + (u^2)^2\right) \sqrt{1 + (u^1)^2 + (u^2)^2}} \mathbf{E}_{\bar{2}}. \tag{5.70}$$

To compute $\nabla_1 \mathbf{E}_{\bar{2}}$ and $\nabla_2 \mathbf{E}_{\bar{2}}$ directly would be a formidable task but because of the symmetry which I will now discuss, we can determine these entities without further computation.

In general, not just for saddle surfaces, $\nabla_i \mathbf{E}_{\bar{j}}$ must be some linear combination of $\mathbf{E}_{\bar{k}}$'s so

$$\nabla_i \mathbf{E}_J = -c_{JK} \left(\boldsymbol{\gamma}_i\right) \mathbf{E}^K. \tag{5.71}$$

(I did not make a mistake!) For a Euclidean space or for a space embedded in a Euclidean space, $\langle \mathbf{E}_K, \mathbf{E}_K \rangle$ is always equal to $+1$ and $\mathbf{E}^K = \mathbf{E}_K$. However in pseudo-Euclidean spaces or in spaces embedded in pseudo-Euclidean spaces, a member of a moving frame $\mathbf{E}_K$ may have the property that $\langle \mathbf{E}_K, \mathbf{E}_K \rangle = -1$. In such a case, one defines $\mathbf{E}^K = -\mathbf{E}_K$. This guarantees that for all cases, $\left\langle \mathbf{E}^K, \mathbf{E}_J \right\rangle = \delta^K_J$. The $\boldsymbol{\gamma}_i$ that appears as an argument of c_{JK} in (5.71) is used to indicate that the intrinsic derivative is being computed along the coordinate curve corresponding to u^i. With this convention, it can be shown that $c_{JK} \left(\boldsymbol{\gamma}_i\right) = -c_{KJ} \left(\boldsymbol{\gamma}_i\right)$.

Theorem 90. *If* $\nabla_i \mathbf{E}_J = -c_{JK}(\boldsymbol{\gamma}_i)\mathbf{E}^K$ *then* $c_{JK}(\boldsymbol{\gamma}_i) = -c_{KJ}(\boldsymbol{\gamma}_i)$.

Proof. *Now* $\langle \mathbf{E}_J, \mathbf{E}_K \rangle = n_{JK}$, *where* $n_{JK} = \pm 1$, *if* $J = K$ *and* $n_{JK} = 0$, *if* $J \neq K$, *so*

$$
\begin{aligned}
0 = -\nabla_i \langle \mathbf{E}_J, \mathbf{E}_K \rangle &= -\langle \nabla_i \mathbf{E}_J, \mathbf{E}_K \rangle - \langle \mathbf{E}_J, \nabla_i \mathbf{E}_K \rangle \\
&= c_{JM}(\boldsymbol{\gamma}_i)\langle \mathbf{E}^M, \mathbf{E}_K \rangle + c_{KM}(\boldsymbol{\gamma}_i)\langle \mathbf{E}_J, \mathbf{E}^M \rangle \\
&= c_{JM}(\boldsymbol{\gamma}_i)\delta_K^M + c_{KM}(\boldsymbol{\gamma}_i)\delta_J^M \\
&= c_{JK}(\boldsymbol{\gamma}_i) + c_{KJ}(\boldsymbol{\gamma}_i)
\end{aligned}
$$

□

It should be noted that Theorem 90 implies that $c_{KK}(\boldsymbol{\gamma}_i) = 0$.

Returning to the saddle surface discussed above, we note that from (5.69) and (5.70)

$$
\begin{aligned}
c_{\bar{1}\bar{2}}(\boldsymbol{\gamma}_1) &= 0 \text{ and} \\
c_{\bar{1}\bar{2}}(\boldsymbol{\gamma}_2) &= \frac{-u^1}{(1+(u^2)^2)\sqrt{1+(u^1)^2+(u^2)^2}}.
\end{aligned}
$$

Thus

$$
\begin{aligned}
\nabla_1 \mathbf{E}_{\bar{2}} &= -c_{\bar{2}\bar{1}}(\boldsymbol{\gamma}_1)\mathbf{E}^{\bar{1}} = c_{\bar{1}\bar{2}}(\boldsymbol{\gamma}_1)\mathbf{E}^{\bar{1}} = 0 \text{ and} \\
\nabla_2 \mathbf{E}_{\bar{2}} &= -c_{\bar{2}\bar{1}}(\boldsymbol{\gamma}_2)\mathbf{E}^{\bar{1}} = c_{\bar{1}\bar{2}}(\boldsymbol{\gamma}_2)\mathbf{E}^{\bar{1}} = \frac{-u^1}{(1+(u^2)^2)\sqrt{1+(u^1)^2+(u^2)^2}}\mathbf{E}_1.
\end{aligned}
$$

It is worth noting that if we had taken advantage of the information available to an extrinsic observer, the computation of $c_{\bar{1}\bar{2}}(\boldsymbol{\gamma}_1)$ and $c_{\bar{1}\bar{2}}(\boldsymbol{\gamma}_2)$ would have been simpler. In that situation, we could have written

$$
\begin{aligned}
\mathbf{E}_{\bar{1}} &= \frac{\boldsymbol{\gamma}_1}{|\boldsymbol{\gamma}_1|} = \frac{\mathbf{e}_1 + \mathbf{e}_3 u^2}{\sqrt{(1+(u^2)^2)}} \text{ so} \\
\frac{\partial}{\partial u^1}\mathbf{E}_{\bar{1}} &= 0 = -c_{\bar{1}\bar{2}}(\boldsymbol{\gamma}_1)\mathbf{E}^{\bar{2}} + h_{\bar{1}}(\boldsymbol{\gamma}_1)\mathbf{N}.
\end{aligned}
$$

Thus $c_{\bar{1}\bar{2}}(\boldsymbol{\gamma}_1) = 0$. Furthermore, since $\boldsymbol{\gamma}_1 = \mathbf{e}_1 + \mathbf{e}_3 u^2$ and $\boldsymbol{\gamma}_2 = \mathbf{e}_2 + \mathbf{e}_3 u^1$, it follows that $\boldsymbol{\gamma}_{12} = -\mathbf{e}_{23}u^2 - \mathbf{e}_{31}u^1 + \mathbf{e}_{12}$. We then have:

$$
\begin{aligned}
\mathbf{E}_{\bar{2}} &= \frac{\boldsymbol{\gamma}_1}{|\boldsymbol{\gamma}_1|}\frac{\boldsymbol{\gamma}_{12}}{|\boldsymbol{\gamma}_{12}|} = \frac{\mathbf{e}_1 + \mathbf{e}_3 u^2}{\sqrt{(1+(u^2)^2)}}\frac{-\mathbf{e}_{23}u^2 - \mathbf{e}_{31}u^1 + \mathbf{e}_{12}}{\sqrt{1+(u^1)^2+(u^2)^2}} \\
&= \frac{-\mathbf{e}_1 u^1 u^2 + \mathbf{e}_2(1+(u^2)^2) + \mathbf{e}_3 u^1}{\sqrt{(1+(u^2)^2)}\sqrt{1+(u^1)^2+(u^2)^2}}
\end{aligned}
$$

Now

$$\begin{aligned}\frac{\partial}{\partial u^2}\mathbf{E}_{\bar{1}} &= \frac{\partial}{\partial u^2}\frac{\mathbf{e}_1 + \mathbf{e}_3 u^2}{\sqrt{(1+(u^2)^2)}} \\ &= \left(\frac{\partial}{\partial u^2}\frac{1}{\sqrt{(1+(u^2)^2)}}\right)(\mathbf{e}_1 + \mathbf{e}_3 u^2) + \frac{\mathbf{e}_3}{\sqrt{(1+(u^2)^2)}} \\ &= -c_{\bar{1}\bar{2}}(\boldsymbol{\gamma}_2)\mathbf{E}^{\bar{2}} + h_{\bar{1}}(\boldsymbol{\gamma}_2)\mathbf{N} \qquad (5.72)\end{aligned}$$

From (5.72)

$$\begin{aligned}c_{\bar{1}\bar{2}}(\boldsymbol{\gamma}_2) &= -\left\langle \mathbf{E}_{\bar{2}}, \frac{\partial}{\partial u^2}\mathbf{E}_{\bar{1}}\right\rangle \\ &= -\left\langle \mathbf{E}_{\bar{2}}, \left(\frac{\partial}{\partial u^2}\frac{1}{\sqrt{(1+(u^2)^2)}}\right)(\mathbf{e}_1 + \mathbf{e}_3 u^2) + \frac{\mathbf{e}_3}{\sqrt{(1+(u^2)^2)}}\right\rangle \\ &= -\left\langle \mathbf{E}_{\bar{2}}, \frac{\mathbf{e}_3}{\sqrt{(1+(u^2)^2)}}\right\rangle \\ &= -\left\langle \frac{-\mathbf{e}_1 u^1 u^2 + \mathbf{e}_2(1+(u^2)^2) + \mathbf{e}_3 u^1}{\sqrt{(1+(u^2)^2)}\sqrt{1+(u^1)^2+(u^2)^2}}, \frac{\mathbf{e}_3}{\sqrt{(1+(u^2)^2)}}\right\rangle \text{ or}\end{aligned}$$

$$c_{\bar{1}\bar{2}}(\boldsymbol{\gamma}_2) = \frac{-u^1}{(1+(u^2)^2)\sqrt{1+(u^1)^2+(u^2)^2}} \qquad (5.73)$$

For two dimensions, there are only two relevant coefficients ($c_{\bar{1}\bar{2}}(\boldsymbol{\gamma}_1)$ and $c_{\bar{1}\bar{2}}(\boldsymbol{\gamma}_2)$). But for higher dimensions, there are many more. As a result, people have found a couple of ways to condense the information stored in these coefficients.

Users of differential forms have observed that

$$\begin{aligned}\nabla_{\mathbf{v}}\mathbf{E}_J &= v^i \nabla_i \mathbf{E}_J = -v^i c_{JK}(\boldsymbol{\gamma}_i)\mathbf{E}^K. \text{ Also} \\ \nabla_{\mathbf{v}}\mathbf{E}_J &= -c_{JK}(\mathbf{v})\mathbf{E}^K. \text{ And thus} \\ c_{JK}(\mathbf{v}) &= c_{JK}(v^i\boldsymbol{\gamma}_i) = v^i c_{JK}(\boldsymbol{\gamma}_i). \qquad (5.74)\end{aligned}$$

This shows us that c_{JK} is a real valued linear function on the n-dimensional vector space spanned by $\boldsymbol{\gamma}_1, \boldsymbol{\gamma}_2, \ldots, \boldsymbol{\gamma}_n$. This implies that there exists a vector $\mathbf{w}_{JK}$ (or 1-form) such that

$$c_{JK}(\mathbf{v}) = \langle \mathbf{w}_{JK}, \mathbf{v}\rangle\,. \qquad (5.75)$$

An explicit formula for $\mathbf{w}_{JK}$ is

$$\mathbf{w}_{JK} = c_{JK}(\boldsymbol{\gamma}_i)\boldsymbol{\gamma}^i = c_{JK}(\mathbf{E}_I)\mathbf{E}^I. \tag{5.76}$$

To check (5.76), note:

$$\begin{aligned}\langle \mathbf{w}_{JK}, \mathbf{v}\rangle &= \left\langle c_{JK}(\boldsymbol{\gamma}_i)\boldsymbol{\gamma}^i, v^m\boldsymbol{\gamma}_m\right\rangle = c_{JK}(\boldsymbol{\gamma}_i)v^m\left\langle \boldsymbol{\gamma}^i, \boldsymbol{\gamma}_m\right\rangle \\ &= c_{JK}(\boldsymbol{\gamma}_i)v^m\delta^i_m = c_{JK}(\boldsymbol{\gamma}_i)v^i = c_{JK}(v^i\boldsymbol{\gamma}_i) \\ &= c_{JK}(\mathbf{v}),\end{aligned}$$

which agrees with (5.74) and (5.75). The $\mathbf{w}_{JK}$'s are known as *connection 1-forms.*

An alternative method of condensing the information that I feel is even better – particularly for higher dimensions is to use *Fock–Ivanenko coefficients*, which I will discuss in the next subsection.

Problem 91. In the derivation of (5.73), I used the fact that

$$\left\langle \mathbf{E}_{\bar{2}}, \left(\frac{\partial}{\partial u^2}\frac{1}{\sqrt{1+(u^2)^2}}\right)\left(\mathbf{e}_1 + \mathbf{e}_3 u^2\right)\right\rangle = 0.$$

Explain why this is true without doing any computations. Why did I know that it would be unnecessary to compute

$$\frac{\partial}{\partial u^2}\frac{1}{\sqrt{1+(u^2)^2}}\,?$$

5.6.2 Gauss Curvature via Fock–Ivanenko Coefficients

Fock–Ivanenko coefficients were first introduced by Vladimir Fock and Dmitrii Ivanenko to make Paul Dirac's equation for the electron compatible with Albert Einstein's theory of general relativity (Fock and Ivanenko 1929a, 1929b; Fock 1929). A Fock–Ivanenko coefficient, $\boldsymbol{\Gamma}_k$, is defined by the equation

$$\boldsymbol{\Gamma}_k = \frac{1}{4}c_{PQ}(\boldsymbol{\gamma}_k)\mathbf{E}^P\mathbf{E}^Q. \tag{5.77}$$

Using Fock–Ivanenko coefficients, we can write

$$\nabla_k \mathbf{E}_I = \boldsymbol{\Gamma}_k\mathbf{E}_I - \mathbf{E}_I\boldsymbol{\Gamma}_k. \tag{5.78}$$

To verify (5.78), we note that from (5.77)

$$\begin{aligned}\boldsymbol{\Gamma}_k\mathbf{E}_I - \mathbf{E}_I\boldsymbol{\Gamma}_k &= \frac{1}{4}c_{PQ}(\boldsymbol{\gamma}_k)\left[\mathbf{E}^P\mathbf{E}^Q\mathbf{E}_I - \mathbf{E}_I\mathbf{E}^P\mathbf{E}^Q\right]\\ &= \frac{1}{4}c_{PQ}(\boldsymbol{\gamma}_k)\left[\mathbf{E}^P\left(\mathbf{E}^Q\mathbf{E}_I + \mathbf{E}_I\mathbf{E}^Q\right) - \left(\mathbf{E}^P\mathbf{E}_I + \mathbf{E}_I\mathbf{E}^P\right)\mathbf{E}^Q\right]\\ &= \frac{1}{4}c_{PQ}(\boldsymbol{\gamma}_k)\left[\mathbf{E}^P 2\delta_I^Q - 2\delta_I^P\mathbf{E}^Q\right]\\ &= \frac{1}{2}c_{PI}(\boldsymbol{\gamma}_k)\mathbf{E}^P - \frac{1}{2}c_{IQ}(\boldsymbol{\gamma}_k)\mathbf{E}^Q \text{ or}\\ &= -c_{IQ}(\boldsymbol{\gamma}_k)\mathbf{E}^Q\\ &= \boldsymbol{\nabla}_k\mathbf{E}_I.\end{aligned}$$

Equation (5.78) can be generalized to products of $\mathbf{E}_J$'s. That is

$$\boldsymbol{\nabla}_k\left(\mathbf{E}_{I_1}\mathbf{E}_{I_2}\cdots\mathbf{E}_{I_p}\right) = \boldsymbol{\Gamma}_k\left(\mathbf{E}_{I_1}\mathbf{E}_{I_2}\cdots\mathbf{E}_{I_p}\right) - \left(\mathbf{E}_{I_1}\mathbf{E}_{I_2}\cdots\mathbf{E}_{I_p}\right)\boldsymbol{\Gamma}_k. \tag{5.79}$$

(See Prob. 93.)

More generally if

$$\mathbf{A} = \frac{1}{p!}A^{I_1 I_2 \ldots I_p}\mathbf{E}_{I_1}\mathbf{E}_{I_2}\cdots\mathbf{E}_{I_p}, \text{ then}$$

$$\begin{aligned}\boldsymbol{\nabla}_k\mathbf{A} = &\frac{1}{p!}\left(\frac{\partial}{\partial u^k}A^{I_1 I_2 \ldots I_p}\right)\mathbf{E}_{I_1}\mathbf{E}_{I_2}\cdots\mathbf{E}_{I_p}\\ &+ \frac{1}{p!}A^{I_1 I_2 \ldots I_p}\boldsymbol{\Gamma}_k\mathbf{E}_{I_1}\mathbf{E}_{I_2}\cdots\mathbf{E}_{I_p} - \frac{1}{p!}A^{I_1 I_2 \ldots I_p}\mathbf{E}_{I_1}\mathbf{E}_{I_2}\cdots\mathbf{E}_{I_p}\boldsymbol{\Gamma}_k.\end{aligned}$$

To represent this last equation in a more concise manner it is useful to define what might be called a *moving frame derivative* ∂_k. Let us define $\partial_k\mathbf{E}_J$ to be zero and ∂_k to be identical to $\partial/\partial u^k$ when it is applied to any real valued coefficient of a product of $\mathbf{E}_J$'s. Thus in the example above where

$$\partial_k\mathbf{A} = \frac{1}{p!}\left(\frac{\partial}{\partial u^k}A^{I_1 I_2 \ldots I_p}\right)\mathbf{E}_{I_1}\mathbf{E}_{I_2}\cdots\mathbf{E}_{I_p},$$

we can write

$$\boldsymbol{\nabla}_k\mathbf{A} = \partial_k\mathbf{A} + \Gamma_k\mathbf{A} - \mathbf{A}\Gamma_k. \tag{5.80}$$

Fock–Ivanenko coefficients can be used to rapidly compute the curvature 2-forms and thereby all components of the Riemann tensor. To get the necessary relation, we observe that

$$\begin{aligned}(\nabla_i\nabla_j - \nabla_j\nabla_i)\,\mathbf{E}_K &= \nabla_i\left(\boldsymbol{\Gamma}_j\mathbf{E}_K - \mathbf{E}_K\boldsymbol{\Gamma}_j\right) - \nabla_j\left(\boldsymbol{\Gamma}_i\mathbf{E}_K - \mathbf{E}_K\boldsymbol{\Gamma}_i\right)\\ &= \left(\nabla_i\boldsymbol{\Gamma}_j\right)\mathbf{E}_K + \boldsymbol{\Gamma}_j\left(\boldsymbol{\Gamma}_i\mathbf{E}_K - \mathbf{E}_K\boldsymbol{\Gamma}_i\right) - \left(\boldsymbol{\Gamma}_i\mathbf{E}_K - \mathbf{E}_K\boldsymbol{\Gamma}_i\right)\Gamma_j\\ &\quad - \mathbf{E}_K\nabla_i\boldsymbol{\Gamma}_j - \left(\nabla_j\boldsymbol{\Gamma}_i\right)\mathbf{E}_K - \Gamma_i\left(\boldsymbol{\Gamma}_j\mathbf{E}_K - \mathbf{E}_K\Gamma_j\right)\\ &\quad + \left(\boldsymbol{\Gamma}_j\mathbf{E}_K - \mathbf{E}_K\Gamma_j\right)\boldsymbol{\Gamma}_i + \mathbf{E}_K\nabla_j\boldsymbol{\Gamma}_i\end{aligned}$$

or restated:

$$\begin{aligned}(\nabla_i\nabla_j - \nabla_j\nabla_i)\,\mathbf{E}_K - &\left(\nabla_i\boldsymbol{\Gamma}_j - \nabla_j\boldsymbol{\Gamma}_i - \Gamma_i\boldsymbol{\Gamma}_j + \boldsymbol{\Gamma}_j\boldsymbol{\Gamma}_i\right)\mathbf{E}_K\\ &- \mathbf{E}_K\left(\nabla_i\boldsymbol{\Gamma}_j - \nabla_j\boldsymbol{\Gamma}_i - \boldsymbol{\Gamma}_i\boldsymbol{\Gamma}_j + \boldsymbol{\Gamma}_j\boldsymbol{\Gamma}_i\right).\end{aligned} \tag{5.81}$$

From (5.61)

$$(\nabla_i\nabla_j - \nabla_j\nabla_i)\,\mathbf{E}_K = \frac{1}{2}\mathbf{R}_{ij}\mathbf{E}_K - \frac{1}{2}\mathbf{E}_K\mathbf{R}_{ij}.$$

From these last two equations, we would like to conclude that

$$\frac{1}{2}\mathbf{R}_{ij} = \nabla_i\boldsymbol{\Gamma}_j - \nabla_j\boldsymbol{\Gamma}_i - \boldsymbol{\Gamma}_i\boldsymbol{\Gamma}_j + \boldsymbol{\Gamma}_j\boldsymbol{\Gamma}_i. \tag{5.82}$$

However suppose

$$\mathbf{A}\mathbf{E}_K - \mathbf{E}_K\mathbf{A} = \mathbf{B}\mathbf{E}_K - \mathbf{E}_K\mathbf{B} \quad \text{for} \quad K = \bar{1}, \bar{2}, \ldots, \bar{n}. \tag{5.83}$$

Does it then follow that $\mathbf{A} = \mathbf{B}$? Clearly if $\mathbf{A} = \mathbf{B}+\alpha\mathbf{I}$, (5.83) would still be satisfied. If $\bar{n}$ is an odd number, we could also add a scalar multiple of the pseudoscalar $\mathbf{E}_{\bar{1}}\mathbf{E}_{\bar{2}}\cdots\mathbf{E}_{\bar{n}}$ to $\mathbf{B}$ without disturbing the validity of (5.83). However if we insist that both $\mathbf{A}$ and $\mathbf{B}$ be 2-vectors, then (5.83) does indeed imply that $\mathbf{A} = \mathbf{B}$. (See Prob. 94.) Thus to verify (5.82), we are left with the task of showing that the right hand side is a 2-vector. Clearly $\nabla_i\boldsymbol{\Gamma}_j$ and $\nabla_j\boldsymbol{\Gamma}_i$ are 2-vectors. I will leave it to you to show that $\boldsymbol{\Gamma}_i\boldsymbol{\Gamma}_j - \boldsymbol{\Gamma}_j\boldsymbol{\Gamma}_i$ must be a 2-vector. (See Prob. 95.)

Using (5.80), we can obtain a slightly more useful form of (5.82). From (5.80), we have

$$\begin{aligned}\nabla_i\boldsymbol{\Gamma}_j &= \partial_i\boldsymbol{\Gamma}_j + \boldsymbol{\Gamma}_i\boldsymbol{\Gamma}_j - \boldsymbol{\Gamma}_j\boldsymbol{\Gamma}_i \text{ and}\\ \nabla_j\boldsymbol{\Gamma}_i &= \partial_j\boldsymbol{\Gamma}_i + \boldsymbol{\Gamma}_j\boldsymbol{\Gamma}_i - \boldsymbol{\Gamma}_i\boldsymbol{\Gamma}_j.\end{aligned}$$

Substituting these last two results into (5.82), we get

$$\frac{1}{2}\mathbf{R}_{ij} = \partial_i\boldsymbol{\Gamma}_j - \partial_j\boldsymbol{\Gamma}_i + \boldsymbol{\Gamma}_i\boldsymbol{\Gamma}_j - \boldsymbol{\Gamma}_j\boldsymbol{\Gamma}_i. \tag{5.84}$$

As a formula for $\mathbf{R}_{ij}$, (5.84) is easier to compute than (5.82) since the moving frame derivative ∂_k does not act on the $\mathbf{E}_K$'s. When using either formula, be careful to note that the two formulas differ in the signs for the last two terms. It is also useful to note that for two-dimensional surfaces, all 2-vectors are scalar multiples of the pseudo-vector. Therefore for two-dimensional surfaces, all 2-vectors commute with one another and the last two terms of (5.82) or (5.84) cancel one another out.

Returning to the saddle surface, we can compute the curvature 2-form with relative ease and with it the Gaussian curvature $R^{12}{}_{12}$.

From (5.69) and (5.70), we know that

$$c_{\bar{1}\bar{2}}(\boldsymbol{\gamma}_1) = 0 \text{ and}$$

$$c_{\bar{1}\bar{2}}(\boldsymbol{\gamma}_2) = \frac{-u^1}{(1+(u^2))\sqrt{1+(u^1)^2+(u^2)^2}}.$$

This means that

$$\boldsymbol{\Gamma}_1 = 0 \text{ and}$$

$$\begin{aligned}\boldsymbol{\Gamma}_2 &= \frac{1}{4} c_{PQ}(\boldsymbol{\gamma}_2)\mathbf{E}^P\mathbf{E}^Q = \frac{1}{4} c_{\bar{1}\bar{2}}(\boldsymbol{\gamma}_2)\mathbf{E}^{\bar{1}}\mathbf{E}^{\bar{2}} + \frac{1}{4} c_{\bar{2}\bar{1}}(\boldsymbol{\gamma}_2)\mathbf{E}^{\bar{2}}\mathbf{E}^{\bar{1}} \\ &= \frac{1}{2} c_{\bar{1}\bar{2}}(\boldsymbol{\gamma}_2)\mathbf{E}^{\bar{1}}\mathbf{E}^{\bar{2}} = \frac{-u^1}{2(1+(u^2)^2)\sqrt{1+(u^1)^2+(u^2)^2}}\mathbf{E}^{\bar{1}}\mathbf{E}^{\bar{2}}.\end{aligned}$$

In this circumstance, (5.84) becomes

$$\begin{aligned}\frac{1}{2}\mathbf{R}_{12} = \partial_1\boldsymbol{\Gamma}_2 &= \frac{-1}{2(1+(u^2)^2)}\left(\frac{\partial}{\partial u^1}\left[u^1\left(1+(u^1)^2+(u^2)^2\right)^{-1/2}\right]\right)\mathbf{E}^{\bar{1}}\mathbf{E}^{\bar{2}} \\ &= \frac{-1}{2(1+(u^2)^2)}\left[\left(1+(u^1)^2+(u^2)^2\right)^{-1/2} - \frac{1}{2}u^1\left(1+(u^1)^2+(u^2)^2\right)^{-3/2}2u^1\right]\mathbf{E}^{\bar{1}}\mathbf{E}^{\bar{2}} \\ &= \frac{-\left(1+(u^1)^2+(u^2)^2\right)^{-3/2}}{2(1+(u^2)^2)}\left[1+(u^1)^2+(u^2)^2-(u^1)^2\right]\mathbf{E}^{\bar{1}}\mathbf{E}^{\bar{2}}.\end{aligned}$$

Or

$$\mathbf{R}_{12} = \frac{-1}{\left(1+(u^1)^2+(u^2)^2\right)^{3/2}}\mathbf{E}^{\bar{1}}\mathbf{E}^{\bar{2}}. \tag{5.85}$$

For this problem, $\mathbf{E}^{\bar{1}}\mathbf{E}^{\bar{2}} = \mathbf{E}_{\bar{1}}\mathbf{E}_{\bar{2}} = \boldsymbol{\gamma}_{12}/\left|\boldsymbol{\gamma}_{12}\right|$. For the saddle surface, $\mathbf{s} = u^1\mathbf{e}_1 + u^2\mathbf{e}_2 + u^1u^2\mathbf{e}_3$, $\boldsymbol{\gamma}_1 = \mathbf{e}_1 + u^2\mathbf{e}_3$, $\boldsymbol{\gamma}_2 = \mathbf{e}_2 + u^1\mathbf{e}_3$, and $\boldsymbol{\gamma}_{12} = -u^2\mathbf{e}_{23} - u^1\mathbf{e}_{31} + \mathbf{e}_{12}$. Thus $|\boldsymbol{\gamma}_{12}| = \sqrt{1+(u^1)^2+(u^2)^2}$. Combining these results, (5.85) becomes

$$\mathbf{R}_{12} = \frac{-1}{[1+(u^1)^2+(u^2)^2]^2}\boldsymbol{\gamma}_{12}. \tag{5.86}$$

Also

$$\mathbf{R}_{12} = \frac{1}{2}R^{jk}{}_{12}\boldsymbol{\gamma}_{jk} = \frac{1}{2}R^{12}{}_{12}\boldsymbol{\gamma}_{12} + \frac{1}{2}R^{21}{}_{12}\boldsymbol{\gamma}_{21}$$

$$= R^{12}{}_{12}\boldsymbol{\gamma}_{12}. \tag{5.87}$$

Comparing (5.86) and (5.87), it follows that the Gaussian curvature for the saddle surface is

$$R^{12}{}_{12} = \frac{-1}{[1+(u^1)^2+(u^2)^2]^2}. \tag{5.88}$$

From the informal discussion of Gaussian curvature presented in Sect. 5.1, it should not be surprising that the Gaussian curvature for the saddle surface is negative.

In closing this section, I will derive a formula for $\boldsymbol{\Gamma}_\alpha$ which is useful for intrinsic computations – particularly if the metric g_{ij} is diagonal.

Theorem 92.

$$\boldsymbol{\Gamma}_\alpha = \frac{1}{4}\boldsymbol{\gamma}^{\lambda\eta}\frac{\partial g_{\alpha\lambda}}{\partial u^\eta} + \frac{1}{4}\boldsymbol{\gamma}^\eta \wedge \partial_\alpha \boldsymbol{\gamma}_\eta. \tag{5.89}$$

Proof. We need to keep in mind that

$$\boldsymbol{\Gamma}_\alpha = \frac{1}{4}c_{JK}(\boldsymbol{\gamma}_\alpha)\mathbf{E}^J\mathbf{E}^K \tag{5.90}$$

where $c_{JK}(\boldsymbol{\gamma}_\alpha)$ is defined by the equation

$$\nabla_\alpha \mathbf{E}_J = -c_{JK}(\boldsymbol{\gamma}_\alpha)\mathbf{E}^K. \tag{5.91}$$

You should also note that several steps below, I will use the fact that if $\mathbf{v}$ is an arbitrary vector then

$$\mathbf{v} = \langle \mathbf{v}, \boldsymbol{\gamma}_\beta\rangle \boldsymbol{\gamma}^\beta = \langle \mathbf{v}, \mathbf{E}_J\rangle \mathbf{E}^J. \tag{5.92}$$

(See Prob. 96.) Since

$$\mathbf{E}_J = \langle \mathbf{E}_J, \boldsymbol{\gamma}_\beta\rangle \boldsymbol{\gamma}^\beta,$$

$$\nabla_\alpha \mathbf{E}_J = \langle \mathbf{E}_J, \boldsymbol{\gamma}_\beta\rangle \nabla_\alpha \boldsymbol{\gamma}^\beta + \left(\frac{\partial}{\partial u^\alpha}\langle \mathbf{E}_J, \boldsymbol{\gamma}_\beta\rangle\right)\boldsymbol{\gamma}^\beta. \text{ And therefore}$$

$$\mathbf{E}^J\nabla_\alpha \mathbf{E}_J = -\mathbf{E}^J\langle \mathbf{E}_J, \boldsymbol{\gamma}_\beta\rangle \Gamma^\beta_{\eta\alpha}\boldsymbol{\gamma}^\eta + \mathbf{E}^J\left(\partial_\alpha\langle \mathbf{E}_J, \boldsymbol{\gamma}_\beta\rangle\right)\boldsymbol{\gamma}^\beta.$$

Using (5.91), (5.92), and (5.31), this last equation becomes

$$-c_{JK}(\boldsymbol{\gamma}_\alpha)\mathbf{E}^J\mathbf{E}^K = -\boldsymbol{\gamma}_\beta\frac{g^{\beta\lambda}}{2}\left[\frac{\partial g_{\alpha\lambda}}{\partial u^\eta}+\frac{\partial g_{\eta\lambda}}{\partial u^\alpha}-\frac{\partial g_{\alpha\eta}}{\partial u^\lambda}\right]\boldsymbol{\gamma}^\eta$$

$$+\left(\partial_\alpha\left(\mathbf{E}^J\left\langle\mathbf{E}_J,\boldsymbol{\gamma}_\beta\right\rangle\right)\right)\boldsymbol{\gamma}^\beta. \text{ That is:}$$

$$c_{JK}(\boldsymbol{\gamma}_\alpha)\mathbf{E}^J\mathbf{E}^K = \frac{1}{2}\boldsymbol{\gamma}^\lambda\boldsymbol{\gamma}^\eta\left[\frac{\partial g_{\alpha\lambda}}{\partial u^\eta}+\frac{\partial g_{\eta\lambda}}{\partial u^\alpha}-\frac{\partial g_{\alpha\eta}}{\partial u^\lambda}\right]-\left(\partial_\alpha\boldsymbol{\gamma}_\beta\right)\boldsymbol{\gamma}^\beta. \tag{5.93}$$

Since the left hand side of (5.93) is a 2-vector, we only need to retain the 2-vector terms on the right hand side. Thus

$$c_{JK}(\boldsymbol{\gamma}_\alpha)\mathbf{E}^J\mathbf{E}^K = \frac{1}{2}\boldsymbol{\gamma}^{\lambda\eta}\left[\frac{\partial g_{\alpha\lambda}}{\partial u^\eta}+\frac{\partial g_{\eta\lambda}}{\partial u^\alpha}-\frac{\partial g_{\alpha\eta}}{\partial u^\lambda}\right]-\left(\partial_\alpha\boldsymbol{\gamma}_\beta\right)\wedge\boldsymbol{\gamma}^\beta. \tag{5.94}$$

Since $\boldsymbol{\gamma}^{\lambda\eta}=-\boldsymbol{\gamma}^{\eta\lambda}$ *and* $\partial g_{\eta\lambda}/\partial u^\alpha=\partial g_{\lambda\eta}/\partial u^\alpha$, $\boldsymbol{\gamma}^{\lambda\eta}\partial g_{\eta\lambda}/\partial u^\alpha=0$, *(5.94) becomes*

$$c_{JK}(\boldsymbol{\gamma}_\alpha)\mathbf{E}^J\mathbf{E}^K = \frac{1}{2}\boldsymbol{\gamma}^{\lambda\eta}\left[\frac{\partial g_{\alpha\lambda}}{\partial u^\eta}-\frac{\partial g_{\alpha\eta}}{\partial u^\lambda}\right]+\boldsymbol{\gamma}^\beta\wedge\partial_\alpha\boldsymbol{\gamma}_\beta.$$

So finally, we have

$$\boldsymbol{\Gamma}_\alpha = \frac{1}{4}\boldsymbol{\gamma}^{\lambda\eta}\frac{\partial g_{\alpha\lambda}}{\partial u^\eta}+\frac{1}{4}\boldsymbol{\gamma}^\beta\wedge\partial_\alpha\boldsymbol{\gamma}_\beta. \tag{5.95}$$

If you review (5.77) and (5.78), you might suspect that the Fock–Ivanenko coefficients might depend on the choice of the orthonormal non-coordinate basis. This is indeed the case and it is reflected in the second term on the R.H.S. of (5.95). □

When the metric g_{jk} is diagonal, this formula for $\boldsymbol{\Gamma}_\alpha$ can be simplified further if we choose an orthonormal non-coordinate basis aligned with the coordinate basis. In particular, we can let

$$\boldsymbol{\gamma}_\beta = \mathbf{E}_B\left|\boldsymbol{\gamma}_\beta\right| = \mathbf{E}_B\left|g_{\beta\beta}\right|^{1/2}.$$

where $B=\beta,\ \beta=1,2,\ldots,$ or n and no sum is intended. Since it is also true that

$$\boldsymbol{\gamma}_\beta = g_{\beta\beta}\boldsymbol{\gamma}^\beta$$

with no sum intended, we have

$$\begin{aligned}\partial_\alpha\boldsymbol{\gamma}_\beta &= \mathbf{E}_B\partial_\alpha\left|g_{\beta\beta}\right|^{1/2} = \mathbf{E}_B\frac{1}{2}\left|g_{\beta\beta}\right|^{-1/2}\frac{\partial\left|g_{\beta\beta}\right|}{\partial u^\alpha}\\ &= \frac{1}{2}\boldsymbol{\gamma}^\beta\frac{g_{\beta\beta}}{\left|g_{\beta\beta}\right|}\frac{\partial\left|g_{\beta\beta}\right|}{\partial u^\alpha} = \frac{1}{2}\boldsymbol{\gamma}^\beta\frac{\partial g_{\beta\beta}}{\partial u^\alpha}.\end{aligned}$$

It then follows that

$$\gamma^{\beta} \wedge \partial_{\alpha}\gamma_{\beta} = \frac{1}{2}\gamma^{\beta} \wedge \gamma^{\beta}\frac{\partial g_{\beta\beta}}{\partial u^{\alpha}} = 0.$$

The first term on the right hand side of (5.89) simplifies more or less automatically. We merely note that in the summation over the λ index, $g_{\alpha\lambda} = 0$ unless $\lambda = \alpha$. Thus if g_{jk} is diagonal, we may write

$$\mathbf{\Gamma}_{\alpha} = \frac{1}{4}\gamma^{\alpha\eta}\frac{\partial g_{\alpha\alpha}}{\partial u^{\eta}}, \tag{5.96}$$

where the η index is summed but the α index is not.

Problem 93. Starting with (5.78) prove the following equation by induction:

$$\nabla_k \left(\mathbf{E}_{I_1}\mathbf{E}_{I_2}\cdots\mathbf{E}_{I_p}\right) = \mathbf{\Gamma}_k \left(\mathbf{E}_{I_1}\mathbf{E}_{I_2}\cdots\mathbf{E}_{I_p}\right) - \left(\mathbf{E}_{I_1}\mathbf{E}_{I_2}\cdots\mathbf{E}_{I_p}\right)\mathbf{\Gamma}_k.$$

Problem 94. Show that any p-vector other than a scalar or multiple of the pseudo vector $\mathbf{E}_{\bar{1}\bar{2}\cdots\bar{n}}$ (when n is odd) will fail to commute with at least one $\mathbf{E}_K$.

Problem 95. Show that if $\mathbf{\Gamma}_1$ and $\mathbf{\Gamma}_2$ are both 2-vectors then $\mathbf{\Gamma}_1\mathbf{\Gamma}_2 - \mathbf{\Gamma}_2\mathbf{\Gamma}_1$ is also a 2-vector. (Suggestion: Use an orthonormal basis or (4.58).)

Problem 96. Show that $\mathbf{v} = \langle \mathbf{v}, \gamma_k \rangle \gamma^k$.

Problem 97. Use the methods of this section to show that for the sphere ${R_{12}}^{12} = 1/r^2$.

Problem 98. TORUS Consider the torus. See Fig. 5.7. The surface is parameterized by the equations

$$\begin{aligned}
x^1(\phi,\theta) &= r\cos\phi = (R + a\cos\theta)\cos\phi,\\
x^2(\phi,\theta) &= r\sin\phi = (R + a\cos\theta)\sin\phi, \text{ and}\\
x^3(\phi,\theta) &= a\sin\theta.
\end{aligned}$$

Where $0 \le \phi < 2\pi$ and $-\pi < \theta \le \pi$. Use the methods of this section to show that

$${R_{12}}^{12} = \frac{\cos\theta}{a(R + a\cos\theta)}.$$

Note that the Gaussian curvature is positive on the outer portion of the surface and negative on the inner portion.

Problem 99. TORUS continued. See Fig. 5.8.

a) Consider the problem of parallel transporting the vector around the perimeter of a rectangular figure on the surface of a torus bounded by $\phi = \phi_0$,

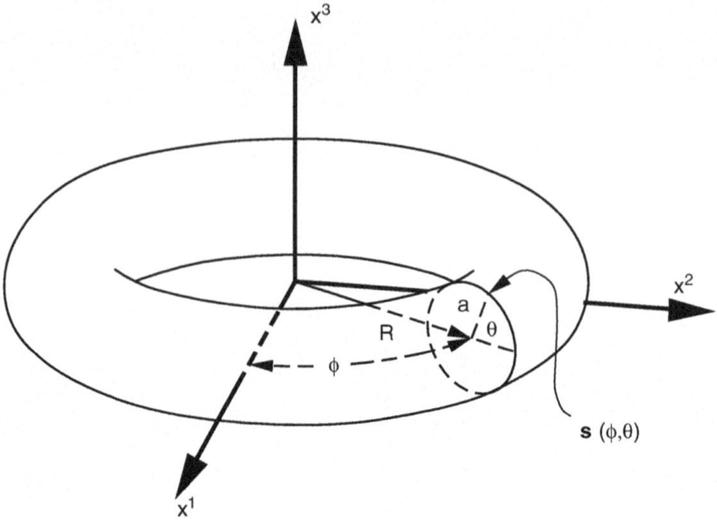

Fig. 5.7 Coordinate system for a torus: $\mathbf{s}(\phi, \theta) = \mathbf{e}_1(R+a\cos\theta)\cos\phi + \mathbf{e}_2(R+a\cos\theta)\sin\phi + \mathbf{e}_3 a\sin\theta$

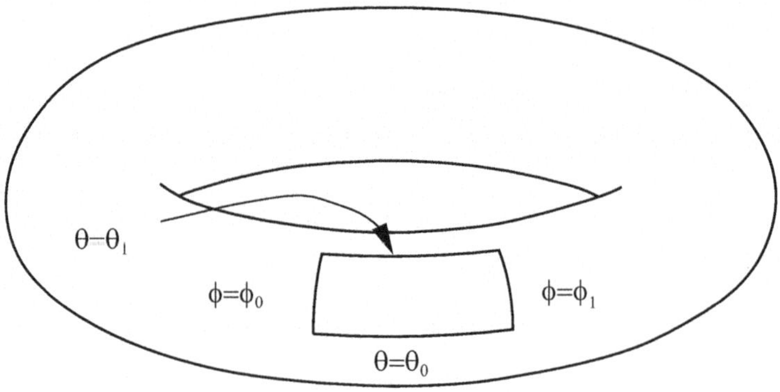

Fig. 5.8 Coordinate quadrilateral on surface of torus

$\phi = \phi_1$, $\theta = \theta_0$, and $\theta = \theta_1$ where $\phi_1 > \phi_0$ and $\theta_1 > \theta_0$. Suppose you start with an initial vector $\mathbf{v}_0 = \mathbf{E}_{\bar{1}} = \boldsymbol{\gamma}_\phi / \left|\boldsymbol{\gamma}_\phi\right|$ at $\phi = \phi_1$ and $\theta = \theta_0$. Parallel transport it along the path $\theta = s/a + \theta_0$ and $\phi = \phi_1$ until you arrive at $\theta = \theta_1$ and $\phi = \phi_1$. Then parallel transport it along the path $\phi = -s/(R + a\cos\theta) + \phi_1$ and $\theta = \theta_1$ until you arrive at $\phi = \phi_0$ and $\theta = \theta_1$. (At this point your parallel transported vector $\mathbf{v}_2$ should equal $\mathbf{E}_{\bar{1}}\cos([\phi_1 - \phi_2]\sin\theta_1) + \mathbf{E}_{\bar{2}}\sin([\phi_1 - \phi_2]\sin\theta_1)$.) Continuing the parallel transport along the remaining two edges, you should return to the original point with $\mathbf{v}_4 = \mathbf{E}_{\bar{1}}\cos([\phi_1 - \phi_2][\sin\theta_1 - \sin\theta_0]) + \mathbf{E}_{\bar{2}}\sin([\phi_1 - \phi_2][\sin\theta_1 - \sin\theta_0])$.

This implies that the parallel transported vector has undergone an apparent rotation through an angle equal to $[\phi_1 - \phi_2]\,[\sin\theta_1 - \sin\theta_0]$.

If the angles θ_0 and θ_1 are chosen so the four sided figure lies entirely in the region where the Gaussian curvature is positive, is the angle of rotation positive? What if θ_0 and θ_1 are chosen so the figure lies entirely in the region where the Gaussian curvature is negative? How can θ_0 and θ_1 be chosen with $\theta_1 \neq \theta_0$ so that the apparent rotation is zero? Does this make sense?

b) It is not hard to see that the area for the four sided figure described above is

$$\int_{\phi_0}^{\phi_1}\int_{\theta_0}^{\theta_1}(r d\phi) a d\theta = \int_{\phi_0}^{\phi_1}\int_{\theta_0}^{\theta_1} a(R + a\cos\theta) d\phi d\theta.$$

Compute this integral and then adjust the definition of (5.4) so that you can compute the Gaussian curvature using this integral and the result of part a. You should get the same result as you did for the computation for ${R_{12}}^{12}$ in Prob. 98.

Problem 100. Suppose the cone is parameterized by the equations

$$x^1(\phi, z) = az\cos\phi,$$
$$x^2(\phi, z) = az\sin\phi, \text{ and}$$
$$x^3(\phi, z) = az.$$

a) Compute the metric tensor.
b) Use (5.31) and the equation of Prob. 88 to compute the Gaussian curvature ${R_{12}}^{12}$.
c) Use the methods of this section to compute ${R_{12}}^{12}$. (To use (5.31) you may need to use the fact that $\boldsymbol{\gamma}^{12} = \mathbf{E}^{\bar{1}\bar{2}}\left|\boldsymbol{\gamma}^{12}\right|$.

Problem 101. ELLIPSOID (See Fig. 5.9. The problem of computing the Gaussian curvature for the ellipsoid from an intrinsic point of view is very difficult (at least for me). However the problem is manageable using an extrinsic approach. The standard equation for the ellipsoid is

$$\frac{x^2}{a^2} + \frac{y^2}{b^2} + \frac{z^2}{c^2} = 1. \tag{5.97}$$

Using a slightly modified version of spherical coordinates, we can write:

$$\mathbf{s} = \mathbf{e}_1 a\cos\phi\sin\theta + \mathbf{e}_2 b\sin\phi\sin\theta + \mathbf{e}_3 c\cos\theta.$$

a) Compute the Gaussian curvature. *Suggestions:* Let $u^1 = \phi$, $u^2 = \theta$, and $\mathbf{E}_{\bar{1}} = \boldsymbol{\gamma}_1/\left|\boldsymbol{\gamma}_1\right|$. (This expedites the computation of $c_{\bar{1}\bar{2}}(\boldsymbol{\gamma}_2)$.) Review the calculations for the saddle surface following Theorem 90 and the content of Prob. 91. Also review the calculations for the saddle surface following (5.84). The expression

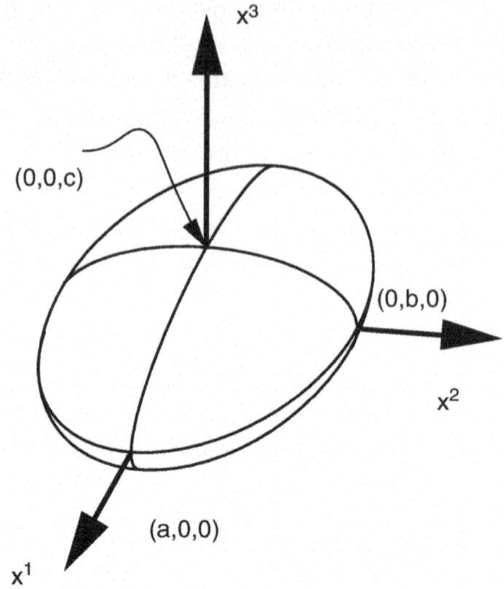

Fig. 5.9 Ellipsoid defined by the equation $\frac{x^2}{a^2}+\frac{y^2}{b^2}+\frac{z^2}{c^2}=1$

$(a^2\sin^2\phi + b^2\cos^2\phi)$ makes a frequent appearance so you can save some writing by replacing that expression by $f(\phi)$. Carrying out the computation, you should get

$$R^{12}{}_{12} = \frac{a^2b^2c^2}{\left[a^2b^2\cos^2\theta + c^2\sin^2\theta\left(a^2\sin^2\phi + b^2\cos^2\phi\right)\right]^2}.$$

This formula is not very enlightening but using the fact that $x = a\cos\phi\sin\theta$, $y = b\sin\phi\sin\theta$, and $z = c\cos\theta$, you will find that

$$R^{12}{}_{12} = \frac{a^2b^2c^2}{\left[\frac{b^2c^2}{a^2}x^2 + \frac{a^2c^2}{b^2}y^2 + \frac{a^2b^2}{c^2}z^2\right]^2}$$
$$= \left(\frac{1}{a^2b^2c^2}\right)\frac{1}{\left[\frac{x^2}{a^4} + \frac{y^2}{b^4} + \frac{z^2}{c^4}\right]^2}.$$

Are these formulas consistent with the Gaussian curvature for a sphere?

b) Suppose $a^2 > b^2 > c^2$. Determine the points on the ellipsoid where the Gaussian curvature is a maximum and where the Gaussian curvature is a minimum. (It may not be necessary but it is not cheating to use calculus.) Are your results consistent with your intuition?

Problem 102. *ONE SHEET HYPERBOLOID*

A one sheet hyperboloid can be represented by the equation

$$\frac{x^2}{a^2} + \frac{y^2}{b^2} - \frac{z^2}{c^2} = 1.$$

One way to parameterize this surface is

$$\mathbf{x}(u^1, u^2) = \mathbf{e}_1 a \cosh u^2 \cos u^1 + \mathbf{e}_2 b \cosh u^2 \sin u^1 + \mathbf{e}_3 c \sinh u^2.$$

Use Fock–Ivanenko coefficients to determine the Gaussian curvature. You should get

$$\boldsymbol{\Gamma}_1 = \frac{a^2 b^2 \sinh u^2}{2f(u^1)\left[a^2 b^2 \sinh^2 u^2 + c^2 f(u^1)\cosh^2 u^2\right]^{1/2}} \mathbf{E}^{\bar{1}}\mathbf{E}^{\bar{2}}, \text{ where}$$

$f(u^1) = a^2 \sin^2 u^1 + b^2 \cos^2 u^1$, $\boldsymbol{\Gamma}_2 = 0$, and

$$K = \frac{-a^2 b^2 c^2}{\left[a^2 b^2 \sinh^2 u^2 + c^2 \cosh^2 u^2 (a^2 \sin^2 u^1 + b^2 \cos^2 u^1)\right]^2}.$$

Or alternatively:

$$K = \frac{-1}{a^2 b^2 c^2 \left[\frac{x^2}{a^4} + \frac{y^2}{b^4} + \frac{z^2}{c^4}\right]^2}.$$

Problem 103. ELLIPTIC PARABOLOID An elliptic paraboloid can be represented by the equation:

$$z = \frac{x^2}{a^2} + \frac{y^2}{b^2}.$$

Show that the Gaussian curvature for the elliptic paraboloid may be written in the form

$$K = R_{12}{}^{12} = \frac{1}{4a^2 b^2 \left[\frac{x^2}{a^4} + \frac{y^2}{b^4} + \frac{1}{4}\right]^2}$$

Problem 104. HYPERBOLIC PARABOLOID A hyperbolic paraboloid can be represented by the equation:

$$z = \frac{x^2}{a^2} - \frac{y^2}{b^2}.$$

a) Show that the Gaussian curvature for the hyperbolic paraboloid may be written in the form:

$$K = R_{12}{}^{12} = \frac{-1}{4a^2 b^2 \left[\frac{x^2}{a^4} + \frac{y^2}{b^4} + \frac{1}{4}\right]^2} \tag{5.98}$$

b) If $a = b = \sqrt{2}$, (5.98) matches (5.88). Why should the Gaussian curvature for this special case of the hyperbolic paraboloid match the Gaussian curvature of the saddle surface?

5.6.3 *The Riemann Tensor for Orthonormal Frames

If you wish to make extensive computations in orthonormal frames, you should be aware of the fact that

$$R^A{}_{BPQ}\mathbf{E}_A \neq (\nabla_P\nabla_Q - \nabla_Q\nabla_P)\mathbf{E}_B.$$

Indeed

$$R^A{}_{BPQ}\mathbf{E}_A = (\nabla_P\nabla_Q - \nabla_Q\nabla_P - f^S{}_{PQ}\nabla_S)\mathbf{E}_B, \tag{5.99}$$

where $f^S{}_{PQ}$ is defined by the relation:

$$f^S{}_{PQ}\partial_S g = (\partial_P\partial_Q - \partial_Q\partial_P)g, \quad \text{where} \tag{5.100}$$

the ∂_A's are the moving frame derivatives defined in the previous section.

To make sense of this, we need to back up and introduce some definitions. The conversion of a coordinate basis to an orthonormal non-coordinate basis and the reverse conversion can be carried out using a matrix W_α^A and its inverse W_A^α. The W_α^A's and the W_A^α's can be defined by the equations

$$\mathbf{E}^A = W_\alpha^A \boldsymbol{\gamma}^\alpha, \quad \text{and} \tag{5.101}$$

$$\mathbf{E}_A = W_A^\alpha \boldsymbol{\gamma}_\alpha. \tag{5.102}$$

I leave it to you to show that (5.101) and (5.102) imply that

$$W_A^\alpha W_\alpha^B = \delta_A^B, \text{ and} \tag{5.103}$$

$$W_A^\alpha W_\beta^A = \delta_\beta^\alpha. \tag{5.104}$$

(See Prob. 106.)

In addition, you should be able to show that

$$\boldsymbol{\gamma}^\alpha = W_A^\alpha \mathbf{E}^A, \text{ and} \tag{5.105}$$

$$\boldsymbol{\gamma}_\alpha = W_\alpha^A \mathbf{E}_A. \tag{5.106}$$

(See Prob. 107.)

We are now in a position to state and prove a theorem:

Theorem 105. *If*

$$R^A{}_{BCD} = W_\alpha^A W_B^\beta W_C^\rho W_D^\eta R^\alpha{}_{\beta\rho\eta}, \textit{ and} \tag{5.107}$$

$$\nabla_A = W_A^\alpha \nabla_\alpha, \textit{ then} \tag{5.108}$$

$$R^A{}_{BPQ}\mathbf{E}_A = (\nabla_P\nabla_Q - \nabla_Q\nabla_P - f^S{}_{PQ}\nabla_S)\mathbf{E}_B, \textit{ where} \tag{5.109}$$

$f^S{}_{PQ}$ is defined by the equation:

$$f^S{}_{PQ}\partial_S g = (\partial_P\partial_Q - \partial_Q\partial_P)g, \tag{5.110}$$

where it is understood that g is an arbitrary scalar valued function of the coordinates.

Proof. The first step of this proof is to show that

$$W_B^\beta(\nabla_\rho\nabla_\eta - \nabla_\eta\nabla_\rho)\boldsymbol{\gamma}_\beta = (\nabla_\rho\nabla_\eta - \nabla_\eta\nabla_\rho)W_B^\beta\boldsymbol{\gamma}_\beta. \tag{5.111}$$

I leave this first step for you to do. (See Prob. 108.)

From this equation, we can infer that

$$W_B^\beta R^\alpha{}_{\beta\rho\eta}\boldsymbol{\gamma}_\alpha = (\nabla_\rho\nabla_\eta - \nabla_\eta\nabla_\rho)\mathbf{E}_B. \tag{5.112}$$

From (5.112), it follows that

$$\begin{aligned} W_B^\beta R^\alpha{}_{\beta\rho\eta}\boldsymbol{\gamma}_\alpha &= \left[(W_\rho^E\nabla_E)(W_\eta^F\nabla_F) - (W_\eta^F\nabla_F)(W_\rho^E\nabla_E)\right]\mathbf{E}_B \\ &= W_\rho^E W_\eta^F\left[\nabla_E\nabla_F - \nabla_F\nabla_E\right]\mathbf{E}_B \\ &\quad + W_\rho^E\left(\nabla_E W_\eta^F\right)\nabla_F\mathbf{E}_B - W_\eta^F\left(\nabla_F W_\rho^E\right)\nabla_E\mathbf{E}_B. \end{aligned}$$

Thus

$$\begin{aligned} W_B^\beta R^\alpha{}_{\beta\rho\eta}\boldsymbol{\gamma}_\alpha &= W_\rho^E W_\eta^F\left[\nabla_E\nabla_F - \nabla_F\nabla_E\right]\mathbf{E}_B \\ &\quad + W_\rho^E\left(\partial_E W_\eta^S\right)\nabla_S\mathbf{E}_B - W_\eta^F\left(\partial_F W_\rho^S\right)\nabla_S\mathbf{E}_B. \end{aligned} \tag{5.113}$$

We note that

$$\begin{aligned} \partial_E W_\eta^S &= \delta_\eta^\alpha\left(\partial_E W_\alpha^S\right) = W_\eta^F W_F^\alpha\left(\partial_E W_\alpha^S\right), \quad \text{and} \\ \partial_F W_\rho^S &= \delta_\rho^\alpha\left(\partial_F W_\alpha^S\right) = W_\rho^E W_E^\alpha\left(\partial_F W_\alpha^S\right). \end{aligned}$$

With these relations, (5.113) becomes

$$\begin{aligned} W_B^\beta R^\alpha{}_{\beta\rho\eta}\boldsymbol{\gamma}_\alpha &= W_\rho^E W_\eta^F\left[\nabla_E\nabla_F - \nabla_F\nabla_E\right]\mathbf{E}_B \\ &\quad + W_\rho^E W_\eta^F\left[W_F^\alpha\left(\partial_E W_\alpha^S\right) - W_E^\alpha\left(\partial_F W_\alpha^S\right)\right]\nabla_S\mathbf{E}_B, \end{aligned}$$

which means:

$$W_B^\beta R^\alpha{}_{\beta\rho\eta}\boldsymbol{\gamma}_\alpha = W_\rho^E W_\eta^F\left[\nabla_E\nabla_F - \nabla_F\nabla_E - f^S{}_{EF}\nabla_S\right]\mathbf{E}_B, \text{ where} \tag{5.114}$$

$$f^S{}_{EF} = -W_F^\alpha\left(\partial_E W_\alpha^S\right) + W_E^\alpha\left(\partial_F W_\alpha^S\right) \tag{5.115}$$

This last equation can be adjusted further. In particular

$$W_F^\alpha \left(\partial_E W_\alpha^S\right) = \partial_E \left(W_F^\alpha W_\alpha^S\right) - \left(\partial_E W_F^\alpha\right) W_\alpha^S$$
$$= -\left(\partial_E W_F^\alpha\right) W_\alpha^S.$$

Adjusting the second term on the R.H.S. of (5.115) in a similar manner, (5.115) becomes

$$f^S{}_{EF} = \left(\partial_E W_F^\alpha\right) W_\alpha^S - \left(\partial_F W_E^\alpha\right) W_\alpha^S$$
$$= \left[\left(\partial_E W_F^\alpha\right) - \left(\partial_F W_E^\alpha\right)\right] W_\alpha^S. \tag{5.116}$$

If we now prove that

$$\left(\partial_E \partial_F - \partial_F \partial_E\right) g = \left[\left(\partial_E W_F^\alpha\right) - \left(\partial_F W_E^\alpha\right)\right] W_\alpha^S \partial_S g,$$

we will know that

$$f^S{}_{EF} \partial_S g = \left(\partial_E \partial_F - \partial_F \partial_E\right) g$$

as previously claimed. At that point we will be near the end of this proof. We start by noting that

$$\left(\partial_E \partial_F - \partial_F \partial_E\right) g = \left[\partial_E W_F^\alpha \partial_\alpha - \partial_F W_E^\beta \partial_\beta\right] g$$
$$= \left[\left(\partial_E W_F^\alpha\right) \partial_\alpha - \left(\partial_F W_E^\beta\right) \partial_\beta\right] g$$
$$+ \left[W_F^\alpha \partial_E \partial_\alpha - W_E^\beta \partial_F \partial_\beta\right] g.$$

That is:

$$\left(\partial_E \partial_F - \partial_F \partial_E\right) g = \left[(\partial_E W_F^\alpha)\partial_\alpha - (\partial_F W_E^\alpha)\partial_\alpha\right] g$$
$$+ \left[W_F^\alpha W_E^\beta \partial_\beta \partial_\alpha - W_E^\beta W_F^\alpha \partial_\alpha \partial_\beta\right] g. \tag{5.117}$$

Since

$$\left(\partial_\alpha \partial_\beta - \partial_\beta \partial_\alpha\right) g = 0,$$

the second term on the R.H.S. of (5.117) is zero. Thus we have

$$\left(\partial_E \partial_F - \partial_F \partial_E\right) g = \left[\left(\partial_E W_F^\alpha\right) - \left(\partial_F W_E^\alpha\right)\right] \partial_\alpha g$$
$$= \left[\left(\partial_E W_F^\alpha\right) - \left(\partial_F W_E^\alpha\right)\right] W_\alpha^S \partial_S g$$

which was our desired result.

Multiplying both sides of (5.114) by $W^{\rho}_{C} W^{\eta}_{D}$ and then carrying out some summations, we have:

$$\begin{aligned} W^{\beta}_{B} W^{\rho}_{C} W^{\eta}_{D} R^{\alpha}{}_{\beta\rho\eta} \boldsymbol{\gamma}_{\alpha} &= \left(W^{\rho}_{C} W^{E}_{\rho}\right)\left(W^{\eta}_{D} W^{F}_{\eta}\right)\left[\nabla_E \nabla_F - \nabla_F \nabla_E - f^{S}{}_{EF} \nabla_S\right] \mathbf{E}_B \\ &= \delta^{E}_{C} \delta^{F}_{D} \left[\nabla_E \nabla_F - \nabla_F \nabla_E - f^{S}{}_{EF} \nabla_S\right] \mathbf{E}_B \\ &= \left[\nabla_C \nabla_D - \nabla_D \nabla_C - f^{S}{}_{CD} \nabla_S\right] \mathbf{E}_B, \text{ or restated:} \end{aligned}$$

$$R^{\alpha}{}_{BCD} \boldsymbol{\gamma}_{\alpha} = \left[\nabla_C \nabla_D - \nabla_D \nabla_C - f^{S}{}_{CD} \nabla_S\right] \mathbf{E}_B. \tag{5.118}$$

I will let you show that

$$R^{\alpha}{}_{BCD} \boldsymbol{\gamma}_{\alpha} = R^{A}{}_{BCD} \mathbf{E}_A.$$

(See Prob. 109.) This result combined with (5.118), completes the proof of our theorem. That is

$$\begin{aligned} R^{A}{}_{BCD} \mathbf{E}_A &= \left[\nabla_C \nabla_D - \nabla_D \nabla_C - f^{S}{}_{CD} \nabla_S\right] \mathbf{E}_B, \text{ where} \\ f^{S}{}_{CD} \partial_S g &= \left(\partial_C \partial_D - \partial_D \partial_C\right) g. \end{aligned}$$

□

Problem 106. Use (5.101) and (5.102) to prove (5.103) and (5.104).

Problem 107. Use (5.101), (5.102), (5.103), and (5.104) to prove (5.105) and (5.106).

Problem 108. Prove (5.111).

Problem 109. Show that

$$R^{\alpha}{}_{BCD} \boldsymbol{\gamma}_{\alpha} = R^{A}{}_{BCD} \mathbf{E}_A.$$

Problem 110. Show that $R^{AB}{}_{AB} = R^{\alpha\beta}{}_{\alpha\beta}$. Also show that for two dimensions, $R^{\bar{1}\bar{2}}{}_{\bar{1}\bar{2}} = R^{12}{}_{12}$.

5.7 *Doing Physics Under Stalin

The pervasive role of Communist ideology and the KGB (the secret police), in Stalin's Russia, had a huge impact on the lives of Fock and Ivanenko. (Note! I am using the convention of applying the label "KGB" for the secret police regardless of the time period. Properly "KGB" (Komitet Gosudarstvennoye Bezopasnosti) or (Committee for State Security) was the title for the secret police only between 1954 and 1991. These are roughly the years between the death of Stalin and the collapse of communist rule.)

During Stalin's rule, party ideologues would quote the writings of Marx, Engels, and Lenin to denounce both relativity and quantum mechanics. Einstein was attacked for using complex mathematics divorced from physical reality and turning physics into a branch of geometry (Vucinich 2001, pp. 22–23). Quantum mechanics was also attacked for its reliance on mathematical symbolism. Furthermore, Max Born's probabilistic interpretation of the wave function and Heisenberg's uncertainty principle seemed to be at odds with the deterministic outlook of Karl Marx.

In 1929, when Fock and Ivanenko introduced what later became known as the "Fock–Ivanenko coefficients", these party ideologues could be safely ignored. However a few years later, it became quite dangerous to express a dissident opinion. And when the Great Terror became most intense between 1936 and 1938, the decision to select someone for execution or a long sentence in a labor camp could depend on the whim of a KGB bureaucrat. In turn many of these KGB bureaucrats eventually became victims of this same wave of terror. To understand the stress imposed on Vladimir Fock and Dmitrii Ivanenko, it is useful to also examine the fates of three other Soviet physicists: George Gamow, Lev Landau, and Matvei Bronstein. Each of the five physicists dealt with Stalin's totalitarianism in his own way. The success of each approach was not entirely in the hands of the individual and the results were mixed.

In the late 1920s, Vladimir Fock was a young professor at the University of Leningrad and the other four were graduate students at the same university.

5.7.1 *George Gamow 1904–1968

George Gamow was born on March 4, 1904. In the summer of 1928, George Gamow was sent abroad to the University of Göttingen in Germany. While there, he discovered that he could use the newly developed theory of quantum mechanics to explain the emission of alpha particles from radioactive elements. This won him recognition both home and abroad. When he returned to Russia in the summer of 1929, he was hailed in *Pravda*, the official newspaper of the Communist Party. After a second stay abroad at Cambridge, England and Copenhagen, Denmark, he returned to Russia in the spring of 1931. He immediately discovered that the political atmosphere had now become quite oppressive. In the previous September, the Nazi party had achieved significant success in national elections. Russian fears stemming from the unfolding events in Germany were well founded. While serving his nine month prison sentence in 1924, Hitler authored his *Mein Kampf* (My Struggle). In that manifesto, he stated that Germany would never be a first rate power without obtaining control of more land. He went on to argue that Russia was the most logical place for Germany to acquire the needed land.

In Russia, there were opportunists who were all too eager to exploit this situation for their own ends. It now became easy to establish an official party line orthodoxy and then denounce those who might disagree with it. At the time of Gamow's return to Russia, ideologues, with encouragement of the Communist Party, were attacking physicists for being receptive to Einstein's etherless theory of relativity and Heisenberg's uncertainty principle. These theories were condemned for being capitalistic heresies that were un-Russian and anti-Marxist.

One of the milder ideologues, Boris Hessen, was responsible for an entry on ether in the 1931 edition of the *Greater Soviet Encyclopedia.* In response, Bronstein, Gamow, Ivanenko, and Landau along with two other young physicists, sent a sarcastic telegram to Hessen:

> Having read your entry on ether started our enthusiastic studies of it. Looking forward to reading about phlogiston. Bronstein, Gamow, Ivanenko, Iamailov, Landau, Chumbadze.
>
> (At one time, "phlogiston" was hypothesized to be a substance released during combustion. This concept was discarded near the end of the eighteenth century when it was discovered that an entity undergoing combustion combined with oxygen and actually *gained* weight.)

All six physicists soon received severe reprimands and both Bronstein and Landau were temporarily barred from lecturing at Polytechnic in Leningrad (Gorelik and Frenkel 1994, p. 51).

In this political atmosphere, obtaining a new passport was nearly impossible. In the summer of 1932, George Gamow and his bride tried to escape Russia by rowing a collapsible boat across the Black Sea to Turkey. On the second day of rowing, a severe storm arose and the consequent winds drove their boat back to the Russian shore.

Fortunately, with the assistance of Niels Bohr, Gamow and his wife were able to obtain passports the following year. Officially Gamow was being permitted to attend the 1933 Solvay Conference in Brussels. He did attend the conference but he never returned to Russia and emigrated to the U.S.

Like Einstein, Gamow could not get the required security clearance to work on the Manhattan project and became a consultant to the U.S. Navy during World War II. Later he did contribute to the development of the hydrogen bomb. From his own work, he became an early advocate of the Big Bang theory for the origin of the universe. He also became famous as a popularizer of science, being the author of many books including *Mr. Tompkins in Wonderland* (1937), *One Two Three – Infinity* (1947), and *Thirty Years that Shook Physics* (1966).

In 1953, after reading the famous paper by Francis Crick and James Watson describing the double helical structure of DNA, Gamow sent Crick a letter outlining a plausible code connecting the structure of DNA with the existence of 20 amino acids. This suggestion stimulated the research that did indeed reveal a code that was not too far off from the kind of code proposed by Gamow. George Gamow died on August 20, 1968.

5.7.2 *Lev Davidovich Landau 1908–1968

Lev Landau was born on January 22, 1908. He eventually became the most prominent of the five physicists. For many years the multivolume series of books organized by L.D. Landau and E.M. Lifshitz was required reading for physics students not only in Russia but also in the United States and other countries. Like Gamow, Landau enjoyed the opportunity to study in Western Europe before Stalin cut down on communication with other countries.

While in Cambridge on a Rockefeller grant, Landau was challenged by discussions he had with Pyotr Kapitza on an anomalous property of the electric conductivity of bismuth in a strong magnetic field (the Kapitza effect). As a result, Landau soon developed a theory of diamagnetism (Gorelik and Frenkel 1994, p. 38).

Unlike Gamow, Landau was a political radical (at least in his younger days) who enthusiastically supported the ideals of the communist revolution. (Most of what follows was extracted from Gennady Gorelik's article in *Scientific American* entitled "The Top-Secret Life of Lev Landau." (Gorelik 1997, pp. 72–77).

Despite his run-in with political authorities for being one of the signatories to the "ether-phlogiston" telegram in 1931, Landau became head of the theoretical division of the Ukrainian Institute of Physics and Technology at Kharkov in 1932. However, a few short years later he was in hot water again. In 1934, the Kharkov institute received a new director with orders to redirect research into applied areas directed toward military and other national priorities. Landau was never the diplomat and he alienated local authorities by trying to preserve a role for pure science.

In 1937, the KGB arrested several scientists working at Kharkov. Before being shot, a couple of Landau's friends "confessed" that Landau was head of some counterrevolutionary organization. Landau felt that he would be safer away from Kharkov and Pyotr Kapitza offered him a position in Moscow.

At the age of 27, Kapitza had emigrated to England in 1921 to work in the Cavendish laboratory with Ernest Rutherford. Initially, Rutherford did not want to employ Kapitza. He told Kapitza that with about 30 people around him, he had no openings. However when Kapitza responded by pointing out that one additional person would be within the usually accepted experimental error, Rutherford relented (Hargittai 2002, pp. 114–115). Kapitza soon achieved an international reputation for his investigations of strong magnetic fields. The Royal Society Mond Laboratory was built in Cambridge specifically for him and was officially opened in February 1933.

From time to time Kapitza returned to Russia to visit his family and maintain professional contacts. However when he did this in 1934, his passport was seized and he was denied permission to return to England by Stalin's order. This was despite the fact that by 1934, Kapitza had obtained dual citizenship in Great Britain and Russia. Kapitza then founded the Institute for Physical Problems in Moscow and the Soviet Government purchased the equipment of the Mond Laboratory with the cooperation of Rutherford.

Within a year of arriving in Moscow, Landau and two friends were arrested on April 28, 1938. According to KGB files made available to Gorelik, Landau was forced to stand for seven hours a day and threatened with transfer to a more oppressive prison. After two months, Landau broke down and wrote a six page confession. Landau signed an oath of secrecy on leaving prison and he never talked about his ordeal. Thus one cannot be sure how much truth can be attributed to this confession. Nonetheless the confession is not consistent with any of the usual KGB scripts. Furthermore it is consistent with Landau's political outlook. For these reasons, Gorelik has concluded that the confession must be taken seriously. According to the confession, Landau and one of the friends who was arrested with him planned to distribute anti-Stalin pamphlets at the 1938 May Day parade. An excerpt of the pamphlet reads as follows:

Comrades!

> The great cause of the October revolution has been evilly betrayed ... Millions of innocent people are thrown in prison, and no one knows when his own turn will be ...
>
> Stalin, with his rabid hatred of socialism, has become like Hitler and Mussolini ....The proletariat of our country that has overthrown the power of the tsar and the capitalists will be able to overthrow a fascist dictator and his clique. ...

It is amazing that Landau was not executed. Two factors saved him. One factor was the replacement of Nikolai Ivanovich Yezhov by Lavrenti Pavlovich Beria as head of the KGB late in 1938 bringing an end to the most extreme aspects of the great terror. Beria may have been as ruthless but he was much more rational and pragmatic. When Germany attacked Russia in 1941, Beria released about 140 former intelligence and security officers from jail to fill personnel slots in his organization (Sudoplatov 1995, p. 127). The second factor was Kapitza. By now Kapitza had invented a new technique for the production of oxygen – vital for metallurgy and therefore industry. For this reason, Kapitza was valued by the government and he used his status to appeal to Molotov to release Landau. Kapitza wrote that he had just made a discovery "in the most puzzling field of modern physics" and that no theorist other than Landau could explain it.

On the eve of May Day 1939, after one year of imprisonment, Landau was freed on bail. Landau was clearly grateful to Kapitza for Kapitza's display of courage on his behalf (Vucinich 1984, p. 174). Presumably, Landau was strongly motivated to give credibility to the claim that Kapitza had made to Molotov. At any rate, within a few months he constructed an explanation for Kapitza's superfluidity. For their work on the phenomenon, both Landau and Kapitza won Nobel Prizes (Landau in 1962 and Kapitza in 1978).

Shortly after Hiroshima, Beria was put in charge of the Russian atomic bomb project and Landau was recruited to participate. Despite his opposition to Stalin, Landau made significant contributions to the development of the bomb and continued to work on the bomb project until Stalin died in 1953. After Stalin died, Landau commented to a friend and pupil, Isaac M. Khalatnikov, "That's it. He's gone. I'm no longer afraid of him, and I won't work on (nuclear weapons) anymore." And he quit the bomb project.

In January 1962, the same year that Landau won the Nobel Prize, he was the victim of a car accident. Although he survived, he suffered brain damage that made it impossible for him to work effectively as a scientist. He died six years later on April 1, 1968.

5.7.3 *Matvei Petrovich Bronstein 1906–1938

Matvei Bronstein was born on December 2, 1906. In the short 31 years of his life, Matvei Petrovich Bronstein accomplished much. We can only speculate what he might have accomplished had he lived longer. According to his KGB file, Bronstein was in detention and undergoing interrogation on his 31st birthday. He had been arrested about four months earlier on August 6, 1937 and would be executed by a firing squad three months later on February 18, 1938 (Gorelik and Frenkel 1994, p. 145 and p. 153).

Matvei and his twin Isidor were only seven years old, when their father was drafted into the Czar's army in 1914 to participate in the war effort against the Germans. Without the father's income to pay for formal schooling, much of Matvei's and Isidor's early schooling occurred at home. They were supplied with good books and Matvei became a voracious reader. He eventually became conversant in Ukrainian, Russian, Georgian, English, French, German, Latin, Greek, Spanish, Hebrew, Turkish, and Japanese. Furthermore he could recite poetry in a good many of these languages (Gorelik and Frenkel 1994, p. 124).

In 1925, he joined a physics club at the local university in Kiev. Although Bronstein had no formal secondary education, the faculty advisor of the club, Petr Tartakovsky, soon recognized him as an exceptionally promising student. Presumably with Tartakovsky's encouragement, Bronstein applied to Russia's most prestigious university – Leningrad University. He entered Leningrad University in 1926. However during the previous year while still only 18 years old, Bronstein had three papers published. Two of these papers were published in *Zeitschrift für Physik,* which at the time was one of the world's most prestigious journals.

In 1929, he successfully attacked a problem in stellar atmospheres whose exact solution had eluded such outstanding physicists as Jeans, Eddington, and Milne (Gorelik and Frenkel 1994, p. 27). The result became known as the "Hopf–Bronstein correlation" (Chandrasekhar 1953, pp. 85 and 95). (Not much later, Hopf obtained the same result independently.)

During the early 1930s, Bronstein realized that a quantum gravity theory would have to be developed to deal with several cosmological problems. Such a theory has not yet been achieved but Bronstein was the first to obtain some useful results (Gorelik and Frenkel 1994, pp. 83–121).

As Stalin's Great Terror intensified, Matvei Bronstein found it increasingly difficult to advocate sanity in a world of ideological lunatics. During his student days at Leningrad University, he had begun writing popular science books to supplement his income (Gorelik and Frenkel 1994, p. 32). He enjoyed the opportunity to stimulate interest in science and continued writing for young and old. His last book was addressed to teenagers with the title *Inventors of Radiotelegraph.* In the spring of 1937, it had been accepted for publication, the editor had done his job, and the printers had done their job. The only task remaining was the binding.

At this point a ideological fanatic was appointed as a new director of the publishing house. The thrust of the book was that the invention of the radiotelegraph was the culmination of a long series of advances in both the theory and the applications of electromagnetism. In this context, it was not surprising that two men, Alexander Popov and Guglielmo Marconi, should independently succeed in achieving the last step in 1895. (Popov did his work in an academic setting and was not the entrepreneur and promoter that Marconi was. As a result, Popov did not receive much attention outside of Russia. However for many Russians, Popov is considered the "true inventor.")

For the new director, the thought that Popov was not the sole inventor of the radiotelegraph was unpatriotic. Presumably for the director, even the title of Bronstein's book was inflammatory. When Bronstein refused to make major changes, the director ordered that tens of thousands of copies ready to be bound should be destroyed. This occurred a few months before Bronstein's arrest (Gorelik and Frenkel 1994, pp. 192–193).

At about the same time Bronstein was waging a similar war on another front. Bronstein and Ivanenko had collaborated on the translation of the first edition of Dirac's *Quantum Mechanics* into Russian. The translation appeared in 1932 – two years after the publication date of the original English version. Presumably Bronstein and Ivanenko were provided with early copies of Dirac's second edition so they could rush a Russian edition into print more quickly. The publication date for the English edition is 1935 and in a letter dated April 21 of the same year, Bronstein informed Dirac that the Russian translation "will appear very soon, in two or three months." (Gorelik and Frenkel 1994, p. 60) However, this publication schedule was severely disrupted. In March of 1935 sometime before Bronstein's letter to Dirac, Ivanenko had been arrested by the KGB as the son of a noble landlord who served as a high ranking official in the Czar's government (Private e-mail communications: Gennady Gorelik June 14, 2002; Alexei Kojevnikov September 8, 2002: and Sergiu Vacaru August 9, 2002). Ivanenko was originally sent to a labor camp but in December he was exiled to Tomsk State University in Siberia.

Although Ivanenko had become a political pariah, Bronstein continued the collaboration. This was despite the fact that their collaboration on the first edition had not gone smoothly (Gorelik and Frenkel 1994, p. 40). On April 11, 1937, nearly two years after his letter to Dirac, Bronstein wrote to Fock:

> Today I signed Dirac's book to be sent to press. Unfortunately, this time I lost the battle I was waging for this book with the scoundrels in the publishing house. First, they insisted that Dymus' name be removed from the front page; to balance things I removed my name as well, though preserved it as the name of the editor. I have the right to do this since I corrected what Dymus had done. Second, they prefaced it with an indecent piece, explaining that Dirac is a villain (Gorelik and Frenkel 1994, p. 60). (Note! "Dymus" was an affectionate nickname for Dmitrii Ivanenko.)

On August 6, 1937, Matvei Bronstein was arrested. What was the reason for the arrest and execution of Bronstein? Matvei's widow, Lydia Korneevna Chukovskaya, once wrote a letter to *The New York Review* that was published on April 12, 1990:

> ...The human mind, unwilling to reconcile itself to the senseless, looks for *reasons* to explain every case. There was only one reason for the terror of 1937: quotas. The authorities

> set themselves the task of arresting a certain number of professors, teachers, dentists, chauffeurs, deaf and dumb people, workers, factory mangers, officers, musicians, actors, film directors, shoe shiners, and many others, and it didn't matter whom they chose in each category. Those who were arrested were tortured until they admitted they were members of counterrevolutionary organizations and denounced others whose names were provided by the investigator. Afterward they were either sent to a camp or shot. If they hadn't confessed they would have met the same fate. Why? To what end? Ask the executioners, I don't know what they'll answer. ...

In February 1938, Bronstein's wife was told that the verdict was ... *ten years of hard labor in a far-away concentration camp without the right to write and receive letters and with all his belongings confiscated.* She enlisted the aid of Fock. As soon as Fock had learned of Bronstein's arrest he showed up at his home to determine what had happened. This took a great deal of courage since he had himself been detained by the KGB twice ... for one day in 1935 and again for a week only six months earlier in February of 1937. The second time he was released only after Kapitza used his influence on Fock's behalf.

After Beria reined in the worst aspects of the Great Terror, Fock composed a letter in March 1939 to the USSR prosecutor Andrei Vishinsky supporting the request of Lydia Chukovskaya that Bronstein's case be reconsidered. This letter was signed by four other prominent physicists. Others also wrote letters on Bronstein's behalf but those efforts stopped in December of 1939 when Lydia learned that Matvei had died.

It was only 20 years later when Khruschev was ruler, that she learned that her efforts had been pointless. Matvei had been executed on February 18, 1938 ... roughly the same time she had been told that he was being sent off to a labor camp (Gorelik and Frenkel 1994, pp. 141, 146).

*5.7.4 *Vladimir Alexandrovich Fock 1898–1974*

Vladimir Alexandrovich Fock was born on December 22, 1898. Among chemists, Fock is best known for his contribution to the Hartree–Fock method for computing approximate wave functions for multi-electron systems (Fock 1930, pp. 126–148). Among physicists, Fock is known for much more. In 1929, Fock and Ivanenko introduced what are now known as the Fock–Ivanenko coefficients (Fock and Ivanenko 1929a, 1929a) also (Fock 1929). This was one of the earliest introduction of a gauge term in a wave equation.

The application of the Fock–Ivanenko coefficients to Dirac's wave equation for the electron makes it invariant under a change of coordinates in curved space. According to the theory of general relativity, this is what is needed to include the effect of gravity. As far as I know the addition of these terms have no measurable consequence because the force of gravity is extremely weak in this context. Nonetheless this approach became important in the development of quantum field theory. The use of gauge terms in wave equations for particles other than the electron eventually became the most effective mathematical method to describe the physical consequences of both the weak and the strong force.

Actually, the 1929 papers were not the first introduction of a gauge term in a quantum mechanical wave equation. The first example was also introduced by Fock. He did that in 1926 to modify Schrödinger's wave equation (Fock 1926, pp. 242–250). Fock obtained this result about a week after he read Schrödinger's first paper. Fock's name is also attached to Fock Space, the Fock representation, and Fock symmetry of the hydrogen atom. Many feel that his name should also be attached to the Klein–Gordon equation and Kaluza–Klein theories. Fock's *Foundations of Quantum Mechanics* was the first book on quantum mechanics written by a Russian. And his book: *The Theory of Space, Time, and Gravitation* was also highly regarded.

In the late 1940s and early 50s, many physicists working at research centers administered by the Soviet Academy of Sciences were besieged by ideological attacks from members of the physics department at Moscow State University. These ideologues hoped to exploit the cold war atmosphere to gain control of Russia's most prestigious research centers. Their assaults were usually composed of roughly equal parts of Marxist theology and an ignorance of modern physics. These assaults did not achieve their intended purpose because many of their intended victims were the same physicists who had developed Russia's atomic bomb and were then working on the hydrogen bomb.

Nonetheless, these ideologues had created an atmosphere in which it was difficult to teach and discuss either quantum mechanics or general relativity in an intelligent and coherent manner.

The mathematical structures introduced with both quantum mechanics and general relativity required difficult interpretations. Fock was an independent thinker who invested a great deal of time and thought on his own interpretations. It seems somewhat idiosyncratic that he insisted that his insight into these interpretations was aided by his study of Marxist theory. Actually his interpretation of the physics embedded in the mathematics was not substantially different from that of Max Born, Werner Heisenberg, and Niels Bohr. However his familiarity with Marxist theory enabled him to be the most effective defender of the integrity of Soviet physics against ideological attacks.

Surely Fock understood the risk of confronting these Marxist ideologues but it appears that he concluded that it was just as risky to ignore them. He was quoted to say, "Cowardice does not influence the probability of arrest." (Aleksandrov 1988, p. 489) During 1952, the last year of Stalin's rule, Fock decided it was time to launch a counter attack. In particular, he decided to make a direct response to some positions that had been taken by A.A. Maksimov in the journal: *Questions of Philosophy.*

Getting a response published in the same journal would not be easy since Maksimov was one of the editors. Fock first wrote a letter to Malenkov complaining that in the process of denying the validity of modern physical theories, Maksimov was slandering Marxist–Leninist philosophy (Pollock 2000, p. 234). (Malenkov would become Stalin's successor as premier one year later.) When this letter was ignored, Fock turned to Kurchatov who was the top science administrator of Russia's atomic bomb project under Beria. Kurchatov forwarded a copy of Fock's article to Beria with a strong endorsement along with a letter of support from eleven of the most important atomic physicists in Russia, including Landau and Andrei Sakharov (Pollock 2000, pp. 234–238).

In their letter, according to Ethan Pollock,

...The authors recognized that philosophers played an important role in the struggle between idealism and materialism, but complained that some philosophers who were ignorant of the foundations of physics were now engaging in attacks on quantum mechanics and relativity. These concepts, the authors emphasized, formed the basis of modern physics and the theoretical foundation for electronic and atomic technology.

They singled out Maksimov's article "Against Reactionary Einsteinians in Physics" as particularly dangerous and anti-scientific. Maksimov's criticisms of Einstein's theory were troublesome because, they claimed, it would be impossible to solve problems of elementary particle physics or atomic power without the use of the theory of relativity. To make matters worse, Maksimov's ignorance allowed him to attack quantum theory by labeling all modern physicists "idealists" (un-Marxist). Furthermore, the authors continued, articles by other philosophers in *Questions of Philosophy* and *The Literary Gazette* indicated that this ignorance was pervasive. ...(Pollock 2000, pp. 235–236).

At the beginning of 1953, after the turning of some bureaucratic wheels, Fock's article was published in *Questions of Philosophy* under the title: "Against Ignorant Criticisms of Modern Physical Theory." Understandably, this was over the objections of Maksimov.

Fock's article marked a turning point in the ideological battle. Stalin died on March 5, 1953 and after that the level of rancor subsided. Two years later, Fock was able to get his book, *The Theory of Space, Time, and Gravitation* published without fear of being subjected to ideological abuse.

Fock continued to be productive to the end of his life. In the final year of his life at age of 76, he nearly completed the second edition of his *Foundations of Quantum Mechanics*. He died on December 27, 1974 and the book was completed by others and appeared in print two years later.

*5.7.5 *Dmitrii Dmitrievich Ivanenko 1904–1994*

Dmitrii Dmitrievich Ivanenko was born on July 9, 1904. The evaluation of Dmitrii Ivanenko both as a physicist and as a person tends to go to extremes. At one extreme is the high regard of the prominent historian of Soviet science, Alexander Vucinich. Discussing some of the ideological divisions that were already occurring before the Great Terror, Vucinich wrote,

> The deep and irreconcilable differences between the two groups broke out in the open in 1928, when the young and fiery Ivanenko told the Sixth Congress of Soviet Physicists that the time had come for theoretical physics to replace philosophy. To be sure, this was a time of profound epistemological inquiries into the theories of quantum mechanics and relativity. But behind Ivanenko's statement was an open rebellion against the ambitious

efforts of Marxist philosophers to hold the upper hand in validating new physical thought. Ivanenko paid a heavy price for his critical comments: though one of the most cited Soviet physicists in Western scientific literature, he never became a member of the USSR Academy of Sciences (Vucinich 2001, p. 54).

Generally in the 1930s, Ivanenko was a respected physicist although Ivanenko's friendship with Landau ended in the early 30s for reasons that now seem to be unknown (Gorelik and Frenkel 1994, p. 54). Whatever the cause of Landau's dislike for Ivanenko, the depth of Landau's feelings were on public display for all to see. In 1956, Freeman Dyson became witness to Landau's bitterness. Under Krushchev, that year, there was enough of a thaw in the cold war that it became possible for American physicists to visit the Soviet Union. In the course of a visit to Moscow, Freeman Dyson met Ivanenko. He and Ivanenko were walking along an institute corridor and came across Landau. Ivanenko started to introduce Freeman but Landau turned around and walked away. Freeman spoke to Landau on other occasions in other settings and got along fine with him (Cherkis 2004, private communication).

As an aside, Sergei Cherkis took some math seminars under my supervision when he was an undergraduate. (He had very little choice since I was the only one teaching upper level math courses at Upsala College for several years before it became bankrupt and went out of business in 1995.) Sergei Cherkis shared an earlier version of "Doing Physics under Stalin" with Freeman Dyson and Freeman Dyson responded with the anecdote of Ivanenko's inability to introduce him to Landau.

Landau was not the only person that Ivanenko alienated. Many others would come to despise Ivanenko. For example, at an international conference on gravitation that occurred in Warsaw, Poland in July 1962, Fock displayed a very open contempt for Ivanenko. During World War I, Fock had served in an artillery unit. As a consequence, Fock suffered from a severe case of deafness. For this reason, he wore a hearing aid that required a large battery that was attached to a belt around his waist. When Ivanenko was about to begin a presentation at the 1962 conference, Fock shut off his hearing aid in a very theatrical manner for all to see (Engelbert Schücking 2002, 2003, Private Communications).

The suggestion of Vucinich that Ivanenko's 1928 speech cost Ivanenko membership in the USSR Academy of Science does not withstand scrutiny. Gamow was elected as a corresponding member in 1932 not long after the ether-phlogiston telegram. Fock was also elected as a corresponding member in 1932 and then elected as a full member in 1939 after he had been arrested twice. And in an unusual action, Landau was elected as a full member to the Academy in 1946 skipping the usual intermediate status as corresponding member.

To substantiate his claim that Ivanenko was "one of the most cited Soviet physicists in Western scientific literature", Vucinich refers his readers to the book, *Inward Bound,* written by the physicist Abraham Pais.

Much of Ivanenko's reputation stems from a paper he wrote in 1932 proposing a new model for the atomic nucleus (Ivanenko 1932, pp. 439–441). Before 1932, it was thought that nuclei were composed solely of protons and electrons. However this model conflicted with the laws of quantum mechanics in several ways. Early in 1932, James Chadwick discovered the neutron. This stimulated some new ideas.

Because of his 1932 paper, Ivanenko is sometimes credited as the first to propose the proton–neutron model of atomic nuclei. Actually the paper proposes that nuclei are composed mostly of alpha particles plus whatever number of protons and neutrons are necessary to give the correct atomic weight and atomic number. According to Pais, others were proposing the same model at the same time (Perrin 1932) and (Auger 1932).

Pais indicates that Ivanenko's real accomplishment was to point out that if the neutron was considered an elementary particle with spin 1/2 then his near proton–neutron model would give the correct spin for the Nitrogen nucleus. This was at a time when no one had proposed that the neutron was an elementary particle and the spin of the neutron was unknown (Pais 1986, pp. 409–411).

At this time, Ivanenko was respected on the world stage although he never had star status. During the late 1920s and early 30s, Ivanenko collaborated with the best Soviet physicists of his time. He collaborated not only with V.A. Fock and M.P. Bronstein but also with future Nobel prize winners L.D. Landau and I.Y.Tamm. Historian Alexei Kojevnikov, who has read the correspondence between Fock and Ivanenko, reports that Fock actively lined up academic positions for Ivanenko and continued to be supportive to him while Ivanenko was in exile (Kojevnikov Sept. 6, 2002, Private Communication).

How did Ivanenko become the object of Fock's contempt? At age 86, Vitaly Ginzburg is one of the few people who can throw some light on this question at this late date. In his book, *The Physics of a Lifetime,* Ginzburg discusses combative encounters with Richard Feynman and Lev Landau with good humor (Ginzburg 2001, pp. 369–370 and pp. 446–447). However he is reluctant to discuss Ivanenko who evokes bad memories.

In 1942, Ivanenko's exile ended and he obtained a position at Moscow State University. Ginzburg reports that it was the perception of many in the physics community that the KGB had persuaded Ivanenko to spy on them (Ginzburg 2003, Private Communication). This is plausible since the KGB frequently recruited informers from their detainees. On the other hand, according to historian Gennady Gorelik, there is no documentary support for this charge and Ivanenko's KGB file is only available to Ivanenko's closest relatives. Defenders of Ivanenko could point out that it was also the practice of the KGB to arrange things so that people who did not know one another from distant regions would be brought together at possible centers of dissent to create an atmosphere of mutual distrust.

At any rate, Ivanenko was no longer the rebel who had attacked the Marxist philosophers in 1928. He did not join with his Moscow State University colleagues in their bizarre attacks against relativity and quantum mechanics. However he did advocate Bohm's intellectually respectable but politically safer alternative to Bohr's interpretation of quantum mechanics (Vucinich 1984, p. 334). More importantly, he enthusiastically joined his university colleagues in a campaign to emphasize the accomplishments of Russian physicists, which would include himself.

The Marxist philosopher B.M. Kedrov published his *Engels and Natural Science* in 1946 and Ivanenko attacked him for not giving broader treatment to Russian contributions (Vucinich 1984, pp. 232–233).

In the fall of 1947, Ginzburg was eagerly expecting to receive the title of professor. At about this same time, Trofim Denisovich Lysenko was making his final move to gain control of all of Russia's agricultural research centers. Lysenko was a charlatan who was able to persuade many party officials including Stalin and later Krushchev that a man with the field experience of a Russian peasant could outperform academically trained geneticists whose minds were polluted by western ideas. His most prestigious opponent Nikolay Ivanovich Vavilov had died in a Soviet prison from malnutrition in 1942.

On October 4, an article appeared in the *Literaturnaya Gazeta (The Literary Gazette)* lambasting the foes of Lysenko. Although it was not completely obvious at the time, it was a list of prominent geneticists who would soon be losing positions for being "cosmopolitan" and advocates of Mendel's gene theory. (The epithet "cosmopolitan" was applied to those who were receptive to "bourgeois" ideas from the west. If the object of the epithet was Jewish there was also the implication that the person was more loyal to Israel than to Mother Russia. It should be noted that Jews in Soviet Russia were officially treated as a separate nationality like Ukrainians or Georgians.)

To Ginzburg's horror, he discovered that his name was added to the list of geneticists who should be condemned for their cosmopolitan behavior. For contrast, the author mentioned Ivanenko as an exemplar to be admired ... a true Russian. On the same day, Ginzburg was notified that he would not be promoted to professor because he was too "cosmopolitan." Since Ivanenko was on the board that made this decision and since the author of the article in *The Literary Gazette* would not normally be aware of Ginzburg's writings, Ginzburg had good reason to believe that Ivanenko was instrumental in both events. Ginzburg remembers the date quite well because it was his 31st birthday! (Ginzburg 2003, Private Communication). (I obtained this story from Prof. Ginzburg through an exchange of e-mail messages and a three way phone conversation for which Sasha Rozenberg acted as translator. A few months later, in October 2003, Prof. Ginzburg was awarded a Nobel Prize in physics.)

(If Ginzburg had been born about ten years earlier, things might have been worse. If my arithmetic is correct, Bronstein, Ivanenko, and Landau all observed their 31st birthdays either in a KGB prison or in a KGB labor camp.)

During the following year in 1948, G.S. Landsberg accused Ivanenko of making citation of his work and that of his students the touchstone of a Soviet physicist's patriotism (Holloway 1994, p. 53).

Also in 1948, Fock was asked to write an official review of a manuscript written by Ivanenko and Sokolov. The topic was quantum gravity. It had been written the year before and was now being submitted for the Stalin Prize. Fock wrote, "Whatever causes compelled the authors to avoid mentioning Bronstein's achievements, their word may not be considered as the construction of the quantum theory of gravitation, for this theory was created by Bronstein 11 years earlier." (Gorelik 1993, p. 313).

Failing to cite Bronstein because he was still "an enemy of the people" would not have excused Ivanenko in the eyes of Fock. Fock knew how Bronstein had battled with the publishers of the translation of Dirac's second edition of *Quantum Mechanics* to give Ivanenko due recognition when Ivanenko was in exile.

Ivanenko's nationalistic attacks on his brother physicists alienated many and he paid for it. During the 30's, Ivanenko could take pride in the fact that he had coauthored some significant papers. However his new enemies were eager to point out that he had published very little without a coauthor. Questions were raised about how much he had actually contributed to those coauthored papers. Furthermore it was recognized that his grasp of mathematics was not as strong as one would normally expect of a theoretical physicist. All this was reflected in a humiliating encounter Ivanenko had with Richard Feynman at the same 1962 conference in Warsaw where Fock shut off his hearing aid.

According to Engelbert Schücking, Ginzburg gave a talk, which was parallel translated by one of Infeld's students. Ginzburg became unhappy with the translation and continued his talk in English to the chagrin of his fellow Russians. After the talk, Ivanenko commented to Feynman that Ginzburg was "like a child."

Feynman who had a high regard for Ginzburg lashed back,

"What have you ever done in physics, Ivanenko?

"I've written a book with Sokolov."

"How do I know what you contributed to it? Ivanenko, what is the integral of e to minus x squared from minus infinity to plus infinity?"

Silence

"Ivanenko, what is one and one?"

The following day, Schücking saw Ivanenko giving Feynman a paper of his and Feynman was rather apologetic saying that he was unable to read Russian. Nonetheless, it is clear that Feynman did not have a high regard for Ivanenko. Actually, Feynman took a dim view of the whole conference. He told Schücking over a drink, "These people are like worms in a bottle crawling over each other."

Fock consistently battled for his students when they encountered difficulties with the authorities. How would Ivanenko react in a similar situation? Ivanenko was given the opportunity to intervene on behalf of one of his students in 1987.

Sergiu Vacaru was doing graduate work from 1984 to 1987 in Moscow under the official supervision of Ivanenko. (Perhaps it should be noted that someone else was overseeing the actual writing of Vacaru's Ph.D. thesis.) For much of this period, Yeltsin was leader of the Communist Party in Moscow and Vacaru along with other students were encouraging him in his efforts to carry out various reforms. However in 1987, the reformers suffered a setback. Yeltsin was ousted and Vacaru along with many others was arrested.

After one month, Sergiu was released from prison but he had become politically undesirable. He had published over 30 papers and had completed all academic requirements for a doctorate. However he soon found that he could not get the required endorsements from communist party functionaries to receive his degree. As far as Sergiu knows, Ivanenko never lifted a finger on his behalf. At any rate, Vacaru had to go elsewhere to get his Ph.D (Vacaru 2002, Private Conversation).

Perhaps it is easy for someone who has lived a painless existence to condemn Ivanenko. However it appears that many who were living under the same oppressive government as Ivanenko felt that he could have been a better person. Dmitrii Ivanenko died on December 30, 1994.

Problem 111. Compute

$$I = \int_{-\infty}^{+\infty} \exp(-x^2)dx.$$

Hint

$$I^2 = \int_{-\infty}^{+\infty} \int_{-\infty}^{+\infty} \exp(-\left(x^2 + y^2\right))dxdy.$$

This integral can be determined by using polar coordinates.

Chapter 6
The Gauss–Bonnet Formula

6.1 The Exterior Derivative and Stokes' Theorem

We can define the *exterior derivative* **d** by the equation:

$$\mathbf{dA} = \boldsymbol{\gamma}^k \wedge \nabla_k \mathbf{A}, \tag{6.1}$$

where **A** is any differentiable Clifford number that is not necessarily index free. For example,

$$\mathbf{d}u^j = \boldsymbol{\gamma}^k \frac{\partial u^j}{\partial u^k} = \boldsymbol{\gamma}^k \delta_k^j = \boldsymbol{\gamma}^j. \tag{6.2}$$

In the context of differential forms, we would write

$$\mathbf{d}u^1 \wedge \mathbf{d}u^2 \wedge \ldots \wedge \mathbf{d}u^p = \mathbf{d}u^1\mathbf{d}u^2 \ldots \mathbf{d}u^p \text{ in place of}$$
$$\boldsymbol{\gamma}^1 \wedge \boldsymbol{\gamma}^2 \wedge \ldots \boldsymbol{\gamma}^p = \boldsymbol{\gamma}^{12\ldots p}$$

and call the resulting product a *p-form*. In this book, I have adopted some of the terminology from the formalism of differential forms because many of the results of differential forms carry over into Clifford algebra in a virtual isomorphic form. One significant difference is that in the formalism of differential forms, 1-forms and tangent vectors span distinct spaces. The analogs in Clifford algebra (the upper and lower index Dirac vectors) span the same space.

The formalism of differential forms is particularly appropriate when one wants to investigate problems without introducing a metric tensor. The definition that I have used to define the exterior derivative in (6.1) depends on the existence of a metric tensor. However, had I restricted the definition of the exterior derivative to Clifford numbers written in the form

$$\mathbf{A} = A_{k_1 k_2 \ldots k_p} \boldsymbol{\gamma}^{k_1 k_2 \ldots k_p},$$

J. Snygg, *A New Approach to Differential Geometry using Clifford's Geometric Algebra*, DOI 10.1007/978-0-8176-8283-5_6,

I could have used a definition that does not depend on the metric. Namely:

$$\mathbf{dA} = \frac{\partial A_{k_1 k_2 \ldots k_p}}{\partial u^j} \boldsymbol{\gamma}^{j k_1 k_2 \ldots k_p}.$$

To see this, we note that

$$\begin{aligned}\nabla_j \left(\boldsymbol{\gamma}^{k_1} \boldsymbol{\gamma}^{k_2} \cdots \boldsymbol{\gamma}^{k_p}\right) = &\left(\nabla_j \boldsymbol{\gamma}^{k_1}\right) \boldsymbol{\gamma}^{k_2} \cdots \boldsymbol{\gamma}^{k_p} \\ &+ \boldsymbol{\gamma}^{k_1} \left(\nabla_j \boldsymbol{\gamma}^{k_2}\right) \boldsymbol{\gamma}^{k_3} \ldots \boldsymbol{\gamma}^{k_p} \\ &+ \boldsymbol{\gamma}^{k_1} \boldsymbol{\gamma}^{k_2} \left(\nabla_j \boldsymbol{\gamma}^{k_3}\right) \boldsymbol{\gamma}^{k_4} \cdots \boldsymbol{\gamma}^{k_p} \\ &+ \cdots + \boldsymbol{\gamma}^{k_1} \boldsymbol{\gamma}^{k_2} \cdots \boldsymbol{\gamma}^{k_{p-1}} \left(\nabla_j \boldsymbol{\gamma}^{k_p}\right).\end{aligned}$$

Then

$$\begin{aligned}\boldsymbol{\gamma}^j \wedge \nabla_j \left(\boldsymbol{\gamma}^{k_1} \boldsymbol{\gamma}^{k_2} \ldots \boldsymbol{\gamma}^{k_p}\right) = &- \Gamma_{\eta j}{}^{k_1} \boldsymbol{\gamma}^j \wedge \boldsymbol{\gamma}^\eta \boldsymbol{\gamma}^{k_2} \boldsymbol{\gamma}^{k_3} \cdots \boldsymbol{\gamma}^{k_p} \\ &- \Gamma_{\eta j}{}^{k_2} \boldsymbol{\gamma}^j \wedge \boldsymbol{\gamma}^{k_1} \boldsymbol{\gamma}^\eta \boldsymbol{\gamma}^{k_3} \cdots \boldsymbol{\gamma}^{k_p} \\ &- \Gamma_{\eta j}{}^{k_3} \boldsymbol{\gamma}^j \wedge \boldsymbol{\gamma}^{k_1} \boldsymbol{\gamma}^{k_2} \boldsymbol{\gamma}^\eta \boldsymbol{\gamma}^{k_4} \ldots \boldsymbol{\gamma}^{k_p} \\ &+ \cdots - \Gamma_{\eta j}{}^{k_p} \boldsymbol{\gamma}^j \wedge \boldsymbol{\gamma}^{k_1} \boldsymbol{\gamma}^{k_2} \ldots \boldsymbol{\gamma}^{k_{p-1}} \boldsymbol{\gamma}^\eta.\end{aligned}$$

This of course implies that

$$\begin{aligned}\boldsymbol{\gamma}^j \wedge \nabla_j \left(\boldsymbol{\gamma}^{k_1 k_2 \ldots k_p}\right) = &-\Gamma_{\eta j}{}^{k_1} \boldsymbol{\gamma}^{j \eta k_2 k_3 \ldots k_p} - \Gamma_{\eta j}{}^{k_2} \boldsymbol{\gamma}^{j k_1 \eta k_3 k_4 \ldots k_p} \\ &- \Gamma_{\eta j}{}^{k_3} \boldsymbol{\gamma}^{j k_1 k_2 \eta k_4 \ldots k_p} + \cdots - \Gamma_{\eta j}{}^{k_p} \boldsymbol{\gamma}^{j k_1 k_2 \ldots k_{p-1} \eta}. \end{aligned} \tag{6.3}$$

It turns out that each term on the right-hand side of (6.3) is zero. For example, consider the second term. The exchange of two numbers in a sequence is an odd permutation, so

$$\boldsymbol{\gamma}^{j k_1 \eta k_3 k_4 \ldots k_p} = -\boldsymbol{\gamma}^{\eta k_1 j k_3 k_4 \ldots k_p}.$$

On the other hand, the Christoffel symbol is symmetric with respect to its two lower indices, so

$$\Gamma_{\eta j}{}^{k_2} \boldsymbol{\gamma}^{j k_1 \eta k_3 k_4 \ldots k_p} = -\Gamma_{j \eta}{}^{k_2} \boldsymbol{\gamma}^{\eta k_1 j k_3 k_4 \ldots k_p}. \tag{6.4}$$

However j and η are dummy indices that can be replaced, respectively, by η and j. If we do this on the right-hand side of (6.4), we have

$$\Gamma_{\eta j}{}^{k_2} \boldsymbol{\gamma}^{j k_1 \eta k_3 k_4 \ldots k_p} = -\Gamma_{\eta j}{}^{k_2} \boldsymbol{\gamma}^{j k_1 \eta k_3 k_4 \ldots k_p}. \tag{6.5}$$

In general, $x = -x$ implies that $x = 0$. Thus, (6.5), implies that

$$\Gamma_{\eta j}{}^{k_2} \boldsymbol{\gamma}^{j k_1 \eta k_3 k_4 \ldots k_p} = 0.$$

Essentially, the same argument applies to any term on the right-hand side of (6.3), so we have

$$\boldsymbol{\gamma}^j \wedge \nabla_j \left(\boldsymbol{\gamma}^{k_1k_2\ldots k_p}\right) = \mathbf{d}\boldsymbol{\gamma}^{k_1k_2\ldots k_p} = 0. \tag{6.6}$$

Now suppose

$$\mathbf{F} = \frac{1}{p!} F_{k_1k_2\ldots k_p} \boldsymbol{\gamma}^{k_1k_2\ldots k_p},$$

then

$$\begin{aligned}\mathbf{dF} &= \boldsymbol{\gamma}^j \wedge \nabla_j \mathbf{F}\\ &= \frac{1}{p!}\left(\frac{\partial}{\partial u^j} F_{k_1k_2\ldots k_p}\right)\boldsymbol{\gamma}^{jk_1k_2\cdots k_p} + \frac{1}{p!} F_{k_1k_2\ldots k_p}\boldsymbol{\gamma}^j \wedge \nabla_j \left(\boldsymbol{\gamma}^{k_1k_2\ldots k_p}\right).\end{aligned}$$

Since the last term is zero, we have

$$\begin{aligned}\mathbf{dF} &= \boldsymbol{\gamma}^j \wedge \nabla_j \mathbf{F} = \frac{1}{p!}\left(\frac{\partial}{\partial u^j} F_{k_1k_2\ldots k_p}\right)\boldsymbol{\gamma}^{jk_1k_2\ldots k_p}, \text{ where}\\ \mathbf{F} &= \frac{1}{p!} F_{k_1k_2\ldots k_p}\boldsymbol{\gamma}^{k_1k_2\ldots k_p}.\end{aligned} \tag{6.7}$$

From (6.7), we see that for a restricted class of Clifford numbers, the exterior derivative operator **d** is meaningful in the absence of a metric. The intrinsic operator definition, that I have used, will allow us to apply **d** to all differentiable Clifford numbers, when the metric is assumed to exist.

A very important and useful result that flows from the formalism of differential forms is a generalized version of Stokes' Theorem:

Theorem 112. *Stokes' Theorem*
Suppose

$$\mathbf{F} = \sum_{k=1}^{n} F_{12\ldots\hat{k}\ldots n}\boldsymbol{\gamma}^{12\ldots\hat{k}\ldots n} = \sum_{k=1}^{n} F_{12\ldots\hat{k}\ldots n}\mathbf{d}u^1\mathbf{d}u^2\ldots\mathbf{d}\hat{u}^k\ldots\mathbf{d}u^n.$$

Then

$$\int_V \mathbf{dF} = \int_{\partial V}\mathbf{F}, \text{ or rewritten, we have} \tag{6.8}$$

$$\begin{aligned}&\int_V \sum_{k=1}^{n}(-1)^{k-1}\frac{\partial}{\partial u^k}F_{12\ldots\hat{k}\ldots n}\mathbf{d}u^1\mathbf{d}u^2\ldots\mathbf{d}u^k\ldots\mathbf{d}u^n\\ &= \int_{\partial V}\sum_{k=1}^{n}F_{12\ldots\hat{k}\ldots n}\mathbf{d}u^1\mathbf{d}u^2\ldots\widehat{\mathbf{d}u^k}\cdots\mathbf{d}u^n.\end{aligned}$$

It should be noted that the circumflex $\wedge$ *is used to indicate an omitted index or other entity. Furthermore,*

$$\begin{aligned}\mathbf{dF} &= \sum_{k=1}^{n} \frac{\partial}{\partial u^k} F_{12\cdots\hat{k}\ldots n} \mathbf{du}^k \mathbf{du}^1 \mathbf{du}^2 \ldots \widehat{\mathbf{du}^k} \ldots \mathbf{du}^n \\ &= \sum_{k=1}^{n} (-1)^{k-1} \frac{\partial}{\partial u^k} F_{12\ldots\hat{k}\ldots n} \mathbf{du}^1 \mathbf{du}^2 \ldots \mathbf{du}^k \ldots \mathbf{du}^n.\end{aligned}$$

The symbol V *designates an* n*-dimensional bounded region while the symbol* ∂V *designates the* $(n-1)$*-dimensional boundary of that same region. The boundary is assumed to have a finite* $(n-1)$*-dimension volume. To compute the integrals, we replace each* $\mathbf{du}^k$ *by* du^k *after ordering the* $\mathbf{du}^k$*'s in increasing order. It is difficult to state the most general conditions for which the theorem is true. Clearly, the two integrals in (6.8) have to exist. The theorem is relatively easy to prove if we require that every coordinate curve passes through the region intersecting the boundary twice, touches the region tangentially (possibly along a connected interval), or misses the region altogether. (This is a slight generalization of an alternate requirement that the region be compact and convex.) One can then generalize the regions for which the theorem is true by fitting together regions of the type just described. This can be done because the contributions of shared boundaries will cancel out. I will prove the theorem below only for 2-dimensional volumes because that is the only case we need in this text. For higher dimensions, see (Flanders 1963) or (Snygg 1997).*

Proof. (2-dimensions only)

What we wish to prove is

$$\int_V \left(\frac{\partial F_2}{\partial u^1} - \frac{\partial F_1}{\partial u^2}\right) du^1 du^2 = \int_{\partial V} \left(F_1 du^1 + F_2 du^2\right). \tag{6.9}$$

To prove (6.9), we merely carry out the obvious integrations on the left-hand side. (Refer to Fig. 6.1.)

$$\begin{aligned}\int_V \frac{\partial F_2(u^1, u^2)}{\partial u^1} du^1 du^2 &= \int_{u^2_{MIN}}^{u^2_{MAX}} F_2(u^1_{MAX}(u^2), u^2) du^2 \\ &\quad - \int_{u^2_{MIN}}^{u^2_{MAX}} F_2(u^1_{MIN}(u^2), u^2) du^2 \\ &= \int_{ABC} F_2 du^2 - \int_{EDC} F_2 du^2 \\ &= \int_{ABC} F_2 du^2 + \int_{CDE} F_2 du^2 \\ &= \int_{ABCDE} F_2 du^2.\end{aligned} \tag{6.10}$$

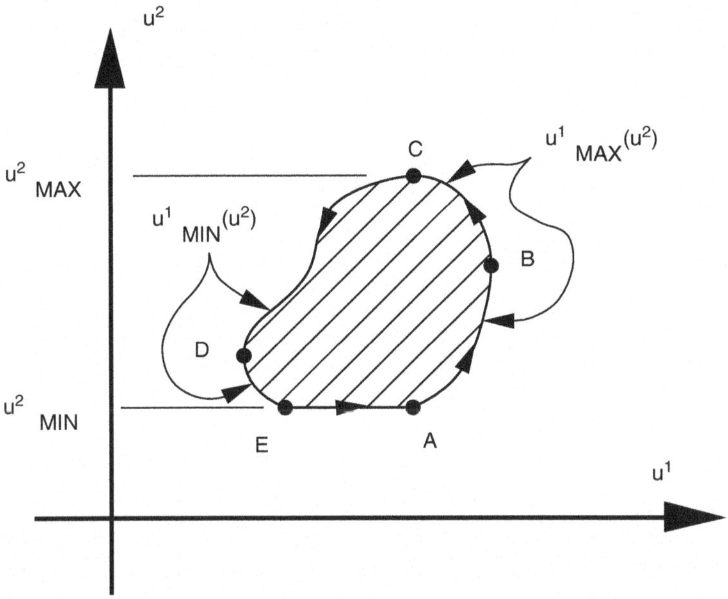

Fig. 6.1 $u^1_{MAX}(u^2) =$ path ABC and $u^1_{MIN}(u^2) =$ path EDC

We note that

$$\int_{EA} F_2 du^2 = \int_{u^2_{MIN}}^{u^2_{MIN}} F_2(u^1(u^2), u^2) du^2. \tag{6.11}$$

The integrand on the right-hand side of (6.11) is not well defined as a function of u^2 since u^1 is not really a function of u^2. Nevertheless, the interval of integration has length zero so the integral is zero. Combining this result with (6.10) gives us

$$\begin{aligned}\int_V \frac{\partial F_2(u^1, u^2)}{\partial u^1} du^1 du^2 &= \int_{ABCDE} F_2 du^2 + \int_{EA} F_2 du^2 \\ &= \int_{ABCDEA} F_2 du^2 = \int_{\partial V} F_2 du^2.\end{aligned} \tag{6.12}$$

Similarly,

$$\begin{aligned}-\int_V \frac{\partial F_1}{\partial u^2} du^1 du^2 = &-\int_{u^1_{MIN}}^{u^1_{MAX}} F_1(u^1, u^2_{MAX}(u^1)) du^1 \\ &+ \int_{u^1_{MIN}}^{u^1_{MAX}} F_1(u^1, u^2_{MIN}(u^1)) du^1\end{aligned}$$

$$= -\int_{DCB} F_1 du^1 + \int_{DEAB} F_1 du^1$$
$$= \int_{BCD} F_1 du^1 + \int_{DEAB} F_1 du^1$$
$$= \int_{\partial V} F_1 du^1 .$$

Combining this result with (6.12), we get our desired result:

$$\int_V \left(\frac{\partial F_2}{\partial u^1} - \frac{\partial F_1}{\partial u^2} \right) du^1 du^2 = \int_{\partial V} \left(F_1 du^1 + F_2 du^2 \right).$$

□

Problem 113. Suppose $\mathbf{F} = \frac{1}{p!} F_{i_1 i_2 \ldots i_p} \gamma^{i_1 i_2 \ldots i_p}$. Show $\mathbf{ddF} = 0$. Does this generalize to other Clifford numbers? For example, if $\mathbf{E}^{\bar{k}}$ is a member of a noncoordinate orthonormal basis, is it true that $\mathbf{ddE}^{\bar{k}} = 0$? These relations are known as the Poincaré lemma by some authors and the converse of the Poincaré lemma by others. (It is the converse of a less trivial lemma.)

Problem 114. Find an example where $\mathbf{d}\boldsymbol{\gamma}_k \neq 0$.

Problem 115. The proof I gave for the 2-dimensional version of Stokes' Theorem applies only to very simply shaped regions described immediately preceding my proof. However, Stokes' Theorem can easily be extended to a wide variety of regions by fitting such simple regions together. This is because the path integrals for the shared boundaries will cancel out.

For the simple regions discussed in the proof, the boundary integral is carried out in essentially a counter-clockwise direction with the interior of the region to the left of the direction of path integration. What happens if several of these simple regions are fitted together to form a region with one or more holes? That is for a boundary associated with a hole, in what direction should one carry out the path integration to maintain the validity of Stokes' Theorem? (Draw pictures to justify your answer.)

6.2 *Curvature via Connection 1-Forms

You should at least skim over the material in Subsect. (5.6.3) before reading this section.

The intent of this section is to give some indication of how computations are carried out using the formalism of differential forms rather than the formalism of Clifford algebra. This section is not a prerequisite for any other section in this text. Thus, you may wish to skip this section on first reading and save it for your summer vacation.

A popular method of computing the components of the Riemann tensor is to compute the curvature 2-forms via the connection 1-forms that were defined by (5.71) and (5.76). In particular,

$$\mathbf{R}_{AB} = \mathbf{d}\mathbf{w}_{AB} + \mathbf{w}_A{}^C \wedge \mathbf{w}_{CB}. \tag{6.13}$$

Now that I have defined the exterior derivative $\mathbf{d}$, I can now derive this formula.

From (5.51),

$$\begin{aligned} R_{AB\alpha\beta}\mathbf{E}^A &= R^{\lambda}{}_{B\alpha\beta}\boldsymbol{\gamma}_\lambda = W_B^\eta R^\lambda{}_{\eta\alpha\beta}\boldsymbol{\gamma}_\lambda \\ &= W_B^\eta \left(\nabla_\alpha\nabla_\beta - \nabla_\beta\nabla_\alpha\right)\boldsymbol{\gamma}_\eta. \end{aligned}$$

You should confirm that it follows from (5.56) that we can rewrite this last equation in the form:

$$\begin{aligned} R_{AB\alpha\beta}\mathbf{E}^A &= \left(\nabla_\alpha\nabla_\beta - \nabla_\beta\nabla_\alpha\right) W_B^\eta\boldsymbol{\gamma}_\eta, \text{ or restated:} \\ R_{AB\alpha\beta}\mathbf{E}^A &= \left(\nabla_\alpha\nabla_\beta - \nabla_\beta\nabla_\alpha\right)\mathbf{E}_B. \end{aligned}$$

From (5.71), this becomes

$$\begin{aligned} R_{AB\alpha\beta}\mathbf{E}^A &= -\nabla_\alpha\left(c_{BD}(\boldsymbol{\gamma}_\beta)\mathbf{E}^D\right) + \nabla_\beta\left(c_{BD}(\boldsymbol{\gamma}_\alpha)\mathbf{E}^D\right) \\ &= -\left(\nabla_\alpha c_{BA}(\boldsymbol{\gamma}_\beta)\right)\mathbf{E}^A + c_{BD}(\boldsymbol{\gamma}_\beta)c^D{}_A(\boldsymbol{\gamma}_\alpha)\mathbf{E}^A \\ &\quad + \left(\nabla_\beta c_{BA}(\boldsymbol{\gamma}_\alpha)\right)\mathbf{E}^A - c_{BD}(\boldsymbol{\gamma}_\alpha)c^D{}_A(\boldsymbol{\gamma}_\beta)\mathbf{E}^A. \end{aligned}$$

This last equation implies

$$\begin{aligned} R_{AB\alpha\beta} &= -\nabla_\alpha c_{BA}(\boldsymbol{\gamma}_\beta) + \nabla_\beta c_{BA}(\boldsymbol{\gamma}_\alpha) \\ &\quad + c_{BD}(\boldsymbol{\gamma}_\beta)c^D{}_A(\boldsymbol{\gamma}_\alpha) - c_{BD}(\boldsymbol{\gamma}_\alpha)c^D{}_A(\boldsymbol{\gamma}_\beta). \end{aligned}$$

Using the fact that $c_{GH} = -c_{HG}$ and $c^G{}_H = -c_H{}^G$ (See Prob. 116), we have

$$\begin{aligned} R_{AB\alpha\beta} &= \nabla_\alpha c_{AB}(\boldsymbol{\gamma}_\beta) - \nabla_\beta c_{AB}(\boldsymbol{\gamma}_\alpha) \\ &\quad + c_A{}^D(\boldsymbol{\gamma}_\alpha)c_{DB}(\boldsymbol{\gamma}_\beta) - c_A{}^D(\boldsymbol{\gamma}_\beta)c_{DB}(\boldsymbol{\gamma}_\alpha). \end{aligned}$$

Now

$$\begin{aligned} \mathbf{R}_{AB} &= \frac{1}{2}R_{AB\alpha\beta}\boldsymbol{\gamma}^{\alpha\beta} \\ &\frac{1}{2}\Big[\boldsymbol{\gamma}^\alpha \wedge \left(\nabla_\alpha c_{AB}(\boldsymbol{\gamma}_\beta)\right)\boldsymbol{\gamma}^\beta + \boldsymbol{\gamma}^\beta \wedge \left(\nabla_\beta c_{BA}(\boldsymbol{\gamma}_\alpha)\right)\boldsymbol{\gamma}^\alpha \\ &\quad + c_A{}^D(\boldsymbol{\gamma}_\alpha)\boldsymbol{\gamma}^\alpha \wedge c_{DB}(\boldsymbol{\gamma}_\beta)\boldsymbol{\gamma}^\beta + c_A{}^D(\boldsymbol{\gamma}_\beta)\boldsymbol{\gamma}^\beta \wedge c_{DB}(\boldsymbol{\gamma}_\alpha)\boldsymbol{\gamma}^\alpha\Big]. \end{aligned} \tag{6.14}$$

We note that

$$\boldsymbol{\gamma}^{\alpha} \wedge \left(\nabla_{\alpha} c_{AB}(\boldsymbol{\gamma}_{\beta})\right) \boldsymbol{\gamma}^{\beta} = \boldsymbol{\gamma}^{\alpha} \wedge \nabla_{\alpha} \left(c_{AB}(\boldsymbol{\gamma}_{\beta})\boldsymbol{\gamma}^{\beta}\right) - c_{AB}(\boldsymbol{\gamma}_{\beta})\boldsymbol{\gamma}^{\alpha} \wedge \nabla_{\alpha}\boldsymbol{\gamma}^{\beta}.$$

From (6.6),

$$\boldsymbol{\gamma}^{\alpha} \wedge \nabla_{\alpha}\boldsymbol{\gamma}^{\beta} = \mathbf{d}\boldsymbol{\gamma}^{\beta} = 0.$$

Also $\mathbf{w}_{AB} = c_{AB}(\boldsymbol{\gamma}_{\beta})\boldsymbol{\gamma}^{\beta}$. With these results, (6.14) becomes

$$\mathbf{R}_{AB} = \mathbf{d}\mathbf{w}_{AB} + \mathbf{w}_{A}{}^{D} \wedge \mathbf{w}_{DB}. \tag{6.15}$$

For two dimensions, $\mathbf{w}_{\bar{1}}{}^{P}\mathbf{w}_{P\bar{2}} = \mathbf{w}_{\bar{1}}{}^{\bar{1}}\mathbf{w}_{\bar{1}\bar{2}} + \mathbf{w}_{\bar{1}}{}^{\bar{2}}\mathbf{w}_{\bar{2}\bar{2}} = 0$. This is true because $\mathbf{w}_{\bar{2}\bar{2}} = 0$ and $\mathbf{w}_{\bar{1}}{}^{\bar{1}} = \pm\mathbf{w}_{\bar{1}\bar{1}} = 0$. Thus, in the two-dimensional case $\mathbf{R}_{\bar{1}\bar{2}} = \mathbf{d}\mathbf{w}_{\bar{1}\bar{2}}$.

Another useful formula is

$$\mathbf{d}\mathbf{E}^{A} = -\mathbf{w}^{A}{}_{M} \wedge \mathbf{E}^{M}. \tag{6.16}$$

To obtain this equation, we note that

$$\mathbf{d}\mathbf{E}^{A} = \boldsymbol{\gamma}^{k} \wedge \nabla_{k}\mathbf{E}^{A} = -\boldsymbol{\gamma}^{k} \wedge c^{A}{}_{M}(\boldsymbol{\gamma}_{k})\mathbf{E}^{M} = -\mathbf{w}^{A}{}_{M} \wedge \mathbf{E}^{M}.$$

Equation(6.16) can frequently used to obtain the connection 1-forms. If we write

$$\mathbf{E}^{A} = W_{\lambda}^{A}\boldsymbol{\gamma}^{\lambda}, \text{ then}$$

$$\mathbf{d}\mathbf{E}^{A} = \boldsymbol{\gamma}^{k} \wedge \nabla_{k}\left(W_{\lambda}^{A}\boldsymbol{\gamma}^{\lambda}\right) = \left(\frac{\partial W_{\lambda}^{A}}{\partial u^{k}}\right)\boldsymbol{\gamma}^{k\lambda}.$$

Knowing the left-hand side of (6.16) enables us to infer the connection 1-forms from the right-hand side by a "guess-and-check" or "trial and error" method discussed in *Gravitation* by Misner et al. (1973, pp. 355–356). I personally think the use of Fock–Ivanenko coefficients is more straight forward and efficient.

Using either method to compute curvature 2–forms, you must be prepared to go back and forth between some coordinate frame and some orthonormal noncoordinate frame.

To compute $\mathbf{d}\mathbf{w}_{AB}$, one generally represents w_{AB} in the form $c_{AB}(\boldsymbol{\gamma}_{\lambda})\boldsymbol{\gamma}^{\lambda}$ rather than the form $c_{AB}(\mathbf{E}_{\lambda})\mathbf{E}^{\lambda}$. Thus the resulting curvature 2-form, $\mathbf{R}_{AB}$ is represented in the form

$$\mathbf{R}_{AB} = \frac{1}{2}R_{ABij}\boldsymbol{\gamma}^{ij}.$$

Therefore to obtain the components of the Riemann tensor in a purely coordinate form or a purely orthogonal noncoordinate form, one must go an extra step. In particular,

$$R_{\alpha\beta ij} = W_{\alpha}^{A}W_{\beta}^{B}R_{ABij}, \text{ or}$$

$$R_{ABPQ} = W_{P}^{i}W_{Q}^{j}R_{ABij}.$$

Of course, the problem of going back and forth between the two types of frames also occurs when one uses Fock–Ivanenko coefficients to carry out the desired computations.

Problem 116. Using the fact that ${c_A}^B = c_{AC} n^{CB}$ demonstrates that ${c_A}^B = -{c^B}_A$.

Problem 117. Show ${R^{AB}}_{AB} = {R^{\alpha\beta}}_{\alpha\beta}$. Also show ${R^{\alpha\beta}}_{\alpha\beta} = {R_{\alpha\beta}}^{\alpha\beta}$ (indices unsummed).

Problem 118. Starting with (6.16),

(a) Show

$$\begin{aligned} \mathbf{ddE}^A &= -\left(\mathbf{dw}^A{}_B + \mathbf{w}^A{}_C \wedge \mathbf{w}^C{}_B\right) \wedge \mathbf{E}^B \\ &= -\mathbf{R}^A{}_B \wedge \mathbf{E}^B. \end{aligned}$$

(b) Show the result of part a) is consistent with the result of Prob. 113. Hint! Refer to (5.54).

6.3 Geodesic Curvature on a 2-dimensional Surface

(You may wish to review my comments on the arc length parameterization of a curve in Sect 5.4.)

The *curvature* of a curve on the 2-dimensional Euclidean plane may be considered to be the rate at which its unit tangent vector $\mathbf{t}(s)$ rotates away from (or toward) a straight line. (See Fig. 6.2.) To discuss the nature of this curvature, it is useful to introduce a second unit vector $\mathbf{n}(s)$ perpendicular to $\mathbf{t}(s)$. It is standard practice to choose the direction of $\mathbf{n}(s)$ so that $\{\mathbf{t}(s), \mathbf{n}(s)\}$ considered as a basis is right handed. That is

$$\begin{aligned} \mathbf{t}(s)\mathbf{n}(s) &= \mathbf{e}_1\mathbf{e}_2 \text{ or} \\ \mathbf{n}(s) &= \mathbf{t}(s)\mathbf{e}_1\mathbf{e}_2. \end{aligned}$$

Thus, if

$$\begin{aligned} \mathbf{t}(s) &= \mathbf{e}_1 \cos\theta(s) + \mathbf{e}_2 \sin\theta(s), \text{ then} \\ \mathbf{n}(s) &= -\mathbf{e}_1 \sin\theta(s) + \mathbf{e}_2 \cos\theta(s). \end{aligned}$$

If curve A in Fig. 6.2 is parameterized in terms of s where s measures arc length, then

$$\frac{d}{ds}\mathbf{t}(s) = (-\mathbf{e}_1 \sin\theta + \mathbf{e}_2 \cos\theta)\frac{d\theta}{ds} = \frac{d\theta}{ds}\mathbf{n}(s) = k(s)\mathbf{n}(s). \tag{6.17}$$

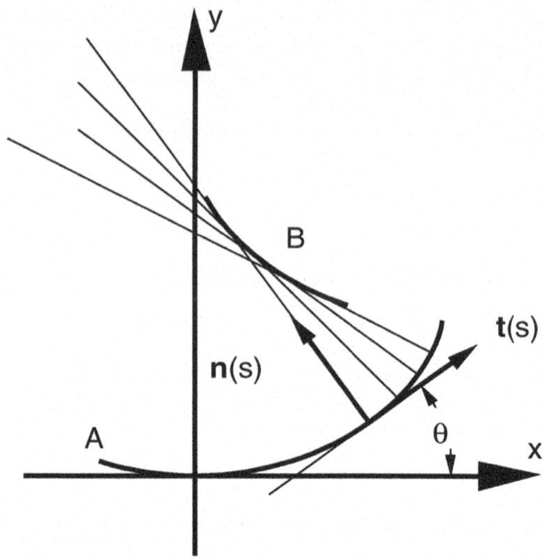

Fig. 6.2 Curve B is the evolute of curve A

Similarly,

$$\frac{d}{ds}\mathbf{n}(s) = (-\mathbf{e}_1 \cos\theta - \mathbf{e}_2 \sin\theta)\,\frac{d\theta}{ds} = -k(s)\mathbf{t}(s). \tag{6.18}$$

In this context, $d\theta/ds$ or $k(s)$ is said to be the *curvature* of curve A. If the curvature $k(s)$ is positive, $\mathbf{t}(s)$ is rotating counterclockwise and if $k(s)$ is negative, $\mathbf{t}(s)$ is rotating clockwise.

The lines normal to curve A have an envelope (curve B) that is said to be the *evolute* of curve A. To determine a formula for curve B, we note that each point on curve B lies somewhere on a line normal to curve A. Thus if $\mathbf{x}(s)$ is a formula for curve A where s is arc length, then a parameterization of curve B is

$$\mathbf{y}(s) = \mathbf{x}(s) + \alpha(s)\mathbf{n}(s), \tag{6.19}$$

where $\alpha(s)$ is yet to be determined. (In this computation, s is a measure of arc length for $\mathbf{x}(s)$ but not for $\mathbf{y}(s)$.) To determine the function $\alpha(s)$, we note that a vector tangent to curve B at $\mathbf{y}(s)$ is normal to curve A at the corresponding point $\mathbf{x}(s)$. Using (6.17) and (6.18), we have

$$\frac{d}{ds}\mathbf{y}(s) = \mathbf{t}(s) - \alpha(s)k(s)\mathbf{t}(s) + \left(\frac{d\alpha(s)}{ds}\right)\mathbf{n}(s). \tag{6.20}$$

Thus,

$$[1 - \alpha(s)k(s)]\,\mathbf{t}(s) = 0 \text{ and}$$

$$\alpha(s) = \frac{1}{k(s)}. \tag{6.21}$$

And therefore

$$\mathbf{y}(s) = \mathbf{x}(s) + \frac{1}{k(s)}\mathbf{n}(s). \tag{6.22}$$

From (6.20) and (6.21), it is clear that if $k(s)$ is constant then $d\mathbf{y}/ds = 0$. In this case, $\mathbf{y}(s)$ becomes a single point, which we can label $\mathbf{a}$. From (6.22), we see

$$|\mathbf{x}(s) - \mathbf{a}| = \frac{1}{|k(s)|}. \tag{6.23}$$

Thus, when the curvature k is constant, then $\mathbf{x}(s)$ is a circle with center $\mathbf{y}(s) = \mathbf{a}$ and radius equal to $1/|k|$.

Therefore in the more general case where $\mathbf{x}(s)$ is not a circle, it is natural to refer to $1/|k(s_0)|$ as the *radius of curvature* for the point $\mathbf{x}(s_0)$. Similarly, it is natural to refer to $\mathbf{y}(s_0)$ as the *center of curvature* for the point $\mathbf{x}(s_0)$.

Another approach, which justifies the same terminology, is to consider the circle that makes the best approximation to the curve $\mathbf{x}(s)$ in the neighborhood of $\mathbf{x}(s_0)$. If $\mathbf{u}(s)$ is an arc length formula for a circle of radius r and center $\mathbf{a}$, then

$$\langle \mathbf{u}(s) - \mathbf{a}, \mathbf{u}(s) - \mathbf{a}\rangle = r^2 \tag{6.24}$$

Our problem now is to determine both r and $\mathbf{a}$ so that the resulting circle is the best approximation to our curve at $\mathbf{x}(s_0)$. To carry out this endeavor, we consider the function

$$f(s) = \langle \mathbf{x}(s) - \mathbf{a}, \mathbf{x}(s) - \mathbf{a}\rangle - r^2. \tag{6.25}$$

If we wanted to obtain the unique circle that passed through the points $\mathbf{x}(s_{-1})$, $\mathbf{x}(s_0)$, and $\mathbf{x}(s_1)$, where $s_{-1} < s_0 < s_1$, we would determine r and $\mathbf{a}$ so that

$$f(s_{-1}) = f(s_0) = f(s_1) = 0.$$

See Fig. 6.3. According to Rolle's Theorem, this implies that there are parameter values t_1 and t_2 such that $s_{-1} < t_1 < s_0 < t_2 < s_1$ and

$$\left.\frac{d}{ds}f(s)\right|_{s=t_1} = \dot{f}(t_1) = \left.\frac{d}{ds}f(s)\right|_{s=t_2} = \dot{f}(t_2) = 0.$$

Applying Rolle's Theorem once more, we know there is a parameter value u such that $t_1 < u < t_2$ and

$$\left.\frac{d^2}{ds^2}f(s)\right|_{s=u} = \ddot{f}(u) = 0.$$

Thus, necessary conditions for the limiting circle obtained when the three points coalesce at $\mathbf{x}(s_0)$ is that

$$f(s_0) = \dot{f}(s_0) = \ddot{f}(s_0) = 0.$$

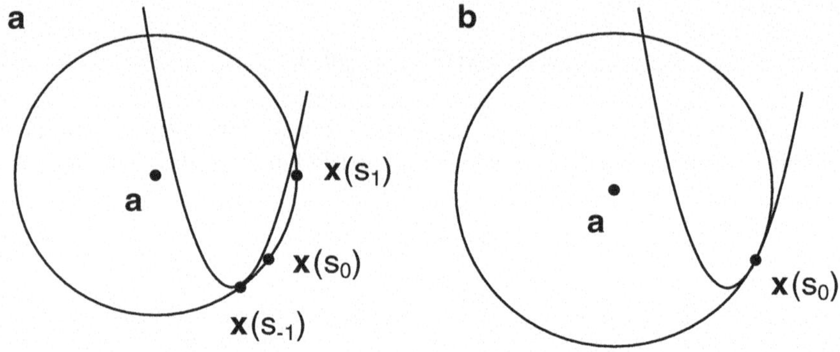

Fig. 6.3 (**a**) Circle passing through three points of parabola. (**b**) Osculating circle for point $\mathbf{x}(s_0)$

(When I carry out the computation below, it will become obvious that these conditions are also sufficient to determine r and $\mathbf{a}$.)

Applying the first condition, we have

$$f(s_0) = \langle \mathbf{x}(s_0) - \mathbf{a}, \mathbf{x}(s_0) - \mathbf{a} \rangle - r^2 = 0.$$

This implies that

$$r = |\mathbf{x}(s_0) - \mathbf{a}| \,. \tag{6.26}$$

Next

$$\dot{f}(s) = 2 \langle \mathbf{t}(s), \mathbf{x}(s) - \mathbf{a} \rangle \,. \tag{6.27}$$

Setting $\dot{f}(s_0) = 0$, we infer that for some value of λ,

$$\mathbf{x}(s_0) - \mathbf{a} = \lambda \mathbf{n}(s_0). \tag{6.28}$$

From (6.27), we have

$$\ddot{f}(s) = 2 \langle k(s)\mathbf{n}(s), \mathbf{x}(s) - \mathbf{a} \rangle + 2 \langle \mathbf{t}(s), \mathbf{t}(s) \rangle \,.$$

Setting $\ddot{f}(s_0) = 0$ and using (6.28), we have

$$2 \langle k(s_0)\mathbf{n}(s_0), \mathbf{x}(s_0) - \mathbf{a} \rangle + 2 \langle \mathbf{t}(s_0), \mathbf{t}(s_0) \rangle = 0, \text{ therefore}$$

$$k(s_0) \langle \mathbf{n}(s_0), \lambda \mathbf{n}(s_0) \rangle + \langle \mathbf{t}(s_0), \mathbf{t}(s_0) \rangle = 0 \text{ and thus}$$

$$k(s_0)\lambda + 1 = 0, \text{ or restated:}$$

$$\lambda = \frac{-1}{k(s_0)} \tag{6.29}$$

Combining this result with (6.26) and (6.28), we have

$$r = |\mathbf{x}(s_0) - \mathbf{a}| = |\lambda| = \frac{1}{|k(s_0)|}, \text{ and thus}$$

$$r = \frac{1}{|k(s_0)|}. \tag{6.30}$$

Combining (6.28) and (6.29), we have

$$\mathbf{a} = \mathbf{x}(s_0) + \frac{1}{k(s_0)}\mathbf{n}(s_0). \tag{6.31}$$

Comparing (6.30) and (6.31) with (6.23) and (6.22) now gives us another justification for the terms "*radius of curvature*" and "*center of curvature*" that I introduced in my discussion of the evolute.

The circle constructed above is referred to as the *osculating* (kissing) *circle*. This is a term coined by Leibniz (1646–1716).

For the intrinsic observer on a 2-dimensional curved surface, it does not make sense to define the curvature of a curve as the rate the unit tangent vector rotates away from (or toward) a straight line. But it does make sense to speak of the rate at which the unit tangent vector rotates away from (or toward) a geodesic. This is called the *geodesic curvature* $k_g(s)$. Instead of having

$$\frac{d}{ds}\mathbf{t}(s) = k(s)\mathbf{n}(s), \text{ we have}$$

$$\nabla_s \mathbf{t}(s) = k_g(s)\mathbf{n}(s), \text{ where} \tag{6.32}$$

as before, the direction of the unit vector $\mathbf{n}(s)$ is chosen so that $\{\mathbf{t}(s), \mathbf{n}(s)\}$ is "right handed." If $\{\mathbf{V}_1(s), \mathbf{V}_2(s)\}$ is a pair of orthonormal right handed vectors that are parallel transported along the curve $\mathbf{x}(s)$, then we can write

$$\mathbf{t}(s) = \mathbf{V}_1(s)\cos\theta(s) + \mathbf{V}_2(s)\sin\theta(s), \text{ and}$$

$$\mathbf{n}(s) = -\mathbf{V}_1(s)\sin\theta(s) + \mathbf{V}_2(s)\cos\theta(s).$$

Since $\nabla_s \mathbf{V}_1(s) = \nabla_s \mathbf{V}_2(s) = 0$, we have

$$\nabla_s \mathbf{t}(s) = \mathbf{n}(s)\frac{d\theta}{ds}, \text{ and} \tag{6.33}$$

$$\nabla_s \mathbf{n}(s) = -\mathbf{t}(s)\frac{d\theta}{ds}. \tag{6.34}$$

Problem 119. For theoretical discussions of curves, it is convenient to use an arc length parameter. However in actual practice, solving the equation $ds/dt = |\mathbf{v}(t)|$

and then obtaining the inverse function $t(s)$ can complicate the computations. However, the desired entities can be computed without using an explicit form of $t(s)$. Show

$$\frac{d}{dt}\mathbf{t}(t) = k(t)\,|\mathbf{v}(t)|\,\mathbf{n}(t), \text{ and} \tag{6.35}$$

$$\frac{d}{dt}\mathbf{n}(t) = -k(t)\,|\mathbf{v}(t)|\,\mathbf{t}(t), \text{ where} \tag{6.36}$$

$\mathbf{v}(t) = \frac{d}{dt}\mathbf{x}(t)$, $\mathbf{t}(t) = \mathbf{v}(t)/\,|\mathbf{v}(t)|$, and $\mathbf{n}(t) = \mathbf{t}(t)\mathbf{e}_1\mathbf{e}_2$.

Problem 120. Consider the parabola $\mathbf{x}(t) = (t, t^2)$. Use the results of Prob. 119 to obtain the curvature and the equation for the evolute. Also show that there is a cusp in the evolute at $(0, \frac{1}{2})$.

Problem 121. Reconsider (6.25). That is

$$f(s) = \langle \mathbf{x}(s) - \mathbf{a}, \mathbf{x}(\mathbf{s}) - \mathbf{a}\rangle - r^2.$$

It was shown that the necessary and sufficient conditions for $\mathbf{a}$ and r to be, respectively, the center and radius of the osculating circle for $s = s_0$ are

$$f(s_0) = \dot{f}(s_0) = \ddot{f}(s_0) = 0.$$

(a) Show $f^{(3)}(s_0) = -2\dot{k}(s_0)/k(s_0)$, where $f^{(3)}(s) = \frac{d^3 f(s)}{ds^3}$. This shows that if neither $\dot{k}(s_0)$ nor $k(s_0)$ is zero, then the osculating circle crosses the original curve $\mathbf{x}(s)$ at the point of tangency. Why? This is contrary to some informal drawings that appear in some text books. It also suggests to me that the name "osculating" (kissing) circle may be inappropriate.
(b) Is there some point on a parabola where the parabola is not crossed by the companion osculating circle? Justify your answer by some computations.

Problem 122. CYCLOID

Consider a wheel in the x-y plane rolling along the x axis without slipping (See Fig. 6.4). The path generated by a point P attached to the wheel is called a *cycloid*. The equation for the cycloid pictured in Fig. 6.4 is

$$\mathbf{x}(\theta) = \mathbf{e}_1 b(\theta - \sin\theta) + \mathbf{e}_2 b(1 - \cos\theta). \tag{6.37}$$

Show that the evolute of this cycloid is another cycloid. In particular, show

$$\begin{aligned}\mathbf{y}(\theta) &= [\mathbf{e}_1 b\pi - \mathbf{e}_2 2b] + [\mathbf{e}_1 b(\phi - \sin\phi) + \mathbf{e}_2 b(1 - \cos\phi)]\,, \text{ where}\\ \phi &= \theta - \pi.\end{aligned} \tag{6.38}$$

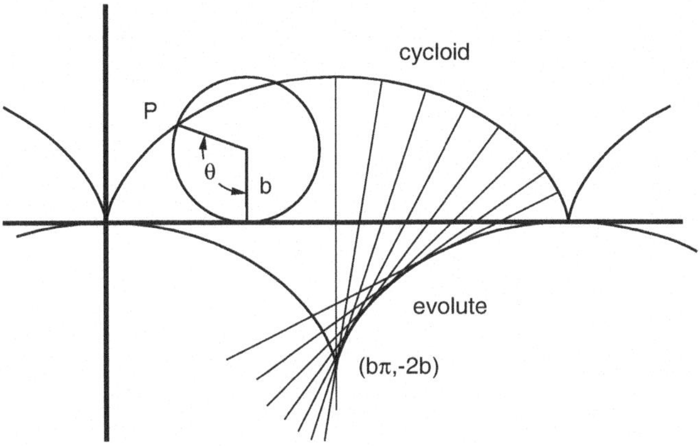

Fig. 6.4 The evolute of a cycloid is another cycloid

Comment:

$$\mathbf{v}(\theta) = \frac{d\mathbf{x}}{d\theta} = \mathbf{e}_1 b(1 - \cos\theta) + \mathbf{e}_2 b \sin\theta$$
$$= 2b\left[\mathbf{e}_1 \sin^2\frac{\theta}{2} + \mathbf{e}_2 \sin\frac{\theta}{2}\cos\frac{\theta}{2}\right].$$

Thus,

$$|\mathbf{v}(\theta)| = 2b\left|\sin\frac{\theta}{2}\right| \quad \text{and}$$
$$\mathbf{t}(\theta) = \frac{\sin\frac{\theta}{2}}{\left|\sin\frac{\theta}{2}\right|}\left[\mathbf{e}_1 \sin\frac{\theta}{2} + \mathbf{e}_2 \cos\frac{\theta}{2}\right]$$

except where θ is an integral multiple of 2π.

Problem 123. Consider the circle of constant latitude on a sphere. Namely

$$\mathbf{x}(\phi) = \mathbf{e}_1 R\cos\phi\sin\theta_0 + \mathbf{e}_2 R\sin\phi\sin\theta_0 + \mathbf{e}_3 R\cos\theta_0.$$

Show that $1/\left|k_g\right|$ is the distance from any point on the circle to the vertex of the tangent cone shown in Fig. 6.5.

6.4 *Huygens' Pendulum Clock and the Cycloid

Note! Some Newtonian physics is used in this section.

Although Galileo Galilei (1564–1642) designed a pendulum clock, he never built one. Due to friction, one must transfer energy to the pendulum to keep it swinging

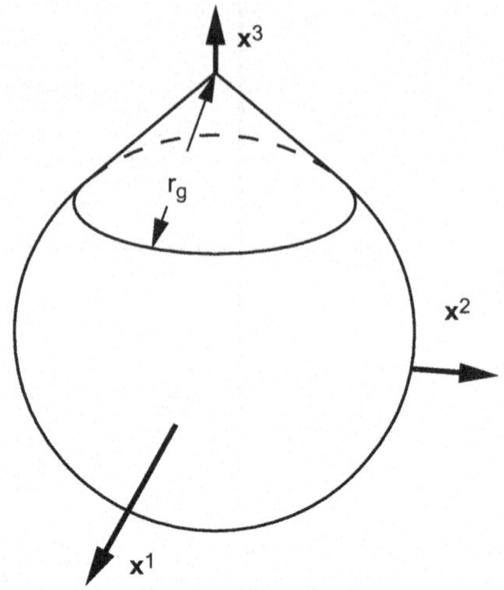

Fig. 6.5 If $r_g = 1/|k_g|$, where k_g is the geodesic curvature, then for a circle of constant latitude on a sphere, r_g is the distance from any point on the circle to to the vertex of the tangent cone

back and forth. That was one problem. A second problem faced by clock designers is that the period of a simple pendulum depends on the maximum angle of deflection from a vertical line (the *amplitude*).

In the winter of 1656–1657, Christiaan Huygens (1629–1695) solved the first problem by designing a clock that transferred the energy of a slowly falling weight to the pendulum by a ratchet mechanism involving a system of gears.

Huygens used two methods to deal with the second problem. It had already been noted by Galileo that for small amplitudes, the period of a pendulum is nearly *isochronous* (independent of amplitude). Huygens also discovered that if he used a bob hung on a "weightless" cord for his pendulum, he could make the period nearly isochronous by having the cord wind around bent pieces of wood or metal plates. (See Fig. 6.6a). This had the effect of shortening the period for large amplitudes. However initially, Huygens did not know what curve he should use for his constraining plates.

Even with his limited knowledge, Huygens and others were able to design fairly accurate clocks. In 1658, one year after Huygens' first clock, Samuel Coster guaranteed church officials in Utrecht that a clock he was building for them would not deviate more than eight minutes in a week. (Bos 1986, p. xi). However, Huygens had much broader interests than building accurate clocks. For example, he is currently best known for his wave theory of light.

Late in 1659, Huygens attacked the problem of measuring the gravitational constant g. This problem was not trivial. Without stop watches, it was impossible to get even an approximate value directly. Galileo may never have dropped anything from the tower of Pisa. Doing so would not have given him very much information. Using inclined planes, Galileo was able to slow down the rate of descent. Doing this, he was able to conclude that for a rolling ball released with zero initial velocity, the formula for the distance traveled down the plane would be $s = \frac{1}{2}ct^2$, where s is

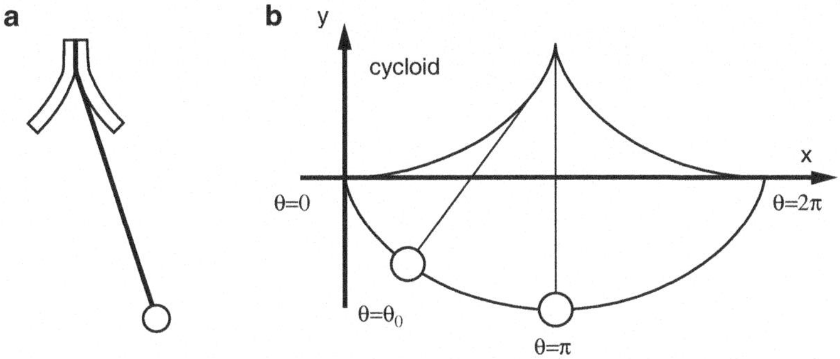

Fig. 6.6 (**a**) Bent plates shorten period of pendulum for large amplitudes. (**b**) A cycloid defined by the equation $\mathbf{x}(\theta) = \mathbf{e}_1 b(\theta - \sin\theta) - \mathbf{e}_2 b(1 - \cos\theta)$. The period is independent of the position of maximum deflection where $\theta = \theta_0$

the distance, t is the time, and c is a constant, which is determined by the angle of inclination. From this information, it was reasonable to infer that the formula for an object in free fall should be $s = \frac{1}{2}gt^2$. However, his measurements using rolling balls could not be directly translated into the situation of a truly free fall. Using simple trigonometry, Galileo could have inferred the value of g by sliding objects down frictionless planes. Without a sophisticated knowledge of Newtonian mechanics, rolling balls cannot be used to determine the value of g. When a ball rolls down an inclined plane, part of the potential energy is converted into the kinetic energy of the spinning motion. Thus, a rolling ball will not arrive at the bottom of an inclined plane as fast as an object sliding down a frictionless plane. Unfortunately for Galileo, frictionless planes were unavailable.

In the hopes of determining an accurate value for g, Huygens started to investigate the mathematical nature of the period of the pendulum at the end of 1659. He had access to some results on centrifugal force that he had derived himself but he did not have the tools of calculus. Newton was not quite 17 and had not yet invented calculus. Using the coordinate free methods of Archimedes with a small amount of algebra, Huygens was unable to solve the problem for a circular arc with an arbitrary amplitude. However, he was able to show that for small amplitudes the period T is approximately $2\pi\sqrt{L/g}$, where L is the length of the "weightless" cord and g is the constant of gravity.

Furthermore, his accomplishments did not stop there. Applying the methods that he was using he developed considerable insight into the problem. He already knew more about the geometry of cycloids than I will discuss in this book. With this knowledge, he was soon able to show that the same formula that was only approximate for the period of a pendulum with a circular arc would be exact if the bob was somehow constrained to follow the path of a cycloid. A discussion of these computations appears in Joella G. Yoder's *Unrolling Time* (1988).

To reconstruct his isochronous result for the cycloid using calculus, consider Fig. 6.6b. We assume that at the point where $\theta = \theta_0$, the bob of the pendulum is at its maximum distance from the point of lowest elevation where $\theta = \pi$. To demonstrate that the path is isochronous, we must show that the period is independent of θ_0. The period T is the time required for the bob to complete an entire cycle. Thus, $(1/4)T$ is the time required for the bob to descend from the point, where $\theta = \theta_0$ to the point where $\theta = \pi$. From the conservation of energy

$$\frac{1}{2}mv^2 + mgy = mgy_0,$$

where y_0 is the value for y at the point where $\theta = \theta_0$. It therefore follows that the speed

$$v = \frac{ds}{dt} = \sqrt{2g(y_0 - y)}.$$

For the cycloid presented in Prob. 122, pictured in Fig. 6.4 and adjusted for Fig. 6.6b, we have

$$\mathbf{x}(\theta) = \mathbf{e}_1 b(\theta - \sin\theta) - \mathbf{e}_2 b(1 - \cos\theta).$$

Thus,

$$\begin{aligned}
\frac{ds}{dt} &= \sqrt{2g(y_0 - y)} \\
&= \sqrt{2g\left[-b(1 - \cos\theta_0) + b(1 - \cos\theta)\right]} \\
&= \sqrt{2gb(\cos\theta_0 - \cos\theta)}.
\end{aligned} \tag{6.39}$$

Furthermore,

$$\frac{ds}{d\theta} = \sqrt{\left(\frac{dx}{d\theta}\right)^2 + \left(\frac{dy}{d\theta}\right)^2} = b\sqrt{(1 - \cos\theta)^2 + (-\sin\theta)^2} = b\sqrt{2(1 - \cos\theta)}$$
$$= b\sqrt{4\sin^2\frac{\theta}{2}}.$$

That is:

$$\frac{ds}{d\theta} = 2b\sin\frac{\theta}{2}. \tag{6.40}$$

From (6.39) and (6.40),

$$\frac{dt}{d\theta} = \frac{ds}{d\theta}\Big/\frac{ds}{dt} = \sqrt{\frac{2b}{g}}\frac{\sin\frac{\theta}{2}}{\sqrt{\cos\theta_0 - \cos\theta}}.$$

Since

$$\cos\theta_0 - \cos\theta = \left(2\cos^2\frac{\theta_0}{2} - 1\right) - \left(2\cos^2\frac{\theta}{2} - 1\right)$$
$$= 2\left(\cos^2\frac{\theta_0}{2} - \cos^2\frac{\theta}{2}\right),$$

it follows that

$$dt = \sqrt{\frac{b}{g}}\frac{\sin\frac{\theta}{2}d\theta}{\sqrt{\cos^2\frac{\theta_0}{2} - \cos^2\frac{\theta}{2}}}.$$

Therefore,

$$\frac{1}{4}T = \sqrt{\frac{b}{g}}\int_{\theta_0}^{\pi}\frac{\sin\frac{\theta}{2}d\theta}{\sqrt{\cos^2\frac{\theta_0}{2} - \cos^2\frac{\theta}{2}}}.$$

Substituting $u = (\cos\frac{\theta}{2})/(\cos\frac{\theta_0}{2})$, we have $du = -\frac{1}{2}(\sin\frac{\theta}{2})d\theta/(\cos\frac{\theta_0}{2})$ and

$$\frac{1}{4}T = -2\sqrt{\frac{b}{g}}\int_1^0\frac{du}{\sqrt{1-u^2}} = 2\sqrt{\frac{b}{g}}\int_0^1\frac{du}{\sqrt{1-u^2}}$$
$$= 2\sqrt{\frac{b}{g}}\arcsin 1 = \pi\sqrt{\frac{b}{g}}. \tag{6.41}$$

Noting that b is the radius of the rolling circle that generates the cycloid, it is clear from Figs. 6.6b and 6.4 that the length L of the cord equals $4b$, so (6.41) becomes

$$T = 2\pi\sqrt{\frac{L}{g}}. \tag{6.42}$$

When Huygens demonstrated that the cycloid is an isochronous path, he was then faced with the problem of how to constrain the bob of a pendulum to stay on the path of a cycloid. It so happens that he had already designed a clock with bent plates, which solved this problem – at least approximately. (See Fig. 6.6a again.)

If he could figure out the correct shape for the bent plates, he would have a solution for his problem. In this context, he invented the concept of *evolute*, and soon discovered that the evolute of a cycloid was another cycloid. (See Prob. 122.)

Soon thereafter he determined the evolutes of other curves such as the parabola. (See Prob. 120.) He was not concerned with the notions of curvature or center of curvature so those concepts are credited to Newton and Leibniz.

Huygens eventually published his results in 1673 in a book entitled *Horologium Oscillatorium* (*The Pendulum Clock*). This work has been translated from Latin into English by Robert J. Blackwell (Huygens 1986).

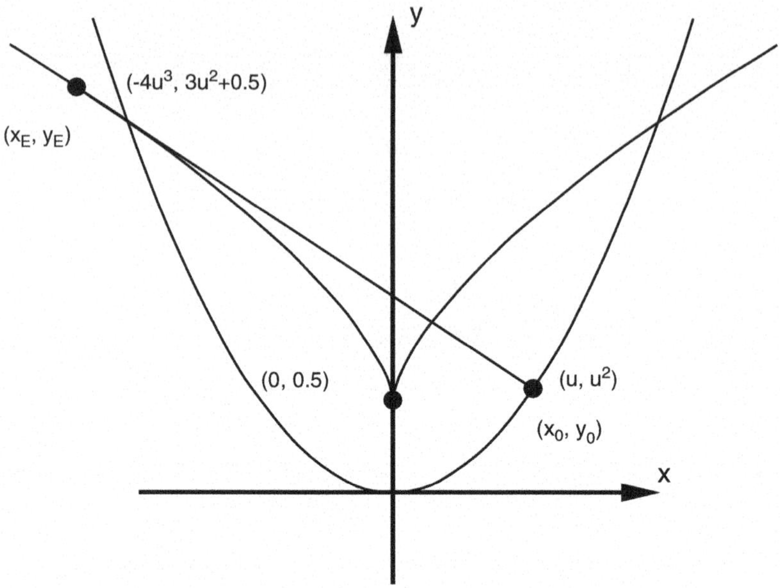

Fig. 6.7 A parabola with its evolute

Fig. 6.8 A simple pendulum

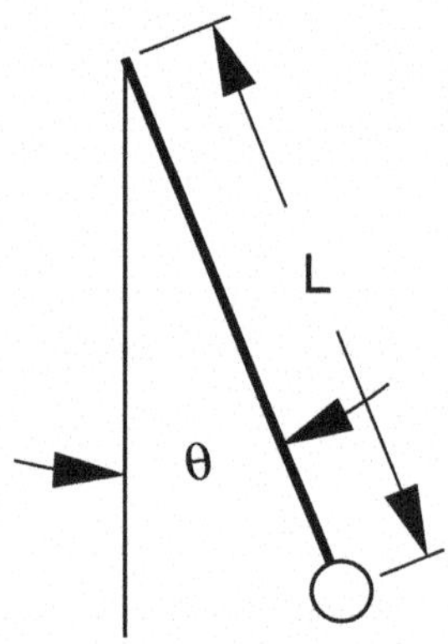

An unexpected consequence of these results was that Huygens now had a method for determining the arc length of a segment of any curve that is the evolute of another curve. For example, consider the evolute of a parabola (See Fig. 6.7). The curved distance from (x_E, y_E) to the cusp at $(0, \frac{1}{2})$ is the same as the straight line distance from (x_E, y_E) to (x_0, y_0) minus $1/2$.

Problem 124. Consider the simple pendulum (See Fig. 6.8). From the conservation of energy,

$$\frac{1}{2}mv^2 - mgL\cos\theta = -mgL\cos\theta_0,$$

where θ_0 is the angle of amplitude (or angle of maximum deflection). Also, $ds = \pm L d\theta$.

(a) Show that

$$\frac{1}{4}T = \sqrt{\frac{L}{2g}}\int_0^{\theta_0} \frac{d\theta}{\sqrt{\cos\theta - \cos\theta_0}}.$$

Equating

$$\begin{aligned}\cos\theta - \cos\theta_0 &= \left(1 - 2\sin^2\frac{\theta}{2}\right) - \left(1 - 2\sin^2\frac{\theta_0}{2}\right)\\ &= 2\sin^2\frac{\theta_0}{2} - 2\sin^2\frac{\theta}{2},\end{aligned}$$

it is not difficult to show that T is an *elliptic integral of the first kind*. See (Magnus et al. 1966, p. 358). Such integrals cannot be expressed in terms of elementary functions. However for small amplitudes,

$$\cos\theta - \cos\theta_0 \approx 1 - \frac{\theta^2}{2} - \left(1 - \frac{\theta_0^2}{2}\right) = \frac{1}{2}\left(\theta_0^2 - \theta^2\right).$$

(b) Use this approximation to show that for small amplitudes

$$T \approx 2\pi\sqrt{\frac{L}{g}}.$$

(c) Can you improve this last approximation with elementary functions?

Problem 125. Consider the equation for the parabola;

$$\mathbf{x}(u) = \mathbf{e}_1 u + \mathbf{e}_2 u^2.$$

(a) Do Prob. 120 if you have not done it before to obtain the equation for the evolute,

$$\mathbf{y}(t) = -4u^3\mathbf{e}_1 + \left(3u^2 + \frac{1}{2}\right)\mathbf{e}_2.$$

Use Huygen's method to obtain the length of the curve of the evolute from the cusp to the point $-4u^3\mathbf{e}_1 + (3u^2 + \frac{1}{2})\mathbf{e}_2$ in terms of u (See Fig. 6.7).

(b) Use the fact that

$$ds = \sqrt{\left(\frac{dx}{du}\right)^2 + \left(\frac{dy}{du}\right)^2}\, du,$$

to replicate the result of part a).

Problem 126. (a) Use Huygen's method to obtain the length of the arc of a cycloid from one cusp to the next in terms of the parameter b, where b is the radius of the rolling circle that generates the cycloid. What is the straight line distance between two cusps in terms of b (See Figs. 6.4 and 6.6b).

(b) Use the fact that

$$ds = \sqrt{\left(\frac{dx}{d\theta}\right)^2 + \left(\frac{dy}{d\theta}\right)^2}\, d\theta,$$

to reproduce the result of part a).

6.5 The Gauss–Bonnet Formula

I begin this section by showing that the informal definition for the Gaussian curvature that I gave in (5.4) is equivalent to the formal definition that I gave in Def. 81. You may recall that the formal definition was

$$K = \frac{1}{2} R_{\alpha\beta}{}^{\alpha\beta} = \frac{1}{2} R^{\alpha\beta}{}_{\alpha\beta} = R_{12}{}^{12} = R^{12}{}_{12}. \tag{6.43}$$

To demonstrate this equivalence, I begin by letting

$$\mathbf{F} = \boldsymbol{\gamma}^k \left\langle \mathbf{E}_{\bar{1}}, \nabla_k \mathbf{E}_{\bar{2}} \right\rangle, \tag{6.44}$$

where the pair $\{\mathbf{E}_{\bar{1}}, \mathbf{E}_{\bar{2}}\}$ forms a right handed orthonormal frame on the region of the surface under consideration. Applying the operator $\mathbf{d}$, we get

$$\mathbf{dF} = \boldsymbol{\gamma}^{jk} \left\langle \nabla_j \mathbf{E}_{\bar{1}}, \nabla_k \mathbf{E}_{\bar{2}} \right\rangle + \boldsymbol{\gamma}^{jk} \left\langle \mathbf{E}_{\bar{1}}, \nabla_j \nabla_k \mathbf{E}_{\bar{2}} \right\rangle. \tag{6.45}$$

Since $\nabla_j \mathbf{E}_{\bar{1}} = -c_{\bar{1}\bar{2}}(\boldsymbol{\gamma}_j)\mathbf{E}^{\bar{2}}$ and $\nabla_k \mathbf{E}_{\bar{2}} = -c_{\bar{2}\bar{1}}(\boldsymbol{\gamma}_k)\mathbf{E}^{\bar{1}}$, it follows that

$$\left\langle \nabla_j \mathbf{E}_{\bar{1}}, \nabla_k \mathbf{E}_{\bar{2}} \right\rangle = c_{\bar{1}\bar{2}}(\boldsymbol{\gamma}_j) c_{\bar{2}\bar{1}}(\boldsymbol{\gamma}_k) \left\langle \mathbf{E}^{\bar{2}}, \mathbf{E}^{\bar{1}} \right\rangle = 0.$$

Thus, (6.45) becomes

$$\mathbf{dF} = \boldsymbol{\gamma}^{jk} \left\langle \mathbf{E}_{\bar{1}}, \nabla_j \nabla_k \mathbf{E}_{\bar{2}} \right\rangle = \boldsymbol{\gamma}^{12} \left\langle \mathbf{E}_{\bar{1}}, \nabla_1 \nabla_2 \mathbf{E}_{\bar{2}} \right\rangle + \boldsymbol{\gamma}^{21} \left\langle \mathbf{E}_{\bar{1}}, \nabla_2 \nabla_1 \mathbf{E}_{\bar{2}} \right\rangle.$$

That is:

$$\begin{aligned} \mathbf{dF} &= \boldsymbol{\gamma}^{12} \left\langle \mathbf{E}_{\bar{1}}, (\nabla_1 \nabla_2 - \nabla_2 \nabla_1) \mathbf{E}_{\bar{2}} \right\rangle \\ &= \boldsymbol{\gamma}^{12} R^A{}_{\bar{2}12} \left\langle \mathbf{E}_{\bar{1}}, \mathbf{E}_A \right\rangle = \boldsymbol{\gamma}^{12} R_{B\bar{2}12} \left\langle \mathbf{E}_{\bar{1}}, \mathbf{E}^B \right\rangle = \boldsymbol{\gamma}^{12} R_{B\bar{2}12} \delta^B_{\bar{1}} = R_{\bar{1}\bar{2}12} \boldsymbol{\gamma}^{12}. \end{aligned} \tag{6.46}$$

Fig. 6.9 $dA = |\gamma_{12}|\, du^1 du^2$

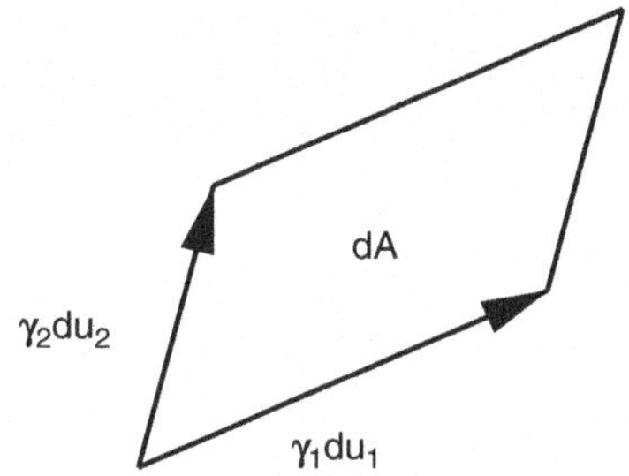

Now

$$R_{\bar{1}\bar{2}12} = R_{\bar{1}\bar{2}12}\left|\mathbf{E}^{\bar{1}}\mathbf{E}^{\bar{2}}\right| = R^{12}{}_{12}\,|\gamma_{12}|\,.$$

Therefore,

$$\mathbf{dF} = R^{12}{}_{12}\,|\gamma_{12}|\,\gamma^{12}.$$

From Stokes' Theorem:

$$\int_A \mathbf{dF} = \int_{\partial A} \mathbf{F}, \tag{6.47}$$

which for our case becomes

$$\int_A R^{12}{}_{12}\,|\gamma_{12}|\,du^1du^2 = \int_{\partial A} \langle \mathbf{E}_{\bar{1}}, \nabla_1 \mathbf{E}_{\bar{2}}\rangle \frac{du^1}{ds} ds + \int_{\partial A} \langle \mathbf{E}_{\bar{1}}, \nabla_2 \mathbf{E}_{\bar{2}}\rangle \frac{du^2}{ds} ds.$$

or restated,

$$\int_A K\,|\gamma_{12}|\,du^1du^2 = \int_{\partial A} \langle \mathbf{E}_{\bar{1}}, \nabla_S \mathbf{E}_{\bar{2}}\rangle\, ds, \tag{6.48}$$

The term $|\gamma_{12}|\,du^1du^2$ that appears on the left-hand side of (6.48) may be interpreted as an infinitesimal area. To see this, we note that the area of a parallelogram is equal to the product of two adjacent sides multiplied by the sine of the angle between them. Thus, the area of the parallelogram in Fig. 6.9 is $|\gamma_1|\,|\gamma_2|\sin\theta du^1du^2$. But

$$\begin{aligned} |\gamma_{12}|^2 &= \langle \gamma_{12}, \gamma_{12}\rangle = \gamma_{12}\gamma_{21} = \langle \gamma_1, \gamma_1\rangle\langle \gamma_2, \gamma_2\rangle - \langle \gamma_1, \gamma_2\rangle\langle \gamma_1, \gamma_2\rangle \\ &= |\gamma_1|^2\,|\gamma_2|^2\,(1-\cos^2\theta) = |\gamma_1|^2\,|\gamma_2|^2 \sin^2\theta, \text{ which implies} \\ |\gamma_{12}| &= |\gamma_1|\,|\gamma_2|\sin\theta. \end{aligned}$$

Thus, (6.48) becomes

$$\int_A K dA = \int_{\partial A} \langle \mathbf{E}_{\bar{1}}, \nabla_S \mathbf{E}_{\bar{2}}\rangle\, ds. \tag{6.49}$$

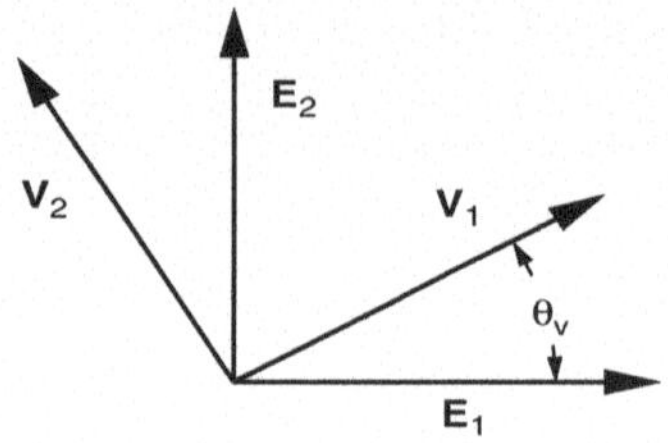

Fig. 6.10 For one who uses $\mathbf{E}_1$ and $\mathbf{E}_2$ to construct a coordinate system for a map of a curved surface, a frame $\{\mathbf{V}_1, \mathbf{V}_2\}$ which is parallel transported around a closed curve will appear to rotate with respect to the map

Note! A review of the calculations above shows that to say $dA = |\boldsymbol{\gamma}_{12}|\, du^1 du^2$ is equivalent to saying that $dA = \sqrt{g} du^1 du^2$. This latter form is more suitable for generalization to higher dimensions.

The significance of the right-hand side of (6.49) is less obvious but it has an even simpler interpretation – at least for a region that is simply connected (no holes). Suppose we consider an orthonormal frame $\{\mathbf{V}_1, \mathbf{V}_2\}$ that is parallel transported around the closed curve. From Fig. 6.10,

$$\mathbf{E}_{\bar{1}} = \mathbf{V}_1 \cos\theta_V - \mathbf{V}_2 \sin\theta_V \text{ and} \tag{6.50}$$

$$\mathbf{E}_{\bar{2}} = \mathbf{V}_1 \sin\theta_V + \mathbf{V}_2 \cos\theta_V. \tag{6.51}$$

(To write down these equations quickly without resorting to a lengthy derivation, it is useful to note that since $\mathbf{E}_2$ is "between" $\mathbf{V}_1$ and $\mathbf{V}_2$, both of its components in the$\{\mathbf{V}_1, \mathbf{V}_2\}$ system must be positive. Also, $\mathbf{E}_1$ must be perpendicular to $\mathbf{E}_2$.)

Since $\mathbf{V}_1$ and $\mathbf{V}_2$ are parallel transported around the curve,

$$\nabla_S \mathbf{V}_1 = \nabla_S \mathbf{V}_2 = 0.$$

Thus,

$$\nabla_S \mathbf{E}_{\bar{2}} = (\mathbf{V}_1 \cos\theta_V - \mathbf{V}_2 \sin\theta_V)\frac{d\theta_V}{ds} = \mathbf{E}_{\bar{1}}\frac{d\theta_V}{ds}.$$

With this result, (6.49) becomes

$$\int_A K dA = \int_{\partial A} \frac{d\theta_V}{ds} ds = \theta, \tag{6.52}$$

where θ is the total angle of rotation that appears to occur to a vector when it is parallel transported around a complete circuit and returned to its original location. (It is assumed that region V is simply connected.) From (6.52), we see that the informal definition for the Gaussian curvature that I gave in (5.9) is consistent with the formal definition of (6.43).

If the boundary of the closed curve is composed of geodesic curves pieced together like the three edges of a geodesic triangle, then the extreme right-hand side of (6.52) can be decomposed into quantities that are easily measured. To obtain this alternative interpretation, consider the unit tangent vector **t** and the unit normal

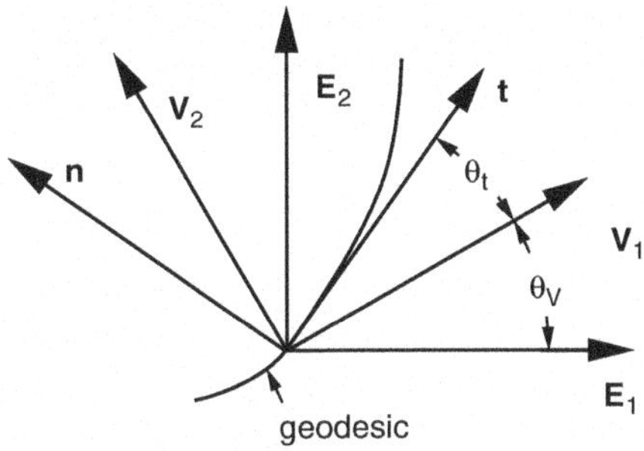

Fig. 6.11 Along a geodesic the angle θ_t remains constant

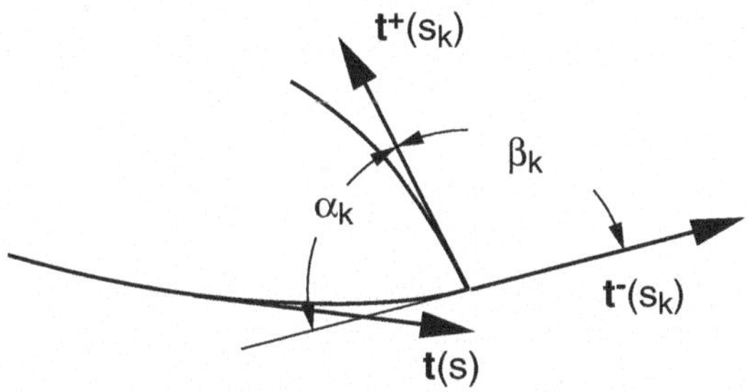

Fig. 6.12 At the kth vertex the unit tangent vector **t** rotates through angle β_k where β_k is the exterior angle for that vertex

vector **n**. See Fig. 6.11. Along a geodesic where the geodesic curvature k_g is zero, the angle θ_t remains constant. However at each vertex, the value of θ_t makes a sudden change. See Fig. 6.12. At the kth vertex θ_t increases by β_k that is the exterior angle of the vertex. (Note! It is possible for β_k to be negative in which case the change in θ_t is not really an increase.) At any rate, when a complete circuit is made,

$$\text{the final value of } \theta_t \text{ minus the initial value of } \theta_t = \sum_k \beta_k . \tag{6.53}$$

Also, when a complete circuit is made,

$$\text{the final value of } (\theta_t + \theta_V) \text{ minus the initial value of } (\theta_t + \theta_V) = 2\pi . \tag{6.54}$$

Furthermore,

$$\text{the final value of } \theta_V \text{ minus the initial value of } \theta_V = \theta, \tag{6.55}$$

where θ is the entity that appears on the right-hand side of (6.52). Subtracting (6.55) from (6.54) gives us

$$\text{the final value of } \theta_t \text{ minus the initial value of } \theta_t = 2\pi - \theta, \text{ or}$$

$$\theta = 2\pi - (\text{the final value of } \theta_t \text{ minus the initial value of } \theta_t) . \tag{6.56}$$

Combining this result with (6.53) and (6.52), we have for simply connected regions bounded by piecewise connected geodesics,

$$\int_A K dA = 2\pi - \sum_k \beta_k. \tag{6.57}$$

where β_k designates the exterior angle at the kth vertex.

Comment! Equation (6.54) is not as trivial as it might first appear. If the boundary curve was allowed to cross itself, then the unit tangent vector might rotate some alternate multiple of 2π with respect to a fixed frame. In his book *Geometry from a Differentiable Viewpoint*, John McCleary (McCleary 1994, p. 173) cites Heinz Hopf's Umlaufsatz (a theorem that says that the tangent along a closed piecewise differentiable curve enclosing a simply connected region turns through 2π (Hopf 1935, pp. 50–62.)). He then goes on to say,

> "The proof of this innocent-sounding result is lengthy and involves ideas that really belong in a course on topology. In order to avoid such a long detour we postpone a sketch of the proof to the end of the chapter."

McCleary presents what he describes as a sketch of the proof on pp. 181–184 of his book. In this book, I will simply omit the proof altogether.

For the case of a geodesic triangle,

$$\sum_k \beta_k = \sum_{k=1}^{3} \beta_k = \sum_{k=1}^{3} (\pi - \alpha_k) = 3\pi - (\alpha_1 + \alpha_2 + \alpha_3),$$

where α_k designates the interior angle of the kth vertex. Substituting this result into (6.57) gives us a result for geodesic triangles that was published by Gauss in 1828 (Gauss 1828). That is

$$\int_A K dA = \alpha_1 + \alpha_2 + \alpha_3 - \pi. \tag{6.58}$$

This is of course a generalization of the result of Thomas Harriot that was discussed at the beginning of Chap. 5. Presumably, Gauss was pleased with the result because he referred to it as "*Theorema Elegantissimum*." (Klingenberg 1978, p. 141).

In 1848, Pierre Ossian Bonnet was able to generalize the result of Gauss to regions that have closed boundaries that are pieced together by finite number of smooth curves, which are not necessarily geodesics (Bonnet 1848). Bonnet was the first to publish this generalized version, but it is plausible that Gauss was aware of the result somewhat earlier (Struik 1988, p. 153). We are now in a position to prove this result, which is now known as the *Gauss–Bonnet formula*.

Theorem 127. *For a 2-dimensional simply connected region bounded by a finite number of twice differentiable paths,*

$$\int_A K dA + \int_{\partial A} k_g ds + \sum_k \beta_k = 2\pi, \tag{6.59}$$

where β_k is the exterior angle of the kth vertex.

Proof. From (6.56),

$$\theta = 2\pi - (\text{the final value of } \theta_t \text{ minus the initial value of } \theta_t).$$

Combining this result with (6.52), we get

$$\int_A KdA + (\text{the final value of } \theta_t \text{ minus the initial value of } \theta_t) = 2\pi, \text{ or}$$

$$\int_A KdA + \int_{\partial A} \frac{d\theta_t}{ds} ds = 2\pi. \tag{6.60}$$

If the unit tangent vector is continuously turning along the entire boundary, then $d\theta_t/ds$ is the geodesic curvature k_g discussed in the last section. In that case, the equation in question becomes

$$\int_A KdA + \int_{\partial A} k_g ds = 2\pi.$$

However if the boundary has vertices, $d\theta_t/ds$ cannot be treated as an ordinary function at those points. Nonetheless, we can still deal with the situation in a rational manner by treating $d\theta_t/ds$ as a distribution function. In particular at the kth vertex, we have

$$\int_{s_k^-}^{s_k^+} \frac{d\theta_t}{ds} ds = \theta_t(s_k^+) - \theta_t(s_k^-) = \beta_k.$$

(See Fig. 6.12.) Equation(6.60) then becomes

$$\int_A KdA + \int_{\partial A} k_g ds + \sum_k \beta_k = 2\pi,$$

where it understood that the vertices are omitted in the calculation of $\int_{\partial A} k_g ds$. □

Problem 128. How should the Gauss–Bonnet formula be adjusted for regions that are not simply connected? (A simply connected region is one with no holes.) Hint! A region with one hole can be considered the difference of two simply connected regions. For Stokes' Theorem, the path integrals along the shared boundary in diagram A of Fig. 6.13 cancel out and the form of Stokes' Theorem remains unchanged for regions that are not simply connected. What makes the Gauss–Bonnet formula different? What is the correct Gauss–Bonnet formula for a region with p holes?

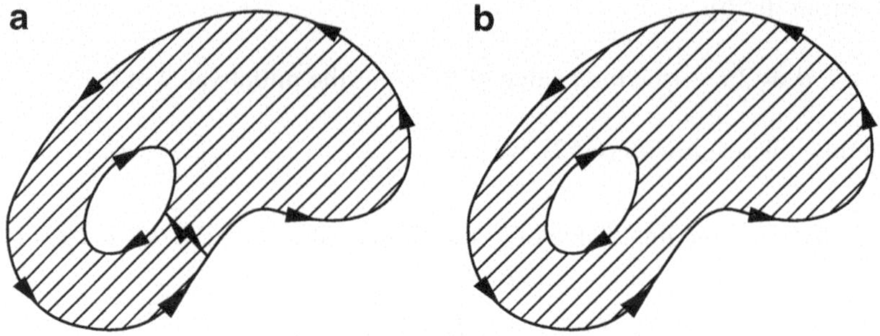

Fig. 6.13 For Stokes' Theorem, you can simply replace the path of integration shown in figure **a** by that shown in figure **b**. For the Gauss–Bonnet Theorem, a further adjustment must be made. What is it?

Fig. 6.14 The pseudosphere

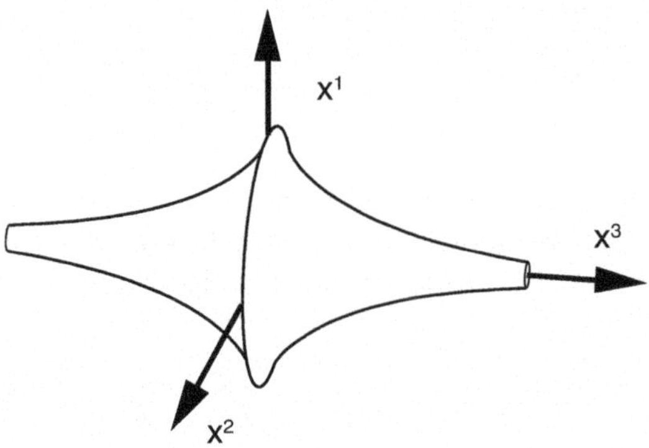

Problem 129. Show $\boldsymbol{\gamma}^k \langle \mathbf{E}_{\bar{1}}.\nabla_k \mathbf{E}_{\bar{2}} \rangle = \mathbf{w}_{12}$, so (6.47) becomes

$$\int_A d\,\mathbf{w}_{12} = \int_{\partial A} \mathbf{w}_{12}.$$

Problem 130. Some authors use the expression R_{1212}/g for the Gaussian curvature where $g = g_{11}g_{22} - g_{12}g_{21}$. Show $R_{1212}/g = R^{12}{}_{12}$.

Problem 131. PSEUDOSPHERE (See Fig. 6.14.)

Consider the surface

$$\mathbf{x}(\phi, \theta) = \mathbf{e}_1 \frac{R\cos\phi}{\cosh\theta} + \mathbf{e}_2 \frac{R\sin\phi}{\cosh\theta} + \mathbf{e}_3 R(\theta - \tanh\theta).$$

Show that the Gaussian curvature $K = R^{12}{}_{12} = -1/R^2$.

Problem 132. Use the Gauss–Bonnet formula to show that on a 2-dimensional surface with negative curvature, you cannot construct a closed figure bounded by only two geodesics.

Problem 133. Use (6.58) to show that on a surface of constant negative Gaussian curvature, there is a least upper bound for the areas of geodesic triangles. Can this least upper bound be realized by an actual geodesic triangle?

6.6 The Interpretation of Curvature 2-Forms as Infinitesimal Rotation Operators

Equation (6.52) can be generalized when ${R_{12}}^{12}$ is not constant – at least for infinitesimal loops. In that case,

$$K dA = {R_{12}}^{12} \, |\boldsymbol{\gamma}_{12}| \, du^1 du^2 = d\theta. \tag{6.61}$$

On the other hand, the rotation operator for such an infinitesimal rotation is

$$\mathbf{R} = \mathbf{I}\cos\frac{d\theta}{2} + \frac{\boldsymbol{\gamma}_{12}}{|\boldsymbol{\gamma}_{12}|}\sin\frac{d\theta}{2} = \mathbf{I} + \frac{1}{2}\frac{\boldsymbol{\gamma}_{12}}{|\boldsymbol{\gamma}_{12}|}d\theta.$$

Using (6.61), this becomes

$$\mathbf{R} = \mathbf{I} + \frac{1}{2}{R_{12}}^{12}\boldsymbol{\gamma}_{12} du^1 du^2 = \mathbf{I} + \frac{1}{4}{R_{12}}^{\alpha\beta}\boldsymbol{\gamma}_{\alpha\beta} du^1 du^2.$$

That is

$$\mathbf{R} = \mathbf{I} + \frac{1}{2}\mathbf{R}_{12} du^1 du^2, \text{ where}$$

$\mathbf{R}_{12}$ is a curvature 2-form. This last equation can also be written in the form:

$$\mathbf{R} = \mathbf{I} + \frac{1}{4}\mathbf{R}_{jk} du^j du^k.$$

It turns out that this last equation is valid in spaces of arbitrary dimension. See (Snygg 1997, pp. 104–110). You should note that the orientation of the rotation depends on the orientation of the infinitesimal loop. If we consider only the lowest order terms, this rotation operator may be factored into rotations each one of which corresponds to an infinitesimal loop in a 2-dimensional coordinate plane. That is

$$\mathbf{R} = \left(\mathbf{I} + \frac{1}{2}\mathbf{R}_{12} du^1 du^2\right)\left(\mathbf{I} + \frac{1}{2}\mathbf{R}_{13} du^1 du^3\right)\cdots\left(\mathbf{I} + \frac{1}{2}\mathbf{R}_{n\,n-1} du^{n-1} du^n\right).$$

Each of these rotation will be quite different. Obviously, the "axis" of rotation for

$$\mathbf{I} + \frac{1}{2}\mathbf{R}_{12} du^1 du^2 = \mathbf{I} + \frac{1}{4}{R_{12}}^{\alpha\beta}\boldsymbol{\gamma}_{\alpha\beta} du^1 du^2$$

will be quite different than the "axis" for

$$\mathbf{I}+\frac{1}{2}\mathbf{R}_{37}du^3du^7 = \mathbf{I}+\frac{1}{4}R_{37}{}^{\alpha\beta}\boldsymbol{\gamma}_{\alpha\beta}du^3du^7.$$

In general, we see that $R_{jk}{}^{\alpha\beta}$ is a measure of the motion of a vector in the α-β plane when it is parallel transported around an infinitesimal loop in the j-k plane.

6.7 *Euler's Theorem for Convex Polyhedrons

The Gauss–Bonnet formula has some significant consequences for closed surfaces (finite surfaces with no boundaries such as the sphere or torus). In Prob. 18, I mentioned that Euler proposed that for any convex polyhedron $F - E + V = 2$, where F is the number of faces, E is the number of edges, and V is the number of vertices. The Gauss–Bonnet formula can not only be used to prove Euler's formula but also a generalized version.

(What follows is an informal version of a proof that appears in Wolfgang Kühnel's *Differential Geometry* (Kühnel 2000, p. 171).)

First suppose the edges of a convex polyhedron are deformed to fit on the surface of a sphere or some topologically equivalent surface such as an ellipsoid. (Clearly, this can be done without changing the number of faces, edges, or vertices.) Then we rewrite (6.59) in terms of interior angles for the jth "polygon" or "face". Since the kth exterior angle of the jth face β_{jk} is the supplement of the corresponding interior angle α_{jk}, we have $\beta_{jk} = \pi - \alpha_{jk}$ and (6.59) becomes

$$\int_{A_j} KdA + \int_{\partial A_j} k_g ds = \sum_k (\alpha_{jk} - \pi) + 2\pi \text{ or}$$

$$\int_{A_j} KdA + \int_{\partial A_j} k_g ds = \sum_k \alpha_{jk} - E(j)\pi + 2\pi, \tag{6.62}$$

where $E(j)$ designates the number of edges associated with the jth face. Now suppose we sum (6.62) over all faces on the sphere (Fig. 6.15). This gives us

$$\sum_j \int_{A_j} KdA + \sum_j \int_{\partial A_j} k_g ds = \sum_{j,k} \alpha_{jk} - \sum_j E(j)\pi + 2\pi F. \tag{6.63}$$

The first thing to observe is that the the total contribution of the geodesic curvature is zero. This is because the contributions of the shared edges of adjacent "polygons" cancel out. (Draw your own picture.) Second, any number of faces may share a common vertex but the sum of the interior angles at any given vertex must add up to 2π. Thus, the total sum of interior angles $\sum_{j,k} \alpha_{jk}$ must be $2\pi V$, where V is the

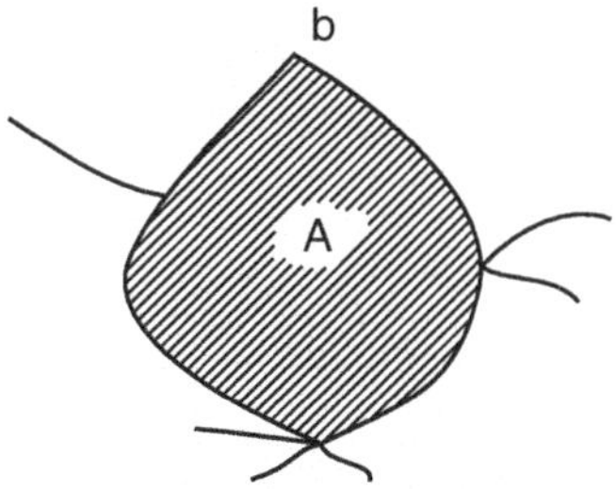

Fig. 6.15 For purposes of this section, area A represents a "pentagon" with five edges and five vertices even though it only has two corners. Note that if corner b was smoothed out, both the number of edges and vertices would be reduced by one and the Euler characteristic $F - E + V$ would remain the same

number of vertices. Furthermore, $\sum_j E(j) = 2E$ because every edge in the sum is counted twice. Thus, (6.63) becomes

$$\int_A K dA = 2\pi(F - E + V). \tag{6.64}$$

The integral on the left-hand side of (6.64) is known as the *total Gaussian curvature* for the closed surface.

For the sphere,

$$\int_A K dA = \frac{1}{R^2}\int_A dA = \frac{1}{R^2}\left(4\pi R^2\right) = 4\pi. \tag{6.65}$$

Combining this with (6.64) gives us

$$F - E + V = 2. \tag{6.66}$$

Equation (6.66) is known as *Euler's Theorem (or formula) for Convex Polyhedrons*. This formula is quite profound, but the consequences of (6.64) are even more profound.

First, a review of the proof shows that the result is not restricted to spheres or ellipsoids. Clearly, it is valid for any smooth closed surface that can be continuously deformed into a sphere. Second, I started the proof by deforming the edges of a convex polyhedron onto a sphere. But I could have started with some lattice already on the sphere. Similarly, I could have started by dividing up a torus into "polygons", and we would still have arrived at (6.64). In the case of the torus, the total Gaussian curvature would have been different so the sum $F - E + V$ would have been different also. In general for any smooth closed surfaces, we can write

$$\int_A K dA = 2\pi\chi(A), \text{ where} \tag{6.67}$$

$$\chi(A) = F - E + V. \tag{6.68}$$

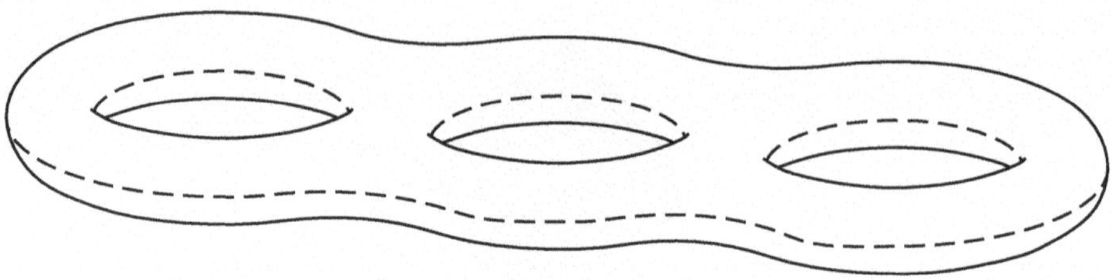

Fig. 6.16 A surface of genus 3 sliced like a bagel

In this context, $\chi(A)$ is known as the *Euler* or *Euler–Poincaré characteristic of the surface.* Surfaces that can be continuously deformed into one another have the same Euler characteristic. For this reason, $\chi(A)$ is regarded as a "topological invariant." The Euler characteristic for a doughnut is the same as that of a teacup with one handle. (Geometers tend to extol themselves for having the insight to see that doughnuts and teacups have a similarity while others dismiss them for "not being able to tell the difference.")

To conclude this chapter, I will mention the notion of genus. The *genus* of a closed surface is equal to the number of holes in the solid covered by the closed surface. Thus, a sphere has genus zero; a torus has genus one; a teacup with two handles has genus two: and a sphere with p-handles has genus p. I leave it to you to show that using the Gauss–Bonnet Theorem,

$$\int_A K dA = 4\pi(1 - p) \text{ or} \tag{6.69}$$

$$F - E + V = 2(1 - p), \text{ where } p \text{ is the genus of } A. \tag{6.70}$$

Problem 134. (a) Use the result of Prob. 98 to show that the total Gaussian curvature for a torus is zero.

(b) Figure out some way of dividing the surface of a torus into "polygons" so that you can use (6.64) to check the result of part (a).

Problem 135. Prove (6.69).

Suggestion: Slice up a version of a multihole torus into two surfaces (See Fig. 6.16). Then use the result of Prob. 128. Alternatively, figure out some way of dividing up the surface of a multihole torus into "polygons" so that you can apply (6.64).

6.8 *Carl Friedrich Gauss and Bernard Riemann

*6.8.1 *Carl Friedrich Gauss 1777–1855*

Since his contributions to mathematics and physics are both deep and wide, Gauss is frequently honored by the use of his name. Most of statistics is based on the

Gaussian distribution. In electrostatics, he is honored by *Gauss' Theorem* (a special case of the Generalized Stokes' Theorem). In this book, we have encountered *Gaussian curvature* and the *Gauss–Bonnet formula*. My math dictionary mentions *Gauss' formulas* (some spherical trigometric formulas). For his contributions to number theory, one encounters *Gaussian integers*. Carl Gauss collaborated with Wilhelm Weber in the study of magnetic fields and both are honored by physicists who use their names for units of magnetic field strengths. There is even a *Gauss crater* on the moon.

Gauss and Weber attained public acclaim in Europe for their invention of the first electromagnetic telegraph in 1833 – two years before Samuel Morse had his telegraph working.

To measure the shape of the earth, Gauss invented the heliotrope and before the introduction of satellite technology his theoretical work was the foundation of modern geodesy (Hall 1970, pp. 89–90). Gauss also had a gift for languages. Along with his native German, he was considered literate in Greek, Latin, English, and French. At the age of 60, he also became literate in Russian. On the other hand, his knowledge of Italian, Spanish, and Swedish was considered to be superficial (Hall 1970, p. 159).

Although Gauss is mainly recognized for his contributions to mathematics, for virtually his entire working life, Gauss made his living as an astronomer.

Gauss did not begin life in promising circumstances. He was born on April 30, 1777 in the city of Braunschweig (also known as Brunswick). Braunschweig was the capital of a duchy that had the same name. Gauss' father was a strict disciplinarian and if his father had had his way, Carl would have joined his father doing odd jobs such as stone mason, canal worker, and gardener.

However, his teachers recognized his extraordinary talent in mathematics and Duke Carl Wilhelm Ferdinand of Braunschweig was persuaded to become his patron when Gauss was 14 years old. He attended the Collegium Carolinum for three years before entering the University of Göttingen in the fall of 1795. During the spring of his first year, at the age of 18, Gauss proved that a regular polygon with 17 sides could be constructed by compass and ruler. To achieve this result, he showed that the largest real part of any of the roots of

$$x^{16} + x^{15} + x^{14} + - - - + x^2 + x + 1 = 0$$

is

$$-\frac{1}{16} + \frac{1}{16}\sqrt{17} + \frac{1}{16}\sqrt{34 - 2\sqrt{17}}$$
$$+\frac{1}{8}\sqrt{17 + 3\sqrt{17} - \sqrt{34 - 2\sqrt{17}} - 2\sqrt{34 + 2\sqrt{17}}}.$$

(Hall 1970, pp. 21–34) One should note that a thousand place decimal expansion would not have demonstrated that construction of the 17-gon is possible by compass and ruler.

Carl Frederich Gauss made few friends while a student but he did make friends with Wolfgang Bolyai (otherwise known as Bolyai Farkas). Wolfgang was a Hungarian who would become a lifetime correspondent with Gauss. After three years at Göttingen, Gauss returned home in Braunschweig hoping the Duke would continue to be his patron in the fall of 1798. It was at least two months before he obtained an audience with the Duke. When he finally saw the Duke, the Duke agreed to continue his financial support of Gauss' studies. Gauss was eager to get some of his mathematical results published but the Duke urged him to get a Ph.D (Dunnington 1955, pp. 34–35). It was agreed that Gauss would obtain a Ph.D. at Helmstedt where Germany's most prominent mathematician Johann Friedrich Pfaff was on the faculty.

Gauss attended no courses at Helmstedt, made substantial use of the library, had some conversations with Prof. Pfaff, and was awarded a Ph.D. on July 16, 1799. This was roughly nine months after he made his first visit to Helmstedt. The degree was awarded *in absentia* without the usual oral examination (Bühler 1981, p. 17).

Unlike most Ph.D's, that of Gauss was a significant piece of work – a proof of the Fundamental Theorem of Algebra. The theorem states that any polynomial with real coefficients can be factored into terms that are either linear or quadratic. For many, this was considered to be proven previously but Gauss demonstrated that previous proofs were inadequate and he introduced a new rigor into the problem.

At about the same time, Gauss was assembling a book on number theory entitled *Disquisitiones arithmeticae* (Arithmetical Investigations), which is considered by some his most significant work. Among other things, it discusses number theory using substantial amounts of modular arithmetic. The printing took longer than Gauss expected but the Duke paid the cost of publication and the book appeared in print in 1801.

Lagrange wrote to the young Gauss, "Your *Disquisitiones* have with one stroke elevated you to the rank of the foremost mathematicians, and the contents of the last section (theory of the equations of circle division) I look on as the most beautiful analytic discovery which has been made for a long time." (Dunnington 1955, p. 44)

Even after Gauss obtained his Ph.D., the Duke continued to support him financially. During these years in Braunschweig, Gauss had no particular duties yet he was extremely productive. Long after Gauss' death, Felix Klein (1849–1925) described Gauss' years in Braunschweig as his *Heldenzeit* (heroic period) (James 2002, p. 62). However, biographer W.K. Bühler states that when Gauss terminated his student days at Göttingen in the fall of 1798, "...he had already developed the basic ideas of nearly all his important mathematical papers, which he was to publish over the next twenty-five years." (Bühler 1981, p. 17)

On January 1, 1801, the Italian astronomer Joseph Piazzi (1746–1826) in Palermo was first to discover an asteroid (Ceres). However after tracking it for only 41 days and 9 degrees of arc, it disappeared behind the sun (Hall 1970, p. 69). It then became a challenge to the best astronomers of Europe to predict where it would reappear. Gauss developed a new approach to the problem and his prediction was quite different from any of the others. Thus when Gauss' prediction proved correct, he achieved fame among the community of astronomers and in September of 1802 he was soon offered the position of director of the observatory in St. Petersburg, Russia (Hall 1970, p. 67). In response, the Duke raised Gauss' stipend and Gauss was soon planning an up-to-date observatory to be built in Braunschweig with the Duke's financial support.

On January 30, 1806, the Duke was sent on a diplomatic mission to St. Petersburg in hopes that the Russians would join Germany as an ally against Napoleon. The mission was a diplomatic failure but the Russians urged the Duke to allow the young astronomer Gauss to accept a second job offer. When the Duke returned to Braunschweig in the spring of 1806, he raised Gauss' stipend again (Dunnington 1955, pp. 78–79). Meanwhile, Gauss had fallen madly in love with Johanna Osthoff, the daughter of a tannery owner in Braunschweig and married her on October 9, 1805.

This was an idyllic time for Gauss but it was about to have an abrupt end. At the age of 71, Duke Carl Wilhelm Ferdinand accepted command of a woefully deficient army to defend against the invasion by Napoleon. The decisive battle of Jena–Auerstädt took place on October 14, 1806. In the course of battle the Duke was mortally wounded by a musket ball that entered above his right eye and carried away his left eye. The Duke was taken back to his palace in Braunschweig and a delegation was sent to Napoleon asking him to allow the Duke to die in peace among family and friends. Napoleon denied the request and on the morning of October 25, Gauss saw a long carriage drawn by two horses leave the castle yard with the Duke. The Duke was taken away out of Napoleon's reach, north to Altoona where the Duke died on November 10 (Dunnington 1955, p. 82).

Gauss now had a wife and an infant son (Joseph). He was 29 and the Duke had been his sole means of support since the age of 14. The following summer, Gauss was offered the position of director of the observatory at the University of Göttingen. Gauss accepted the offer and he held that position for the rest of his life.

During his lifetime, he published approximately 150 papers. However, the majority of his work was not published until after his death. He was writing for posterity and was reluctant to publish material that had loose ends or could plausibly be polished and made more elegant. It is also conceivable that he sometimes withheld preliminary results in hopes that he would achieve priority for more significant results.

Many times young mathematicians would send Gauss some of their results with the hope of getting his endorsement. Often Gauss would respond that their accomplishment was significant, although he had achieved the same result 20 or 30 years earlier. This happened so often that some began to suspect that Gauss was not being honest about his claims. Although priority is generally given to the one who publishes first, Gauss' behavior tended to undercut the accomplishments of others and Lagrange was particularly outraged by this. Only after his death did it become clear that most of the time Gauss was at least being truthful. (His treatment of Grassmann was an exception.) Gauss' reluctance to publish had another consequence that angered mathematicians other than Lagrange. It was disappointing for anyone to discover that he had spent months or years repeating what Gauss had done earlier when they could have been advancing the cutting edge of mathematics.

Soon after Gauss died, the Royal Scientific Society of Göttingen set out to organize his papers and publish them with annotations. This project took seven decades between 1863 and 1933 and was carried out by ten German scientists, each an expert in his own field. The result was a twelve volume set of books known as Gauss' *Collected Works* (Hall 1970, p. 164).

Gauss never published any of his extensive work on non-Euclidean geometry. The circumstances surrounding this fact tells us a lot about Gauss' personality.

Non-Euclidean geometry evolved from a 2000-year long struggle to deal with the concept of parallel lines in a logical manner. When Euclid (~300 BC) laid down his axioms and postulates, it was clear that he was uncomfortable with his last postulate that is sometimes known as the axiom of parallelism. In an equivalent form, this postulate stated by Proclus about 700 years later is "In the plane, given a straight line and a point not on this line, there exists one and only one straight line passing through the given point that does not intersect the original given line."

Euclid tried to prove as much as he could without using this postulate. Furthermore without using this postulate, he was able to show that the remaining axioms and postulates implied there would be at least one nonintersecting line passing through the given point (Kárteszi 1987, pp. 14–15). Proclus also tried unsuccessfully to prove that the axiom of parallelism was unnecessary.

It was only during Gauss' lifetime that Johann Bolyai, Nikolai Lobachevsky, and Gauss himself made the critical breakthrough. Indeed, they discovered that one could replace the axiom of parallelism by an axiom (or postulate) that stated, "In a plane, given a straight line passing and a given point not on this line, there exist more than one straight line that does not intersect the original given line." If one makes this replacement, one gets a geometry in which the sum of the interior angles of a triangle is less than 180^0 and similar triangles are also congruent. About 13 years after Gauss died, Eugenio Beltrami made the observation that if one identified the geodesics of a surface of constant negative curvature with "straight lines", one is dealing with non-Euclidean geometry (Beltrami 1868). Later, it was observed that if one not only adjusted the axiom of parallelism but also another postulate of Euclid, one would get another form of non-Euclidean geometry in which "straight lines" would correspond to the geodesics on the surface of a sphere.

During the eighteenth century, interest in the axiom of parallelism had been stimulated by a controversial assertion by the philosopher Immanuel Kant (1724–1804) (Kárteszi 1987, p. 15). As a result, many tried to derive the axiom from the remaining axioms and postulates. Of course, these efforts all ended in failure. Others tried to prove the postulate by contradiction. That is assume that the more than one line passing through the given point is nonintersecting and arrive at some totally absurd result.

Some of these mathematicians got some interesting results. They derived some consequences that were seemingly peculiar but not entirely illogical. These people included Giovanni Girolamo Saccheri (1667–1733) and Johann Heinrich Lambert (1728–1777) – both before the Gauss era. Along the same lines, two lawyers: Ferdinand Karl Schweikart (1780–1859) and his nephew Franz Adolf Taurinus (1794–1874) achieved some results that they shared in personal correspondence with Gauss. From the responding letters to Schweikart in 1819 and to Taurinus in 1824, it is clear that Gauss had made a great deal of progress in developing a full-blown non-Euclidean geometry (Dunnington 1955, pp. 180–182).

Nonetheless, Gauss was reluctant to publish his results. His friend Friedrich Wilhelm Bessel urged him to publish this material and in a famous letter of reply written on January 27, 1829, Gauss wrote, "Perhaps it will not happen during my lifetime, since I fear the Boeotian's cries if I were to express my opinion clearly" (Bonola 1955, p. 67) and (Hall 1970, pp. 112–113). (The Boeotians were a Greek tribal group who were reputed by their enemies to be dimwits.) In the 1824 letter to Taurinus, Gauss expressing similar fears had urged Taurinus to keep the letter private.

Many historians take the position that Gauss' fears of a negative outcry were unjustified. However in Gauss' time, the study of geometry was considered to be the study of God's creation. Even Gauss posed the question, "What is the true geometry?" In 1816, Gauss had used tough language when he reviewed two articles on the parallel postulate in a scientific journal. In return, he was subjected to vulgar attack (Hall 1970, p. 113).

Those who think that Gauss' prestige would have made him immune from the kind of abuse suffered by Einstein in pre-Nazi Germany or by Fock in Stalin's Russia should consider a diatribe cited by Gauss biographer Waldo Dunnington (1955, p. 274). The following was written roughly 20 years after Gauss' death when his private math papers had become public. Referring to Gauss, Eugen Karl Dühring (1833–1921) wrote:

> His megalomania rendered it impossible for him to take exception to any tricks that the deficient parts of his own brain played on him, particularly in the realm of geometry. Thus he arrived at a pretentiously mystical denial of Euclid's axioms and theorems, and proceeded to set up the foundations of an apocalyptic geometry not only of nonsense but of absolute stupidity . . . They are abortive products of the deranged mind of a mathematical professor, whose mania for greatness proclaims them as new and superhuman truths! The mathematical delusions and deranged ideas in question are the fruits of a veritable *paranoia geometrica.*

It would appear from this passage that Dühring was a raving lunatic. Dühring was a contentious anti-Semite. Nevertheless, his contemporaries took his writings seriously. His following as a political philosopher was so substantial that Frederick Engels felt compelled to write an entire book to refute some of his writings (Anti-Dühring (1878)). According to Michael Monastyrsky, one of the Riemann's biographers, Dühring's name (but not his work) became wellknown in the Soviet Union since Engel's work was required reading (Monastyrsky 1998, p. 78).

More relevant for the history of mathematics is the fact that in 1872 Eugen Karl Dühring was awarded the Benecke Prize by the philosophical faculty of the University of Göttingen. This prestigious prize was awarded for a 513-page history of mechanics.

Most of this work of Dühring is a scholarly treatment of the contributions of such figures as Archimedes, Leonardo da Vinci, Galileo, Descartes, Newton, Fermat, the Bernoullis, Leibnitz, L'Hopital, d'Alembert, Lagrange, Carnot, Euler,

Hamilton, and Joule. However, when he gets to Gauss and Riemann, he expresses utter contempt for the notion of non-Euclidean geometry. To support his ridicule, he cites a letter that Gauss wrote to Schumacher on July 12, 1831. Dühring indicates that Gauss wrote that in a non-Euclidean geometry, one would have equilateral triangles with unequal angles whose sum would be less than 180^0 (Dühring 1873, pp. 488–489). In a later edition (Dühring 1877, p. 459), Dühring published the relevant passage. It is clear that what Gauss actually wrote was that for an equilateral triangle, each angle would be unequal to 60^0 (not unequal to one another). This distinction may have been too subtle for Dühring but that did not prevent Dühring from increasing the harshness of his invective toward Gauss as the years went by.

Why would the philosophy faculty of the University of Göttingen endorse the views of someone attacking Gauss who was probably the most prestigious member that the faculty at their university ever had? It so happens that when non-Euclidean geometry was becoming mainstream mathematics, the leading figure resisting this development was a philosopher by the name of Rudolf Hermann Lotze (1817–1881) (Szenassy 1987, p.234). Presumably, it was not a coincidence that Lotze was a member of the philosophy department at the University of Göttingen at the time the Benecke prize was awarded to Dühring. Furthermore, he had been a member of the same faculty since 1844, roughly 10 years before Gauss died. So much for faculty collegiality.

How did non-Euclidean geometry eventually gain acceptance? It did not occur overnight. In January 1832, Gauss received a letter from his old friend Wolfgang Bolyai. One has good reason to believe that Wolfgang would have been quite worried about what kind of response he would get. Wolfgang had written a letter to Gauss 16 years earlier without getting a reply.

In that earlier letter, Wolfgang had asked Gauss to take Johann (then 13) into his home as an apprentice mathematician. In the same letter, Wolfgang also mentioned that he had suffered a financial set back in 1811 because of a currency devaluation. Gauss was not the nurturing type. Two of his own sons emigrated to America to escape his control – one at the age of 19 and the other at the age of 24. Thus, Gauss would probably not have been eager to accept the young Johann into his home in the best of circumstances. When he sent his request to Gauss, Wolfgang did not know what the situation was in Gauss' home. At the time of their previous correspondence in 1808, Gauss had two children. Meanwhile, Gauss had remarried after his first wife had died and his second wife was now expecting her third child. Furthermore, Gauss was then in the process of making arrangements for his mother to move in.

Having not heard from Gauss in 1816, Wolfgang Bolyai was writing to Gauss in 1832 seeking an evaluation of a paper written by his son Johann (also known as Bolyai János). The paper was a fairly complete discussion of the foundations of hyperbolic non-Euclidean geometry. The paper was not easy reading and it arrived at another bad time for Gauss. Gauss' second wife Frederica (Minna) had died on September 12, 1831 after a 14-year battle with tuberculosis. This was only a few months before receiving the Bolyai paper.

Gauss eventually responded to the January letter on March 6, 1832. The responding letter contains lines that have become infamous,

> Now something about the work of your son. If I begin by saying that I must not praise him, surely, you will be startled for a moment; but I cannot do otherwise; praising him would mean praising myself; because all the contents of the work, the way followed by your son, and the results he obtained agree almost from beginning to end with the meditations I have been engaged in partly for 30–35 years already. This extremely surprised me indeed (Kárteszi 1987, p. 34).

At first glance, these appear to be self-serving comments by an insensitive egomaniac. As noted above, this was the kind of remark received by other young mathematicians seeking Gauss' approval. Certainly, it was not what Johann Bolyai wanted to hear. However, one must keep in mind that this letter was addressed to Johann's father and that puts it in a somewhat different context. Actually, Wolfgang was elated by the response. He wrote to his son, "Gauss' answer respecting your work is very fine and redounds to the honor of our fatherland and nation. A good friend says it would be a great satisfaction."

Why did Wolfgang accept Gauss' response so favorably? The educational background of Wolfgang and his son Johann differed enough to give them divergent outlooks. Wolfgang and therefore his son came from a noble family but Wolfgang suffered financial reversals that made it impossible to give his son the same education that he had received. Wolfgang had attended the University of Göttingen, which had a fine library and where mathematicians were doing cutting edge research. Wolfgang's job as professor of mathematics, physics, and chemistry at the Reformed College of Marosvásárhely did not pay well and Wolfgang did not have the financial resources to send his son to Göttingen. As a result, Johann obtained most of his mathematical education from his father and then chose a career as an officer in the Hungarian army.

Wolfgang had spent hours on end over a period of many years trying to derive the parallel axiom from the other Euclidean axioms. Once or twice he thought he had been successful but Gauss was able to detect flaws in his "proofs." When his son wrote to his father in 1820 that he was taking up the problem, Wolfgang tried to discourage him.

> Do not waste even one hour's time or that problem. It does not lead to any result; instead it will come to poison all of your life. For hundreds of years hundreds of the world's foremost geometers have cogitated without having succeeded in proving the parallel axiom, as long as they refrained from taking some new axiom as help. I believe that I myself have investigated all conceivable ideas in this connection (Hall 1970, p. 113).

His father's advice was ignored and in November of 1823 Johann informed his father, "From nothing I have created a new world." (Hall 1970, p. 113) In February 1825, Johann visited his father but was unable to convince his father that his non-Euclidean structure was not flawed (Kárteszi 1987, p. 33). Nonetheless, Wolfgang recognized that if his son was correct he had achieved an historically important accomplishment. On the other hand, from letters he had received from Gauss in 1799 and 1804, Wolfgang had good reason to believe that his son's work was probably not quite as original as his son thought it was. He therefore urged his son to get his results written up and published as soon as possible. Johann, referring to his father, commented,

> He advised me that, if I was really successful, I should speedily make a public announcement and that for two reasons. One reason is that the idea might easily pass to someone else who would then publish it. Another reason - and one that seems valid enough - is that when the time is ripe for certain things, these things appear in different places in the manner of violets coming to light in early spring. And since scientific striving is like a war of which one does not know when it will be replaced by peace one must, if possible, win; for here preeminence comes to him who is first (Meschkowski 1964, pp. 33–34).

Fortunately, Wolfgang was completing a two-volume mathematical work in Latin for his college (Szenassy 1987, p. 221). In 1829, the printing was authorized and Johann's results were added as an appendix. Wolfgang saw to it that the appendix was printed before the rest of the work.

When he mailed the "Appendix" to Gauss, he was presumably worried about two things: Had Gauss already published some of the same material and was his son's work logically correct? On both counts Gauss' response told Wolfgang what he wanted to hear. The lines of Gauss' response following those already quoted were:

> It had been my intention to publish nothing of my own work during my life; by the way, I have noted down only a small portion so far. Most people do not even have a right sense of what this matter depends on, and I have met only few to accept with particular interest what I told them. One needs a strong feeling of what in fact is missing and, as to this point, the majority of people lack it. On the other hand, I had planned to write down everything in the course of time so that at least it would not vanish with me some day.
> Thus I was greatly surprised that now I can save myself this trouble, and I am very glad that it is just my good old friend's son who so wonderfully outmatched me. ... (Kárteszi 1987, pp. 34–35).

Wolfgang was probably also elated to see himself referred to as "old friend." Since Gauss never responded to Wolfgang's move-in proposal for his son, he had not heard from Gauss since 1808, which was more than 20 years before (Schmidt and Stäckel 1899).

Wolfgang's concerns about getting Johann's priority established turned out to be well founded. At approximately the same time that Wolfgang was urging his son to get his work published, Nikolai Ivanovich Lobachevsky published a lengthy dissertation that covered roughly the same material. Since Lobachevsky's paper was written in Russian, it received essentially no attention in western Europe. In 1837, Lobachevsky published a paper in French. (Lobachevsky 1837). In 1840, Lobachevsky did receive some limited attention when he had a 61-page book published in Berlin under the title *Geometrische Untersuchungen zur Theorie der Parallelinien* (*Geometrical Investigations on the Theory of Parallels*).

However, neither Johann Bolyai nor Nikolai Lobachevsky received the kind of attention that they deserved during their lifetimes. Most of those who were aware of their work were inclined to treat their expositions as mathematical curiosities, which were probably logically flawed.

Gauss is frequently criticized for not publicly drawing attention to their work and thereby giving them the prominence that they deserved. However without being willing to face the cries of the "dimwit Boeotians", there may have been little that Gauss could do. In 1842, Gauss was able to get Lobachevsky elected as a corresponding member of the Royal Society of Sciences in Göttingen. The official letter of notification sent to Lobachevsky makes no mention of non-Euclidean geometry. It is plausible that Gauss was avoiding a confrontation with fellow members of the Göttingen faculty. (This is my personal speculation that may be on thin ice since it was about two years before Rudolf Hermann Lotze joined the Göttingen faculty. But it is unlikely the Benecke prize was awarded to Dühring by a one person committee.) Lobachevsky would have known the reason for the honor even if some members of the Göttingen faculty might not have. After all there was only one reason for Lobachevsky to receive such an honor. According to Dunnington, Gauss wrote Lobachevsky a personal letter of congratulations (Dunnington 1955, p. 187). The Russian historian, G. E. Izotov is convinced that no such letter was sent (Izotov 1993, p. 9). However, Dunnington describes the contents of a return letter dated June 1843 that Lobachevsky sent to Gauss (Dunnington 1955, p. 187).

In either case, the honor had little practical effect. Four years later, Lobachevsky was forced into early retirement from his position as Rector at the University of Kazan (Dunnington 1955, p. 187).

Some years before, Johann Bolyai encountered a circumstance that was not completely dissimilar. A few months after Johann Bolyai received a copy of Gauss' response to his father, Johann applied for a furlough from the Hungarian army so that he could develop a more elaborate version of his "Appendix." The mathematicians who evaluated his application were not able to understand it. And when their attention was drawn to Gauss' letter, the favorable remarks were attributed to the early friendship between Gauss and Johann's father (Szenassy 1987, p. 225).

Had Gauss been willing to do battle with the "Boeotians" would things have gone better for Johann Bolyai and Nicholas Lobachevsky? Several historians have pointed out that it was only after Gauss' death in 1855 when his private correspondence was published, that the mathematical world began to take non-Euclidean geometry seriously. However, besides the posthumous publication of Gauss' papers, there was another factor that set the stage for acceptance of non-Euclidean geometry in the mathematical community.

Geometer Michael Spivak once wrote, "The single most important work in the history of differential geometry is Gauss' paper of 1827 *Disquisitiones generales circa superficies curvas* (in Latin)." (Spivak 1970, p.3A-1). (In the succeeding pages of his book, Spivak gives tips on how to read an English translation of the work (Gauss 1965).) The short book cited by Spivak includes an early version of the Gauss–Bonnet theorem that stimulated substantial research into curved surfaces. For example, Ferdinand Minding (1806–1885), inventor of the pseudosphere, was able to show that if one took some of the formulas one encounters in spherical trigonometry and formally replaced R (the radius of the sphere) by iR, one would obtain formulas valid on a surface with constant negative curvature $-R^2$ (Minding 1840, pp. 323–327).

Lobachevsky and Bolyai had obtained the same results in their non-Euclidean geometry. As a matter of fact, in his early papers, Lobachevsky called this geometry "imaginary."

Eventually, it was observed that at least for two dimensions, one can logically identify the geodesics on a surface of constant negative curvature with the "straight lines" in the non-Euclidean geometry of Bolyai and Lobachevsky. Thus, the mathematical community was much more receptive to non-Euclidean geometry by the time Gauss' papers were published. Presumably, the community of philosophers were less familiar with surfaces of negative curvature and were thus less receptive to any drastic alternative to the axiom of parallelism.

Unfortunately, the acceptance of non-Euclidean geometry in the mathematical community came too late to benefit either Bolyai or Lobachevsky. Both died only a few years after Gauss. Nicholai Lobachevsky died in 1856 and Johann Bolyai died in 1860.

In April 1831, a few months before Gauss' second wife died, Wilhelm Eduard Weber (1804–1891) arrived in Göttingen to assume the position of professor of physics. Gauss had recommended Weber for the position, and although Gauss was almost 27 years older, they soon became close friends and collaborators.

They worked jointly in the study of magnetism and, as a result, they are both honored by having units of magnetism named after them. Together, they stimulated an international project to measure the daily fluctuation of the earth's magnetic field. However, political events would soon interfere with their relationship.

From 1714 to 1837, Hanover and Great Britain shared the same king. As a matter of fact between 1714 and 1760, Great Britain had a succession of two kings (George I and George II) who did not speak English. Neither one of them had a deep interest in the affairs of Great Britain. By default, the power of prime ministers and parliament grew under their reigns.

George III (grandson of George II) became king in 1760. George III had a reputation for being intellectually slow but he was much more actively engaged in governing than his two predecessors. His approach was sufficiently tyrannical to provoke 13 American colonies to revolt.

George III was succeeded by two of his sons (George IV and then William IV). It is possible that William IV learned something from the mistake of his father. At any rate when faced with political turmoil in Hanover, he granted the people of Hanover a more democratic constitution in 1833.

When William IV died in 1837, Victoria became Queen of Great Britain. However, the law of succession in Hanover did not permit a female ruler. At this point, Hanover separated from Great Britain and Ernest Augustus who was the fifth son of George III now became King of Hanover at the age of 66. Although he did not speak German, Ernest Augustus was determined to make an impact. That he did. The faculty at Göttingen soon got an indication of what was to come. In September of 1837, the university was celebrating its 100th anniversary. The new king showed up for the festivities but when a monument to William IV was unveiled, Ernest Augustus showed his displeasure by turning away (Dunnington 1955, p. 196).

Before the end of the year, Ernest Augustus revoked the constitution that his brother had agreed to four years earlier. Seven prominent members of the Göttingen faculty signed a petition in protest. In the eyes of the new king, this act was a serious affront. The fact that the University of Göttingen had been founded by his great grandfather 100 years before may have made the protest more reprehensible. At any rate, the seven professors were summarily fired. Two of the professors were particularly close to Gauss. One was his son-in-law Georg Heinrich August Ewald (1803–1875) who was a professor of oriental languages. Another was Wilhelm Weber.

Although Gauss did not try to defend his son-in-law, he tried to use his contacts to reverse the king's decision to fire Weber. In particular, he wrote to his friend Alexander von Humboldt indicating that the continuance of his whole scientific activity depended on Weber's staying in Göttingen. Humboldt was not able to gain direct access to the king but two members of the court informed Humboldt that the king would not reconsider his decision (Dunnington 1955, p. 200). Later at a banquet the king told Humboldt, "With my money I can buy as many ballet dancers, whores, and professors as I wish." (Hall 1970, p. 158)

Without employment, Weber stayed in Göttingen until 1843 when he accepted a position as professor of physics at Leipzig. Meanwhile, Ewald obtained a position in Tübingen. In 1848, when disorders forced the royalty of Europe into temporary retreat, Weber and Ewald were able to return to Göttingen. But by then Gauss was 71 and too old to pick up where he had left off with Weber.

In 1854, Gauss was diagnosed as having an enlarged heart and he died the following year on February 23, 1855 at the age of 77.

*6.8.2 *Georg Friedrich Bernhard Riemann 1826–1866*

Bernhard Riemann like William Clifford died from tuberculosis at a relatively young age. Nonetheless, he had a tremendous impact on the development of mathematics.

Today, his name is attached to many concepts that he developed. These include the *Riemann curvature tensor*, the *Riemann integral*, *Riemann surfaces*, *Riemann curvature*, and *Riemannian geometry*. In the year 2000, Clay Mathematics Institute (www.claymath.org) announced that the Institute would award a $1 million prize to anyone who proved the *Riemann hypothesis*.

Generally, his contributions were characterized by a great deal of originality – going beyond problems that were being attacked by his contemporaries. It is believed that some of the originality in his approach to mathematics stemmed from his study of physics. Indeed, during his career he wrote about the theory of gasses, fluid dynamics, heat, light, magnetism, and acoustics. He himself stated that the laws of physics were his greatest interest (James 2002, pp. 188–189).

Bernhard Riemann was born on September 17, 1826 in the rural village of Breselenz, near the city of Dannenberg in the kingdom of Hanover. His father was a well-educated Lutheran pastor who had served as a lieutenant in the Napoleonic Wars of 1812–1814. Bernhard was the second child in a family with two boys and four girls. The local countryside was poverty stricken and Bernhard's family suffered from malnutrition (Motz and Weaver 1993, p. 235). Bernhard's own health was never very good and he died about two months before his 40th birthday. Nonetheless, he not only outlived both of his parents but he also outlived all his siblings except for one sister.

Until he was ten years old, Bernhard's only teacher was his father. At that time, a teacher from a local school was recruited to supplement Bernhard's education in mathematics. In 1840 at the age of 14, Bernhard was sent to the Lyceum in Hanover (the city) where he could live with his grandmother. Apparently, he was unhappy there so when his grandmother died after Bernhard had been in Hanover for two years, he transferred to Johaneum Gymnasium in Lüneburg. He excelled in mathematics but he struggled in most of his other subjects. Schmalfus, the director of the gymnasium recognized Bernhard's mathematical talents and on one occasion he lent him Legendre's book on the theory of numbers. Bernhard read the 900-page cutting edge book in six days. On the other hand, his teachers became concerned because his written assignments were not being completed when due. As a result in 1844, Schmalfus requested G.H. Seffer, the Hebrew teacher, to take Bernhard into his house as a boarder at a reduced rate so that Bernhard would get closer supervision (Laugwitz 1999, pp. 7–8).

In the spring of 1846 at the age of 19, Riemann entered Göttingen University. Due to the dismissal of the "Göttingen Seven" in 1837 by Ernest Augustus, Göttingen University had lost a great deal of prestige and was not the best place to pursue a degree in mathematics. Although Gauss was on the faculty, he was now 69 and generally taught only low level courses addressed to his astronomy students. However, following the wishes of his father, Bernhard was pursuing a financially secure career in the ministry and Göttingen University was the only university in the sphere of the Hanover church (Laugwitz 1999, p. 18). On the other hand, it was not long before Bernhard was able to persuade his father to allow him to change into mathematics.

After one year at Göttingen, Bernhard transferred to Berlin University, which was staffed by the most prominent mathematicians of Germany. In Berlin, Bernhard was able to learn from such outstanding mathematicians as Carl Gustav Jacob Jacobi (1804–1851), Jakob Steiner (1796–1863), Peter Gustav Legeune-Dirichlet (1805–1859), and Ferdinand Gotthold Max Eisenstein (1823–1852). It is believed that Dirichlet had a particularly strong impact on Riemann's approach to mathematics.

Little is known of Riemann's political outlook but the uprisings of 1848 occured while he was in Berlin. When King Frederick William's personal safety appeared to be threatened by a mob, Riemann joined a corps of fellow students to protect the royal palace. He remained on guard duty from 9AM on March 24 until 1PM the following day (Laugwitz 1999, p. 3).

In 1849, Riemann returned to Göttingen where Gauss became his Ph.D. supervisor. Also, Wilhelm Weber had returned to Göttingen from Leipzig during Riemann's time in Berlin and Riemann became his assistant for 18 months.

At the end of November 1851, Riemann submitted his doctoral thesis: "Grundlagen für eine allgemeine theorie der funktionen einer veränderlichen complexe grösse" (Foundations for a general theory of functions of one complex variable)

(James 2002, p. 184). Gauss' official report to the Philosophical Faculty of the University of Göttingen stated, "The dissertation submitted by Herr Riemann offers convincing evidence of the author's through and penetrating investigations in those parts of the subject treated in the dissertation of a creative, active truly mathematical mind, and of a gloriously fertile originality." (Spivak 1970, Vol. 2 pp. 4A-1, 4A-2)

In his dissertation, Riemann introduced what is now known as "Riemann surfaces." These surfaces enable mathematicians to comprehend and deal with functions of complex variables having branch points with much greater insight than would otherwise be possible.

To obtain the position of lecturer at a university in Germany at that time, Riemann faced two more hurdles: a *Habilitationschrift* and a *Habilitationsvortrag*. For the first requirement, Riemann presented a series of lectures in which he introduced what is now known as the *Riemann integral* and used that concept to advance the theory of Fourier series. For the second requirement, Riemann proposed three possible topics – two on electricity and one on geometry. According to legend, Riemann was surprised when Gauss persuaded the decision making council to choose the topic on geometry.

After Riemann had prepared his talk in the spring of 1854 he had trouble getting a date set for its delivery because of Gauss' fragile health. (Gauss' heart condition was getting worse and he would die early the following year.) At one point, Gauss asked Riemann to delay his talk until August in the hopes that his health would improve (James 2002, p. 185). Then Gauss changed his mind and Riemann gave his talk to the Göttingen faculty on June 10, 1854.

In his lecture, Riemann introduced the notion of a metric in the context of an n-dimensional space. He thus laid down the foundation of what is now known as *Riemannian geometry*. In 1970, Michael Spivak wrote, "Although the lecture was not published until 1866, the ideas within it proved to be the most influential in the entire history of differential geometry." (Spivak 1970, Vol. II, p. 4A-1) (It is worth noting that Spivak included an English translation of Riemann's lecture in his comprehensive work on differential geometry (Spivak 1970, Vol. II, pp. 4A-4 to 4A-20).)

Dedekind, a colleague of Riemann, who became his first biographer, wrote that for Gauss the lecture "surpassed all his expectations with great astonishment, and on the way back from the faculty meeting he spoke to Wilhelm Weber, with the greatest appreciation, and with an excitement rare for him, about the depth of the ideas presented by Riemann." (Spivak 1970, Vol. II, pp. 4A-3)

This lecture won little or no recognition for Riemann during his lifetime for the simple reason that the contents were not published until two years after his death (Riemann 1868). Nonetheless, Riemann gained rapid recognition during the next few years.

When Gauss died, his position was given to Dirichlet who held it for four years until 1859 when he died. In 1859, the chair that had been occupied by Gauss was awarded to Riemann who then became a full professor at the age of 32. That same year Riemann was elected a corresponding member of the Berlin Academy of Sciences. The recommendation for the honor signed by Kummer, Borchardt, and Weirstrass includes the passage:

> We considered it our duty to turn the attention of the Academy to our colleague whom we recommend not as a young talent who shows promise, but rather as a fully mature and independent investigator in our area of science, whose progress he has promoted in significant measure (Monastyrsky 1999, p. 63).

In 1862, Bernhard Riemann married Elise Koch. However that fall, he had a severe cold followed by the onset of tuberculosis. Riemann and his wife spent most of his last four years in Italy where the warmer climate seemed better for his health. During the periods of remission, he had a significant influence on the mathematical community in Italy.

Riemann spent the last month of his life in the village of Selasca on Lake Maggiore with his wife and three-year-old daughter Ida. He died on July 20, 1866. His last words to his wife were, "Kiss our child." (Monastyrsky 1999, p. 76)

Problem 136. In his habilitationsvortrag, Riemann presented only one formula. That formula was for the infintesimal arclength for a particular metric that he introduced. Namely:

$$ds = \frac{1}{1 + \frac{\alpha}{4}\sum (x^j)^2}\sqrt{\sum (dx^j)^2}. \tag{6.71}$$

(a) Show that in the 2-dimensional case, the Gaussian curvature for this metric is

$$K = R^{12}{}_{12} = \alpha.$$

(b) Show that in the n-dimensional case, the Riemann tensor for this metric is

$$R^{jk}{}_{mn} = \alpha\delta^{jk}_{mn}.$$

Chapter 7
Some Extrinsic Geometry in E^n

7.1 The Frenet Frame

A *curve* in E^n is a one parameter mapping $\mathbf{x}(t)$ from some open interval in E^1 to E^n. Generally, a "curve"is considered a function so that a change in parameterization results in a "different curve"even though the *path* (or *trace*) remains the same.

I do not promise to be consistent with this terminology but the reader should be sensitive to the fact that some concepts such as "speed", "velocity", and "acceleration" depend on the choice of parameter. Other concepts such as "curvature" and "torsion"are independent of the choice of function to describe a given path. The entity $\mathrm{d}\mathbf{x}/\mathrm{d}t$ may be considered the *velocity* of a point following a path. One could consider a point moving along a path at different speeds, speeding up on some segments and slowing down on other segments. One could even consider a point moving along a smooth path in some bizarre fashion – coming to a stop and then resuming its motion in the original direction or even reversing direction from time to time. Such a parameterization can be regarded as a stupid choice for a mathematician who wishes to study the shape of the path.

It is generally preferable to have a parameterization for which $\mathrm{d}\mathbf{x}/\mathrm{d}t \neq 0$ along the entire path. This is frequently possible. (See Problem 139.) However this is not always possible. If the path has a sharp corner or cusp, then it will not have a nonzero tangent vector at that point so one cannot have a well-defined $\mathrm{d}\mathbf{x}/\mathrm{d}t$ that is not a zero vector. Any point for which $\mathrm{d}\mathbf{x}/\mathrm{d}t \neq 0$ is said to be *regular*. Most of the theory of curves, presented in this text and others is devoted to line segments containing only regular points. On the other hand, Ian R. Porteous points out in the introduction to his *Geometric Differentiation for the Intelligence of Curves and Surfaces* that cusps and other singularities are sometimes the most interesting aspects of a curve or surface (Porteous 1994, pp. ix–xii).

Nevertheless, it must be said that cusps and other singularities are generally isolated so not much differential geometry of curves can be learned without studying regular points. To study the shape of a path, it is useful (at least for theoretical discussions) to have the point move at a constant unit speed. If s represents the arc

J. Snygg, *A New Approach to Differential Geometry using Clifford's Geometric Algebra*,
DOI 10.1007/978-0-8176-8283-5_7,

length of the path from some arbitrarily chosen point, then

$$\left|\frac{d\mathbf{x}}{ds}\right| = 1.$$

If one already has in place a parameter t for which $|d\mathbf{x}/dt| > 0$, then (at least in theory) one can determine an alternate parameter s for which $|d\mathbf{x}/ds| = 1$. One simply determines a function $s(t)$ for which

$$\frac{ds}{dt} = \left|\frac{d\mathbf{x}}{dt}\right|.$$

Then

$$\frac{d\mathbf{x}}{ds} = \frac{d\mathbf{x}}{dt}\frac{dt}{ds}.$$

So

$$\left|\frac{d\mathbf{x}}{ds}\right| = \left|\frac{d\mathbf{x}}{dt}\right|\left|\frac{dt}{ds}\right| = \left|\frac{ds}{dt}\frac{dt}{ds}\right| = 1.$$

Note! One can also reverse the direction of motion and still have an arc length or unit speed parameterization. That is you can have $ds/dt = -|d\mathbf{x}/dt|$. I will ignore that possibility below so I do not have to consider different cases. If you wish to deal with that possibility, just let $t = -\bar{t} + c$ and then all formulas below will be valid if you replace t by $\bar{t}$.

Example 137. Consider the *circular helix* (spiral) (See Fig. 7.1):

$$\begin{aligned}\mathbf{x}(t) &= \mathbf{e}_1 a \cos t + \mathbf{e}_2 a \sin t + \mathbf{e}_3 bt \\ \frac{d\mathbf{x}}{dt} &= -\mathbf{e}_1 a \sin t + \mathbf{e}_2 a \cos t + \mathbf{e}_3 b, \text{ so} \\ \left|\frac{d\mathbf{x}}{dt}\right| &= \sqrt{a^2 \sin^2 t + a^2 \cos^2 t + b^2} = \sqrt{a^2 + b^2} = \frac{ds}{dt}.\end{aligned} \tag{7.1}$$

Thus,

$$s = \left(\sqrt{a^2 + b^2}\right) t + c,$$

where c is an arbitrary constant. If we choose c to be zero, then (7.1) becomes

$$\mathbf{x}(s) = \mathbf{e}_1 a \cos \frac{s}{\sqrt{a^2 + b^2}} + \mathbf{e}_2 a \sin \frac{s}{\sqrt{a^2 + b^2}} + \mathbf{e}_3 \frac{bs}{\sqrt{a^2 + b^2}}. \tag{7.2}$$

Fig. 7.1 Circular helix

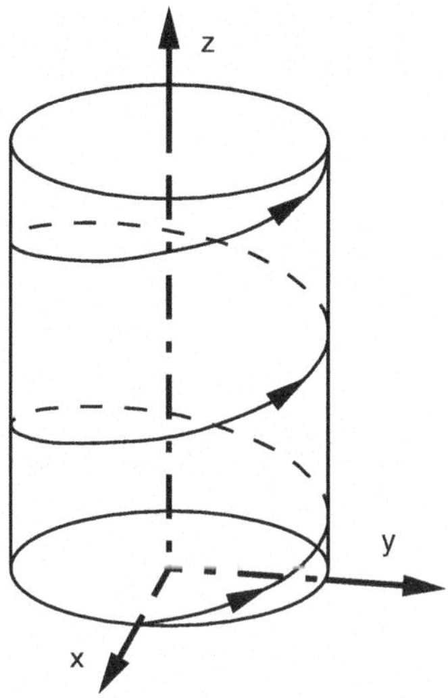

To continue our discussion, it is useful to assume that except at isolated points, a unit speed curve $\mathbf{x}(s)$ in an n-dimensional Euclidean space E^n can be differentiated n times. In addition, the curve is truly n-dimensional only if the set

$$\left\{\frac{\mathrm{d}\mathbf{x}}{\mathrm{d}s}, \frac{\mathrm{d}^2\mathbf{x}}{\mathrm{d}s^2}, \ldots, \frac{\mathrm{d}^n\mathbf{x}}{\mathrm{d}s^n}\right\}$$

spans an n-dimensional space. This is equivalent to saying

$$\frac{\mathrm{d}\mathbf{x}}{\mathrm{d}s} \wedge \frac{\mathrm{d}^2\mathbf{x}}{\mathrm{d}s^2} \wedge \cdots \wedge \frac{\mathrm{d}^n\mathbf{x}}{\mathrm{d}s^n} \neq 0.$$

To study 2-dimensional surfaces, it was useful to introduce the concept of a moving frame as we did in Sect. 5.6.1 of Chap. 5. For a given surface, many alternative moving frames can be used. With the possible exception of surfaces of revolution, there is no particular moving frame that can be considered intrinsic.

However for a curve, there is an intrinsic moving frame known as the *Frenet frame*. For a surface, there is an arbitrariness in the choice of coordinates, and then there is another arbitrariness in the order of the resulting tangent vectors. By contrast, for a curve, there is a natural basis. Namely

$$\left\{\frac{\mathrm{d}\mathbf{x}}{\mathrm{d}s}, \frac{\mathrm{d}^2\mathbf{x}}{\mathrm{d}s^2}, \ldots, \frac{\mathrm{d}^n\mathbf{x}}{\mathrm{d}s^n}\right\}.$$

Furthermore, there is a natural sequence for the members of this basis so there is a natural or intrinsic frame resulting from the Gram–Schmidt process. Namely let

$$\mathbf{E}_1 = \frac{\mathrm{d}\mathbf{x}}{\mathrm{d}s}, \tag{7.3}$$

$$\mathbf{E}_k\mathbf{E}_{k-1}\cdots\mathbf{E}_1 = \frac{\frac{\mathrm{d}^k\mathbf{x}}{\mathrm{d}s^k}\wedge\frac{\mathrm{d}^{k-1}\mathbf{x}}{\mathrm{d}s^{k-1}}\wedge\cdots\wedge\frac{\mathrm{d}\mathbf{x}}{\mathrm{d}s}}{\left|\frac{\mathrm{d}^k\mathbf{x}}{\mathrm{d}s^k}\wedge\frac{\mathrm{d}^{k-1}\mathbf{x}}{\mathrm{d}s^{k-1}}\wedge\cdots\wedge\frac{\mathrm{d}\mathbf{x}}{\mathrm{d}s}\right|}, \text{ for } k = 2,3,\ldots,n-1, \tag{7.4}$$

or

$$\mathbf{E}_k = \frac{\frac{\mathrm{d}^k\mathbf{x}}{\mathrm{d}s^k}\wedge\frac{\mathrm{d}^{k-1}\mathbf{x}}{\mathrm{d}s^{k-1}}\wedge\cdots\wedge\frac{\mathrm{d}\mathbf{x}}{\mathrm{d}s}}{\left|\frac{\mathrm{d}^k\mathbf{x}}{\mathrm{d}s^k}\wedge\frac{\mathrm{d}^{k-1}\mathbf{x}}{\mathrm{d}s^{k-1}}\wedge\cdots\wedge\frac{\mathrm{d}\mathbf{x}}{\mathrm{d}s}\right|}\mathbf{E}_1\mathbf{E}_2\cdots\mathbf{E}_{k-1},$$

$$\text{for } k = 2,3,\ldots,n-1. \tag{7.5}$$

We should note that an equivalent approach is to have

$$\mathbf{E}_k\mathbf{E}_{k-1}\cdots\mathbf{E}_1 = \frac{\frac{d^k\mathbf{x}}{\mathrm{d}s^k}\wedge\mathbf{E}_{k-1}\wedge\cdots\wedge\mathbf{E}_2\wedge\mathbf{E}_1}{\left|\frac{d^k\mathbf{x}}{\mathrm{d}s^k}\wedge\mathbf{E}_{k-1}\wedge\cdots\wedge\mathbf{E}_2\wedge\mathbf{E}_1\right|},$$

$$\text{for } k = 2,3,\ldots,n-1, \tag{7.6}$$

and

$$\mathbf{E}_k = \frac{\frac{d^k\mathbf{x}}{ds^k}\wedge\mathbf{E}_{k-1}\wedge\cdots\wedge\mathbf{E}_2\wedge\mathbf{E}_1}{\left|\frac{\mathrm{d}^k\mathbf{x}}{\mathrm{d}s^k}\wedge\mathbf{E}_{k-1}\wedge\cdots\wedge\mathbf{E}_2\wedge\mathbf{E}_1\right|}\mathbf{E}_1\mathbf{E}_2\cdots\mathbf{E}_{k-1},$$

$$\text{for } k = 2,3,\ldots,n-1. \tag{7.7}$$

We could use (7.5) or (7.7) for $k = n$. However if we did that, the resulting frame could change from right handed to left handed or vice versa on opposite sides of some high order inflection point, where one of the derivatives $\mathrm{d}^j\mathbf{x}/\mathrm{d}s^j = 0$. It is preferable to require that the frame be right handed at all points, where

$$\frac{\mathrm{d}^n\mathbf{x}}{\mathrm{d}s^n}\wedge\frac{\mathrm{d}^{n-1}\mathbf{x}}{\mathrm{d}s^{n-1}}\wedge\cdots\wedge\frac{\mathrm{d}\mathbf{x}}{\mathrm{d}s}\neq 0.$$

That is

$$\mathbf{E}_n\mathbf{E}_{n-1}\cdots\mathbf{E}_1 = \mathbf{e}_n\mathbf{e}_{n-1}\cdots\mathbf{e}_1, \text{ or}$$

$$\mathbf{E}_n = \mathbf{e}_{n\cdots 21}\mathbf{E}_1\mathbf{E}_2\cdots\mathbf{E}_{n-1}. \tag{7.8}$$

Using the same reasoning used in the proof of Theorem 90 in Sect. 5.6.1 of Chap. 5, we know that

$$\frac{\mathrm{d}\mathbf{E}_j}{\mathrm{d}s} = \omega_{jk}\mathbf{E}^k, \text{ where } \omega_{jk} = -\omega_{kj}\,. \tag{7.9}$$

Furthermore, since our vectors lie in a Euclidean space:

$$\mathbf{E}^k = \mathbf{E}_k \text{ for } k = 1,2,\ldots,n\,. \tag{7.10}$$

We will now show that

$$\omega_{jk} = 0 \text{ if } k > j + 1. \tag{7.11}$$

We first note that

$$\{\mathbf{E}_1, \mathbf{E}_2, \ldots, \mathbf{E}_k\} \text{ and } \left\{ \frac{\mathrm{d}\mathbf{x}}{\mathrm{d}s}, \frac{\mathrm{d}^2\mathbf{x}}{\mathrm{d}s^2}, -, \frac{\mathrm{d}^k\mathbf{x}}{\mathrm{d}s^k} \right\}$$

span the same space. This implies that $\mathrm{d}\mathbf{E}_j/\mathrm{d}s$ is in the space spanned by

$$\left\{ \frac{\mathrm{d}\mathbf{x}}{\mathrm{d}s}, \frac{\mathrm{d}^2\mathbf{x}}{\mathrm{d}s^2}, -, \frac{\mathrm{d}^j\mathbf{x}}{\mathrm{d}s^j}, \frac{\mathrm{d}^{j+1}\mathbf{x}}{\mathrm{d}s^{j+1}} \right\},$$ which is equivalent to the space spanned by

$$\{\mathbf{E}_1, \mathbf{E}_2, \ldots, \mathbf{E}_j, \mathbf{E}_{j+1}\} \text{ or } \{\mathbf{E}^1, \mathbf{E}^2, \ldots, \mathbf{E}^j, \mathbf{E}^{j+1}\}.$$

Therefore,

$$\frac{\mathrm{d}\mathbf{E}_j}{\mathrm{d}s} = \omega_{j1}\mathbf{E}^1 + \omega_{j2}\mathbf{E}^2 + \cdots + \omega_{jj+1}\mathbf{E}^{j+1} \text{ for } j = 1, 2, \ldots, n-1 \tag{7.12}$$

and

$$\frac{\mathrm{d}\mathbf{E}_n}{\mathrm{d}s} = \omega_{n1}\mathbf{E}^1 + \omega_{n2}\mathbf{E}^2 + \cdots + \omega_{nn}\mathbf{E}^n. \tag{7.13}$$

Thus, $\omega_{jk} = 0$ for $k > j + 1$.

However since $\omega_{jk} = -\omega_{kj}$, knowing that $\omega_{jk} = 0$ for $k > j + 1$ implies that $\omega_{jk} = 0$ if $k < j - 1$. Furthermore since $\omega_{jk} = -\omega_{kj}$, we know that $\omega_{jj} = 0$. Now dropping the zero terms in (7.12) and (7.13), we have

$$\frac{\mathrm{d}\mathbf{E}_1}{\mathrm{d}s} = \omega_{12}\mathbf{E}^2, \tag{7.14}$$

$$\frac{\mathrm{d}\mathbf{E}_j}{\mathrm{d}s} = \omega_{jj-1}\mathbf{E}^{j-1} + \omega_{jj+1}\mathbf{E}^{j+1} \text{ for } j = 2, 3, \ldots, n-1 \ (j \text{ not summed}), \text{ and} \tag{7.15}$$

$$\frac{\mathrm{d}\mathbf{E}_n}{\mathrm{d}s} = \omega_{nn-1}\mathbf{E}^{n-1}. \tag{7.16}$$

If we define

$$k_j = \omega_{jj+1}, \text{ we then have}$$

$$\frac{\mathrm{d}\mathbf{E}_1}{\mathrm{d}s} = k_1\mathbf{E}_2, \tag{7.17}$$

$$\frac{\mathrm{d}\mathbf{E}_j}{\mathrm{d}s} = -k_{j-1}\mathbf{E}_{j-1} + k_j\mathbf{E}_{j+1} \text{ for } j = 2, 3, \ldots, n-1, \text{ and} \tag{7.18}$$

$$\frac{\mathrm{d}\mathbf{E}_n}{\mathrm{d}s} = -k_{n-1}\mathbf{E}_{n-1}. \tag{7.19}$$

These equations are usually summarized in matrix form:

$$\frac{d}{ds}\begin{bmatrix} \mathbf{E}_1 \\ \mathbf{E}_2 \\ \mathbf{E}_3 \\ - \\ - \\ - \\ \mathbf{E}_n \end{bmatrix} = \begin{bmatrix} 0 & k_1 & 0 & -- & - & 0 \\ -k_1 & 0 & k_2 & -- & - & 0 \\ 0 & -k_2 & 0 & -- & - & 0 \\ - & - & - & --- & - & - \\ - & - & - & --- & - & - \\ - & - & - & --- & 0 & k_{n-1} \\ 0 & 0 & 0 & -- & -k_{n-1} & 0 \end{bmatrix}\begin{bmatrix} \mathbf{E}_1 \\ \mathbf{E}_2 \\ \mathbf{E}_3 \\ - \\ - \\ - \\ \mathbf{E}_n \end{bmatrix}. \tag{7.20}$$

The Frenet frame is defined so that all the k_j's with the possible exception of k_{n-1} are positive. In the next section, I will demonstrate that this is a consequence of (7.4) and (7.5) or (7.6) and (7.7). This is a matter of convenience and has no geometric significance. If you chose different signs for the $\mathbf{E}_j$'s, the k_j's would have different signs. Other authors point out that if the k_j's for $j < n - 1$ were not all positive, it would be a simple matter to correct the situation as you compute the frame inductively. That is, adjust the sign of $\mathbf{E}_2$ to make k_1 positive. Then adjust the sign of $\mathbf{E}_3$ to make k_2 positive, and then continue until you have adjusted the sign of $\mathbf{E}_{n-1}$ to make k_{n-2} positive. At the last step, you adjust the sign of $\mathbf{E}_n$ to make the Frenet frame right-handed so k_{n-1} may be positive or negative. (This makes it possible to distinguish a *right handed curve* from a left handed one.)

The entity k_1 is generally known as the *curvature* of the path. The rest of the k_j's are generally known as *higher order curvatures*. Johan Gerretsen (Gerretsen 1962, p. 75) refers to the higher order curvatures as *torsions*. On the other hand, Wolfgang Kühnel (Kühnel 2002, p. 26) reserves the word *torsion* for k_{n-1}. Most authors of elementary texts on differential geometry restrict themselves to three dimensions. In that context, k_1 is usually designated by the Greek letter kappa (κ) and k_2 is designated by the Greek letter tau (τ). The numerical value of kappa is referred to as "the curvature"and the numerical value of tau is referred to as "the torsion."

Additionally for three-dimensional curves, $\mathbf{E}_1$ is designated by $\mathbf{t}$ and referred to as the *tangent vector*. $\mathbf{E}_2$ is designated by $\mathbf{n}$ and referred to as the *normal vector*. Finally, $\mathbf{E}_3$ is designated by $\mathbf{b}$ and referred to as the *binormal vector*.

The matrix (7.20) summarizes a set of equations that are usually referred to as the *Frenet equations* or the *Serret–Frenet equations*. However Ian R. Porteus (Porteus 1994, p. 116) notes;

> "Space curves were originally known as *curves of double curvature*, first studied by Alexis-Claude Clairaut (1731) when in his teens. The date of the papers of Serret (1851) and Frenet (1852) is surprisingly late. It seems, however, that they were the last to discover the equations attributed to them, for these are to be found in a book by Carl Eduard Senff (1831) of the University of Dorpat (now Tartu in Estonia), who attributed them to his teacher Martin Bartels. Indeed the concept of torsion is already explicit in a paper by Michel-Ange Lancret (1806), a pupil of Monge, –."

Author's Note! Although Martin Bartels (1769–1833) never won recognition as a first rank mathematician, he had a significant impact on the history of non-Euclidean geometry as a teacher. In 1786, in Braunschweig, Germany, a 17-year-old Bartels was assigned the task of exposing the young Gauss to mathematics far more advanced than that usually presented to a 9 year old. It is believed that Bartels influenced the decision of the Duke of Braunschweig to subsidize Gauss' education.

A little over 20 years later, in Russia, at Kazan State University, Bartels' had a student who originally planned to pursue a career in medicine. That student was Lobachevsky who was soon persuaded to change his career plans.

Martin Bartels corresponded with Gauss and at one time it was speculated that Bartels may have transmitted some of Gauss' results on non-Euclidean geometry to Lobachevsky. However, scholars who have studied the correspondence between Gauss and Bartels are convinced that Gauss never revealed those results to Bartels (O'Connor and Robertson: Lobachevsky).

What appears likely is that Bartels directed the attention of both Gauss and Lobachevsky to the problem of the parallel postulate. This is consistent with the early interest that Gauss had in the problem. It is also consistent with the content of a history of mathematics course that Lobachevsky took from Bartels (Laptev 1992).

The matrix (7.20) can be rewritten in terms of Clifford algebra. Namely

$$\frac{\mathrm{d}\mathbf{E}_k(s)}{\mathrm{d}s} = -\mathbf{M}(s)\mathbf{E}_k(s) + \mathbf{E}_k(s)\mathbf{M}(s), \text{ where}$$

$$\mathbf{M}(s) = \frac{1}{2}\sum_{j=1}^{n-1} k_j(s)\mathbf{E}_j(s)\mathbf{E}_{j+1}(s)\,. \tag{7.21}$$

(See Problem 145.)

I will refer to $\mathbf{M}$ as the *Frenet 2-vector*. It turns out that the Frenet 2-vector is related to a rotation operator $\mathbf{R}(s)$. Since $\mathbf{E}_j(s)$ does not change its length, each $\mathbf{E}_j(s)$ undergoes a rotation as s changes. Thus, we can write

$$\mathbf{E}_j(s) = \mathbf{R}^{-1}(s)\mathbf{E}_j(0)\mathbf{R}(s). \tag{7.22}$$

From this equation;

$$\frac{\mathrm{d}\mathbf{E}_j(s)}{\mathrm{d}s} = \frac{\mathrm{d}\mathbf{R}^{-1}(s)}{\mathrm{d}s}\mathbf{E}_j(0)\mathbf{R}(s) + \mathbf{R}^{-1}(s)\mathbf{E}_j(0)\frac{\mathrm{d}\mathbf{R}(s)}{\mathrm{d}s}\,. \tag{7.23}$$

From (7.22),

$$\mathbf{E}_j(0) = \mathbf{R}(s)\mathbf{E}_j(s)\mathbf{R}^{-1}(s).$$

So (7.23) becomes

$$\frac{\mathrm{d}\mathbf{E}_j(s)}{\mathrm{d}s} = \frac{\mathrm{d}\mathbf{R}^{-1}(s)}{\mathrm{d}s}\mathbf{R}(s)\mathbf{E}_j(s) + \mathbf{E}_j(s)\mathbf{R}^{-1}(s)\frac{\mathrm{d}\mathbf{R}(s)}{\mathrm{d}s}. \tag{7.24}$$

Now since

$$\mathbf{R}^{-1}\mathbf{R} = \mathbf{I}, \text{ it follows that}$$

$$\frac{\mathrm{d}\mathbf{R}^{-1}}{\mathrm{d}s}\mathbf{R} + \mathbf{R}^{-1}\frac{\mathrm{d}\mathbf{R}}{\mathrm{d}s} = \mathbf{0} \text{ or } \frac{\mathrm{d}\mathbf{R}^{-1}}{\mathrm{d}s}\mathbf{R} = -\mathbf{R}^{-1}\frac{\mathrm{d}\mathbf{R}}{\mathrm{d}s}, \text{ so}$$

Equation (7.24) becomes

$$\frac{\mathrm{d}\mathbf{E}_j(s)}{\mathrm{d}s} = -\mathbf{R}^{-1}\frac{\mathrm{d}\mathbf{R}}{\mathrm{d}s}\mathbf{E}_j(s) + \mathbf{E}_j(s)\mathbf{R}^{-1}\frac{\mathrm{d}\mathbf{R}}{\mathrm{d}s}.$$

Comparing this with (7.21) gives us

$$\mathbf{M}(s) = \mathbf{R}^{-1}(s)\frac{\mathrm{d}\mathbf{R}(s)}{\mathrm{d}s} \quad \text{or}$$

$$\frac{\mathrm{d}\mathbf{R}(s)}{\mathrm{d}s} = \mathbf{R}(s)\mathbf{M}(s). \tag{7.25}$$

From the theory of differential equations, it is known that (7.20) is always solvable although not necessarily in terms of elementary functions. A solver has the freedom to choose the initial Frenet frame:

$$\mathbf{E}_j(0) \text{ for } j = 1, 2, \ldots, n \text{ and } \mathbf{x}(0).$$

As a consequence, any information about the shape of a curve is stored in the Frenet equations.

Example 138.

As an example let us consider the 3-dimensional case for which both the curvature and torsion are constant. We will use a Clifford algebra approach. For the scalar version of (7.25), the solution is essentially trivial. If

$$\frac{\mathrm{d}R(s)}{\mathrm{d}s} = R(s)M(s), \text{ then}$$

$$\frac{\mathrm{d}R}{R} = M\mathrm{d}s, \text{ and } \ln R = \int M\mathrm{d}s. \tag{7.26}$$

$$\text{Thus, } R(s) = R(0)\exp\int_0^s M(u)\mathrm{d}u\,. \tag{7.27}$$

However if we are dealing with matrices or Clifford numbers, the situation is more complicated. To see this, suppose

$$\mathbf{A}(s) = \int_0^s \mathbf{M}(u)\mathrm{d}u \text{ and } \frac{\mathrm{d}}{\mathrm{d}s}\mathbf{A}(s) = \mathbf{M}(s). \tag{7.28}$$

Then

$$\exp\int_0^s \mathbf{M}(u)\mathrm{d}u = \exp\mathbf{A}(s) = \sum_{k=0}^{\infty}\frac{1}{k!}(\mathbf{A}(s))^k \text{ , and}$$

$$\frac{\mathrm{d}}{\mathrm{d}s}\exp\mathbf{A}(s) = \mathbf{M} + \frac{1}{2!}[\mathbf{MA} + \mathbf{AM}] + \frac{1}{3!}\left[\mathbf{M}(\mathbf{A})^2 + \mathbf{AMA} + (\mathbf{A})^2\mathbf{M}\right] + \cdots.$$

If $\mathbf{M}$ commutes with $\mathbf{A}$, then

$$\frac{\mathrm{d}}{\mathrm{d}s}\exp\mathbf{A}(s) = \left(\mathbf{I}+\mathbf{A}+\frac{1}{2!}(\mathbf{A})^2+\cdots\right)\mathbf{M}$$
$$= [\exp\mathbf{A}(s)]\,\mathbf{M}(s)\,. \tag{7.29}$$

However if $\mathbf{M}$ does not commute with $\mathbf{A}$, then

$$\frac{\mathrm{d}}{\mathrm{d}s}\exp\mathbf{A}(s) \neq [\exp\mathbf{A}(s)]\,\mathbf{M}\,.$$

On the other hand, if $\mathbf{M}$ is a constant matrix or Clifford number, then $\mathbf{A}(s)=\mathbf{M}s$ and so $\mathbf{A}$ and $\mathbf{M}$ commute. If the curvatures are constant, then the Frenet 2-vector

$$\mathbf{M}(s) = \frac{1}{2}\sum_{j=1}^{n-1}k_j\mathbf{E}_j(s)\mathbf{E}_{j+1}(s) = \frac{1}{2}\sum_{j=1}^{n-1}k_j\mathbf{E}_j(0)\mathbf{E}_{j+1}(0) = \mathbf{M}(0)\,.$$

(See Problem 146.)

For the 3-dimensional case, the Frenet 2-vector

$$\mathbf{M}(s) = \frac{1}{2}\left[\kappa\mathbf{t}(s)\mathbf{n}(s)+\tau\mathbf{n}(s)\mathbf{b}(s)\right] = \frac{1}{2}\left[\kappa\mathbf{t}(0)-\tau\mathbf{b}(0)\right]\mathbf{n}(0)\,.$$

Because of the form of the Frenet 2-vector $\mathbf{M}$, it is useful to introduce an alternate orthonormal frame. Namely:

$$\bar{\mathbf{E}}_1 = \frac{\kappa\mathbf{t}(0)-\tau\mathbf{b}(0)}{\sqrt{\kappa^2+\tau^2}}, \tag{7.30}$$

$$\bar{\mathbf{E}}_2 = \mathbf{n}(0), \text{ and} \tag{7.31}$$

$$\bar{\mathbf{E}}_3 = \frac{\tau\mathbf{t}(0)+\kappa\mathbf{b}(0)}{\sqrt{\kappa^2+\tau^2}}. \tag{7.32}$$

In this circumstance

$$\mathbf{M}(s) = \mathbf{M}(0) = \frac{\sqrt{\kappa^2+\tau^2}}{2}\bar{\mathbf{E}}_1\bar{\mathbf{E}}_2$$

and the solution of (7.25) becomes

$$\mathbf{R}(s) = \mathbf{R}(0)\exp\left[\frac{\sqrt{\kappa^2+\tau^2}}{2}\bar{\mathbf{E}}_1\bar{\mathbf{E}}_2 s\right]. \tag{7.33}$$

From (7.22), it is clear that $\mathbf{R}(0) = \mathbf{I}$. Since $(\bar{\mathbf{E}}_1\bar{\mathbf{E}}_2)^2 = -\mathbf{I}$, $\bar{\mathbf{E}}_1\bar{\mathbf{E}}_2$ behaves algebraically like $\sqrt{-1}$. Therefore, (7.33) becomes

$$\mathbf{R}(s) = \exp\left[\frac{\sqrt{\kappa^2+\tau^2}}{2}\bar{\mathbf{E}}_1\bar{\mathbf{E}}_2 s\right]$$

$$= \mathbf{I}\cos\frac{\sqrt{\kappa^2+\tau^2}}{2}s + \bar{\mathbf{E}}_1\bar{\mathbf{E}}_2\sin\frac{\sqrt{\kappa^2+\tau^2}}{2}s\,. \tag{7.34}$$

(See Problem 147.)

To compute **t**(s), we note that from (7.22):

$$\mathbf{t}(s) = \mathbf{R}^{-1}(s)\mathbf{t}(0)\mathbf{R}(s)\,. \tag{7.35}$$

From (7.30) and (7.32),

$$\mathbf{t}(0) = \frac{\kappa}{\sqrt{\kappa^2+\tau^2}}\bar{\mathbf{E}}_1 + \frac{\tau}{\sqrt{\kappa^2+\tau^2}}\bar{\mathbf{E}}_2\,. \tag{7.36}$$

Combining (7.34) and (7.36), we have

$$\mathbf{R}^{-1}(s)\mathbf{t}(0)$$

$$= \exp\left[-\frac{\sqrt{\kappa^2+\tau^2}}{2}\bar{\mathbf{E}}_1\bar{\mathbf{E}}_2 s\right]\left(\frac{\kappa}{\sqrt{\kappa^2+\tau^2}}\bar{\mathbf{E}}_1 + \frac{\tau}{\sqrt{\kappa^2+\tau^2}}\bar{\mathbf{E}}_3\right)$$

$$= \left(\mathbf{I}\cos\frac{\sqrt{\kappa^2+\tau^2}}{2}s - \bar{\mathbf{E}}_1\bar{\mathbf{E}}_2\sin\frac{\sqrt{\kappa^2+\tau^2}}{2}s\right)\left(\frac{\kappa}{\sqrt{\kappa^2+\tau^2}}\bar{\mathbf{E}}_1 + \frac{\tau}{\sqrt{\kappa^2+\tau^2}}\bar{\mathbf{E}}_3\right). \tag{7.37}$$

We note that

$$\bar{\mathbf{E}}_1\bar{\mathbf{E}}_2\bar{\mathbf{E}}_1 = -\bar{\mathbf{E}}_1\left(\bar{\mathbf{E}}_1\bar{\mathbf{E}}_2\right) \text{ and } \bar{\mathbf{E}}_1\bar{\mathbf{E}}_2\bar{\mathbf{E}}_3 = \bar{\mathbf{E}}_3\left(\bar{\mathbf{E}}_1\bar{\mathbf{E}}_2\right)\,.$$

Thus, (7.37) becomes

$$\mathbf{R}^{-1}(s)\mathbf{t}(0) = \frac{\kappa}{\sqrt{\kappa^2+\tau^2}}\bar{\mathbf{E}}_1\left(\mathbf{I}\cos\frac{\sqrt{\kappa^2+\tau^2}}{2}s + \bar{\mathbf{E}}_1\bar{\mathbf{E}}_2\sin\frac{\sqrt{\kappa^2+\tau^2}}{2}s\right)$$

$$+ \frac{\tau}{\sqrt{\kappa^2+\tau^2}}\bar{\mathbf{E}}_3\left(\mathbf{I}\cos\frac{\sqrt{\kappa^2+\tau^2}}{2}s - \bar{\mathbf{E}}_1\bar{\mathbf{E}}_2\sin\frac{\sqrt{\kappa^2+\tau^2}}{2}s\right)$$

$$= \frac{\kappa}{\sqrt{\kappa^2+\tau^2}}\bar{\mathbf{E}}_1\exp\left[\frac{\sqrt{\kappa^2+\tau^2}}{2}\bar{\mathbf{E}}_1\bar{\mathbf{E}}_2 s\right]$$

$$+ \frac{\tau}{\sqrt{\kappa^2+\tau^2}}\bar{\mathbf{E}}_3\exp\left[-\frac{\sqrt{\kappa^2+\tau^2}}{2}\bar{\mathbf{E}}_1\bar{\mathbf{E}}_2 s\right]\,. \tag{7.38}$$

Using (7.34) and (7.38), (7.35) becomes

$$\mathbf{t}(s)=\frac{1}{\sqrt{\kappa^2+\tau^2}}\left[\kappa\bar{\mathbf{E}}_1\exp\left(\frac{\sqrt{\kappa^2+\tau^2}}{2}\bar{\mathbf{E}}_1\bar{\mathbf{E}}_2 s\right)+\tau\bar{\mathbf{E}}_3\exp\left(-\frac{\sqrt{\kappa^2+\tau^2}}{2}\bar{\mathbf{E}}_1\bar{\mathbf{E}}_2 s\right)\right]$$

$$\cdot\exp\left(\frac{\sqrt{\kappa^2+\tau^2}}{2}\bar{\mathbf{E}}_1\bar{\mathbf{E}}_2 s\right)\quad\text{and therefore,}$$

$$\begin{aligned}\mathbf{t}(s)&=\frac{\kappa\bar{\mathbf{E}}_1}{\sqrt{\kappa^2+\tau^2}}\exp\left(\sqrt{\kappa^2+\tau^2}\bar{\mathbf{E}}_1\bar{\mathbf{E}}_2 s\right)+\frac{\tau\bar{\mathbf{E}}_3}{\sqrt{\kappa^2+\tau^2}}\\&=\frac{\kappa\bar{\mathbf{E}}_1}{\sqrt{\kappa^2+\tau^2}}\left[\cos\left(\sqrt{\kappa^2+\tau^2}s\right)+\bar{\mathbf{E}}_1\bar{\mathbf{E}}_2\sin\left(\sqrt{\kappa^2+\tau^2}s\right)\right]+\frac{\tau\bar{\mathbf{E}}_3}{\sqrt{\kappa^2+\tau^2}}\,.\end{aligned}$$

Thus,

$$\mathbf{t}(s)=\frac{\kappa}{\sqrt{\kappa^2+\tau^2}}\left[\bar{\mathbf{E}}_1\cos\left(\sqrt{\kappa^2+\tau^2}s\right)+\bar{\mathbf{E}}_2\sin\left(\sqrt{\kappa^2+\tau^2}s\right)\right]+\frac{\tau\bar{\mathbf{E}}_3}{\sqrt{\kappa^2+\tau^2}}\,. \tag{7.39}$$

To get $\mathbf{x}(s)$, we observe that

$$\begin{aligned}\mathbf{x}(s)&=\int_0^s\mathbf{t}(u)\mathrm{d}u+\mathbf{x}(0)\\&=\frac{\kappa\bar{\mathbf{E}}_1}{(\kappa^2+\tau^2)}\sin\left(\sqrt{\kappa^2+\tau^2}s\right)\\&\quad-\frac{\kappa\bar{\mathbf{E}}_2}{(\kappa^2+\tau^2)}\left[\cos\left(\sqrt{\kappa^2+\tau^2}s\right)-1\right]+\frac{\tau\bar{\mathbf{E}}_3 s}{\sqrt{\kappa^2+\tau^2}}+\mathbf{x}(0)\,.\end{aligned}\tag{7.40}$$

If we let $\bar{\mathbf{E}}_1=\mathbf{e}_2$, $\bar{\mathbf{E}}_2=-\mathbf{e}_1$, $\bar{\mathbf{E}}_3=\mathbf{e}_3$, and $\mathbf{x}(0)=-\kappa\bar{\mathbf{E}}_2/\left(\kappa^2+\tau^2\right)=\kappa\mathbf{e}_1/\left(\kappa^2+\tau^2\right)$, we get

$$\begin{aligned}\mathbf{x}(s)&=\frac{\kappa\mathbf{e}_1}{(\kappa^2+\tau^2)}\cos\left(\sqrt{\kappa^2+\tau^2}s\right)\\&\quad+\frac{\kappa\mathbf{e}_2}{(\kappa^2+\tau^2)}\sin\left(\sqrt{\kappa^2+\tau^2}s\right)+\frac{\tau\mathbf{e}_3 s}{\sqrt{\kappa^2+\tau^2}}\,.\end{aligned}\tag{7.41}$$

This is identical to (7.2) for the circular helix. (See Problem 148.)

Problem 139. Suppose $\mathbf{x}(t) = \mathbf{e}_1 t^3 + \mathbf{e}_2 t^6$. Clearly

$$\left.\frac{d\mathbf{x}(t)}{dt}\right|_{t=0} = 0.$$

Find an alternate parameterization for the same path such that all its points are regular. (What curve is this?)

Problem 140. Suppose $\mathbf{x}(s) = \mathbf{e}_1 \cosh t + \mathbf{e}_2 \sinh t + \mathbf{e}_3 t$. Determine ds/dt and an arc length parameterization for $\mathbf{x}$.

Problem 141. Using the arc length parameterization for the helix

$$\mathbf{x}(s) = \mathbf{e}_1 a \cos \frac{s}{\sqrt{a^2+b^2}} + \mathbf{e}_2 a \sin \frac{s}{\sqrt{a^2+b^2}} + \mathbf{e}_3 \frac{bs}{\sqrt{a^2+b^2}},$$

determine the Frenet apparatus $\mathbf{t}(s)$, $\mathbf{n}(s)$, $\mathbf{b}(s)$, κ, and τ. That is $\mathbf{E}_1$, $\mathbf{E}_2$, $\mathbf{E}_3$, k_1, and k_2. What happens when $\tau = 0$?

Problem 142. Suppose $\mathbf{x}(s)$ is a curve in E^3 that is not a straight line. From your knowledge of geometry, it should be clear that

$$\begin{aligned}&\mathbf{x}(s) \text{ lies in a plane}\\ &\Leftrightarrow \exists \text{ a constant vector } \mathbf{N} \text{ s.t. } \langle \mathbf{N}, \mathbf{x}(s) - \mathbf{x}(0)\rangle = 0.\end{aligned}$$

a. Taking derivatives of the equation above and using whatever arguments you find suitable show that a necessary and sufficient condition for $\mathbf{x}(s)$ to lie in a plane is that $\tau(s) = 0$.
b. What is the relationship between $\mathbf{N}$ and the Frenet frame?

Problem 143. Consider the curve $\mathbf{x}(t) = \mathbf{e}_1 4\cos t + \mathbf{e}_2(5 - 5\sin t) + \mathbf{e}_3(-3\cos t)$.

a. Find an arc length parameterization for $\mathbf{x}$.
b. Compute the Frenet apparatus and show that this curve is a circle.
c. Find the center and radius.
d. Rewrite the equation for the curve in terms of the basis $\{\bar{\mathbf{e}}_1, \bar{\mathbf{e}}_2, \bar{\mathbf{e}}_3\}$, where $\bar{\mathbf{e}}_1 = \mathbf{t}(0)$, $\bar{\mathbf{e}}_2 = \mathbf{n}(0)$, and $\bar{\mathbf{e}}_3 = \mathbf{b}(0)$.
e. Sketch the curve in this new basis.

Problem 144. Consider the curve $\mathbf{x}(s) = \mathbf{e}_1 \left[\frac{(1+s)^{3/2}}{3}\right] + \mathbf{e}_2 \left[\frac{(1-s)^{3/2}}{3}\right] + \mathbf{e}_3 \frac{s}{\sqrt{2}}$ defined for $-1 < s < 1$. Show that $\mathbf{x}(s)$ is an arc length parameterization and compute the Frenet apparatus.

Problem 145. Verify (7.21).

Problem 146. Suppose $\mathbf{M}(s) = \frac{1}{2}\sum_{j=1}^{n-1} k_j \mathbf{E}_j(s)\mathbf{E}_{j+1}(s)$, where the k_j's are constant. Use the Frenet equations to show that $d\mathbf{M}(s)/ds = 0$ and thus $\mathbf{M}(s) = \mathbf{M}(0)$.

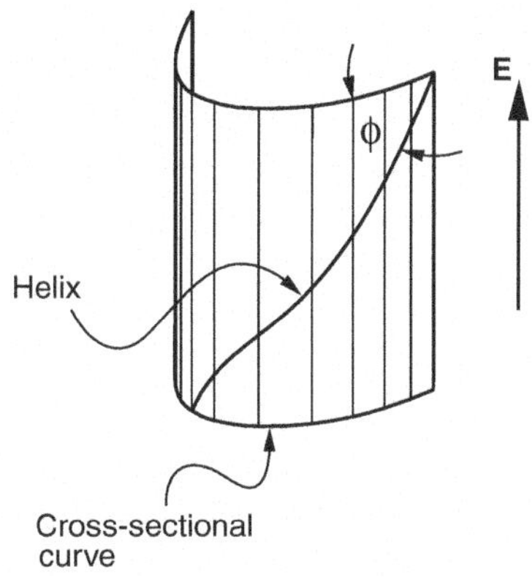

Fig. 7.2 A non-circular helix

Problem 147. Using the fact that

$$\exp \mathbf{A} = \sum_{k=0}^{\infty} \frac{1}{k!} (\mathbf{A})^k,$$

verify (7.34).

Problem 148. Use the formulas you derived in Problem 141 for κ and τ to show that (7.41) is identical to (7.2).

Problem 149. *Helices* are odd dimensional curves with certain constraints on the curvatures. In particular, suppose $k_1, k_2, \ldots, k_{2r}$ are the curvatures of a $(2r+1)$-dimensional curve. Then the curve is said to be a generalized helix if each ratio k_{2j-1}/k_{2j} is constant for $j = 1, 2, \ldots, r$.

a. Show that for a helix, the vector

$$\mathbf{E} = \left(\sum_{j=0}^{r} \alpha_{2j+1} \mathbf{E}_{2j+1}(s) \right) \Big/ \sqrt{\sum_{k=0}^{r} (\alpha_{2k+1})^2} \tag{7.42}$$

is a constant unit vector, if

$$\alpha_1 = 1 \text{ and } \alpha_{2j+1} = \frac{k_1}{k_2} \frac{k_3}{k_4} \cdots \frac{k_{2j-1}}{k_{2j}} \text{ for } j = 1, \ldots, r.$$

b. The vector $\mathbf{E}$ of (7.42) may be referred to as the *axis of the helix*. How does the axis $\mathbf{E}$ relate to the vector $\bar{\mathbf{E}}_3$ of (7.32) and (7.40)?

Problem 150. Suppose that $\mathbf{x}(s)$ is an arc length curve in the $\mathbf{e}_1\mathbf{e}_2$ plane with tangent vector $\mathbf{t}(s)$, normal vector $\mathbf{n}(s)$, and curvature $\kappa(s)$. Suppose that ϕ is a constant angle. (See Fig. 7.2.) In addition, suppose that the curve $\bar{\mathbf{x}}(\bar{s})$ is an arc length curve, where

$$\bar{\mathbf{x}}(\bar{s}(s)) = \mathbf{x}(s) + \mathbf{e}_3 s \tan \phi.$$

a. Determine the function $\bar{s}(s)$ assuming that it is increasing.
b. Assuming $\kappa(s) > 0$ and ϕ is in the first quadrant, compute the Frenet apparatus $\bar{\mathbf{t}}$, $\bar{\mathbf{n}}$, $\bar{\mathbf{b}}$, $\bar{\kappa}$, and $\bar{\tau}$ for $\bar{\mathbf{x}}(\bar{s})$ in terms of $\mathbf{t}$, $\mathbf{n}$, $\mathbf{e}_3$, κ, and ϕ. (To calculate $\bar{\mathbf{b}}$, note that the direction of $\mathbf{n}$ is chosen so that $\mathbf{tn} = \mathbf{e}_1\mathbf{e}_2$ and therefore $\bar{\mathbf{b}}\bar{\mathbf{n}}\bar{\mathbf{t}} = \mathbf{e}_{321} = \mathbf{e}_3\mathbf{nt}$.)
c. Is $\bar{\mathbf{x}}(\bar{s})$ an example of a helix as defined in Problem 149? Use the formulas that you derived in part b) to get a formula for the vector $\mathbf{E}$ of (7.42) in terms of $\mathbf{t}$, $\mathbf{n}$, and $\mathbf{e}_3$. Does this answer surprise you? It should not.

The curve $\mathbf{x}(s)$ is designated as the *cross-sectional curve* of the helix $\bar{\mathbf{x}}(\bar{s})$. For 3-dimensions, we can use essentially any 2-dimensional curve for a cross-sectional curve. For higher dimensions, the situation is considerably more complicated. Helices in 3-dimensions are sometimes referred to as *curves of constant slope*. Why?

Note! We did not have to require that $\kappa(s) > 0$ or require that ϕ be in the first quadrant. I did not want to burden you with the problem of sorting out different cases. However, you may wish to do that. The case discussed above may be regarded as a right-handed helix. How could you get a left-handed helix without drastically changing the approach above?

Problem 151. Consider the curve in Problem 140.

a. Show that the curve in Problem 140 is a helix as defined in Problem 149. Is it right handed or left handed?
b. Determine the axis $\mathbf{E}$.
c. Sketch the cross-sectional curve.

Problem 152. Consider the curve in Problem 144.

a. Show that the curve in Problem 144 is a helix as defined in Problem 149. Is it right handed or left handed?
b. Determine the axis $\mathbf{E}$.
c. Sketch the cross-sectional curve.

Problem 153. Suppose the circle in Problem 143 is the cross-sectional curve of a circular helix.

a. What would be the axis $\mathbf{E}$?
b. Construct a general formula for a right-handed helix for which the circle in Problem 143 is a cross-sectional curve.
c. Adjust your formula in part (b) to get a general formula for a left handed helix with the same cross-sectional curve.

Problem 154. Suppose

$$\mathbf{M}(s) = \frac{1}{2}\sum_{j=1}^{n-1} k_j(s)\mathbf{E}_j(s)\mathbf{E}_{j+1}(s) \text{ and } \mathbf{A}(s) = \int_0^s \mathbf{M}(u)du.$$

a. Is it true that

$$\frac{\mathrm{d}}{\mathrm{d}s}\exp \mathbf{A}(s) = [\exp \mathbf{A}(s)]\,\mathbf{M}(s) \tag{7.43}$$

for a 3-dimensional helix? Is it true for higher dimensional helices?
b. Formulate a sufficient condition for (7.43) to be valid in any dimension.

Problem 155. If you reverse the direction of the parameterization do you change the right or left handedness of a curve? (Your answer may depend on the dimension of the curve.)

Problem 156. Suppose that $\mathrm{d}^k s/\mathrm{d}t^k$ is well defined for $k = 1, 2, \ldots, n$ and $\mathrm{d}s/\mathrm{d}t \neq 0$. Show that the set of vectors

$$\left\{\frac{\mathrm{d}\mathbf{x}}{\mathrm{d}t}, \frac{\mathrm{d}^2\mathbf{x}}{\mathrm{d}t^2}, \ldots, \frac{\mathrm{d}^k\mathbf{x}}{\mathrm{d}t^k}\right\}$$

spans the same space as

$$\left\{\frac{\mathrm{d}\mathbf{x}}{\mathrm{d}s}, \frac{\mathrm{d}^2\mathbf{x}}{\mathrm{d}s^2}, \ldots, \frac{\mathrm{d}^k\mathbf{x}}{\mathrm{d}s^k}\right\}$$

for $k = 1, 2, \ldots, n$. (As a consequence, applying the Gram–Schmidt process to either basis in the normal order will result in the same frame. Thus, it is not necessary to use an arc length parameterization to obtain the Frenet frame.)

7.2 *Arbitrary Speed Curves with Formulas

Virtually, the entire theory of curves is based on unit speed parameterizations. In theory, it is always possible to introduce a unit speed parameterization for a curve $\mathbf{x}(t)$ on any interval of the domain for which $\mathrm{d}\mathbf{x}/\mathrm{d}t \neq 0$. However for many (if not most) cases, it is a computational nightmare to actually carry out a unit speed parameterization. The good news is that you do not have to.

In this section, I will introduce generalized versions of (7.3), (7.5), (7.7), and (7.8). I will demonstrate that these formulas result in the conventional signs for the curvatures as I claimed for the unit speed versions in the last section. In addition, I will derive a formula for the curvatures.

Before proceeding further, you should do Problem 156 if you have not already done so.

In this section, I will assume that $\mathrm{d}s/\mathrm{d}t$ is both well defined and positive on whatever interval is under consideration.

Theorem 157. *Given a curve* $\mathbf{x}(t)$*, the following formulas are valid for the Frenet apparatus:*

$$\mathbf{E}_k = \frac{\frac{\mathrm{d}^k\mathbf{x}}{\mathrm{d}t^k} \wedge \mathbf{E}_{k-1} \wedge \cdots \wedge \mathbf{E}_2 \wedge \mathbf{E}_1}{\left|\frac{\mathrm{d}^k\mathbf{x}}{\mathrm{d}t^k} \wedge \mathbf{E}_{k-1} \wedge \cdots \wedge \mathbf{E}_2 \wedge \mathbf{E}_1\right|}\mathbf{E}_1\mathbf{E}_2 \cdots \mathbf{E}_{k-1}, \tag{7.44}$$

for $k = 1, 2, \ldots, n-1$,

where it is understood that

$$\mathbf{E}_1 = \frac{\frac{d\mathbf{x}}{dt}}{\left|\frac{d\mathbf{x}}{dt}\right|}. \tag{7.45}$$

Also,

$$\mathbf{E}_n = \mathbf{e}_{n\cdots 21}\mathbf{E}_1\mathbf{E}_2\cdots\mathbf{E}_{n-1}. \tag{7.46}$$

We can define the coefficient A_k by the equation:

$$\frac{d^k\mathbf{x}}{dt^k} = A_k\mathbf{E}_k + \sum_{j=1}^{k-1}\alpha_{kj}\mathbf{E}_j, \text{ or equivalently} \tag{7.47}$$

$$A_j = \left|\frac{d^j\mathbf{x}}{dt^j}\wedge\mathbf{E}_{j-1}\wedge\cdots\wedge\mathbf{E}_2\wedge\mathbf{E}_1\right| \tag{7.48}$$

for $j = 1, 2, \ldots, n-1$, where it is understood that

$$A_1 = \left|\frac{d\mathbf{x}}{dt}\right| = \frac{ds}{dt}. \tag{7.49}$$

For $j = n$,

$$A_n = \left(\frac{d^n\mathbf{x}}{dt^n}\wedge\mathbf{E}_{n-1}\wedge\cdots\wedge\mathbf{E}_2\wedge\mathbf{E}_1\right)\mathbf{e}_{12\cdots n}. \tag{7.50}$$

The curvatures can be written in terms of these coefficients. Namely

$$k_j = \frac{A_{j+1}}{A_jA_1}. \tag{7.51}$$

Proof. From the result of Problem 156, we know that the sets

$$\left\{\frac{d\mathbf{x}}{dt}, \frac{d^2\mathbf{x}}{dt^2}, \ldots, \frac{d^k\mathbf{x}}{dt^k}\right\} \text{ and } \left\{\frac{d\mathbf{x}}{ds}, \frac{d^2\mathbf{x}}{ds^2}, \ldots, \frac{d^k\mathbf{x}}{ds^k}\right\}$$

span the same space for $k = 1, 2, \ldots, n$. This implies that

$$\frac{\frac{d^k\mathbf{x}}{dt^k}\wedge\frac{d^{k-1}\mathbf{x}}{dt^{k-1}}\wedge\cdots\wedge\frac{d\mathbf{x}}{dt}}{\left|\frac{d^k\mathbf{x}}{dt^k}\wedge\frac{d^{k-1}\mathbf{x}}{dt^{k-1}}\wedge\cdots\wedge\frac{d\mathbf{x}}{dt}\right|} = \pm\frac{\frac{d^k\mathbf{x}}{ds^k}\wedge\frac{d^{k-1}\mathbf{x}}{ds^{k-1}}\wedge\cdots\wedge\frac{d\mathbf{x}}{ds}}{\left|\frac{d^k\mathbf{x}}{ds^k}\wedge\frac{d^{k-1}\mathbf{x}}{ds^{k-1}}\wedge\cdots\wedge\frac{d\mathbf{x}}{ds}\right|}. \tag{7.52}$$

A review of your computations in Problem 156 should reveal that we must choose the positive sign in (7.52) if ds/dt is positive. This means that in the Gram–Schmidt computation,

$$\mathbf{E}_1 = \frac{\frac{d\mathbf{x}}{dt}}{\left|\frac{d\mathbf{x}}{dt}\right|}, \text{ and} \tag{7.53}$$

$$\mathbf{E}_k\mathbf{E}_{k-1}\cdots\mathbf{E}_1 = \frac{\frac{\mathrm{d}^k\mathbf{x}}{\mathrm{d}t^k}\wedge\frac{\mathrm{d}^{k-1}\mathbf{x}}{\mathrm{d}t^{k-1}}\wedge\cdots\wedge\frac{\mathrm{d}\mathbf{x}}{\mathrm{d}t}}{\left|\frac{\mathrm{d}^k\mathbf{x}}{\mathrm{d}t^k}\wedge\frac{\mathrm{d}^{k-1}\mathbf{x}}{\mathrm{d}t^{k-1}}\wedge\cdots\wedge\frac{\mathrm{d}\mathbf{x}}{\mathrm{d}t}\right|}, \text{ and thus} \tag{7.54}$$

$$\mathbf{E}_k = \frac{\frac{\mathrm{d}^k\mathbf{x}}{\mathrm{d}t^k}\wedge\frac{\mathrm{d}^{k-1}\mathbf{x}}{\mathrm{d}t^{k-1}}\wedge\cdots\wedge\frac{\mathrm{d}\mathbf{x}}{\mathrm{d}t}}{\left|\frac{\mathrm{d}^k\mathbf{x}}{\mathrm{d}t^k}\wedge\frac{\mathrm{d}^{k-1}\mathbf{x}}{\mathrm{d}t^{k-1}}\wedge\cdots\wedge\frac{\mathrm{d}\mathbf{x}}{\mathrm{d}t}\right|}\mathbf{E}_1\mathbf{E}_2\cdots\mathbf{E}_{k-1} \tag{7.55}$$

for $k = 2,\ldots,n-1$.

Equivalently, we can write

$$\mathbf{E}_k\mathbf{E}_{k-1}\cdots\mathbf{E}_1 = \frac{\frac{\mathrm{d}^k\mathbf{x}}{\mathrm{d}t^k}\wedge\mathbf{E}_{k-1}\wedge\cdots\wedge\mathbf{E}_2\wedge\mathbf{E}_1}{\left|\frac{\mathrm{d}^k\mathbf{x}}{\mathrm{d}t^k}\wedge\mathbf{E}_{k-1}\wedge\cdots\wedge\mathbf{E}_2\wedge\mathbf{E}_1\right|}, \text{ so} \tag{7.56}$$

$$\mathbf{E}_k = \frac{\frac{\mathrm{d}^k\mathbf{x}}{\mathrm{d}t^k}\wedge\mathbf{E}_{k-1}\wedge\cdots\wedge\mathbf{E}_2\wedge\mathbf{E}_1}{\left|\frac{\mathrm{d}^k\mathbf{x}}{\mathrm{d}t^k}\wedge\mathbf{E}_{k-1}\wedge\cdots\wedge\mathbf{E}_2\wedge\mathbf{E}_1\right|}\mathbf{E}_1\mathbf{E}_2\cdots\mathbf{E}_{k-1} \tag{7.57}$$

for $k = 1,2,\ldots,n-1$.

And, as before

$$\mathbf{E}_n\mathbf{E}_{n-1}\cdots\mathbf{E}_1 = \mathbf{e}_{n\cdots 21} \text{ so}$$
$$\mathbf{E}_n = \mathbf{e}_{n\cdots 21}\mathbf{E}_1\mathbf{E}_2\cdots\mathbf{E}_{n-1}. \tag{7.58}$$

Since the sets

$$\left\{\frac{\mathrm{d}\mathbf{x}}{\mathrm{d}t},\frac{\mathrm{d}^2\mathbf{x}}{\mathrm{d}t^2},\ldots,\frac{\mathrm{d}^k\mathbf{x}}{\mathrm{d}t^k}\right\} \text{ and } \{\mathbf{E}_1,\mathbf{E}_2,\ldots,\mathbf{E}_k\}$$

span the same space for $k = 1,2,\ldots,n$, (7.47) makes sense and therefore it can be used to define A_k.

If we replaced $\mathrm{d}^k\mathbf{x}/\mathrm{d}t^k$ in the numerator on the right hand side of (7.56) by the formula of (7.47), we get

$$A_k = \left|\frac{\mathrm{d}^k\mathbf{x}}{\mathrm{d}t^k}\wedge\mathbf{E}_{k-1}\wedge\cdots\wedge\mathbf{E}_2\wedge\mathbf{E}_1\right|, \text{ for } k = 2,3,\ldots,n-1.$$

From (7.47) and (7.53), it is clear that

$$A_1 = \left|\frac{\mathrm{d}\mathbf{x}}{\mathrm{d}t}\right| = \frac{\mathrm{d}s}{\mathrm{d}t}.$$

For $k = n$, we note that

$$A_n \mathbf{e}_{n\cdots 21} = (A_n \mathbf{E}_n) \wedge \mathbf{E}_{n-1} \wedge \cdots \wedge \mathbf{E}_1 = \frac{d^n \mathbf{x}}{dt^n} \wedge \mathbf{E}_{n-1} \wedge \cdots \wedge \mathbf{E}_2 \wedge \mathbf{E}_1, \text{ and thus}$$

$$A_n = \left(\frac{d^n \mathbf{x}}{dt^n} \wedge \mathbf{E}_{n-1} \wedge \cdots \wedge \mathbf{E}_2 \wedge \mathbf{E}_1 \right) \mathbf{e}_{12\cdots n}. \tag{7.59}$$

To get the formula for the curvatures, we note that since

$$\frac{d^j \mathbf{x}}{dt^j} = A_j \mathbf{E}_j + \sum_{p=1}^{j-1} \alpha_{jp} \mathbf{E}_p, \text{ it follows that} \tag{7.60}$$

$$\frac{d^{j+1} \mathbf{x}}{dt^{j+1}} = A_j k_j \mathbf{E}_{j+1} \frac{ds}{dt} + \sum_{p=1}^{j} \beta_{jp} \mathbf{E}_p \tag{7.61}$$

On the other hand, we observe that from (7.60),

$$\frac{d^{j+1} \mathbf{x}}{dt^{j+1}} = A_{j+1} \mathbf{E}_{j+1} + \sum_{p=1}^{j} \alpha_{j+1p} \mathbf{E}_p, \text{ for } j = 1, 2, \ldots, n-1. \tag{7.62}$$

Using the fact that

$$\frac{ds}{dt} = A_1,$$

it follows from (7.61), and (7.62) that

$$k_j = \frac{A_{j+1}}{A_j A_1} \text{ for } j = 1, 2, \ldots, n-1. \tag{7.63}$$

I have now verified all the equations listed in this theorem.

From (7.48), and (7.50), we see that all of the k_j's are positive with the possible exception of k_{n-1}.

Example 158. Consider the curve,

$$\mathbf{x}(t) = 6t\mathbf{e}_1 + 3\sqrt{2}t^2 \mathbf{e}_2 + 2t^3 \mathbf{e}_3. \tag{7.64}$$

It follows that

$$\frac{d\mathbf{x}}{dt} = 6\mathbf{e}_1 + 6\sqrt{2}t\mathbf{e}_2 + 6t^2 \mathbf{e}_3, \tag{7.65}$$

and therefore

$$A_1 = \left| \frac{d\mathbf{x}}{dt} \right| = \frac{ds}{dt} = 6\sqrt{1 + 2t^2 + t^4} = 6(t^2 + 1). \tag{7.66}$$

Thus,

$$\mathbf{E}_1 = \frac{\mathrm{d}\mathbf{x}}{\mathrm{d}t} \Big/ \left|\frac{\mathrm{d}\mathbf{x}}{\mathrm{d}t}\right| = \frac{\mathbf{e}_1 + \sqrt{2}t\mathbf{e}_2 + t^2\mathbf{e}_3}{t^2+1}. \tag{7.67}$$

From (7.65),

$$\frac{\mathrm{d}^2\mathbf{x}}{\mathrm{d}t^2} = 6\sqrt{2}\mathbf{e}_2 + 12t\mathbf{e}_3. \tag{7.68}$$

And

$$\begin{aligned}\frac{\mathrm{d}^2\mathbf{x}}{\mathrm{d}t^2} \wedge \mathbf{E}_1 &= 6\left(\sqrt{2}\mathbf{e}_2 + 2t\mathbf{e}_3\right) \wedge \left(\frac{\mathbf{e}_1 + \sqrt{2}t\mathbf{e}_2 + t^2\mathbf{e}_3}{t^2+1}\right)\\ &= 6\frac{\mathbf{e}_{23}(\sqrt{2}t^2 - 2\sqrt{2}t^2) + \mathbf{e}_{31}\,(2t) + \mathbf{e}_{12}(-\sqrt{2})}{t^2+1}, \text{ or}\end{aligned}$$

$$\frac{\mathrm{d}^2\mathbf{x}}{\mathrm{d}t^2} \wedge \mathbf{E}_1 = 6\frac{-\sqrt{2}t^2\mathbf{e}_{23} + 2t\mathbf{e}_{31} - \sqrt{2}\mathbf{e}_{12}}{t^2+1}. \tag{7.69}$$

It then follows that

$$A_2 = \left|\frac{\mathrm{d}^2\mathbf{x}}{\mathrm{d}t^2} \wedge \mathbf{E}_1\right| = \frac{6}{t^2+1}\sqrt{2t^4 + 4t^2 + 2} = 6\sqrt{2}. \tag{7.70}$$

Furthermore,

$$\mathbf{E}_2\mathbf{E}_1 = \frac{\frac{\mathrm{d}^2\mathbf{x}}{\mathrm{d}t^2} \wedge \mathbf{E}_1}{\left|\frac{\mathrm{d}^2\mathbf{x}}{\mathrm{d}t^2} \wedge \mathbf{E}_1\right|} = \frac{-t^2\mathbf{e}_{23} + \sqrt{2}t\mathbf{e}_{31} - \mathbf{e}_{12}}{t^2+1}, \text{ so} \tag{7.71}$$

$$\begin{aligned}\mathbf{E}_2 &= \frac{\frac{\mathrm{d}^2\mathbf{x}}{\mathrm{d}t^2} \wedge \mathbf{E}_1}{\left|\frac{\mathrm{d}^2\mathbf{x}}{\mathrm{d}t^2} \wedge \mathbf{E}_1\right|}\mathbf{E}_1\\ &= \left(\frac{-t^2\mathbf{e}_{23} + \sqrt{2}t\mathbf{e}_{31} - \mathbf{e}_{12}}{t^2+1}\right)\left(\frac{\mathbf{e}_1 + \sqrt{2}t\mathbf{e}_2 + t^2\mathbf{e}_3}{t^2+1}\right)\\ &= \frac{\mathbf{e}_1(-\sqrt{2}t^3 - \sqrt{2}t) + \mathbf{e}_2(-t^4+1) + \mathbf{e}_3(\sqrt{2}t^3 + \sqrt{2}t)}{(t^2+1)^2}\end{aligned}$$

Or simplified,

$$\mathbf{E}_2 = \frac{-\sqrt{2}t\mathbf{e}_1 + (-t^2+1)\mathbf{e}_2 + \sqrt{2}t\mathbf{e}_3}{t^2+1}. \tag{7.72}$$

To get $\mathbf{E}_3$, we note that

$$\mathbf{E}_3 = \mathbf{e}_{321}\mathbf{E}_1\mathbf{E}_2.$$

So using (7.71), we have

$$\mathbf{E}_3 = \mathbf{e}_{321}\left(\frac{t^2\mathbf{e}_{23} - \sqrt{2}t\mathbf{e}_{31} + \mathbf{e}_{12}}{t^2+1}\right) \quad \text{or}$$

$$\mathbf{E}_3 = \frac{t^2\mathbf{e}_1 - \sqrt{2}t\mathbf{e}_2 + \mathbf{e}_3}{t^2+1}. \tag{7.73}$$

To compute A_3, we need to compute $\mathrm{d}^3\mathbf{x}/\mathrm{d}t^3$. From (7.68)

$$\frac{\mathrm{d}^3\mathbf{x}}{\mathrm{d}t^3} = 12\mathbf{e}_3.$$

Using (7.50) and (7.71), we have

$$\begin{aligned} A_3 &= \left(\frac{\mathrm{d}^3\mathbf{x}}{\mathrm{d}t^3} \wedge \mathbf{E}_2 \wedge \mathbf{E}_1\right)\mathbf{e}_{123} \\ &= \left(12\mathbf{e}_3 \wedge \frac{-t^2\mathbf{e}_{23} + \sqrt{2}t\mathbf{e}_{31} - \mathbf{e}_{12}}{t^2+1}\right)\mathbf{e}_{123} \\ &= \frac{-12\mathbf{e}_{123}}{t^2+1}\mathbf{e}_{123} \quad \text{and thus} \end{aligned}$$

$$A_3 = \frac{12}{t^2+1}. \tag{7.74}$$

Using (7.63), (7.66), (7.70), and (7.74), we have

$$k_1 = \frac{A_2}{A_1A_1} = \frac{6\sqrt{2}}{36(t^2+1)^2} = \frac{\sqrt{2}}{6(t^2+1)^2} \quad \text{and} \tag{7.75}$$

$$k_2 = \frac{A_3}{A_2A_1} = \frac{12}{(t^2+1)6\sqrt{2}6(t^2+1)} = \frac{\sqrt{2}}{6(t^2+1)^2}. \tag{7.76}$$

We now see that since k_1/k_2 is a constant, we are dealing with a generalized helix. If $\mathbf{x}(t)$ is a helix, we can project out the cross-sectional curve $\mathbf{y}(t)$ passing through $\mathbf{x}(t_0)$. In particular, if $\mathbf{E}$ is the unit axis vector, then

$$\mathbf{y}(t) = \mathbf{x}(t) - \langle \mathbf{x}(t) - \mathbf{x}(t_0), \mathbf{E}\rangle\, \mathbf{E} \tag{7.77}$$

However if the cross-sectional curve $\mathbf{y}(t)$ does not lie in a coordinate plane or a plane parallel to a coordinate plane, simply projecting out the cross-sectional curve is not very enlightening if you wish to know what the cross-sectional curve looks like. For the curve in this example,

$$\mathbf{E} = \frac{k_2\mathbf{E}_1 + k_1\mathbf{E}_2}{\sqrt{(k_1)^2 + (k_2)^2}} = \frac{\mathbf{e}_1 + \mathbf{e}_2}{\sqrt{2}}.$$

Thus, it should be clear that the cross-sectional curve for this example does not lie in a coordinate plane. If we rotate our curve so that axis of the helix $\mathbf{E}$ aligns with $\mathbf{e}_3$ (or introduce a new basis so that $\mathbf{E} = \bar{\mathbf{e}}_3$), then the cross-sectional curve will lie in some plane parallel to the $\mathbf{e}_1\mathbf{e}_2$ plane (or $\bar{\mathbf{e}}_1\bar{\mathbf{e}}_2$ plane). One way to do this is to jump ahead, do Problem 160, and then take advantage of the result. In particular, introduce a new basis:

$$\bar{\mathbf{e}}_1 = \frac{k_1\mathbf{E}_1(0) - k_2\mathbf{E}_3(0)}{\sqrt{(k_1)^2 + (k_2)^2}}, \tag{7.78}$$

$$\bar{\mathbf{e}}_2 = \mathbf{E}_2(0), \text{ and} \tag{7.79}$$

$$\bar{\mathbf{e}}_3 = \frac{k_2\mathbf{E}_1(0) + k_1\mathbf{E}_3(0)}{\sqrt{(k_1)^2 + (k_2)^2}} = \mathbf{E}. \tag{7.80}$$

For our example, $\mathbf{E}_1(0) = \mathbf{e}_1$, $\mathbf{E}_2(0) = \mathbf{e}_2$, $\mathbf{E}_3(0) = \mathbf{e}_3$, and $k_1 = k_2$. Therefore,

$$\begin{aligned}
\bar{\mathbf{e}}_1 &= \frac{\mathbf{e}_1 - \mathbf{e}_3}{\sqrt{2}}, \\
\bar{\mathbf{e}}_2 &= \mathbf{e}_2, \text{ and} \\
\bar{\mathbf{e}}_3 &= \frac{\mathbf{e}_1 + \mathbf{e}_3}{\sqrt{2}}.
\end{aligned}$$

With a little algebra, it is not difficult to show that these equations imply that

$$\begin{aligned}
\mathbf{e}_1 &= \frac{\bar{\mathbf{e}}_1 + \bar{\mathbf{e}}_3}{\sqrt{2}}, \\
\mathbf{e}_2 &= \bar{\mathbf{e}}_2, \text{ and} \\
\mathbf{e}_3 &= \frac{-\bar{\mathbf{e}}_1 + \bar{\mathbf{e}}_3}{\sqrt{2}}.
\end{aligned}$$

Using these formulas, (7.64) becomes

$$\mathbf{x}(t) = \sqrt{2}\left[(-t^3 + 3t)\bar{\mathbf{e}}_1 + 3t^2\bar{\mathbf{e}}_2 + (t^3 + 3t)\bar{\mathbf{e}}_3\right]. \tag{7.81}$$

In this form, we know that the cross-sectional curve lies in the $\bar{\mathbf{e}}_1\bar{\mathbf{e}}_2$ so it is easy to plot. (See Problem 161.) Because of the choice of parameterization, (7.81) does not look like the equation of a helix. Let us try to introduce a unit speed parameterization.

$$\frac{d\mathbf{x}}{dt} = \sqrt{2}\left[(-3t^2 + 3)\bar{\mathbf{e}}_1 + 6t\bar{\mathbf{e}}_2 + (3t^2 + 3)\bar{\mathbf{e}}_3\right]. \text{ So}$$

$$\frac{d\mathbf{x}}{dt} = 3\sqrt{2}\left[(-t^2 + 1)\bar{\mathbf{e}}_1 + 2t\bar{\mathbf{e}}_2 + (t^2 + 1)\bar{\mathbf{e}}_3\right].$$

This implies that

$$\frac{ds}{dt} = \left|\frac{d\mathbf{x}}{dt}\right| = 3\sqrt{2}\sqrt{(-t^2+1)^2 + 4t^2 + (t^2+1)^2}$$
$$= 3\sqrt{2}\sqrt{2(t^4+2t^2+1)} = 6(t^2+1).$$

Finally, this means that

$$s = 6\left(\frac{t^3}{3} + t\right) = 2(t^3 + 3t) + \text{a possible constant}.$$

If we let the constant be zero, then the coefficient of $\bar{\mathbf{e}}_3$ on the right-hand side of (7.81) would be $(\sqrt{2}/2)s$, which is the kind of simple answer you would expect for a helix. However, writing the coefficients of $\bar{\mathbf{e}}_1$ and $\bar{\mathbf{e}}_2$ in terms of s and then trying to compute the Frenet apparatus from the resulting functions would not be an easy task (at least for me).

For the sake of completeness, I will now write down the Frenet equations adjusted for an arbitrary speed parameterization. Since

$$\frac{d\mathbf{E}_j}{dt} = \frac{d\mathbf{E}_j}{ds}\frac{ds}{dt}, \text{ we have}$$

$$\frac{d}{ds}\begin{bmatrix} \mathbf{E}_1 \\ \mathbf{E}_2 \\ \mathbf{E}_3 \\ - \\ - \\ - \\ \mathbf{E}_n \end{bmatrix} = \begin{bmatrix} 0 & k_1 v & 0 & - & - & - & 0 \\ -k_1 v & 0 & k_2 v & - & - & - & 0 \\ 0 & -k_2 v & 0 & - & - & - & 0 \\ - & - & - & - & - & - & - \\ - & - & - & - & - & - & - \\ - & - & - & - & - & 0 & k_{n-1} v \\ 0 & 0 & 0 & - & - & -k_{n-1} v & 0 \end{bmatrix} \begin{bmatrix} \mathbf{E}_1 \\ \mathbf{E}_2 \\ \mathbf{E}_3 \\ - \\ - \\ - \\ \mathbf{E}_n \end{bmatrix}, \tag{7.82}$$

where

$$v = \frac{ds}{dt}.$$

How should one modify (7.21) and/or (7.22) for the equation using the Frenet 2-vector?

Problem 159. Show that

$$A_k A_{k-1} \cdots A_1 = \left|\frac{d^k\mathbf{x}}{dt^k} \wedge \frac{d^{k-1}\mathbf{x}}{dt^{k-1}} \wedge \cdots \wedge \frac{d\mathbf{x}}{dt}\right|, \text{ for } k = 1, 2, \ldots, n-1, \text{ and}$$

$$A_n A_{n-1} \cdots A_1 = \left(\frac{d^n\mathbf{x}}{dt^n} \wedge \frac{d^{n-1}\mathbf{x}}{dt^{n-1}} \wedge \cdots \wedge \frac{d\mathbf{x}}{dt}\right)\mathbf{e}_{12\cdots n}.$$

Problem 160.

a. Construct an argument that verifies (7.77)
b. Suppose $\mathbf{x}(t)$ has the Frenet frame $\mathbf{E}_1$, $\mathbf{E}_2$, and $\mathbf{E}_3$ and $\mathbf{y}(t)$ is the cross-sectional curve passing through $\mathbf{x}(t_0)$ with the Frenet $\bar{\mathbf{E}}_1$, $\bar{\mathbf{E}}_2$, and $\bar{\mathbf{E}}_3$. Using (7.77) and the fact that

$$\mathbf{E} = \frac{k_2\mathbf{E}_1 + k_1\mathbf{E}_3}{\sqrt{(k_1)^2 + (k_2)^2}}, \text{ show that}$$

$$\bar{\mathbf{E}}_1 = \frac{k_1\mathbf{E}_1 - k_2\mathbf{E}_3}{\sqrt{(k_1)^2 + (k_2)^2}}, \tag{7.83}$$

$$\bar{\mathbf{E}}_2 = \mathbf{E}_2, \text{ and} \tag{7.84}$$

$$\bar{\mathbf{E}}_3 = \frac{k_2\mathbf{E}_1 + k_1\mathbf{E}_3}{\sqrt{(k_1)^2 + (k_2)^2}} = \mathbf{E}. \tag{7.85}$$

c. Suppose $\bar{k}_1$ is the curvature and $\bar{k}_2$ is the torsion for $\mathbf{y}(t)$. Show that

$$\bar{k}_1 = \sqrt{(k_1)^2 + (k_2)^2} \text{ and } \bar{k}_2 = 0. \tag{7.86}$$

(Showing that $\bar{k}_2 = 0$, verifies the fact that $\mathbf{y}(t)$ lies in a plane.)

Problem 161. Plot the cross-sectional curve for the helix of (7.81). Is the curve symmetric with respect to either axis? Does it cross itself?

Problem 162. Consider the curve $\mathbf{x}(t) = \mathbf{e}_1 \cosh t + \mathbf{e}_2 \sinh t + \mathbf{e}_3 t$. Compute the Frenet apparatus and verify the fact that it is a helix. Describe or plot the cross-sectional curve. Can you construct formulas for the curvature and the torsion in terms of the arc length distance from the point where $t = 0$.

Problem 163. Consider the curve $\mathbf{x}(t) = 2t\mathbf{e}_1 + t^2\mathbf{e}_2 + (t^3/3)\mathbf{e}_3$. Compute the Frenet apparatus and verify the fact that it is a helix. Find the limiting values of $\mathbf{E}_1$, $\mathbf{E}_2$, and $\mathbf{E}_3$ as $t \to -\infty$ and $t \to +\infty$. Plot the cross-sectional curve.

7.3 Ruled Surfaces and Developable Surfaces

A 2-dimensional surface embedded in the $(n+1)$-dimensional Euclidean space E^{n+1} is said to be a *ruled surface* if it contains a one-parameter family of straight lines, which can be chosen as coordinate curves on the surfaces. The straight lines are said to be *generators* of the ruled surface.

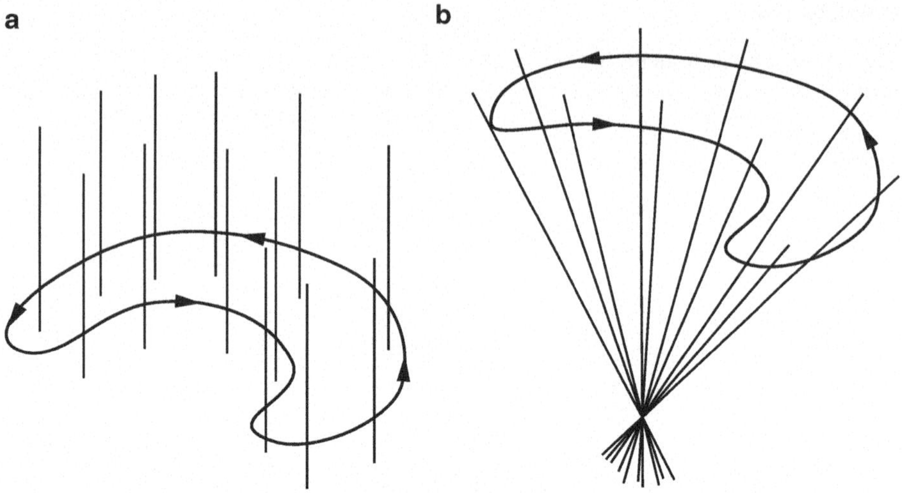

Fig. 7.3 Two ruled surfaces. (**a**) Cylinder. (**b**) Cone

A ruled surface may be thought of as a surface swept out by a continuously moving straight line. Immediate examples that should spring up in your mind are cylinders (not necessarily circular) and cones (not necessarily symmetrical). (See Fig. 7.3a, b.)

Suppose the straight line glides along the curve $\mathbf{x}(s)$ and at each point of the curve, the direction of the straight line is indicated by the unit vector $\mathbf{v}(s)$. In that circumstance, the equation for the surface can be written in the form:

$$\mathbf{y}(u,t) = \mathbf{x}(u)+t\mathbf{v}(u). \tag{7.87}$$

The curve $\mathbf{x}(u)$ is known as the *directrix* of the ruled surface. It should be clear that the coordinate t measures the distance along a straight line generator away from the point $\mathbf{x}(u)$ on the directrix. For $\mathbf{y}(u,t)$ to truly represent a 2-dimensional surface, it is generally required that the coordinate Dirac matrices ($\boldsymbol{\gamma}_u = \partial\mathbf{y}/\partial u$ and $\boldsymbol{\gamma}_t = \partial\mathbf{y}/\partial t = \mathbf{v}(u)$) be linearly independent. However, an important class of ruled surfaces, known as *tangential developables* are swept out by the tangent lines to some smooth directrix. For this class of surfaces, the directrices become edges (*edges of regression*). On such an edge, the condition that $\boldsymbol{\gamma}_u \wedge \boldsymbol{\gamma}_t \neq 0$ is violated. (Actually unless the torsion of the edge of regression is zero, two surfaces are swept out – one by the forward ray of the tangent line and the other by the trailing ray. The two surfaces meet and form a cusp at the edge or regression. An example of this is illustrated in Fig. 7.6.)

Another anomaly occurs at the vertex of a cone, where $\boldsymbol{\gamma}_u \wedge \boldsymbol{\gamma}_t$ is undefined. For cones, the most transparent form of the equation results from using the vertex in place of a directrix curve. In that case, $\mathbf{x}(u)$ becomes a fixed point $\mathbf{p}$ (the vertex) and the equation becomes

$$\mathbf{y}(u,t) = \mathbf{p}+t\mathbf{v}(u). \tag{7.88}$$

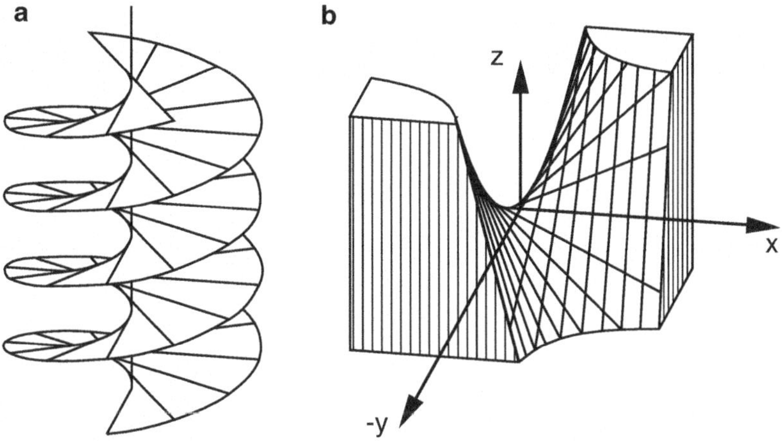

Fig. 7.4 (**a**) Helicoid. (**b**) Saddle surface

At the vertex where $t = 0$, $\mathbf{y}$ is independent of u so neither $\boldsymbol{\gamma}_u$ nor $\boldsymbol{\gamma}_t$ is well defined. When (7.88) is used to represent a cone, you may wish to choose u so that it is an arc length parameter for $\mathbf{v}$.

Cones and cylinders are obvious examples of ruled surfaces. However, there are many other classic examples. For example, suppose you imagine a straight line L attached orthogonally to an axis. If the line L moves along the axis at a constant speed while rotating about the axis at a constant rate, the resulting surface swept out by the line L is said to be a *helicoid*. (See Fig. 7.4a.)

Some ruled surfaces have two sets of generators. Such surfaces are said to be *doubly ruled*. An example of a doubly ruled surface is the saddle surface $z = xy$ (or $x^3 = x^1x^2$), which we have encountered before. (See Fig. 7.4b.) One set of generators cut across the x-axis. For that set of generators, we can parameterize the surface by the equation:

$$\mathbf{y}(u,t) = \mathbf{x}(u) + t\mathbf{v}(u), \text{ where} \tag{7.89}$$

$$\mathbf{x}(u) = \mathbf{e}_1 u \text{ and } \mathbf{v}(u) = \frac{\mathbf{e}_2 + \mathbf{e}_3 u}{\sqrt{1+u^2}}. \tag{7.90}$$

Alternatively, one can use another set of generators that cross the y-axis. (Look at Fig. 7.4b again and Prob. 164.) For me, a less obvious example of a doubly ruled surface is the *one sheet hyperboloid*. (See Fig. 7.5.) The one sheet hyperboloid can be defined by the equation

$$\frac{x^2}{a^2} + \frac{y^2}{b^2} - \frac{z^2}{c^2} = 1. \tag{7.91}$$

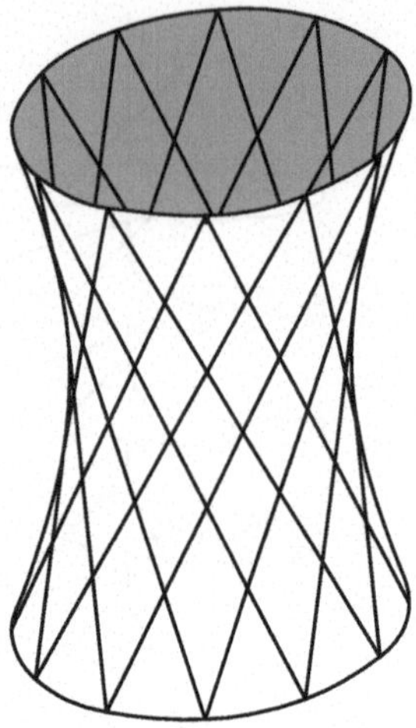

Fig. 7.5 One sheet hyperboloid

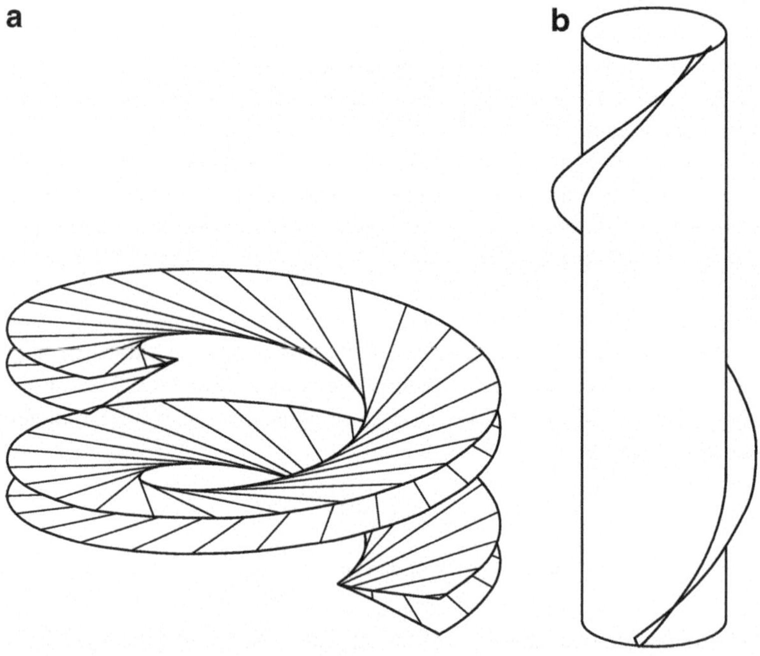

Fig. 7.6 (**a**) Developable helicoid. (**b**) Strake

For the one sheet hyperboloid, one can write:

$$\mathbf{y}(u,t) = \mathbf{x}(u) + t\mathbf{v}(u), \text{ where} \tag{7.92}$$

$$\mathbf{x}(u) = \mathbf{e}_1 a \cos u + \mathbf{e}_2 b \sin u, \text{ and} \tag{7.93}$$

$$\mathbf{v}(u) = -\mathbf{e}_1 a \sin u + \mathbf{e}_2 b \cos u + \mathbf{e}_3 c. \tag{7.94}$$

(In this case, u is not an arc length parameter for $\mathbf{x}$ and $\mathbf{v}$ has not been normalized.) By symmetry, it is not too difficult to show that a one sheet hyperboloid has a second set of generators.

Because of the double set of generators, one sheet hyperboloids are used for nuclear plant cooling towers. Straight steel beams can be used for primary structural members.

In closing this section, I should comment on what is a *developable surface*. In a general context, a developable surface is the envelope of a one parameter family of planes. This approach is outlined in Gerretsen's text (1962, pp. 87-88) for m-dimensional surfaces in $(n+1)$-dimensional Euclidean spaces. For 2-dimensional surfaces, a developable surface may be characterized as a ruled surface with zero Gaussian curvature. In the next chapter (Theorem 233 and Corollary 234), it is shown that at least locally these surfaces are either cylinders, cones, or tangential developables.

Problem 164.

(a) Demonstrate that the formulas of (7.89) and (7.90) are consistent with the equation $z = xy$ (or $x^3 = x^1x^2$).
(b) Write down another equation for the surface $z = xy$, using a second set of generators.
(a) Verify the fact that (7.92), (7.93), and (7.94) are consistent with (7.91).
(b) Construct an alternate version of (7.94) to demonstrate that the one sheet hyperboloid is doubly ruled.

Problem 165. Consider an arbitrary cone represented by the equation:

$$\mathbf{y}(u^1, u^2) = \mathbf{p}+u^1\mathbf{v}(u^2),$$

where $\mathbf{p}$ is a constant vector and $\mathbf{v}(u^2)$ is a unit vector. (See Fig. 7.3b.)

Compute the Gaussian curvature using Fock–Ivanenko coefficients and show that it is zero.

Note! $\langle \mathbf{v}, \dot{\mathbf{v}} \rangle = 0$, where $\dot{\mathbf{v}} = \mathrm{d}\mathbf{v}/\mathrm{d}u^2$. Why?

Problem 166. Consider an arbitrary cylinder represented by the equation:

$$\mathbf{y}(u^1, u^2) = \mathbf{x}(u^2) + u^1\mathbf{v},$$

where u^2 is an arc length parameter for the curve $\mathbf{x}$ and $\mathbf{v}$ is a constant vector of unit length. See Fig. 7.3a. (You can assume that $\boldsymbol{\gamma}_2 \wedge \mathbf{v} \neq 0$, where $\boldsymbol{\gamma}_2 = \dot{\mathbf{x}}(u^2) = \mathrm{d}\mathbf{x}/\mathrm{d}u^2$.) Compute the Gaussian curvature using Fock–Ivanenko coefficients and show that it is zero. Note! When constructing an orthonormal frame for this surface, you may find it useful to use the equality:

$$(\mathbf{b} \wedge \mathbf{a})\,\mathbf{a} = \frac{1}{2}\,[\mathbf{ba} - \mathbf{ab}]\,\mathbf{a} = \left[\mathbf{ba} - \frac{1}{2}\,(\mathbf{ba} + \mathbf{ab})\right]\mathbf{a} \text{ or}$$

$$(\mathbf{b} \wedge \mathbf{a})\,\mathbf{a} = \mathbf{b}\,\langle \mathbf{a}, \mathbf{a} \rangle - \langle \mathbf{a}, \mathbf{b} \rangle\,\mathbf{a}. \tag{7.95}$$

Problem 167 (*Developable Helicoid* and *STRAKE*). Consider the developable helicoid. This is the surface generated by the tangents to a circular helix. Actually, a developable helicoid consists of two surfaces – one swept out by the tangent ray in the forward direction and the other swept out by the tangent ray in the trailing direction. The two surfaces meet at the circular helix forming a sharp cusp known as the *edge of regression*. (See Fig. 7.6a.) For a circular helix, a unit speed parameterization is

$$\mathbf{x}(s) = \mathbf{e}_1 a \cos \frac{s}{\sqrt{a^2+b^2}} + \mathbf{e}_2 a \sin \frac{s}{\sqrt{a^2+b^2}} + \mathbf{e}_3 \frac{bs}{\sqrt{a^2+b^2}}. \tag{7.96}$$

a) Determine the unit tangent vector $\mathbf{E}_1 = \mathbf{t}$ so that you can explicitly write

$$\mathbf{y}(u^1, u^2) = \mathbf{x}(u^2) + u^1 \mathbf{t}(u^2)$$

in terms of u^1, u^2, $\mathbf{e}_1$, $\mathbf{e}_2$, and $\mathbf{e}_3$.

b) Use Fock–Ivanenko coefficients to show that the Gaussian curvature is zero.

c) Because a developable helicoid has zero Gaussian curvature, it can be made out of flat material. (The material has to be bent but not stretched.) Suppose the helix of (7.96) is drawn (perhaps with chalk) on a circular cylinder of radius a. Determine the radius of curvature that you would use to cut a developable helicoid from flat material so that it could be attached to the cylinder along the chalk line without stretching. In the context of industrial applications, the developable helicoid is known as a *strake*. (See Fig. 7.6b.)

A circular cylinder is not rigid since it can be bent without stretching so that the cross-sectional curve becomes something other than a circle. Suppose a circular cylinder was distorted so the cross-sectional curve was no longer a circle. How would this change the design of a strake to be attached to the same chalk line mentioned above? (You may wish to consult Prob. 160. Why does adding a properly designed strake to a circular cylinder make it rigid?

Problem 168 (*Tangential Developable*). Some features of the developable helicoid of Prob. 167 are also true for the more general *tangential developable surfaces*. If u^2 is an arc length parameter for a curve $\mathbf{x}(u^2)$ with nonzero curvature, then the corresponding tangential developable surface can be represented in the form:

$$\mathbf{y}(u^1, u^2) = \mathbf{x}(u^2) + u^1 \mathbf{E}_1(u^2), \text{ where}$$

$$\mathbf{E}_1(u^2) = \frac{d\mathbf{x}}{du^2} = \dot{\mathbf{x}}.$$

If k_2 is nonzero, as in the developable helicoid, $\mathbf{x}(u^2)$ is an edge of regression, where the surface swept out by the tangent ray in the forward direction meets with the surface swept out by the tangent ray in the trailing direction in a sharp cusp. Now suppose the tangential developable is embedded in E^3. Then our equations become

$$\mathbf{y}(u^1, u^2) = \mathbf{x}(u^2) + u^1\mathbf{t}(u^2), \text{ where}$$

$$\mathbf{t}(u^2) = \frac{d\mathbf{x}}{du^2} = \dot{\mathbf{x}}.$$

a) Suppose $\mathbf{E}_{\bar{1}}$ and $\mathbf{E}_{\bar{2}}$ form a moving frame for the a tangential developable surface. Show that if we let $\mathbf{E}_{\bar{1}} = \mathbf{t}(u^2)$ we can use (7.95) to show that $\mathbf{E}_{\bar{2}} = \mathbf{n}(u^2)$. This of course implies that $\mathbf{N} = \mathbf{b}(u^2)$, where $\mathbf{N}$ is normal to the surface.
b) Use Fock–Ivanenko coefficients to show that the Gaussian curvature is zero.

7.4 *Archimedes' Screw

The helicoid is the basis for the Archimedes Screw. If a helicoid is encased in a circular cylinder and tilted, it can be used as a pump. In some versions, the helicoid turns while the cylindrical casing remains stationary during operation. In other versions, the casing is attached to the helicoid and the cylinder is rotated to get the desired pumping action.

If one takes a copy of a helicoid and rotates it 180^0, one ends up with two helicoids equally spaced along a common axis. This "two bladed" design is the one most commonly used for Archimedes Screws but, according to Chris Rorres (Professor Emeritus of Mathematics at Drexel University), Archimedes Screws with as many as eight blades have been constructed.

In ancient times, the Archimedes Screw was used to irrigate lands in Egypt, Greece, and Rome. It was also used to pump water from mines. When the Netherlands wished to expand its boundaries, it would build dikes to enclose an area covered by shallow sea water. Wind powered Archimedes Screws would then be used to drain the enclosed area.

In modern times, manually operated Archimedes Screws continue to be used for irrigation in third world countries. This should not be surprising What may be more surprising is that for some applications, modern engineers have not been able to devise a superior design for a pump. Because of its clog free properties, the Archimedes Screw is the preferred pump for the most advanced water treatment plants.

Traditionally, the invention of the Archimedes Screw has been attributed to Archimedes (287–212BC). However, this has now become a matter of dispute among present-day historians. To paraphrase a section of the relevant Wikipedia web site: The Assyriologist Stephanie Dalley contends it was invented in the sixth century BC during the time of King Sennachrib (Dalley and Oleson 2003). However, John Peter Oleson states that there is a "total lack of any literary and archaeological evidence for the existence of the water-screw before ca. 250BC." (Oleson 1984, p.292)

7.5 Principal Curvatures

7.5.1 The Normal and Geodesic Curvature Vectors

In this section, we will deal with n-dimensional surfaces embedded in an $(n+1)$-dimensional Euclidean space. Such a surface is said to be a *hypersurface*. A metric tensor g_{ij} is said to be *positive definite* if $v^i g_{ij} v^j$ is positive unless $v^i = 0$ for $i = 1, 2, ---, n$. I will restrict the discussion in this section and most of the book to metrics that are positive definite.

In this section, we will also deal with curves that lie in hypersurfaces from an extrinsic point of view. That is we will not restrict ourselves to using measurements taken in the n-dimensional surface. We will use measurements taken in all $(n+1)$-dimensions.

A point on the hypersurface $\mathbf{x}$ is determined by n parameters, so

$$\mathbf{x}(u^1, u^2, ---, u^n) = \mathbf{e}_j x^j(u^1, u^2, ---.u^n), \text{ where} \tag{7.97}$$

j is summed from 1 to $n+1$. If s is an arc length parameter for some curve on the surface, then

$$\mathbf{x}(s) = \mathbf{e}_j x^j(u^1(s), u^2(s), ---, u^n(s)), \text{ and} \tag{7.98}$$

$$\mathbf{E}_1 = \frac{\mathrm{d}}{\mathrm{d}s}\mathbf{x}(s) = \mathbf{e}_j \frac{\partial x^j}{\partial u^k}\frac{\mathrm{d}u^k}{\mathrm{d}s} = \boldsymbol{\gamma}_k \frac{\mathrm{d}u^k}{\mathrm{d}s}, \text{ where} \tag{7.99}$$

the $\boldsymbol{\gamma}_k$'s are coordinate Dirac matrices that span the n-dimensional plane that is tangent to the surface at the given point. If we now took the intrinsic derivative of $\mathbf{E}_1$ with respect to s, we would get some vector in the same tangent plane. This would be appropriate for an intrinsic observer restricted to taking all measurements in the n-dimensional hypersurface.

However, an extrinsic observer would most likely see the tangent vector $\mathbf{E}_1$ moving in the extra dimension. For example, consider the 2-dimensional sphere of radius R embedded in E^3. That is

$$\mathbf{x}(\theta, \phi) = \mathbf{e}_1 R\cos\phi\sin\theta + \mathbf{e}_2 R\sin\phi\sin\theta + \mathbf{e}_3 R\cos\theta. \tag{7.100}$$

The plane tangent to the sphere is spanned by $\boldsymbol{\gamma}_\theta$ and $\boldsymbol{\gamma}_\phi$, where

$$\boldsymbol{\gamma}_\theta = \frac{\partial\mathbf{x}}{\partial\theta} = \mathbf{e}_1 R\cos\phi\cos\theta + \mathbf{e}_2 R\sin\phi\cos\theta - \mathbf{e}_3 R\sin\theta, \text{ and}$$

$$\boldsymbol{\gamma}_\phi = \frac{\partial\mathbf{x}}{\partial\phi} = -\mathbf{e}_1 R\sin\phi\sin\theta + \mathbf{e}_2 R\cos\phi\sin\theta.$$

A third vector $\mathbf{N}$ is normal to the surface. In particular,

$$\mathbf{N} = \mathbf{e}_1\cos\phi\sin\theta + \mathbf{e}_2\sin\phi\sin\theta + \mathbf{e}_3\cos\theta.$$

If we considered the great circle passing through the two poles with $\phi = \phi_0$, we would have

$$\mathbf{x}(s) = \mathbf{e}_1 R\cos\phi_0 \sin\frac{s}{R} + \mathbf{e}_2 R\sin\phi_0 \sin\frac{s}{R} + \mathbf{e}_3 R\cos\frac{s}{R}.$$

Now

$$\mathbf{E}_1(s) = \frac{\mathrm{d}\mathbf{x}(s)}{\mathrm{d}s} = \mathbf{e}_1 \cos\phi_0 \cos\frac{s}{R} + \mathbf{e}_2 \sin\phi_0 \cos\frac{s}{R} - \mathbf{e}_3 \sin\frac{s}{R} = \frac{\boldsymbol{\gamma}_\theta}{R}.$$

This vector clearly lies in the tangent plane spanned by $\boldsymbol{\gamma}_\theta$ and $\boldsymbol{\gamma}_\phi$. However,

$$\frac{\mathrm{d}\mathbf{E}_1}{\mathrm{d}s} = -\mathbf{e}_1\frac{1}{R}\cos\phi \sin\frac{s}{R} - \mathbf{e}_2\frac{1}{R}\sin\phi \sin\frac{s}{R} - \mathbf{e}_3\frac{1}{R}\cos\frac{s}{R} = -\frac{1}{R}\mathbf{N}.$$

An intrinsic observer would not observe anything in the direction $\mathbf{N}$. An intrinsic observer would see that

$$\nabla_s \mathbf{E}_1 = 0 \text{ and}$$

then conclude that the path in question is a geodesic – the closest thing to a straight line in his or her world. On the other hand, for an extrinsic observer

$$\frac{\mathrm{d}\mathbf{E}_1}{\mathrm{d}s} = k_1\mathbf{E}_2 = -\frac{1}{R}\mathbf{N}.$$

Thus, the extrinsic observer would conclude that our curve in question was curved with constant curvature $k_1 = 1/R$. That observer would also determine that $\mathbf{E}_2$ has the direction of a vector pointing from a given point on the curve toward the center of the sphere.

With this motivation, let us see how we can attack the more general problem on an n-dimensional hypersurface. As before, a point on the hypersurface is determined by n parameters, so we have

$$\mathbf{x}(u^1, u^2, ---, u^n) = \mathbf{e}_j x^j(u^1, u^2, ---, u^n).$$

The tangent plane is spanned by the n Dirac coordinate matrices

$$\boldsymbol{\gamma}_k = \frac{\partial \mathbf{x}}{\partial u^k} = \mathbf{e}_j \frac{\partial x^j}{\partial u^k}.$$

To get a normal vector $\mathbf{N}$, we can demand that

$$\frac{\mathbf{N}\boldsymbol{\gamma}_{n\cdots 21}}{|\boldsymbol{\gamma}_{n\cdots 21}|} = \mathbf{e}_{n+1}\mathbf{e}_n \cdots \mathbf{e}_1, \text{ or restated,}$$

$$\mathbf{N} = \mathbf{e}_{n+1}\mathbf{e}_n \cdots \mathbf{e}_1 \frac{\boldsymbol{\gamma}_{12\cdots n}}{|\boldsymbol{\gamma}_{12\cdots n}|}. \tag{7.101}$$

Fig. 7.7 Möbius strip

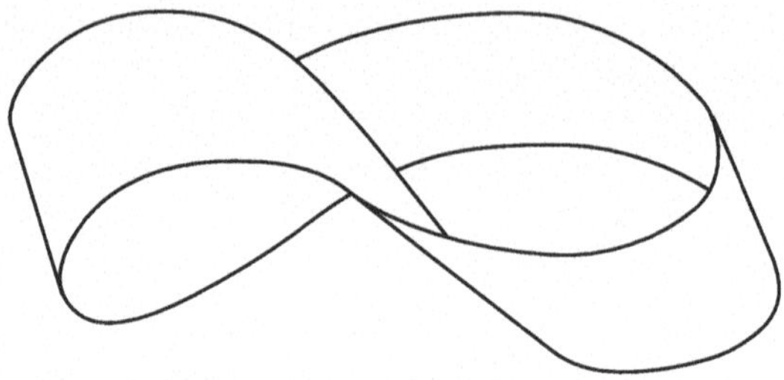

Given a coordinate system, **N** is well defined by (7.101). However, a change in the coordinate system may reverse the direction of **N**. For example, if we relabel the parameter u^1 by u^2 and the parameter u^2 by u^1, the direction of **N** will be reversed. There is no intrinsic way of selecting what is "up" and what is "down."

Because of this sign ambiguity, we have to be a little careful in the interpretation of some entities. Regardless of which direction we choose for **N**, we have one reasonable expectation. If we take some round trip journey, we expect **N** to be pointing in the same direction when we return to the point of origin as it was when we began our journey. However, there exist nonorientable surfaces for which this is not true. The most familiar example is the Möbius strip. (See Fig. 7.7.) If we restrict ourselves to local as opposed to global results, we can ignore the possibility of nonorientable surfaces. That is what I will do in this book.

Now let us return to our curve that lies in the n-dimensional hypersurface. Namely

$$\mathbf{x}(s) = \mathbf{e}_j x^j(u^1(s), u^2(s), ---, u^n(s)). \tag{7.102}$$

If s is an arc length parameter, then

$$\mathbf{E}_1(s) = \frac{\mathrm{d}\mathbf{x}}{\mathrm{d}s} = \mathbf{e}_j \frac{\partial x^j}{\partial u^k}\frac{\mathrm{d}u^k}{\mathrm{d}s} = \boldsymbol{\gamma}_k \frac{\mathrm{d}u^k}{\mathrm{d}s}. \tag{7.103}$$

The unit tangent vector $\mathbf{E}_1(s)$ lies in the plane spanned by the $\boldsymbol{\gamma}_k$'s that is the plane tangent to the hypersurface at the given point.

However, when we compute

$$\frac{\mathrm{d}\mathbf{E}_1(s)}{\mathrm{d}s} = k_1 \mathbf{E}_2(s),$$

we cannot expect $\mathbf{E}_2(s)$ to lie in the same plane. This is because, when we compute $\partial\boldsymbol{\gamma}_k/\partial u^j$, we have two components. First, we have the component that lies in the tangent plane "visible" to the intrinsic observer. That is

$$\nabla_j \boldsymbol{\gamma}_k = \Gamma^m_{jk}\boldsymbol{\gamma}_m.$$

Second, we now have an additional component in the normal direction **N**. Thus,

$$\frac{\partial}{\partial u^j}\boldsymbol{\gamma}_k = \Gamma^m_{jk}\boldsymbol{\gamma}_m + h_{jk}\mathbf{N}. \tag{7.104}$$

When I first introduced the intrinsic derivative $\nabla_j \boldsymbol{\gamma}_k$, I noted that $\Gamma^m_{jk} = \Gamma^m_{kj}$. It is also true that

$$h_{jk} = h_{kj} \tag{7.105}$$

for much the same reason. That is

$$\frac{\partial}{\partial u^j}\boldsymbol{\gamma}_k = \frac{\partial^2}{\partial u^j\,\partial u^k}\mathbf{x}(u^1, u^2, ---, u^n) = \frac{\partial^2}{\partial u^k\,\partial u^j}\mathbf{x}(u^1, u^2, ---, u^n) = \frac{\partial}{\partial u^k}\boldsymbol{\gamma}_j.$$

The h_{jk}'s are known as members of the *second fundamental form*. (The g_{jk}'s are known as members of the *first fundamental form*.)

Like the members of the metric tensor, the members of the second fundamental form transform under a coordinate transformation as members of a tensor. (See Prob. 176.) It is not difficult to show that

$$\frac{\partial \mathbf{N}}{\partial u^j} = -h_{jk}\boldsymbol{\gamma}^k. \tag{7.106}$$

To verify (7.106), we first note that since $\langle \mathbf{N}, \mathbf{N}\rangle = 1$,

$$\frac{\partial}{\partial u^j}\langle \mathbf{N}, \mathbf{N}\rangle = 2\left\langle \mathbf{N}, \frac{\partial \mathbf{N}}{\partial u^j}\right\rangle = 0.$$

This implies that $\partial \mathbf{N}/\partial u^j$ lies in the tangent plane and is thus a linear combination of the $\boldsymbol{\gamma}_k$'s or $\boldsymbol{\gamma}^k$'s. Remember that the $\boldsymbol{\gamma}_k$'s and the $\boldsymbol{\gamma}^k$'s span the same plane since

$$\boldsymbol{\gamma}_k = g_{kj}\boldsymbol{\gamma}^j \text{ and } \boldsymbol{\gamma}^k = g^{kj}\boldsymbol{\gamma}_j.$$

With this thought in mind, we can write

$$\frac{\partial \mathbf{N}}{\partial u^j} = \alpha_{jk}\boldsymbol{\gamma}^k, \text{ where} \tag{7.107}$$

the α_{jk}'s have not yet been determined. Now since $\langle \mathbf{N}, \boldsymbol{\gamma}_k\rangle = 0$, we have

$$\begin{aligned} 0 = \frac{\partial}{\partial u^j}\langle \mathbf{N}, \boldsymbol{\gamma}_k\rangle &= \left\langle \frac{\partial \mathbf{N}}{\partial u^j}, \boldsymbol{\gamma}_k\right\rangle + \left\langle \mathbf{N}, \frac{\partial \boldsymbol{\gamma}_k}{\partial u^j}\right\rangle \\ &= \alpha_{jm}\langle \boldsymbol{\gamma}^m, \boldsymbol{\gamma}_k\rangle + \left\langle \mathbf{N}, \Gamma^m_{jk}\boldsymbol{\gamma}_m\right\rangle + h_{jk}\langle \mathbf{N}, \mathbf{N}\rangle \\ &= \alpha_{jm}\delta^m_k + h_{jk}, \text{ and thus} \end{aligned}$$

$$\alpha_{jk} = -h_{jk}.$$

Furthermore, from (7.107)

$$\frac{\partial \mathbf{N}}{\partial u^j} = -h_{jk}\boldsymbol{\gamma}^k. \tag{7.108}$$

Now we can reexamine the nature of $\partial \mathbf{E}_1/\partial s$. We note that

$$\mathbf{E}_1(s) = \boldsymbol{\gamma}_j \frac{\mathrm{d}u^j}{\mathrm{d}s}, \text{ so}$$

$$\begin{aligned}\frac{\mathrm{d}\mathbf{E}_1}{\mathrm{d}s} &= \frac{\mathrm{d}\boldsymbol{\gamma}_j}{\mathrm{d}s}\frac{\mathrm{d}u^j}{\mathrm{d}s} + \boldsymbol{\gamma}_j \frac{\mathrm{d}^2u^j}{\mathrm{d}s^2}\\ &= \frac{\partial \boldsymbol{\gamma}_j}{\partial u^k}\frac{\mathrm{d}u^k}{\mathrm{d}s}\frac{\mathrm{d}u^j}{\mathrm{d}s} + \boldsymbol{\gamma}_j \frac{\mathrm{d}^2u^j}{\mathrm{d}s^2}.\end{aligned}$$

From this last equation and (7.104), it follows that

$$\frac{\mathrm{d}\mathbf{E}_1}{\mathrm{d}s} = \boldsymbol{\gamma}_m\left(\Gamma^m_{kj}\frac{\mathrm{d}u^k}{\mathrm{d}s}\frac{\mathrm{d}u^j}{\mathrm{d}s} + \frac{d^2u^m}{ds^2}\right) + h_{kj}\frac{\mathrm{d}u^k}{\mathrm{d}s}\frac{\mathrm{d}u^j}{\mathrm{d}s}\mathbf{N}, \text{ and thus} \tag{7.109}$$

$$\frac{\mathrm{d}\mathbf{E}_1}{\mathrm{d}s} = \mathbf{k}_G + \mathbf{k}_N, \text{ where} \tag{7.110}$$

$$\mathbf{k}_G = \boldsymbol{\gamma}_m\left(\Gamma^m_{kj}\frac{\mathrm{d}u^k}{\mathrm{d}s}\frac{\mathrm{d}u^j}{\mathrm{d}s} + \frac{d^2u^m}{ds^2}\right), \text{ and} \tag{7.111}$$

$$\mathbf{k}_N = h_{kj}\frac{\mathrm{d}u^k}{\mathrm{d}s}\frac{\mathrm{d}u^j}{\mathrm{d}s}\mathbf{N} = k_N\mathbf{N}. \tag{7.112}$$

The vector $\mathbf{k}_N$ is the *normal curvature vector*. It does not depend on which direction we choose to be "up." From (7.108), we see that switching the direction of **N** results in a change in the signs of the h_{jk}'s, and thus there is no change in the direction of $\mathbf{k}_N$.

It should be mentioned in passing that the scalar k_N that appears in (7.112) is known as the *normal curvature* and its sign does depend on the direction chosen for **N**.

The vector $\mathbf{k}_G$ is known as the *geodesic curvature vector* or *tangential curvature vector*. In the context of a 2-dimensional surface, we encountered the magnitude of this vector in the Gauss-Bonnet formula.

I will now focus on the normal curvature vector. Essentially, all information that is available to the extrinsic observer but denied to the intrinsic observer is stored in the members of the second fundamental form. From (7.108), we see that once a coordinate system is chosen for the hypersurface and the direction of **N** is chosen, the h_{jk}'s are determined. The vector $\mathbf{E}_1(s)$ is a unit vector and from (7.103), its direction determines the $\mathrm{d}u^k/\mathrm{d}s$'s. From (7.112), we see that this information is sufficient to determine the normal curvature vector. Thus, we see (or should see) that the normal curvature vector $\mathbf{k}_N$ for a curve $\mathbf{x}(s)$ is entirely determined by the direction of its tangent vector $\mathbf{E}_1$ and not by any other detail of the curve.

7.5.2 The Weingarten Map or Shape Operator

Since $\mathbf{k}_N$ depends only on the direction of $\mathbf{E}_1$, it is useful to investigate the possibility that some tangent directions are more interesting than others. In fact that is the case. To pursue this investigation, it is useful to introduce the *Weingarten map* or *shape operator*. We note that $\mathrm{d}\mathbf{N}/\mathrm{d}s$ gives us the change in the direction of $\mathbf{N}$ when we move in the direction of $\mathbf{E}_1$. From (7.108), we see that when we move in any direction in the tangent plane, there is a corresponding change in the direction of $\mathbf{N}$, which is also in the tangent plane. The Weingarten map or shape operator $S(\mathbf{v})$ is simply the explicit expression of this correspondence. Namely

$$S(\mathbf{E}_1) = -\frac{\mathrm{d}\mathbf{N}}{\mathrm{d}s}. \tag{7.113}$$

Looking at the components, we have

$$S(\mathbf{E}_1) = S\left(\frac{\mathrm{d}u^k}{\mathrm{d}s}\boldsymbol{\gamma}_k\right) = -\frac{\mathrm{d}\mathbf{N}}{\mathrm{d}s} = -\frac{\mathrm{d}u^k}{\mathrm{d}s}\frac{\mathrm{d}\mathbf{N}}{\mathrm{d}u^k} = \frac{\mathrm{d}u^k}{\mathrm{d}s}h_{kj}\boldsymbol{\gamma}^j. \tag{7.114}$$

Thus, if we require the shape operator to be a linear operator, then

$$S\left(\frac{\mathrm{d}u^k}{\mathrm{d}s}\boldsymbol{\gamma}_k\right) = \frac{\mathrm{d}u^k}{\mathrm{d}s}S(\boldsymbol{\gamma}_k) = \frac{\mathrm{d}u^k}{\mathrm{d}s}h_{kj}\boldsymbol{\gamma}^j, \text{ and}$$

$$S(\boldsymbol{\gamma}_k) = h_{kj}\boldsymbol{\gamma}^j. \tag{7.115}$$

The h_{jk}'s may be described as the *tensor components of the shape operator*. From (7.115), it is easy to see how to generalize (7.113) for vectors that are not necessarily of unit length. If $\mathbf{v}$ is any vector in the tangent plane, then

$$S(\mathbf{v}) = S(v^j\boldsymbol{\gamma}_j) = v^k h_{kj}\boldsymbol{\gamma}^j = v^k h_k{}^m \boldsymbol{\gamma}_m. \tag{7.116}$$

From (7.110)) and (7.112),

$$\frac{\mathrm{d}\mathbf{E}_1}{\mathrm{d}s} = k_N\mathbf{N} + \mathbf{k}_G.$$

Since $\langle\mathbf{N},\mathbf{k}_G\rangle = 0$,

$$k_N = \left\langle\mathbf{N},\frac{\mathrm{d}\mathbf{E}_1}{\mathrm{d}s}\right\rangle = \frac{\mathrm{d}}{\mathrm{d}s}\langle\mathbf{N},\mathbf{E}_1\rangle - \left\langle\frac{\mathrm{d}\mathbf{N}}{\mathrm{d}s},\mathbf{E}_1\right\rangle = -\left\langle\frac{\mathrm{d}\mathbf{N}}{\mathrm{d}s},\mathbf{E}_1\right\rangle.$$

From (7.113) and this last equation,

$$k_N = \langle S(\mathbf{E}_1),\mathbf{E}_1\rangle. \tag{7.117}$$

7.5.3 Principal Directions and Curvatures

Equation (7.117) suggests to some that $\mathbf{E}_1$ would have an "interesting" direction if

$$S(\mathbf{E}_1) = \lambda \mathbf{E}_1, \text{ where} \tag{7.118}$$

λ is some real scalar. In that case, $\lambda = k_N$. In such a situation, the direction of $\mathbf{E}_1$ is said to be a *principal direction* and k_N is said to be a *principal curvature*. In general if A is a linear operator and

$$A(\mathbf{v}) = \lambda \mathbf{v},$$

$\mathbf{v}$ is said to be an *eigenvector* and λ is said to be an *eigenvalue*.

Why are principal directions and principal curvatures particularly interesting? Most of the "niceness" of principal directions follows from the fact that the shape operator is *symmetric*. What do we mean by symmetric?

Definition 169. A linear operator A is said to be symmetric if $\langle \mathbf{w}, A(\mathbf{v})\rangle = \langle A(\mathbf{w}), \mathbf{v}\rangle$ for all vectors $\mathbf{v}$ and $\mathbf{w}$.

Given a coordinate system, a linear operator can be represented by a tensor. In particular,

$$A(\mathbf{v}) = A(v^j \boldsymbol{\gamma}_j) = v^j A_j{}^k \boldsymbol{\gamma}_k = v^j A_{jm} \boldsymbol{\gamma}^m = v_i A^i{}_m \boldsymbol{\gamma}^m = v_i A^{ij} \boldsymbol{\gamma}_j. \tag{7.119}$$

Corresponding to any linear operator A is the *transpose operator* A^T defined by the equation:

$$\left\langle A^T(\mathbf{w}), \mathbf{v}\right\rangle = \langle \mathbf{w}, A(\mathbf{v})\rangle \tag{7.120}$$

for arbitrary $\mathbf{w}$ and $\mathbf{v}$.

In general, it can be said that

$$A^T(\mathbf{v}) = A^T(v^j \boldsymbol{\gamma}_j) = v^j A^k{}_j \boldsymbol{\gamma}_k = v^j A_{kj} \boldsymbol{\gamma}^k = v_i A_k{}^i \boldsymbol{\gamma}^k = v_i A^{ji} \boldsymbol{\gamma}_j. \tag{7.121}$$

(See Prob. 177.)

From Definition 169 and (7.120), we see that

$$A \text{ is symmetric } \Leftrightarrow A = A^T. \tag{7.122}$$

Reviewing (7.119) and (7.121), we see that

$$A = A^T \Leftrightarrow A_j{}^k = A^k{}_j. \tag{7.123}$$

Thus for a symmetric operator, it makes sense to designate either $A_j{}^k$ or $A^k{}_j$ by A^k_j.

It should also be noted that from (7.119) and (7.121),

$$A = A^T \Leftrightarrow A_{jk} = A_{kj} \text{ or } A^{jk} = A^{kj}.$$

The following two theorems indicate two important properties of real symmetric linear operators.

Theorem 170. *For a positive definite metric, the eigenvalues of a real symmetric linear operator A are real.*

Proof. Suppose

$$A(\mathbf{v}) = \lambda \mathbf{v}. \tag{7.124}$$

Since A is real,

$$A(\mathbf{v}^*) = \lambda^* \mathbf{v}^*, \text{ where} \tag{7.125}$$

$\mathbf{v}^*$ is the complex conjugate of $\mathbf{v}$ and λ^* is the complex conjugate of λ. From (7.124) and (7.125),

$$\langle \mathbf{v}^*, A(\mathbf{v})\rangle = \lambda \langle \mathbf{v}^*, \mathbf{v}\rangle, \text{ and} \tag{7.126}$$

$$\langle A(\mathbf{v}^*), \mathbf{v})\rangle = \lambda^* \langle \mathbf{v}^*, \mathbf{v}\rangle. \tag{7.127}$$

Since A is symmetric,

$$\langle \mathbf{v}^*, A(\mathbf{v})\rangle = \langle A(\mathbf{v}^*), \mathbf{v})\rangle . \text{ So}$$

$$\lambda \langle \mathbf{v}^*, \mathbf{v}\rangle = \lambda^* \langle \mathbf{v}^*, \mathbf{v}\rangle. \tag{7.128}$$

If $\mathbf{v} = \mathbf{p} + i\mathbf{q}$ where $\mathbf{p}$ and $\mathbf{q}$ are real, then

$$\begin{aligned} \langle \mathbf{v}^*, \mathbf{v}\rangle &= (p^j - iq^j)g_{jk}(p^k + iq^k) = p^j g_{jk} p^k + q^j g_{jk} q^k \\ &= \langle \mathbf{p}, \mathbf{p}\rangle + \langle \mathbf{q}, \mathbf{q}\rangle > 0. \end{aligned}$$

Using this result along with (7.128), we have

$$\lambda = \lambda^*.$$

Thus, λ is real □

Theorem 171. *Suppose A is a real, linear, and symmetric operator. Also, suppose $A(\mathbf{v}) = \lambda \mathbf{v}$, $A(\mathbf{u}) = \mu \mathbf{u}$, and $\lambda \neq \mu$. Then $\langle \mathbf{u}, \mathbf{v}\rangle = 0$.*

Proof.

$$\mu \langle \mathbf{u}, \mathbf{v}\rangle = \langle A(\mathbf{u}), \mathbf{v}\rangle = \langle \mathbf{u}, A(\mathbf{v})\rangle = \lambda \langle \mathbf{u}, \mathbf{v}\rangle.$$

Thus,

$$(\mu - \lambda) \langle \mathbf{u}, \mathbf{v}\rangle = 0.$$

And therefore,

$$\langle \mathbf{u}, \mathbf{v}\rangle = 0.$$

□

Now let us return to the problem of determining the eigenvalues and eigenvectors of the shape operator. From (7.105), we know that the shape operator is symmetric. Furthermore, from (7.116),

$$S(\mathbf{v}) = S(v^j \boldsymbol{\gamma}_j) = v^k h_k^m \boldsymbol{\gamma}_m.$$

In addition,

$$\lambda \mathbf{v} = \lambda v^j \boldsymbol{\gamma}_j.$$

Thus,

$$S(\mathbf{v}) = \lambda \mathbf{v} \Leftrightarrow h_k^j v^k = \lambda v^j. \tag{7.129}$$

This can be written in matrix form:

$$\begin{bmatrix} h_1^1 & h_2^1 & \cdot & \cdot & \cdot & h_n^1 \\ h_1^2 & h_2^2 & \cdot & \cdot & \cdot & h_n^2 \\ \cdot & \cdot & \cdot & \cdot & \cdot & \cdot \\ \cdot & \cdot & \cdot & \cdot & \cdot & \cdot \\ \cdot & \cdot & \cdot & \cdot & \cdot & \cdot \\ h_1^n & h_2^n & \cdot & \cdot & \cdot & h_n^n \end{bmatrix} \begin{bmatrix} v^1 \\ v^2 \\ \cdot \\ \cdot \\ \cdot \\ v^n \end{bmatrix} = \lambda \begin{bmatrix} v^1 \\ v^2 \\ \cdot \\ \cdot \\ \cdot \\ v^n \end{bmatrix}. \tag{7.130}$$

That is

$$\begin{bmatrix} h_1^1 - \lambda & h_2^1 & \cdot & \cdot & \cdot & h_n^1 \\ h_1^2 & h_2^2 - \lambda & \cdot & \cdot & \cdot & h_n^2 \\ \cdot & \cdot & \cdot & \cdot & \cdot & \cdot \\ \cdot & \cdot & \cdot & \cdot & \cdot & \cdot \\ \cdot & \cdot & \cdot & \cdot & \cdot & \cdot \\ h_1^n & h_2^n & \cdot & \cdot & \cdot & h_n^n - \lambda \end{bmatrix} \begin{bmatrix} v^1 \\ v^2 \\ \cdot \\ \cdot \\ \cdot \\ v^n \end{bmatrix} = \begin{bmatrix} 0 \\ 0 \\ \cdot \\ \cdot \\ \cdot \\ 0 \end{bmatrix}.$$

Thus to determine the principal curvatures, we need to solve the characteristic equation:

$$\det \begin{bmatrix} h_1^1 - \lambda & h_2^1 & \cdot & \cdot & \cdot & h_n^1 \\ h_1^2 & h_2^2 - \lambda & \cdot & \cdot & \cdot & h_n^2 \\ \cdot & \cdot & \cdot & \cdot & \cdot & \cdot \\ \cdot & \cdot & \cdot & \cdot & \cdot & \cdot \\ \cdot & \cdot & \cdot & \cdot & \cdot & \cdot \\ h_1^n & h_2^n & \cdot & \cdot & \cdot & h_n^n - \lambda \end{bmatrix} = 0. \tag{7.131}$$

Example 172. The saddle surface.

One example worth discussing is the saddle surface $z = xy$, or

$$\mathbf{x}(u^1, u^2) = \mathbf{e}_1 u^1 + \mathbf{e}_2 u^2 + \mathbf{e}_3 u^1 u^2. \tag{7.132}$$

For this surface,

$$\boldsymbol{\gamma}_1 = \frac{\partial \mathbf{x}}{\partial u^1} = \mathbf{e}_1 + \mathbf{e}_3 u^2, \tag{7.133}$$

$$\boldsymbol{\gamma}_2 = \frac{\partial \mathbf{x}}{\partial u^2} = \mathbf{e}_2 + \mathbf{e}_3 u^1, \text{ and} \tag{7.134}$$

$$\boldsymbol{\gamma}_{12} = -\mathbf{e}_{23} u^2 - \mathbf{e}_{31} u^1 + \mathbf{e}_{12}. \tag{7.135}$$

If

$$\mathbf{N}\frac{\boldsymbol{\gamma}_{21}}{|\boldsymbol{\gamma}_{21}|} = \mathbf{e}_{321}, \text{ then}$$

$$\mathbf{N} = \mathbf{e}_{321}\frac{\boldsymbol{\gamma}_{12}}{|\boldsymbol{\gamma}_{12}|} = \frac{-\mathbf{e}_1 u^2 - \mathbf{e}_2 u^1 + \mathbf{e}_3}{[1 + (u^1)^2 + (u^2)^2]^{1/2}}.$$

To get the covariant components of the shape operator, we note that

$$\begin{aligned}\frac{\partial \mathbf{N}}{\partial u^1} &= \frac{-\mathbf{e}_2}{[1 + (u^1)^2 + (u^2)^2]^{1/2}} + \left(-\mathbf{e}_1 u^2 - \mathbf{e}_2 u^1 + \mathbf{e}_3\right)\frac{\partial}{\partial u^1}\frac{1}{[1 + (u^1)^2 + (u^2)^2]^{1/2}} \\ &= -h_{11}\boldsymbol{\gamma}^1 - h_{12}\boldsymbol{\gamma}^2.\end{aligned}$$

Thus,

$$h_{11} = -\left\langle \boldsymbol{\gamma}_1, \frac{\partial \mathbf{N}}{\partial u^1}\right\rangle = 0, \text{ and} \tag{7.136}$$

$$h_{12} = -\left\langle \boldsymbol{\gamma}_2, \frac{\partial \mathbf{N}}{\partial u^1}\right\rangle = \frac{1}{[1 + (u^1)^2 + (u^2)^2]^{1/2}}. \tag{7.137}$$

Similarly,

$$h_{21} = -\left\langle \boldsymbol{\gamma}_1, \frac{\partial \mathbf{N}}{\partial u^2}\right\rangle = \frac{1}{[1 + (u^1)^2 + (u^2)^2]^{1/2}}, \text{ and}$$

$$h_{22} = -\left\langle \boldsymbol{\gamma}_2, \frac{\partial \mathbf{N}}{\partial u^2}\right\rangle = 0.$$

To get the mixed tensor components of the shape operator, we need to compute the contravariant components of the metric tensor.

$$\begin{bmatrix} g_{11} & g_{12} \\ g_{21} & g_{22} \end{bmatrix} = \begin{bmatrix} \langle \boldsymbol{\gamma}_1, \boldsymbol{\gamma}_1\rangle & \langle \boldsymbol{\gamma}_1, \boldsymbol{\gamma}_2\rangle \\ \langle \boldsymbol{\gamma}_2, \boldsymbol{\gamma}_1\rangle & \langle \boldsymbol{\gamma}_2, \boldsymbol{\gamma}_2\rangle \end{bmatrix} = \begin{bmatrix} 1 + (u^2)^2 & u^1 u^2 \\ u^1 u^2 & 1 + (u^1)^2 \end{bmatrix}.$$

Therefore,

$$\begin{bmatrix} g^{11} & g^{12} \\ g^{21} & g^{22} \end{bmatrix} = \frac{1}{[1+(u^1)^2+(u^2)^2]} \begin{bmatrix} 1+(u^1)^2 & -u^1u^2 \\ -u^1u^2 & 1+(u^2)^2 \end{bmatrix}.$$

Thus,

$$h_1^1 = h_{11}g^{11} + h_{12}g^{21} = \frac{-u^1u^2}{[1+(u^1)^2+(u^2)^2]^{3/2}},$$

$$h_2^1 = h_{21}g^{11} + h_{22}g^{21} = \frac{1+(u^1)^2}{[1+(u^1)^2+(u^2)^2]^{3/2}},$$

$$h_1^2 = h_{11}g^{12} + h_{12}g^{22} = \frac{1+(u^2)^2}{[1+(u^1)^2+(u^2)^2]^{3/2}}, \text{ and}$$

$$h_2^2 = h_{21}g^{12} + h_{22}g^{22} = \frac{-u^1u^2}{[1+(u^1)^2+(u^2)^2]^{3/2}}.$$

To get the principal curvatures, we have to solve the characteristic equation:

$$\det\begin{bmatrix} h_1^1-\lambda & h_2^1 \\ h_1^2 & h_2^2-\lambda \end{bmatrix} = \left(\frac{-u^1u^2}{[1+(u^1)^2+(u^2)^2]^{3/2}} - \lambda\right)^2 - \frac{\left(1+(u^1)^2\right)\left(1+(u^2)^2\right)}{[1+(u^1)^2+(u^2)^2]^3} = 0.$$

It then follows that

$$\lambda = \frac{-u^1u^2 \pm \sqrt{(1+(u^1)^2)(1+(u^2)^2)}}{[1+(u^1)^2+(u^2)^2]^{3/2}}. \tag{7.138}$$

You should note that one of the principal curvatures is positive and the other is negative.

Another example worthy of discussion is an n-dimensional surface of revolution.

Example 173. The n-dimensional surface of revolution.

An n-dimensional surface of revolution can be defined by the equation

$$\mathbf{x}(u^1,u^2,---,u^n) = r(u^n)\mathbf{w}(u^1,u^2,---,u^{n-1}) + \mathbf{e}_{n+1}u^n, \text{ where}$$

$$\mathbf{w}(u^1,u^2,---,u^{n-1}) = \mathbf{e}_1 \cos u^1 \cos u^2 \cos u^3 \cdots \cos u^{n-1} +$$

$$\mathbf{e}_2 \sin u^1 \cos u^2 \cos u^3 \cdots \cos u^{n-1} +$$

$$\mathbf{e}_3 \sin u^2 \cos u^3 \cdots \cos u^{n-1} + - - -+$$
$$\mathbf{e}_{n-1} \sin u^{n-2} \cos u^{n-1} +$$
$$\mathbf{e}_n \sin u^{n-1}. \tag{7.139}$$

From these equations, we have

$$\boldsymbol{\gamma}_k = \frac{\partial \mathbf{x}}{\partial u^k} = r \frac{\partial \mathbf{w}}{\partial u^k}, \text{ for } k = 1, 2, - - -, n-1,$$
$$\boldsymbol{\gamma}_n = \frac{\partial \mathbf{x}}{\partial u^n} = \dot{r}\mathbf{w} + \mathbf{e}_{n+1}, \text{ and}$$
$$\mathbf{N} = \frac{\mathbf{w} - \dot{r}\mathbf{e}_{n+1}}{[1 + (\dot{r})^2]^{1/2}}, \text{ where}$$
$$\dot{r} = \frac{\mathrm{d}r}{\mathrm{d}u^n}.$$

(You should convince yourself that $\langle \mathbf{w}, \mathbf{w} \rangle = 1$ and $\langle \boldsymbol{\gamma}_j, \mathbf{N} \rangle = 0$.) We note that for $k = 1, 2, - - -, n-1$:

$$\frac{\partial \mathbf{N}}{\partial u^k} = \frac{1}{[1 + (\dot{r})^2]^{1/2}} \frac{\partial \mathbf{w}}{\partial u^k} = \frac{1}{r\,[1 + (\dot{r})^2]^{1/2}} \boldsymbol{\gamma}_k = -h_k^j \boldsymbol{\gamma}_j. \text{ Also}$$
$$\frac{\partial \mathbf{N}}{\partial u^n} = -\frac{\ddot{r}\,[\dot{r}\mathbf{w} + \mathbf{e}_{n+1}]}{[1 + (\dot{r})^2]^{3/2}} = \frac{-\ddot{r}}{[1 + (\dot{r})^2]^{3/2}} \boldsymbol{\gamma}_n = -h_n^j \boldsymbol{\gamma}_j.$$

From these past two equations, it is clear that the matrix representing the shape operator is diagonal so we can obtain the principal curvatures by reading off the diagonal elements. In particular,

$$\lambda_k = h_k^k = \frac{-1}{r(u^n)\,[1 + (\dot{r}(u^n))^2]^{1/2}} \quad \text{for } k = 1, 2, - - -, n-1 \text{ and} \tag{7.140}$$
$$\lambda_n = h_n^n = \frac{\ddot{r}}{[1 + (\dot{r})^2]^{3/2}}. \tag{7.141}$$

Definition 174.

A curve $\mathbf{x}(t)$ is said to be a *line of curvature* if

$$\frac{\mathrm{d}\mathbf{x}(t)}{\mathrm{d}t} = \mathbf{v}(t), \text{ where}$$

$\mathbf{v}(t)$ has a principal direction.

Fig. 7.8 Catenoid

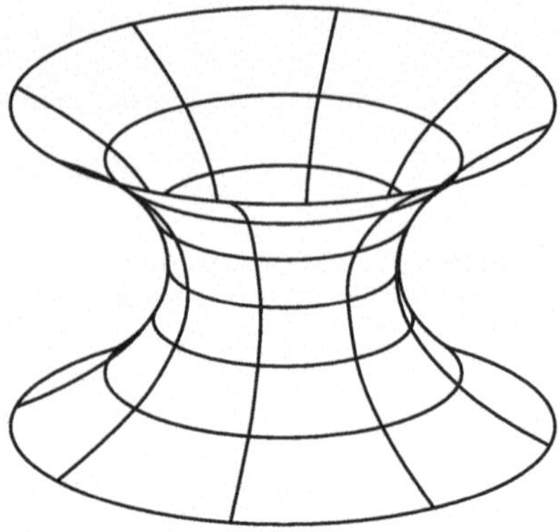

($\mathbf{v}(t)$ does not have to be normalized unless you insist that t be an arc length parameter.) Since

$$S(\boldsymbol{\gamma}_k) = -\frac{\partial \mathbf{N}}{\partial u^k} = h^j_k \boldsymbol{\gamma}_j, \text{ we see}$$

that $\boldsymbol{\gamma}_k$ has a principal direction if the matrix $\left[h^j_k\right]$ is diagonal. That is the case for this example. Furthermore,

$$\frac{\partial \mathbf{x}}{\partial u^k} = \boldsymbol{\gamma}_k.$$

Therefore for a surface of revolution with the parameterization used above, the lines of curvature coincide with the coordinate curves. For a 2-dimensional hypersurface of revolution, (7.139) becomes

$$\mathbf{x}(u^1, u^2) = r(u^2)(\mathbf{e}_1 \cos u^1 + \mathbf{e}_2 \sin u^1) + \mathbf{e}_3 u^2.$$

In this context, the circle coordinate curves for u^1 are said to be *parallels* and the coordinate curves for u^2 are said to be *meridians*. (See Fig. 7.8.)

In closing this section, I will prove another one of Euler's many theorems.

Theorem 175. *Euler's Theorem for Normal Curvature. (Version I)*

Suppose $\{\mathbf{v}(1), \mathbf{v}(2), - - -, \mathbf{v}(n)\}$ *is an orthonormal basis at a given point on a hypersurface such that each member is a unit vector having a principal direction. Suppose* $\mathbf{v}$ *is an arbitrary unit vector. Then the normal curvature* k_N *associated with* $\mathbf{v}$ *is*

$$k_N = \sum_{k=1}^{n} \lambda(k) \cos^2 \theta(k), \text{ where} \tag{7.142}$$

$$\cos \theta(k) = \langle \mathbf{v}, \mathbf{v}(k) \rangle. \tag{7.143}$$

Proof. In Sect. 10.1.2, I will show that the set of eigenvectors for the shape operator spans the n-dimensional plane tangent to the hypersurface at the given point.

We have already shown that a principal direction corresponding to one principal curvature is orthogonal to a principal direction corresponding to a different principal curvature. If the dimension of principal directions corresponding to a given principal curvature is more than one, then one can use the Gram–Schmidt process to find an orthonormal basis that spans such a space. Thus, the orthonormal basis described above can always be constructed.

Now suppose

$$\mathbf{v} = \sum_{k=1}^{n} \alpha_k \mathbf{v}(k). \tag{7.144}$$

(You should demonstrate that $\alpha_k = \langle \mathbf{v}, \mathbf{v}(k) \rangle$.) We then have

$$S(\mathbf{v}) = \sum_{k=1}^{n} \alpha_k S(\mathbf{v}(k)) = \sum_{k=1}^{n} \langle \mathbf{v}, \mathbf{v}(k) \rangle \, S(\mathbf{v}(k)) = \sum_{k=1}^{n} \langle \mathbf{v}, \mathbf{v}(k) \rangle \, \lambda(k) \mathbf{v}(k).$$

From (7.117),

$$k_N = \langle \mathbf{v}, S(\mathbf{v}) \rangle = \sum_{k=1}^{n} \lambda_k \, \langle \mathbf{v}, \mathbf{v}(k) \rangle^2 = \sum_{k=1}^{n} \lambda(k) \cos^2 \theta(k). \tag{7.145}$$

□

It should be remarked if there are n distinct principal curvatures, then the orthonormal basis used in Euler's Theorem is unique and then the terms on the right-hand side of (7.145) are also unique. However if the dimension for the space of principal directions belonging to any of the principal curvatures is more than one, then at least some of the $\mathbf{v}(k)$'s are not unique and therefore some of the terms on the right-hand side of (7.145) are not unique either.

In Sect. 10.1, I will elaborate on the ideas introduced in this section.

Problem 176. From (7.108),

$$\frac{\partial \mathbf{N}}{\partial u^j} = -h_{jm} \boldsymbol{\gamma}^m, \text{ and thus}$$

$$h_{jk} = -\left\langle \frac{\partial \mathbf{N}}{\partial u^j}, \boldsymbol{\gamma}_k \right\rangle.$$

Use this fact to show that the h_{jk}'s transform under a change of coordinates as members of a tensor.

Problem 177. Prove that either (7.121) is correct or I made a mistake.

Problem 178. Tangential Developables

In Prob. 168, I indicated that a tangential developable could be written in the form:

$$\mathbf{y}(u^1, u^2) = \mathbf{x}(u^2) + u^1 \mathbf{E}_1(u^2).$$

If we reparameterize the surface, we can represent the same surface in the form:

$$\mathbf{y}(u^1, u^2) = \mathbf{x}(u^2) + \left(u^1 - u^2\right) \mathbf{E}_1(u^2), \text{ where} \tag{7.146}$$

u^2 is the arc length parameter for the curve $\mathbf{x}(u^2)$.

(a) Consider a tangential developable that lies in E^3. Suppose the curvature of $\mathbf{x}(u^2)$ is $k_1(u^2)$ and the torsion is $k_2(u^2)$. Using the coordinate system of (7.146), compute h_1^1, h_1^2, h_2^1, and h_2^2 in terms of u^1, u^2, k_1, and k_2.
(b) Compute the principal curvatures for tangential developables in terms of u^1, u^2, k_1, and k_2. What happens when the torsion k_2 is zero?
(c) Show that the coordinate curves for a tangential developable as parameterized in (7.146) are also lines of curvature.

Problem 179. Helicoid

The *helicoid* was described in Sect. 8.3. Repeating myself, we note that if a line L moves along the axis at a constant speed while rotating about the axis at a constant rate, the resulting surface swept out by the line L is said to be a helicoid. (See Fig. 7.4a.) From this description, it should be clear that if we choose $\mathbf{e}_3$ for our axis, then one representation of the surface is

$$\mathbf{y}(t, \theta) = \mathbf{e}_3 a\theta + t\left(\mathbf{e}_1 \cos\theta + \mathbf{e}_2 \sin\theta\right). \tag{7.147}$$

For our current purpose, it is useful to change these coordinates. In particular, let

$$\theta = u, \text{ and}$$
$$t = a \sinh(v).$$

Equation (7.147) then becomes

$$\mathbf{x}(u, v) = a \sinh(v) \left[\mathbf{e}_1 \cos(u) + \mathbf{e}_2 \sin(u)\right] + \mathbf{e}_3 a u. \tag{7.148}$$

(a) Compute h_u^u, h_u^v, h_v^u, and h_v^v and then determine the principal curvatures. For the principal curvatures, I got

$$\lambda = \pm \frac{1}{a \cosh^2(v)}.$$

(b) Is this consistent with where you think the surface would be relatively flat?

Problem 180. Catenoid

A catenoid is a surface of revolution generated by rotating a catenary about an appropriate axis. (See Fig. 7.8.) In particular, one representation of a catenoid is: $x^2 + y^2 = a^2 \cosh^2(z/a)$. Another representation is:

$$\mathbf{x}(u, v) = -\mathbf{e}_1 a \sin(u) \cosh(v) + \mathbf{e}_2 a \cos(u) \cosh(v) + \mathbf{e}_3 a v. \tag{7.149}$$

Determine the principal curvatures in terms of this coordinate system. Compare your answers with those in Prob. 179. What distinguishes a catenoid from a helicoid?

Problem 181. Pseudosphere

Consider the pseudosphere defined by

$$\mathbf{x}(u^1, u^2) = \mathbf{e}_1 \frac{R\cos u^2}{\cosh u^1} + \mathbf{e}_2 \frac{R\sin u^2}{\cosh u^1} + \mathbf{e}_3 (u^1 - \tanh u^1).$$

Compute h_1^1, h_2^1, h_1^2, and h_2^2. Then show that the product of the principal curvatures at any given point is $-R^2$.

Problem 182. n-Dimensional Sphere

A special case of the n-dimensional surface of revolution of Example 173 is the n-dimensional sphere. For the sphere

$$r(u^n) = \left[R^2 - (u^n)^2\right]^{1/2}.$$

Use this equation and (7.140) and (7.141) to compute the principal curvatures for the n-dimensional sphere. Is your answer something that you could reasonably expect? What would happen if the sign of $\mathbf{N}$ was changed?

Problem 183. Reviewing (7.143) and (7.144), you will note that

$$\mathbf{v} = \sum_{k=1}^{n} \mathbf{v}(k) \cos\theta(k).$$

The cosines that appear in this formula are known as *direction cosines*.

(a) Show that

$$\sum_{k-1}^{n} \cos^2\theta(k) = 1.$$

(b) In view of (7.145), knowing the values of the principal curvatures tells you what about the minimum and maximum values of the normal curvature? Does this change if we reverse the direction of $\mathbf{N}$?

Problem 184. In the line designated by (7.129), it was stated that

$$S(\mathbf{v}) = \lambda\mathbf{v} \Leftrightarrow h_k^j v^k = \lambda v^j.$$

Show that

$$S(\mathbf{v}) = \lambda\mathbf{v} \Leftrightarrow v_j h_k^j = \lambda v_k.$$

Can (7.130) be modified so that one has a row eigenvector instead of a column eigenvector? How can you interpret the entries of the row eigenvector?

7.6 *Leonhard Euler 1707–1783

Leonhard Euler was probably the most prolific mathematician of all time. Time and again his name has cropped up again and again in this book. Euler was probably more pre-eminent in his time than Gauss was in his time. Since 1911, a publishing project has been underway to amass all his work. The 74th volume was published in 2004 and the project is not yet complete. A substantial amount of his work was achieved after he became totally blind at the age of 61.

He was such a dominating force during his lifetime that we still use much of his notation today. He is responsible for the following notations:

$f(x)$ for functional notation,

e for the base of the natural logarithm,

a, b, c for the sides of triangle ABC,

s for the semiperimeter of a triangle,

$\sum$ for the summation sign, and

i for the imaginary unit $\sqrt{-1}$.

(Eves 1953, p. 359) Although he did not originate the use of the symbol π, Euler is responsible for its acceptance (Boyer 1968, p.484).

Leonhard Euler was born in Basel, Switzerland on April 15, 1707. His father Paul Euler had attended lectures by Jacob Bernoulli at the University of Basel. Furthermore, Paul Euler and Jacob's younger brother Johann had both lived in Jacob Bernoulli's house while undergraduates at Basel (J J O'Connor and E F Robertson, Leonhard Euler).

Nonetheless, Paul became a Lutheran minister and he wanted his son Leonard to follow in his footsteps. At the age of 13, Leonhard entered the University of Basel in the fall of 1720 and by 1723 he received a master's degree in philosophy. In the fall of the same year, he joined the Department of Theology with the intent of fulfilling his father's wish. Although Leonhard remained a devout Calvinist throughout his life, he did not have his heart in theology as a profession.

Meanwhile, Leonhard had been studying mathematics on his own and meeting with Johann Bernoulli on Saturday afternoons. Johann had replaced his older brother Jacob on the Basel faculty when Jacob died in 1705. Leonhard, with Johann's support, was able to obtain the consent of Leonhard's father for him to pursue a career in mathematics. When Euler completed his studies at the University of Basel in 1726, the prospects of a job for a mathematician in Switzerland were grim.

In the fall of the previous year, 1725, two sons of Jacob Bernoulli – Nikolaus and Daniel went to Russia to take positions at the newly organized St. Petersburg Academy of Sciences. These two Bernoullis persuaded the authorities at the Academy to offer Euler a position in physiology. Euler delayed accepting the position because he had to familiarize himself with the topic and he hoped to obtain a position in physics that had opened up at the University of Basel. When it became clear that he would not get the physics position at Basel, Euler moved to St. Petersburg. Once he arrived, Daniel Bernoulli and Jacob Hermann (a distant relative of Euler) persuaded the administration to give him an adjunct position in mathematics. (A position in mathematics had become open because Nikolas Bernoulli had died of hectic fever (Dictionary of Scientific Biography, Vol. II 1970, p. 57) shortly before the Academy offered Euler the position in physiology.) Getting the position in mathematics was good for Euler. However on the day Euler arrived in Russia, Catherine I died (Muir 1961, p. 164). That was bad for Euler.

Political and religious winds in Russia could change direction at any time and the consequences could be severe. The St. Petersburg Academy was established by Peter the Great in 1725 – a few months before he died. For much of Russia's history, times have been turbulent and this period was no exception. In 1682 when Peter was nearing his tenth birthday, the Kremlin where Peter resided was invaded by members of a mutinous militia. Government advisors and household members were tossed out into the street where they were dismembered. This mutiny was the result of machinations by the maternal family of Peter's 25-year-old half sister Sophia who was in the process of seizing power from Peter's mother. At the end of the mutiny, there would be two underage czars (Peter and his half brother Ivan) while Sophia became the ruler of Russia acting as the regent for the two boys (De Jonge 1938, pp. 36-55).

Sophia tried to become Czarina, but at the age of 17, Peter was able to outmaneuver her and become essentially the sole ruler of Russia. (Peter's older half-brother Ivan was feeble minded and nearly blind and therefore played only a ceremonial role.) Sofia was imprisoned in a convent. The convent was palatial but others were punished more severely. Her foreign minister was sent into exile to the north. A close adviser, who had offended the Patriarch of the Russian Orthodox Church by showing western sympathies, was executed. (Authorities in the Russian Orthodox Church would later regret their support of Peter.)

During his reign, Peter devoted much of his efforts to introducing Western European technology into what was essentially a medieval society. That was not an easy task. Anything foreign was suspect in Russia. Before Peter became Czar, foreign visitors to Moscow were required to live in a specified neighborhood. An attempt by a Russian citizen to leave Russia was an offence punishable by death (De Jonge 1938, p. 33).

In 1675 Peter's father, Czar Aleksey Mikhailovich, decreed:

> That no man should adopt the habits of Germans and other foreigners, that they should not cut their hair short, nor trim their clothes, nor wear foreign coats or hats, nor should they allow their followers to do so. And should anyone in their following cut their hair short and wear dress after the foreign fashion, they would incur the wrath of the czar and lose their rank. (De Jonge 1938, p. 24).

Edicts would become quite different while Peter was Czar. In 1698, Peter returned from a tour of Europe shortly after a revolt in support of Sophia had been suppressed in his absence. The day after his return, many nobles and others assembled in the Kremlin to welcome Peter home and demonstrate their loyalty. After some social preliminaries, Peter went about the room with a pair of scissors cutting off the noble's beards one by one. This was Peter's way of introducing Western manners and modes of dress into Russia (De Jonge 1938, p. 128).

This act did not endear Peter to many who treasured the old ways. Many traditional Russians took the view that cutting their beards jeopardized their chances of getting to heaven. In 1700, under instructions from Peter, the citizens of Kamyshin, a town in southern Russia shaved off their beards. Neighboring cossacks responded by running down those who had shaved and cutting off their heads (De Jonge 1938, p. 139).

For Peter, the introduction of Western ways was a consuming passion. His eldest son, Aleksey, was in line to succeed Peter as Czar. When it became clear that Aleksey would reverse much of Peter's reforms if given the chance, Peter had him killed in 1718. (It is not clear whether the son died under torture or from a more formal beheading.) (De Jonge 1938, p. 214).

Peter founded St. Petersburg on the Gulf of Finland at the mouth of the Neva river in 1703. The site chosen by Peter the Great for this new city was inauspicious. It was swampland subject to frequent flooding. Nonetheless, after nine years of building, St. Petersburg became the new capital of Russia. It was Peter's intent to connect Russia's future to that of the rest of Europe. The establishment of the St. Petersburg Academy was a break from the past. Until the regency of Sophia, Moscow did not have a single university (Lindsey 1990, p. 15). Peter's own education was rudimentary. His prose style, spelling, and handwriting showed a lack of discipline (Lindsey 1998, p. 3).

Peter died before the Academy got off the ground but his German born widow, Catherine I, continued his policies. Thus, the Academy got off to a good start before she died two years later. Given the chance, her lover and confidant Alexander Menshikov would have continued strong support for the Academy. Relying on a forged will that he attributed to Catherine I, he was able to get himself declared regent so that he could rule in behalf of the 12-year-old czar, Peter II. This situation lasted for a few weeks but Menshikov was unpopular because of his corruption and pro-German policies. Menshikov tried to isolate the young Czar and get him married off to his daughter. However, Peter II was able to get out the word that he neither approved of Menshikov nor his policies. As a result, Menshikov was soon arrested and sent off to Siberia where he died two years later.

The authority to rule was turned over to Peter II who cut back the funds for the Academy. Euler now found himself in a very bad financial situation. To make ends meet, he became a part-time medical officer in the Russian navy and seriously considered joining full time (Kramer 1981, pp. 214-215).

Fortunately for Euler and the mathematical community, this situation lasted for only about two or three years. Peter II died of small pox in 1730 before reaching his 15th birthday. Anna Ivanova became Czarina by promising to move the capital back to Moscow and to accede to the authority of a self-appointed council, which was anti-Western. However, she soon renounced her promise and became an absolute autocrat. Under her rule, funding was restored to the Academy and Euler was able to give up his part-time job with the navy.

When Daniel Bernoulli returned to Switzerland in 1733, Euler was awarded the Chair of Mathematics. That same year Euler married the daughter of a Swiss painter living in Russia. During their long married life, they had 13 children although only five survived to young adulthood. He remained in St. Petersburg for 14 years until the Czarina Anna died in 1740. During this period, Euler became recognized as the most outstanding mathematician in Europe. He prepared 80–90 papers for publication and actually published 55 works including the 2-volume *Mechanics*. One achievement that attracted admiration was his solution of the "Basel Problem." It had been posed more than 90 years earlier and remained unsolved until Euler did so in 1738. The problem was to sum the series:

$$\sum_{n-1}^{\infty} \frac{1}{n^2}.$$

(Dunham 1999, p. xxii) Euler's answer was $\pi^2/6$. During the same year, Euler became blind in his right eye – probably the result of an eye infection.

When Czarina Anna Ivanovna died in 1740, the political climate in Russia became ominous for the Western Europeans at the St. Petersburg Academy. Czarina Anna Ivanovna had been good to the Academy but she had ruled with an iron hand. Many dissidents had been exiled to Siberia or remote villages. Many others had been executed.

In October, right after Anna's death, a Count Biron got himself named regent for the infant Czar Ivan VI. Less than a month later in November 1740, he was arrested and sent to Siberia. Then Anna Leopoldovna, the mother of the Czar, was named regent. However about one year later in November 1741, she was ousted by Elizabeth who was a daughter of Peter the Great. Anna Leopoldovna and her son would spend the rest of their lives in prison.

As these events were unfolding in Russia, other events were unfolding in Prussia and Austria. Frederick William died on May 31, 1740 and his eldest son, Frederick II, became King of Prussia. This new king would become known as Frederick the Great. Frederick the Great would never introduce democracy but as soon as he became king he admonished provincial administrators who were deemed to be overly repressive and replaced at least one. He also abolished the use of torture in criminal investigations (Asprey 1986, p. 145). One of his first acts was to revive a science academy that had been originated by his grandfather and become moribund under his father.

Euler was invited to become one of the first members of what would soon be named the Académie Royale des Sciences et des Belles Lettres de Berlin. (Frederick the Great preferred the use of French.) Euler eagerly accepted and on June 19, 1741, he and his family were on their way out of St. Petersburg.

Frederick the Great had established a reputation as a man of letters before he became King. Nonetheless, Euler should have had reason to believe that life in Berlin might not be idyllic. Less than five months after Frederick the Great succeeded his father, Emperor Charles VI of the Holy Roman Empire died on

October 20, 1740. His daughter Maria Theresa became ruler of Austria but since she was a woman, many Electors of outlying territories would not accept her as the legitimate ruler of the Empire. Since Maria Theresa was only 23, and not properly prepared for her role, the rulers of Europe saw this situation as an opportunity to expand the boundaries of their countries at the expense of Austria.

Prussia had the largest standing army and was first to strike. Frederick II invaded Upper and Lower Silesia on December 16, 1740 – less than a month after the death of Emperor Charles VI. The weather soon became too severe to continue so Frederick returned to Berlin leaving his troops in control of the invaded territory.

Meanwhile, Frederick persuaded Maupertuis to come to Berlin and tried to get him to assume the presidency of the newly revived academy. Maupertuis was reluctant to assume the office but he did agree to accompany Frederick as an intellectual dinner companion to the war front when hostilities resumed in the spring of 1741. The idea of going to the war front with King Frederick II was ill considered. On April 10 at Mollwitz, Frederick's army became engaged in a fierce battle with the Austrians. It soon became evident that the Prussians faced what looked like certain defeat. The Prussian general Schwerin persuaded Frederick to leave the battlefield. Frederick was reluctant to do so but it would be bad enough to lose a battle, it would have been worse to lose a kingdom.

Frederick left the battlefield with a small escort that included Maupertuis. After a ride of about 33 miles (53 kilometers), Frederick discovered that his route of escape was cut off by the Austrians. Forced to return to the battlefield, Frederick joyfully learned that his army had turned the tide and had won a decisive battle. This battle would win Prussia new respect and enable Frederick to bargain with his neighbors with far more clout than heretofore possible.

However, when Frederick's party was returning to the battlefield, Maupertuis was captured by the Austrians (Asprey 1986, pp. 201-203). He was well treated, taken to Vienna and soon released and returned to Berlin. Presumably, this experience suggested to Maupertuis that the intellectual life in the cafés and salons of Paris was preferable to the intellectual life at Frederick's dinner table. By June, he was back in Paris (O'Connor: Maupertuis, p. 3). Frederick was unable to persuade Maupertuis to become President of his academy until five years later when the war was about to end.

The capture of Maupertuis occurred about two months before Euler left St. Petersburg. But Euler was clearly more fearful of the situation in Russia.

When Euler was presented to Frederick's mother, she tried to engage him in conversation but Euler was essentially closemouthed. One day she asked him the reason for this and Euler replied, "Madam, it is because I have just come from a country where every person who speaks is hanged." (Turnbull 1993, p. 110).

We can assume that Euler knew that people could get executed in Prussia also. King Frederick II had nearly been executed by his own father. Frederick the Great's grandfather, Frederick I, had tried to emulate Louis XIV and had nearly driven Prussia into bankruptcy. His son Frederick William over reacted, lived a frugal existence, enjoyed life in the army barracks, and scorned all art, all books, all paintings, and all science (Asprey 1986, p. 13).

At age five, the future Frederick the Great was given a uniform and required to learn a Prussian parade drill that required 54 movements (Asprey 1986, p. 15). His tutors were instructed to teach him the Lutheran catechism without the Calvin heresy of predestination. He was to be taught French and German but no Latin. Frederick William also instructed his son's tutors to instill a disgust for plays and operas.

Many of these instructions were ignored and Frederick was taught Greek, Italian and even some Latin. He enjoyed the Roman classics and got satisfaction in playing the flute (Asprey 1986, pp.18-19). He would eventually acquire a library of nearly 4,000 books that had to be kept hidden from his father. The library included works by Descarte, Boyle, Locke, and Voltaire (Asprey 1986, p. 35).

Frederick William became disturbed by the direction of his son's development. Thus when Frederick became a young teenager, his father supervised his son's education more closely and saw to it that it would have a narrow focus on military science and religion (Asprey 1986, p. 23).

In 1730, at age 18, Frederick rebelled and tried to escape to England where he hoped to get protection from King George II who was his mother's brother. Since his suspicious father saw to it that his son was closely watched, this attempt failed.

Having been caught, Frederick may have reflected on the fate of his maternal grandmother, Sophia Dorothea. She had died four years earlier while imprisoned in a castle. Sophia was the mother of both young Frederick's mother and George II of England. (It was George II who would establish the University of Göttingen in 1737.) Like Frederick she also tried to escape from a painful situation. As a pawn of international diplomacy, she had been chosen to be the wife of George Louis, the second Elector of Hanover, and she resented the infidelities of her husband. When her plot to escape with Count Philip von Königsmark was discovered, Philipp was presumably executed. She would spend the remaining 32 years of her life in prison. Her husband would become King George I of England.

Her son vehemently protested the treatment of his mother but as a mere prince he could do nothing about it. When he ascended to the throne of England as George II in 1727, it was too late. His mother had died seven months earlier (Durant 1965, pp. 89-94).

How would Frederick William deal with the rebellious Frederick? In a similar situation, Peter the Great had convened a court that condemned his son to death. Frederick William also convened a military court with the expectation of a similar result.

The court sentenced Lieutenant von Katte, a confidant of Frederick, to life in prison and ruled that they did not have the authority to judge Frederick. That would be left to Frederick William (Asprey 1986, p. 69). When the court refused to change its decisions, Frederick William intervened and ruled that his son would be forced to watch his friend Katte beheaded.

This choreography may have been inspired by Peter the Great. In 1698, when Peter returned from his trip to Western Europe, more than 1,000 participants of the mutiny against him were executed. To drive home a point, three of the ring leaders were hung outside his step sister's convent window. The corpses were left there for five months (Hughes 1990, p. 255).

To further satisfy his vengeance against his son, Frederick William had his son's library auctioned off and his flute confiscated.

Before he was beheaded, Katte wrote a letter to Frederick pleading him to reconcile his differences with his father. During the following ten years leading up to the death of Frederick William, a reconciliation did take place after a fashion. Frederick eventually reached the conclusion that for Prussia to survive, he would have to become a knowledgeable military commander. His father eventually gave Frederick his own residency where he could advance his education in his own way and communicate with leaders of the Enlightenment, which included Voltaire.

As rulers, Frederick the Great of Prussia and later Catherine the Great of Russia became known as "enlightened despots." They were enlightened because they enjoyed reading and discussing the works of Voltaire and his friends. They were despots because they never implemented the ideas that would later inspire some revolutionaries to write the American Constitution and Bill of Rights.

Diplomacy in the age of Frederick the Great was the art of deceit. Kings would not try to directly persuade their neighboring kings to pursue some policy change. Instead, they would bribe the advisors of their fellow kings to do the persuading for them. A king at war would wait for a timely moment to double-cross his current allies to cut a favorable deal with his current enemies. (During the American Revolution, France supplied critically needed money and arms to the cause. At the pivotal battle of Yorktown, the decisive factors for American success were French troops and naval support. The French navy prevented the British from delivering badly needed supplies to their ground troops. However, when given the chance in 1783, American diplomats ignored French interests and negotiated separately with Great Britain to gain American independence on more favorable terms than would have been possible otherwise.)

Somehow in this context of ever-changing political alignments, Euler was a successful survivor. During most of his twenty five years in Berlin, Euler was able to collect stipends from both Prussia and Russia. Before moving to Prussia, Euler had made himself useful to Russia. According to the entry on Euler in the Dictionary of Scientific Biography written by A. P. Youschkevitch:

> "From 1733 on, he successfully worked with Delisle on maps in the department of geography. From the middle of the 1730s he studied problems of shipbuilding and navigation, which were especially important to the rise of Russia as a great sea power. He joined various technological committees and engaged in testing scales, fire pumps, saws, and so forth." (Youschkevitch 1971, p. 469).

After moving to Prussia, Euler did not devote his talents completely to ivory tower mathematics. In the same biographical entry cited above, A. P. Youschkevitch wrote,

> "The king also charged Euler with practical problems, such as the project in 1749 of correcting the level of the Finow Canal, which was built in 1744 to join the Havel and Oder. At that time he also supervised the work on pumps and pipes of the hydraulic system at Sans Souci, the royal summer residence.

In 1749 and again in 1763 he advised on the organization of state lotteries and was a consultant to the government on problems of insurance, annuities, and widows' pensions." (Ibid., p. 470).

In 1742, Benjamin Robins advanced the theory of ballistics in a very substantial way with his publication of *New Principles of Gunnery*. Three years later Euler translated this work into German adding a lengthly commentary.

As an aside, Larry D'Antonio, Frederick Rickey, and Sandro Caparrinni (all experts on Euler) have informed me that the collected papers of Napoleon contain notes that Napoleon took on a French translation of the Robins-Euler treatise. These were taken when Napoleon was a 16-year-old student at the École Militaire in 1784-1785. Napoleon was a good student. He crammed the contents of a two-year program into a single year and then passed a comprehensive exam administered by the French mathematician, Pierre-Simon Laplace. Of course, later Napoleon would become the most preeminent artillery officer of all time in the western world.

While Euler was living in Prussia, about half of his output was published in Latin by the St. Petersburg Academy while the other half was published in French by the Berlin Academy (Ibid., pp. 470-471).

He was able to benefit from his good relations with Russia even when Russia and Prussia were fighting one another during the Seven Years War (1756-1763). In 1760, when the Russian army invaded Berlin two Russian soldiers were assigned to protect Euler and his household. Despite this measure, Euler's country home was looted (Kramer 1981, pp. 216-217). When the Russian general was informed, he immediately arranged for Euler to be properly compensated for his losses. An additional four thousand florins was delivered to Euler, when Empress Elizabeth learned what had happened (Turnbull 1993, p. 111).

Euler was not as well treated by Frederick II. Frederick wanted an intellectual dinner companion as much or more than a scientist who would advance the welfare of Prussia. Euler was also too religious for Frederick's taste. Rather than dining with Frederick, Euler ate at home. Rather than discussing philosophy or listening to poetry with Frederick at the King's residence, Euler would assemble his family at home each night and read a chapter of the Bible (Dunham 1999, p. xxv).

During the Seven Years War, the position of President of the Academy became open when Maupertuis died in 1759. Euler became acting president, but he was never given the full title of President. Furthermore, Euler's management decisions were frequently overruled by Frederick. It became clear to Euler that he was appreciated more in Russia. During Euler's 25-year stay in Prussia, events at the top political ranks in Russia continued to unfold in their usual violent ways. Empress Elizabeth died on Christmas day in 1761 near the end of the Seven Years War. She was succeeded by Czar Peter III. At this time, Frederick the Great was in desperate straits. The King had regained control of Berlin but the boundaries of Prussia had shrunk under the combined onslaught of Austria, France, and Russia. Frederick continued to fight on but it appeared he had no chance of restoring the boundaries of Prussia that had existed at the beginning of the war.

Fortunately for Prussia, Czar Peter III was an incompetent who wished to display his admiration for King Frederick. One of Peter III's first official acts was to order a halt to the hostilities with Prussia. Furthermore, he ordered the return of all territorial gains that Russia had attained during the war. Since thousands of soldiers had died to achieve these territorial gains, many Russians were appalled by their new czar. Czar Peter III ruled for about six months before he was overthrown and killed with the tacit approval of his wife Catherine.

As Empress, Peter's widow became known as Catherine the Great. She did not go back to war with Prussia, but she set out to enhance the prestige of Russia in other ways. In 1766, Catherine the Great persuaded Euler to return to St. Petersburg. Although only five of Euler's children had survived to adulthood, his household now consisted of 18 persons. When this group arrived in St. Petersburg, the Empress presented Euler with a furnished house supplied with a royal cook (Kramer 1981, p. 217).

Furthermore, Euler's sons were well treated. In particular, Johann Albrecht became the Chair of Physics in 1766 and permanent secretary of the St. Petersburg Academy in 1769. Euler's youngest son Christoph had become an officer in the Prussian army, and he now became an officer in the Russian army eventually attaining the rank of major-general in artillery (Youschkevitch 1971, p. 472).

During his first stay in Russia, Euler had become blind in his right eye. Within a year after his return, he also became blind in his left eye at the age of 60. This did not slow him down. Several books and 400 research papers were written during the last 17 years of his life after he became totally blind (Kline 1981, p. 429). He was still doing mathematics on September 18, 1783 when he died of a brain hemorrhage at the age of 76.

7.7 *Involutes

Involutes played a prominent role in the early development of differential geometry. However, this section is not a prerequisite for anything else in this text. Thus, you may wish to skip this section.

In Sect. 6.3, we discussed the concept of evolute (at least for the 2-dimensional Euclidean plane). A concept that is complementary is that of the *involute*. Intuitively, an involute is the path of the end point of a string that is unwrapped from a curve while being kept taut. See Fig. 7.9. If $\mathbf{x}(s)$ is the arc length representation of a curve in the 2-dimensional Euclidean plane, then the involute $\mathbf{y}(s)$ is defined by the equation:

$$\mathbf{y}(s) = \mathbf{x}(s) - (s - s_0)\mathbf{t}(s), \tag{7.150}$$

where s is the arc length parameter for $\mathbf{x}$ but not necessarily for $\mathbf{y}$. In Fig. 7.9, $\mathbf{y}(s)$ is an involute of $\mathbf{x}(s)$ and $\mathbf{x}(s)$ is the evolute of $\mathbf{y}(s)$. The definition of an involute remains essentially unchanged for any finite dimension. In particular,

$$\mathbf{y}(s) = \mathbf{x}(s) - (s - s_0)\mathbf{E}_1(s), \tag{7.151}$$

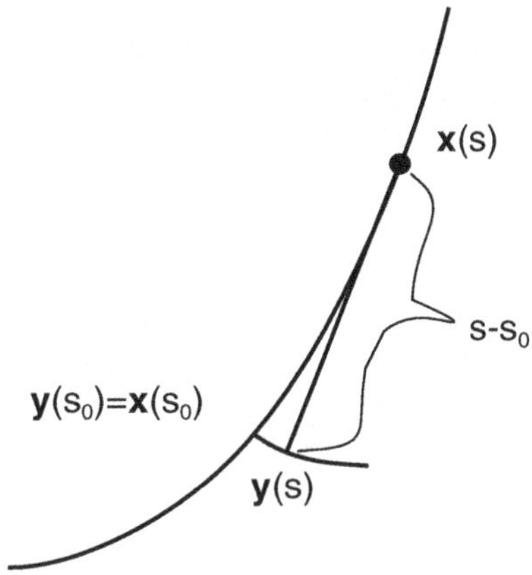

Fig. 7.9 Involute

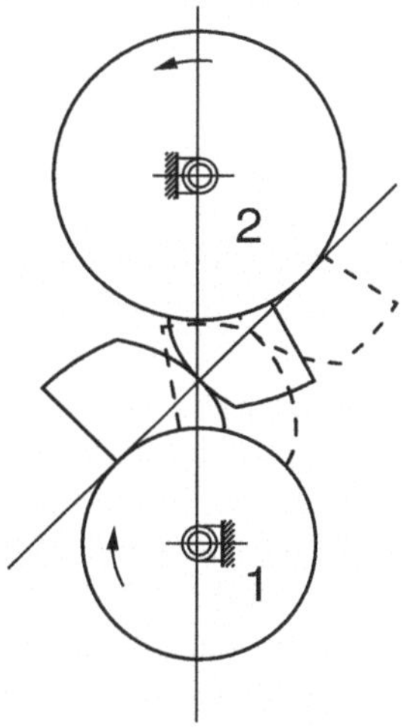

Fig. 7.10 Involutes of circles are used to manufacture gears

where s is again the arc length parameter for $\mathbf{x}(s)$ and $\mathbf{E}_1(s)$ is the first member of the basis for the Frenet apparatus. By contrast, for higher dimensions the definition of an evolute becomes more complex than that used in Sect. 6.3 and I will not discuss it. If you wish to read more on this topic, I recommend Struik (1988) for three dimensions and Gerretsen (1962) for n-dimensions.

(Comment: Comparing (7.151) with (7.146) in Prob. 178, we see that for a tangential developable surface, involutes of the edge of regression are lines of curvature.)

Since one may unwind our imaginary string starting from any point on a given curve, the involute of a given curve is not unique. We can choose any value of s_0 in (7.150) that corresponds to a point on the curve.

An interesting application of the concept of involute is in mechanical engineering (Faires 1960, pp.179-224). See Fig. 7.10 and Prob. 186.

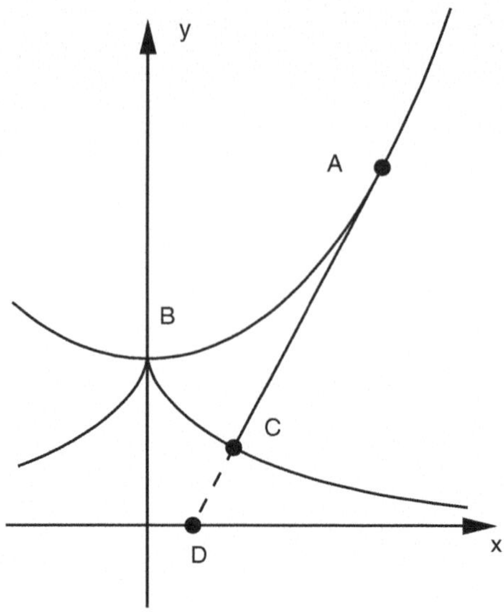

Fig. 7.11 The involute of a catenary is a tractrix

Problem 185. In Fig. 7.9, the path of the involute corresponding to a given value of s_0 is shown for $s > s_0$. Draw the involute for the same value of s_0 for $s < s_0$.

Problem 186. (a) Show that if wheel 1 in Fig. 7.10 rotates clockwise at a constant angular velocity, then wheel 2 will rotate counterclockwise at a constant angular velocity as long as the involutes of the two circles retain mutual contact.

(b) If we had a nonslip surfaces, we could replace gears by circular cylinders with no teeth. What would be the radii of the two nonslip cylinders that would achieve the same effect as the two "gears" shown in Fig. 7.10. (Your formulas should use the variables R_1, R_2, and D, where R_1 is the radius of circle 1 shown in Fig. 7.10, R_2 is the radius of circle 2, and D is the distance between the centers of the two circles.

(c) What is the ratio of the angular velocities in terms of R_1 and R_2? Can this ratio necessarily be duplicated with actual gears?

Problem 187. Show that one of the involutes of a *catenary* is a *tractrix*. The catenary is defined by the equation:

$$\mathbf{x}(\phi) = \mathbf{e}_1 R\phi + \mathbf{e}_2 R \cosh \phi,$$

while the tractrix is defined by the equation:

$$\mathbf{y}(\phi) = \mathbf{e}_1 R(\phi - \tanh \phi) + \mathbf{e}_2(1/\cosh \phi).$$

(See Fig. 7.11.) This result was obtained by Johann Bernoulli in 1691 (Stillwell 2002, p.321). The tractrix is the curve used to generate the surface of revolution that gives us the pseudosphere. See Fig. 6.14 and Prob. 131.

Problem 188. In polar coordinates, a logarithmic spiral may be written in the form $r = a\exp(b\theta)$, where $-\infty < \theta < +\infty$, and both a and b are positive constants. We can also add in the origin where $r = 0$ to the path.

(a) Show that if we unwrap our imaginary string from the origin, the resulting involute is another logarithmic spiral. This result was obtained by Jakob Bernoulli in 1692 (Stillwell 2002, p. 321).
(b) Show that if we adjust the value of b in the equation for the logarithmic spiral, the involute described in part a is a curve congruent to the original logarithmic spiral that has undergone a rotation.

7.8 Theorema Egregium

At this point, it should be clear that the concept of principal curvature is pretty basic. On the other hand, determination of the principal curvatures at a point would be impossible for an intrinsic observer living on a 2-dimensional surface and deprived of access to any higher dimension. A geometer constrained to take measurements on a 2-dimensional surface would not be able to distinguish a truly flat surface from a cylinder, cone, or tangential developable. On the other hand, the intrinsic geometer could determine the Gaussian curvature on her (or his) surface.

Suppose we consider a geometer who had been studying principal curvatures on 2-dimensional surfaces in a 3-dimensional world. We might imagine a phone conversation between this geometer with another geometer living in a 2-dimensional world without access to the third dimension.

After some opening formalities, the conversation might go something like this:

Dr. 3-D I am curious about how things are in your world. What are your curvatures?
Dr. 2-D We live in a simple world. We do not have to worry about some vector in a crazy third dimension rocking forward, backward, and even sidewise in some drunken state. We have only one curvature. I am confident that our curvature is better than either one of your curvatures.
Dr. 3-D How can you say that your world is simple? From what you say, your derivatives do not commute. We keep things simple by telling our first year calculus students that derivatives commute.
Dr. 2-D It sounds to me that you live in a complicated world populated by simple minded mathematicians.

What is the source of this lack of meaningful communication? Having access to the third dimension should be an advantage but how can the extrinsic observer interpret the Gaussian curvature of the intrinsic observer in terms of the principal curvatures? The answer lies in one aspect of the phone conversation above. In one world

$$\left(\nabla_i \nabla_j - \nabla_j \nabla_i\right) \gamma_k \neq 0. \tag{7.152}$$

In the other

$$\left(\frac{\partial^2}{\partial u^i \partial u^j} - \frac{\partial^2}{\partial u^j \partial u^i}\right) \gamma_k = 0. \tag{7.153}$$

In the first (intrinsic world),

$$(\nabla_i \nabla_j - \nabla_j \nabla_i) \boldsymbol{\gamma}_k = R^{\alpha}{}_{kij} \boldsymbol{\gamma}_{\alpha}. \tag{7.154}$$

In the extrinsic world,

$$\begin{aligned}
\left(\frac{\partial^2}{\partial u^i \partial u^j} - \frac{\partial^2}{\partial u^j \partial u^i} \right) \boldsymbol{\gamma}_k &= \frac{\partial}{\partial u^i} \left(\Gamma^{\alpha}_{jk} \boldsymbol{\gamma}_{\alpha} + h_{jk} \mathbf{N} \right) - \frac{\partial}{\partial u^j} \left(\Gamma^{\alpha}_{ik} \boldsymbol{\gamma}_{\alpha} + h_{ik} \mathbf{N} \right) \\
&= \left(\frac{\partial}{\partial u^i} \Gamma^{\alpha}_{jk} \right) \boldsymbol{\gamma}_{\alpha} + \Gamma^{\alpha}_{jk} \Gamma^{\beta}_{i\alpha} \boldsymbol{\gamma}_{\beta} + \Gamma^{\alpha}_{jk} h_{i\alpha} \mathbf{N} + \left(\frac{\partial}{\partial u^i} h_{jk} \right) \mathbf{N} \\
&\quad - h_{jk} h^{\alpha}_i \boldsymbol{\gamma}_{\alpha} - \left(\frac{\partial}{\partial u^j} \Gamma^{\alpha}_{ik} \right) \boldsymbol{\gamma}_{\alpha} - \Gamma^{\alpha}_{ik} \Gamma^{\beta}_{j\alpha} \boldsymbol{\gamma}_{\beta} - \Gamma^{\alpha}_{ik} h_{j\alpha} \mathbf{N} \\
&\quad - \left(\frac{\partial}{\partial u^j} h_{ik} \right) \mathbf{N} + h_{ik} h^{\alpha}_j \boldsymbol{\gamma}_{\alpha} \\
&= \left[\frac{\partial}{\partial u^i} \Gamma^{\alpha}_{jk} - \frac{\partial}{\partial u^j} \Gamma^{\alpha}_{ik} + \Gamma^{\beta}_{jk} \Gamma^{\alpha}_{i\beta} - \Gamma^{\beta}_{ik} \Gamma^{\alpha}_{j\beta} \right] \boldsymbol{\gamma}_{\alpha} \\
&\quad - \left[h_{jk} h^{\alpha}_i - h_{ik} h^{\alpha}_j \right] \boldsymbol{\gamma}_{\alpha} \\
&\quad + \left[\frac{\partial}{\partial u^i} h_{jk} - \frac{\partial}{\partial u^j} h_{ik} + \Gamma^{\alpha}_{jk} h_{i\alpha} - \Gamma^{\alpha}_{ik} h_{j\alpha} \right] \mathbf{N}.
\end{aligned}$$

And therefore,

$$\begin{aligned}
0 = \left(\frac{\partial^2}{\partial u^i \partial u^j} - \frac{\partial^2}{\partial u^j \partial u^i} \right) \boldsymbol{\gamma}_k &= R^{\alpha}{}_{kij} \boldsymbol{\gamma}_{\alpha} - \left[h^{\alpha}_i h_{jk} - h^{\alpha}_j h_{ik} \right] \boldsymbol{\gamma}_{\alpha} \\
&\quad + \left[\frac{\partial}{\partial u^i} h_{jk} - \frac{\partial}{\partial u^j} h_{ik} + \Gamma^{\alpha}_{jk} h_{i\alpha} - \Gamma^{\alpha}_{ik} h_{j\alpha} \right] \mathbf{N}.
\end{aligned} \tag{7.155}$$

Since the coefficient of $\boldsymbol{\gamma}_{\alpha}$ must be zero, we have

$$\begin{aligned}
R^{\alpha}{}_{kij} &= h^{\alpha}_i h_{jk} - h^{\alpha}_j h_{ik}, \text{ or} \\
R^{\alpha k}{}_{ij} &= h^{\alpha}_i h^k_j - h^{\alpha}_j h^k_i.
\end{aligned} \tag{7.156}$$

For a 2-dimensional surface embedded in E^3, the Gaussian curvature

$$K = R^{12}{}_{12} = h^1_1 h^2_2 - h^1_2 h^2_1. \tag{7.157}$$

To compute the principal curvatures, we need to solve a characteristic equation. Namely,

$$\det\begin{bmatrix} h^1_1-\lambda & h^1_2 \\ h^2_1 & h^2_2-\lambda \end{bmatrix} = 0, \text{ which implies that}$$

$$(\lambda)^2 - \lambda(h^1_1 + h^2_2) + h^1_1 h^2_2 - h^1_2 h^2_1 = 0. \tag{7.158}$$

As a consequence,

$$\lambda = \frac{h^1_1 + h^2_2 \pm \sqrt{\left(h^1_1 + h^2_2\right)^2 - 4\left(h^1_1 h^2_2 - h^1_2 h^2_1\right)}}{2}. \tag{7.159}$$

If we let

$$A = \sqrt{\left(h^1_1 + h^2_2\right)^2 - 4\left(h^1_1 h^2_2 - h^1_2 h^2_1\right)},$$

then the product of the two roots is

$$\begin{aligned} \lambda_+\lambda_- &= \frac{1}{4}\left[h^1_1 + h^2_2 + A\right]\left[h^1_1 + h^2_2 - A\right] \\ &= \frac{1}{4}\left[\left(h^1_1 + h^2_2\right)^2 - (A)^2\right] \\ &= h^1_1 h^2_2 - h^1_2 h^2_1. \end{aligned} \tag{7.160}$$

And thus from (7.157),

$$\lambda_+\lambda_- = R^{12}{}_{12} = K. \tag{7.161}$$

Historically, the product $\lambda_+\lambda_-$ was the definition of Gaussian curvature. What was historically significant was the discovery by Gauss that the product of the two principal curvatures is an intrinsic entity. Gauss was quite proud of his discovery that for a 2-dimensional surface embedded in the 3-dimensional Euclidean space E^3, this product is an intrinsic entity. He named this result the "Theorema Egregium," which has been variously translated from the Latin as the "Remarkable Theorem,"the "Outstanding Theorem,"or the "Excellent Theorem."

Gauss published this result as part of a substantial work on curved surfaces entitled, *Disquisitiones generales circas superficies curvas* (General investigations of curved surfaces) (Gauss 1828). The special case of geodesic triangles for the Gauss–Bonnet formula is also included in this work. This treatise appeared about 26 years before Riemann presented his *habilitationsvortrag.* So Gauss did not prove that $\lambda_+\lambda_- = R^{12}{}_{12}$. But he derived an equivalent formula showing that $\lambda_+\lambda_-$ could be computed from members of the metric tensor and their first and second derivatives.

Now that we have seen the relation between Gaussian curvature and the principal curvatures, it may be useful to give a quick overview. We should remind ourselves that the signs of the principal curvatures depend on the chosen direction of **N** but the geometry does not.

If the Gaussian curvature at a point P is positive, the two principal curvatures have the same sign and the neighborhood of P resembles the surface of an ellipsoid or the portion of a torus facing away from the center. If the Gaussian curvature at a point P is negative, the two principal curvatures have the opposite signs and the neighborhood of P resembles a saddle surface or the portion of the torus facing toward the center. If the Gaussian curvature is zero, at least one of the principal curvatures is zero.

7.9 *The Gauss Map

Another way of understanding the nature of Gaussian curvature arises from the *Gaussian spherical mapping*. At each point on a smooth surface, we have a normal vector **N**. If we relocate **N** without changing its direction to the origin, the nose (the pointed end) will determine a point on the surface of a unit sphere. If we allow **N** to roam over the surface, an image will form on the unit sphere. This is the Gaussian spherical mapping.

The image of a circular cylinder will be a great circle. The image of a plane will be a point. The image of an ellipsoid will be the whole sphere. And the image points of a torus will cover the unit sphere twice.

The area of the image will be

$$\int \left| \frac{\partial \mathbf{N}}{\partial u^1} \wedge \frac{\partial \mathbf{N}}{\partial u^2} \right| \mathrm{d}u^1 \mathrm{d}u^2 = \int \left| -h_1^\alpha \boldsymbol{\gamma}_\alpha \wedge -h_2^\beta \boldsymbol{\gamma}_\beta \right| \mathrm{d}u^1 \mathrm{d}u^2. \tag{7.162}$$

We note that

$$\begin{aligned} -h_1^\alpha \boldsymbol{\gamma}_\alpha \wedge -h_2^\beta \boldsymbol{\gamma}_\beta &= h_1^\alpha h_2^\beta \boldsymbol{\gamma}_{\alpha\beta} = h_1^1 h_2^1 \boldsymbol{\gamma}_{11} + h_1^1 h_2^2 \boldsymbol{\gamma}_{12} + h_1^2 h_2^1 \boldsymbol{\gamma}_{21} + h_1^2 h_2^2 \boldsymbol{\gamma}_{22} \\ &= (h_1^1 h_2^2 - h_1^2 h_2^1) \boldsymbol{\gamma}_{12} = K \boldsymbol{\gamma}_{12}. \end{aligned}$$

Also

$$|\boldsymbol{\gamma}_{12}|^2 = \boldsymbol{\gamma}_{12} \boldsymbol{\gamma}_{21} = g_{11} g_{22} - g_{12} g_{21} = g.$$

So (7.162) becomes

$$\int \left| \frac{\partial \mathbf{N}}{\partial u^1} \wedge \frac{\partial \mathbf{N}}{\partial u^2} \right| \mathrm{d}u^1 \mathrm{d}u^2 = \int |K| \sqrt{g} \mathrm{d}u^1 \mathrm{d}u^2.$$

Actually, it is more useful to compute an oriented area:

$$\mathbf{A}_G = \int \left(\frac{\partial \mathbf{N}}{\partial u^1} \wedge \frac{\partial \mathbf{N}}{\partial u^2} \right) \mathrm{d}u^1 \mathrm{d}u^2 = \int \frac{\boldsymbol{\gamma}_{12}}{|\boldsymbol{\gamma}_{12}|} K \sqrt{g} \mathrm{d}u^1 \mathrm{d}u^2.$$

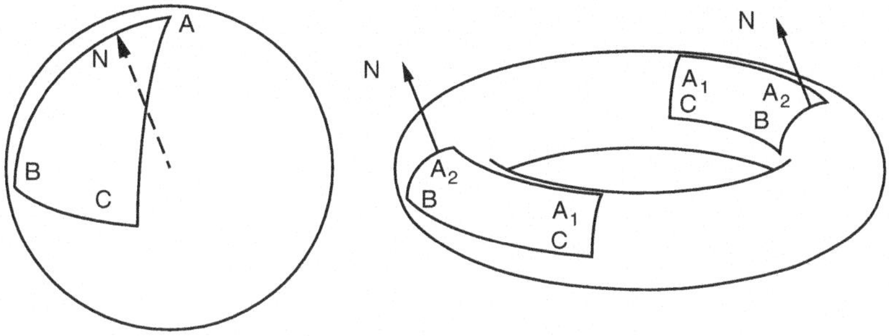

Fig. 7.12 The gauss map from a torus

And thus

$$\mathbf{A}_G = \frac{\gamma_{12}}{|\gamma_{12}|} \int K\sqrt{g}\mathrm{d}u^1\mathrm{d}u^2. \tag{7.163}$$

(To factor out the γ_{12} may not make sense to the extrinsic observer but it does to the intrinsic observer.)

We encountered this same integral (without the $\gamma_{12}/|\gamma_{12}|$ factor) when the Gauss–Bonnet theorem was discussed. Now it has a geometric interpretation. We first note that the oriented area of the pre-image area is

$$A = \frac{\gamma_{12}}{|\gamma_{12}|} \int \sqrt{g}\mathrm{d}u^1\mathrm{d}u^2. \tag{7.164}$$

Thus for small regions, the Gaussian curvature is the ratio of the two areas.

One may ask, if K is negative, the corresponding area on the image sphere must be interpreted as being negative. How is that possible? The answer is where K is positive, counterclockwise loops are mapped into counterclockwise loops. And where K is negative, counterclockwise loops are mapped into clockwise loops. (See Fig. 7.12.)

For the torus, the image sphere is covered twice, once in a positive sense (by the portion of the torus facing away from the center) and once in a negative sense (by the portion of the torus facing toward the center). The integral curvature for the entire torus is zero – a result we got in the form of (6.69) by other means.

To bring this section to a close, I would like to mention the *third fundamental form*. On the surface of the unit image sphere for the Gaussian map, the directed distance corresponding to the coordinate u^1 is

$$h_1^\alpha \gamma_\alpha \mathrm{d}u^1 = \bar{\gamma}_1 \mathrm{d}u^1$$

and the directed distance for u^2 is

$$h_2^\beta \gamma_\beta \mathrm{d}u^2 = \bar{\gamma}_2 \mathrm{d}u^2.$$

Thus, the metric for the "Gauss area" is

$$c_{ij} = \langle \bar{\gamma}_i, \bar{\gamma}_j \rangle = h_i^\alpha h_j^\beta \langle \gamma_\alpha, \gamma_\beta \rangle = h_i^\alpha h_j^\beta g_{\alpha\beta}. \tag{7.165}$$

With this definition, we can now prove the following theorem:

Theorem 189. *If g_{ij} is the first fundamental form, h_{ij} is the second fundamental form, and c_{ij} is the third fundamental form, then*

$$c_{ij} - 2Hh_{ij} + Kg_{ij} = 0, \text{ where} \tag{7.166}$$

K is the Gaussian curvature and H is the mean curvature. That is

$$H = \frac{\lambda_+ + \lambda_-}{2}.$$

Proof. Equation (7.166) follows almost immediately from the Cayley–Hamilton theorem that implies that the matrix $\left[h^i_{\ j}\right]$ satisfies its characteristic equation. From (7.158), the characteristic equation is

$$(\lambda)^2 - \lambda(h_1^1 + h_2^2) + (h_1^1 h_2^2 - h_1^2 h_2^1) = 0.$$

From (7.159),

$$h_1^1 + h_2^2 = 2H.$$

From (7.160) and (7.161),

$$h_1^1 h_2^2 - h_1^2 h_2^1 = K.$$

Using these relations, the characteristic equation for $\left[h^i_{\ j}\right]$ is

$$(\lambda)^2 - 2H\lambda + K = 0.$$

From the Cayley–Hamilton theorem

$$h_\alpha^\beta h_j^\alpha - 2Hh_j^\beta + K\delta_j^\beta = 0.$$

If we now multiply this equation by $g_{i\beta}$ and then sum over β, we have our desired result:

$$c_{ij} - 2Hh_{ij} + Kg_{ij} = 0.$$

□

Problem 190. One equation for a right circular cone is

$$\mathbf{x}(t,\phi) = \mathbf{p} + t\mathbf{v}(\phi), \text{ where}$$

$$\mathbf{v}(\phi) = \mathbf{e}_1 \sin\alpha\cos\phi + \mathbf{e}_2 \sin\alpha\sin\phi + \mathbf{e}_3\cos\alpha.$$

Determine the image of this surface under the Gauss spherical map. Is your result consistent with the comment following (7.164)?

Problem 191. Consider a closed surface with two holes. How many times would the sphere be covered by the Gaussian spherical map? What would be the integral curvature for the surface?

7.10 *Isometries

7.10.1 *Isometry of Surfaces with Constant Gaussian Curvature

In Chap. 5, I pointed out that if a sheet of paper was bent without stretching, the Gaussian curvature would remain zero. It should have been clear from that discussion that an intrinsic observer restricted to taking local measurements on the surface of the paper would not be able to determine the fact that the paper had been bent (or unbent). In the same section, I presented (5.4):

$$K(\mathbf{x}) = \lim_{d \to 0} \theta/(\text{Area of loop}).$$

I presented this equation as an informal definition of Gaussian curvature. Since this definition uses a limit, it can be used to determine the Gaussian curvature at a point whether or not the Gaussian curvature is constant. In Chap. 5, I commented that (5.4) may not be well defined. I presented that definition because it served as a useful lead in to the Gauss–Bonnet theorem. However, better alternatives exist for the intrinsic observer who is not told what the metric for his or her surface is. (See Prob. 194.) For an astute mathematician, the question arises, "Is there information beyond the Gaussian curvature that can be determined by an intrinsic observer that can be used to distinguish one surface from another?" As you might guess, the answer is no for surfaces of zero Gaussian curvature. What is less intuitive (for me) is that the answer is also no for any surface of constant curvature. On the other hand if you are very astute, you might realize that the intrinsic observer could also determine how quickly or slowly the Gaussian curvature varies from point to point.

These questions are closely related to another question. "Given two metric tensors, do they represent different surfaces or do they correspond to different coordinate systems of what are essentially identical surfaces?" If a change in coordinates maps one metric tensor onto another metric tensor, the surfaces represented by the two metric tensors are said to be *isometric*. To determine whether or not two surfaces are isometric is not trivial. Certainly, an informal comparison of the helicoid and the cycloid does not reveal the fact that they are isometric. (See Fig. 7.14 where surface B is a helicoid and surface C is a cycloid.) Are there tests that can be used to determine whether two metrics represent isometric surfaces? Could these tests be used by communicating intrinsic observers confined to two 2-dimensional surfaces to determine whether or not they live on identical surfaces?

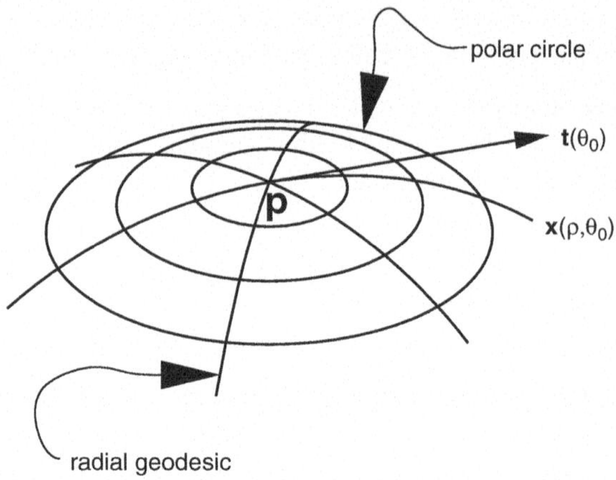

Fig. 7.13 Geodesic polar coordinates

To pursue these questions further, I will first introduce you to the idea of a *geodesic polar parametrization.* In general, this parametrization can be carried out in a neighborhood of any given point **p**. Given a point **p** on a 2-dimensional surface, one can consider the set of all possible unit length tangent vectors located at that point. For each unit vector, there exists a unique geodesic originating at point **p**. If we use an arc length parametrization, then the tangent vector of the geodesic at the point **p** will coincide with the unit tangent vector associated with the geodesic.

We can identify each of the unit tangent vectors by the angle θ, where θ is the angle measured from some arbitrarily selected unit tangent vector. Since the geodesics originating from point **p** are uniquely identified with their initial tangent vectors, the angle θ can also be used to label each geodesic. In this way, we get our geodesic polar parametrization for our surface. That is

$$\mathbf{x} = \mathbf{x}(\rho, \theta), \text{ where}$$

$\mathbf{x}(\rho, \theta_0)$ is the geodesic identified with the initial tangent vector $\mathbf{t}(\theta_0)$ and ρ is the arc length distance measured from the point **p** along the geodesic. (See Fig. 7.13.) As we shall see below, using the arguments of Barrett O'Neill, this is an orthogonal coordinate system (O'Neill 1997, pp. 378-379).

Since ρ is an arc length parameter, $\boldsymbol{\gamma}_\rho$ has unit length where

$$\boldsymbol{\gamma}_\rho = \frac{\partial \mathbf{x}}{\partial \rho}.$$

Thus,

$$g_{\rho\rho} = \langle \boldsymbol{\gamma}_\rho, \boldsymbol{\gamma}_\rho \rangle = \left\langle \frac{\partial \mathbf{x}}{\partial \rho}, \frac{\partial \mathbf{x}}{\partial \rho} \right\rangle = 1. \tag{7.167}$$

Since $\mathbf{x}(\rho, \theta)$ is a geodesic for a constant value of θ, it follows that

$$\nabla_\rho \boldsymbol{\gamma}_\rho = 0 \text{ and therefore } \frac{\partial \boldsymbol{\gamma}_\rho}{\partial \rho} \text{ is orthogonal to the surface.}$$

Now

$$g_{\rho\theta} = g_{\theta\rho} = \langle \boldsymbol{\gamma}_\rho, \boldsymbol{\gamma}_\theta \rangle = \left\langle \frac{\partial \mathbf{x}}{\partial \rho}, \frac{\partial \mathbf{x}}{\partial \theta} \right\rangle, \text{ and}$$

$$\frac{\partial g_{\rho\theta}}{\partial \rho} = \left\langle \frac{\partial^2 \mathbf{x}}{\partial \rho^2}, \frac{\partial \mathbf{x}}{\partial \theta} \right\rangle + \left\langle \frac{\partial \mathbf{x}}{\partial \rho}, \frac{\partial^2 \mathbf{x}}{\partial \rho \partial \theta} \right\rangle. \tag{7.168}$$

Since $\partial^2 \mathbf{x}/\partial^2 \rho$ is orthogonal to the tangent plane, the first term on the right-hand side of (7.168) is zero. Therefore,

$$\frac{\partial g_{\rho\theta}}{\partial \rho} = \left\langle \frac{\partial \mathbf{x}}{\partial \rho}, \frac{\partial^2 \mathbf{x}}{\partial \rho \partial \theta} \right\rangle = \frac{1}{2}\frac{\partial}{\partial \theta}\left\langle \frac{\partial \mathbf{x}}{\partial \rho}, \frac{\partial \mathbf{x}}{\partial \rho} \right\rangle = \frac{1}{2}\frac{\partial}{\partial \theta}(1) = 0.$$

This implies that $g_{\rho\theta}$ is independent of ρ, and thus

$$g_{\rho\theta}(\rho, \theta) = g_{\rho\theta}(0, \theta).$$

For small values of ρ, our surface resembles E^2 and thus our metric approximates ordinary polar coordinates. That is

$$\begin{aligned}(\mathrm{d}s)^2 &= g_{\rho\rho}(\mathrm{d}\rho)^2 + 2g_{\rho\theta}\mathrm{d}\rho\mathrm{d}\theta + g_{\theta\theta}(\mathrm{d}\theta)^2 \\ &\approx (\mathrm{d}\rho)^2 + \rho^2(\mathrm{d}\theta)^2.\end{aligned} \tag{7.169}$$

Thus,

$$\lim_{\rho \to 0} g_{\rho\theta}(\rho, \theta) = g_{\rho\theta}(0, \theta) = 0.$$

Since

$$\begin{aligned}&g_{\rho\theta}(\rho, \theta) = g_{\rho\theta}(0, \theta), \text{ it follows that} \\ &g_{\rho\theta}(\rho, \theta) = 0.\end{aligned}$$

Summarizing, we have

$$\begin{bmatrix} g_{\rho\rho} & g_{\rho\theta} \\ g_{\theta\rho} & g_{\theta\theta} \end{bmatrix} = \begin{bmatrix} 1 & 0 \\ 0 & G(\rho, \theta) \end{bmatrix}. \tag{7.170}$$

(This metric is not valid for a neighborhood of point **p** that includes points where some of the geodesics originating from point **p** intersect one another. Thus for a given point **p**, this metric may be valid for values of ρ less than some positive number.

It is not difficult to obtain a simple formula for the Gaussian curvature in terms of the function $G(\rho, \theta)$. Since the metric is diagonal, we can use (5.96) for the Fock–Ivanenko coefficients. Namely

$$\mathbf{\Gamma}_\alpha = \frac{1}{4}\boldsymbol{\gamma}^{\alpha\eta}\frac{\partial g_{\alpha\alpha}}{\partial u^\alpha}.$$

$$\text{Thus } \mathbf{\Gamma}_\rho = \frac{1}{4}\boldsymbol{\gamma}^{\rho\theta}\frac{\partial g_{\rho\rho}}{\partial\theta} = 0 \text{ and} \tag{7.171}$$

$$\mathbf{\Gamma}_\theta = \frac{1}{4}\boldsymbol{\gamma}^{\theta\rho}\frac{\partial g_{\theta\theta}}{\partial\rho} = -\frac{1}{4}\gamma^{\rho\theta}\frac{\partial G(\rho,\theta)}{\partial\rho}. \tag{7.172}$$

For this metric, it is clear that

$$\boldsymbol{\gamma}^\rho = \boldsymbol{\gamma}_\rho = \mathbf{E}_1 \text{ and} \tag{7.173}$$

$$\boldsymbol{\gamma}^\theta = \frac{1}{G}\boldsymbol{\gamma}_\theta = \frac{1}{\sqrt{G}}\mathbf{E}_2. \tag{7.174}$$

To get the curvature 2-form, we can use (5.84). Namely

$$\frac{1}{2}\mathbf{R}_{\rho\theta} = \partial_\rho\mathbf{\Gamma}_\theta = -\frac{1}{4}\mathbf{E}_1\mathbf{E}_2\frac{\partial}{\partial\rho}\frac{1}{\sqrt{G}}\frac{\partial G}{\partial\rho}. \tag{7.175}$$

However,

$$\frac{1}{\sqrt{G}}\frac{\partial G}{\partial\rho} = 2\frac{\partial}{\partial\rho}\sqrt{G}, \text{ so (7.175) becomes}$$

$$\frac{1}{2}\mathbf{R}_{\rho\theta} = -\frac{1}{2}\mathbf{E}_1\mathbf{E}_2\frac{\partial^2\sqrt{G}}{\partial\rho^2}. \tag{7.176}$$

We note that

$$\mathbf{R}_{\rho\theta} = \frac{1}{2}R^{ij}{}_{\rho\theta}\boldsymbol{\gamma}_{ij} = R^{\rho\theta}{}_{\rho\theta}\boldsymbol{\gamma}_{\rho\theta}, \text{ where}$$

the ρ and θ indices in the last term are unsummed. Using (7.173) and (7.174), we have

$$K = R^{\rho\theta}{}_{\rho\theta} = -\frac{1}{\sqrt{G}}\frac{\partial^2}{\partial\rho^2}\sqrt{G}. \tag{7.177}$$

We are now in a position to ask the question, "If the Gaussian curvature is known, can we determine the geodesic polar parametrization?" It certainly can be done if the Gaussian curvature is constant. The following theorem is due to Ferdinand Minding (1806-1885) (Minding 1839).

Theorem 192. *Minding's Theorem: All 2-dimensional surfaces with the same constant Gaussian curvatures are isometric. (They share the same metric tensor or equivalent metrics that can be converted into one another by a change of coordinates.)*

Proof. K=0. From (7.177),

$$\frac{\partial^2 \sqrt{G}}{\partial^2 \rho} = 0 \ \textit{and thus}$$

$$\sqrt{G} = A(\theta)\rho + B(\theta).$$

From (7.169), it is clear that if $G(\rho, \theta)$ *can be expanded as a Taylor's series in* ρ*, then*

$$G(\rho, \theta) = g_{\theta\theta} = \rho^2 + \textit{ higher powers of } \rho.$$

Thus for our case

$$\sqrt{G(\rho, \theta)} = \rho \ \textit{and} \ g_{\theta\theta} = \rho^2.$$

This means that

$$(\mathrm{d}s)^2 = (\mathrm{d}\rho)^2 + \rho^2(\mathrm{d}\theta)^2, \ \textit{which} \tag{7.178}$$

is the metric corresponding to the use of polar coordinates for the flat Euclidean plane E^2*. If you choose to change the coordinates so that*

$$x = \rho\cos\theta \ \textit{and} \ y = \rho\sin\theta,$$

you would have the more usual metric. Namely

$$(\mathrm{d}s)^2 = (\mathrm{d}x)^2 + (\mathrm{d}y)^2.$$

Case 2) K>0. In this case, (7.177) becomes

$$\frac{\partial^2 \sqrt{G}}{\partial^2 \rho} = -K\sqrt{G}.$$

The general solution for this equation is

$$\sqrt{G} = A(\phi)\sin(\sqrt{K}\rho) + B(\phi)\cos(\sqrt{K}\rho).$$

(I have relabeled the angle variable for a reason that will become apparent.) As before

$$G(\rho, \theta) = g_{\theta\theta} = \rho^2 + \textit{ higher powers of } \rho.$$

Since

$$\sin(\sqrt{K}\rho) = \sqrt{K}\rho + \textit{ higher powers of } \rho \ \textit{ and}$$

$$\cos(\sqrt{K}\rho) = 1 + \textit{ higher powers of } \rho, \textit{ it follows}$$

that

$$B(\phi) = 0 \ \textit{and} \ A(\phi) = \frac{1}{\sqrt{K}}, \ \textit{so}$$

$$g_{\phi\phi} = G(\rho, \phi) = \frac{1}{K} \sin^2(\sqrt{K}\rho).$$

This implies that

$$(\mathrm{d}s)^2 = (\mathrm{d}\rho)^2 + \frac{1}{K} \sin^2(\sqrt{K}\rho)(\mathrm{d}\phi)^2. \tag{7.179}$$

We should note that for spherical coordinates on the surface of a sphere of radius R, we have

$$(\mathrm{d}s)^2 = R^2(\mathrm{d}\theta)^2 + R^2 \sin^2(\theta)(\mathrm{d}\phi)^2. \tag{7.180}$$

We see that these metrics are equivalent if we note that for a sphere, $K = 1/R^2$ and replace the distance from point **p** *by $R\theta$, which is the distance from the North pole! Thus, we see that all surfaces with constant Gaussian curvature are isometric to the surface of a sphere. Finally, we have*

Case 3) $K < 0$. In this case,

$$\frac{\partial^2 \sqrt{G}}{\partial \rho^2} = |K| \sqrt{G} \ \textit{and}$$

$$\sqrt{G} = A(\theta)\sinh(\sqrt{|K|}\rho) + B(\theta)\cosh(\sqrt{|K|}\rho).$$

Using the same kind of arguments we used for Case 2) we have

$$g_{\theta\theta} = G(\rho, \theta) = \frac{1}{|K|} \sinh^2(\sqrt{|K|}\rho). \tag{7.181}$$

□

7.10.2 *Isometries for Surfaces with Nonconstant Gaussian Curvature

To what extent can the result derived for surfaces of constant curvature be generalized to surfaces of nonconstant curvature? Suppose you could determine the geodesic polar coordinates for two points on their respective surfaces. It would then be an easy matter to determine whether the two surfaces are isometric in the neighborhoods of the two given points. You would simply see whether the metric tensors matched. (You would have to allow for the fact that the reference angles in the two coordinate systems might have to be adjusted to get a match.) However generally, it is difficult or impossible to get the explicit formulas for the geodesics that would be necessary for this procedure. Let us consider another approach, which will at least give us some insight into the problem.

Suppose we consider two geometers on the opposite end of a phone line. Each lives in a 2-dimensional world and they wish to determine whether they live in worlds that are isometric to one another.

At the beginning of the conversation they agree to measure their Gaussian curvatures. (Previously, some third party has given each of them a measuring stick of the same length.) If their Gaussian curvatures agree, then they know it is possible they live in isometric worlds. If their Gaussian curvatures differ, they may decide to seek out a pair of points where their Gaussian curvatures agree. One of the geometers could move in a direction such that the Gaussian curvature is increasing and the other could move in a direction of decreasing Gaussian curvatures. If the lower bound of the Gaussian curvature of one world is not greater than the upper bound of the Gaussian curvature of the other, then eventually each could find a point in their respective worlds where their Gaussian curvatures agree.

Can it now be said that at least in their respective neighborhoods that they live in isometric worlds? The answer is no! It is not even sufficient that they each have a coordinate system that gives the same formula for the Gaussian curvature in their respective neighborhoods. Consider one of the geometers. What further information could he or she elicit besides the Gaussian curvature at his or her location. If the Gaussian curvature is not a local maximum or minimum, then there would be some direction the geometer could move so that the Gaussian curvature did not change. Suppose the surface was defined by the formula

$$\mathbf{x} = \mathbf{x}(u^1, u^2) \text{ and}$$

the curve passing through values of equal K was

$$\mathbf{x}(\mathbf{t}) = \mathbf{x}(u^1(t), u^2(t)), \text{ then}$$

$$\frac{\mathrm{d}K}{\mathrm{d}t} = \frac{\partial K}{\partial u^\alpha}\frac{\mathrm{d}u^\alpha}{\mathrm{d}t} = \nabla_\alpha K \frac{\mathrm{d}u^\alpha}{\mathrm{d}t} = 0.$$

The vector $\boldsymbol{\gamma}^\alpha \nabla_\alpha K = \nabla K$ is known as the *gradient* of K. Clearly, this vector is orthogonal to $\boldsymbol{\gamma}_\beta \frac{\mathrm{d}u^\beta}{\mathrm{d}t}$ that has the direction of a path with constant K. For the two surfaces to be isometric, it does not matter whether the components of the gradient of the two surfaces agree at a given pair of points. In different coordinates, the coordinates of a vector have different meanings. What is necessary for the two surfaces to be isometric is that the magnitudes of the two gradients agree. Thus, a necessary condition for the two worlds to be locally isometric is

$$|\nabla K|^2 = g_{\alpha\beta}\frac{\partial K}{\partial u^\alpha}\frac{\partial K}{\partial u^\beta} = \acute{g}_{\alpha\beta}\frac{\partial K}{\partial \acute{u}^\alpha}\frac{\partial K}{\partial \acute{u}^\beta} = \left|\acute{\nabla} K\right|^2, \text{ where} \tag{7.182}$$

$g_{\alpha\beta}$ is the metric tensor for the (u^1, u^2) coordinate system in one world and $\acute{g}_{\alpha\beta}$ is the metric tensor for the $(\acute{u}^1, \acute{u}^2)$ coordinate system in the other world.

If (7.182) is satisfied, then the two geometers will discover that if they each move in the direction of increasing Gaussian curvature orthogonal to the direction of constant Gaussian curvature, the rate of change will be the same for both of them. The agreement of the Gaussian curvatures and their gradients in the neighborhoods of two respective points are clearly necessary conditions for an isometry. Also, they are almost but not quite sufficient conditions. Conditions that are both necessary and sufficient are presented by Eisenhart in his book *An Introduction to Differential Geometry* (Eisenhart 1947, pp. 155-169).

In closing this section, I would like to point out that the notion of a gradient is an important concept in mathematics. With that thought in mind, I will prove the following theorem:

Theorem 193. *The direction of the gradient ∇f of the scalar function $f(u^1, u^2, ---, u^n)$ is the direction of maximum increase or maximum decrease of f.*

Proof. Suppose $\mathbf{x}(u^1(s), u^2(s), ---, u^n(s))$ is the arc length parameterized path through some given point $\mathbf{p}$, that is the path of maximum increase or decrease for the function f. Then the maximum rate of change for f would be

$$\frac{\mathrm{d}f}{\mathrm{d}s} = \frac{\partial f}{\partial u^j}\frac{\mathrm{d}u^j}{\mathrm{d}s}. \tag{7.183}$$

Thus, we are seeking, the vector

$$\gamma_k \frac{\mathrm{d}u^k}{\mathrm{d}s}, \text{ that}$$

maximizes the magnitude of df/ds. If we set

$$\frac{\mathrm{d}u^k}{\mathrm{d}s} = \alpha^k, \text{ then}$$

our task is to find the maximum or minimum of

$$\frac{\partial f}{\partial u^k}\alpha^k \text{ subject to the constraint that}$$

$$g_{ik}\alpha^i\alpha^k = 1.$$

Using the method of Lagrange multipliers, we let

$$G(\lambda, \alpha^1, \alpha^2, ---, \alpha^n) = \frac{\partial f}{\partial u^k}\alpha^k - \lambda(g_{ik}\alpha^i\alpha^k - 1).$$

We then require

$$\frac{\partial G}{\partial \alpha^j} = \frac{\partial f}{\partial u^k}\delta^k_j - \lambda(g_{ik}\delta^i_j\alpha^k + g_{ik}\alpha^i\delta^k_j) = \frac{\partial f}{\partial u^j} - 2\lambda\alpha_j = 0.$$

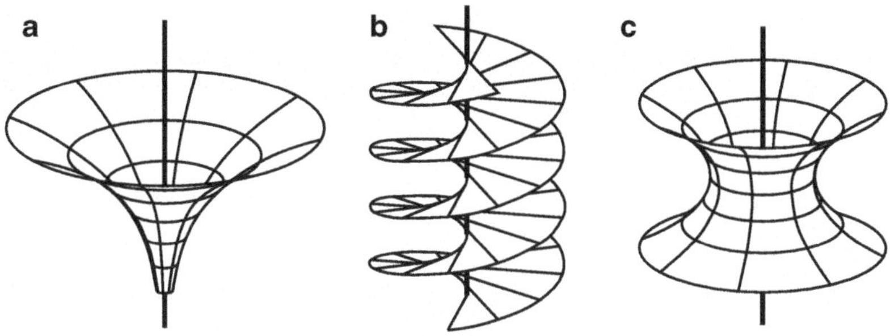

Fig. 7.14 Surfaces **b** and **c** are isometric. With an appropriate coordinate system, surface **a** has the same Gaussian curvature but is not isometric to the other two surfaces

This implies that

$$\gamma_k \frac{\mathrm{d}u^k}{\mathrm{d}s} = \gamma_k \alpha^k = \gamma^j \alpha_j = \frac{1}{2\lambda}\gamma^j \frac{\partial f}{\partial u^j} = \frac{1}{2\lambda}\nabla f.$$

This is what we set out to prove. □

Problem 194. For a flat surface, the equation for the circumference of a circle is $2\pi\rho$.

(a) Use (7.179) and (7.181) to obtain formulas for the circumference of a circle when the Gaussian curvature is constant and $K > 0$ and when $K < 0$.
(b) Use the results of part a) to construct a limit that can be used to determine the Gaussian curvature at a point (whether constant or not).
(c) Determine the appropriate formulas for the areas of circles on surfaces of constant Gaussian curvature. Then use that result to construct a limit that could be used to determine the Gaussian curvature at a point. Which limit, the one from part b) or c), would be more useful?

Problem 195. Consider the three surfaces in Fig. 7.14. Coordinate systems that can be used for the three surfaces are

$$\mathbf{a}\text{: } \mathbf{x}(u,\phi) = \mathbf{e}_1 au\cos\phi + \mathbf{e}_2 au\sin\phi + \mathbf{e}_3 a\ln u.$$

$$\mathbf{b}\text{: } \mathbf{x}(u,\phi) = \mathbf{e}_1 au\cos\phi + \mathbf{e}_2 au\sin\phi + \mathbf{e}_3 a\phi.$$

$$\mathbf{c}\text{: } \mathbf{x}(u,\phi) = \mathbf{e}_1 a\cosh u\cos\phi + \mathbf{e}_2 a\cosh u\sin\phi + \mathbf{e}_3 au.$$

(Surface **b** is a helicoid and surface **c** is a catenoid.)

(a) For surfaces **a** and **b**, determine $\boldsymbol{\gamma}_u$ and $\boldsymbol{\gamma}_\phi$. Then use the results to determine the metric tensors for the two surfaces. (You should get different answers.)
(b) Use Fock–Ivanenko coefficients to determine formulas for the Gaussian curvatures of surfaces **a** and **b**. (You should get the same answer.)

(c) Demonstrate the fact that surfaces **a** and **b** are not isometric.

(d) Replace the "u" in the equation for surface **b** by "$\sinh u$" and then obtain the metric tensor for the helicoid in the new coordinate system. Obtain the metric tensor for the catenoid and thereby show that the helicoid and catenoid are isometric. (Surfaces that are not only isometric but can also be bent into one another without stretching are said to be *applicable*. The helicoid and catenoid are applicable. (See Figs. 11.2 and 11.3.).

Chapter 8
*Non-Euclidean (Hyperbolic) Geometry

8.1 *Early Developments

You should be forewarned that a prerequisite for this chapter is a strong familiarity with the basic manipulations of complex numbers – multiplication, the polar representation, and the notion of complex conjugate. The non-Euclidean geometry of Bolyai and Lobachevsky eventually became known as *hyperbolic geometry* because the ordinary trigonometric functions sine and cosine that appear in formulas for the surface of a sphere are replaced by the hyperbolic functions sinhϕ and coshϕ for surfaces of constant negative Gaussian curvature.

To get a taste of hyperbolic geometry, consider Fig. 8.1a. This is a slight modification of a diagram that appears in Dr. George Bruce Halsted's English translation of Lobachevsky's *Geometrische Untersuchungen zur Theorie der Parallellinien* (*Geometric Investigations on the Theory of Parallels*). The original work was a 61-page book published in Berlin in 1840. The 1891 translation is included as an 35-page appendix in H. S. Carslaw's translation of Roberto Bonola's *Non-Euclidean Geometry* (Bonola 1955). Nicholai Lobachevsky observed that if more than one straight line passing through point A failed to intersect line BC, then there would be two lines HH' and KK' that would approach BC asymptotically in opposite directions. Furthermore if line AD was perpendicular to BC, then the angle ϕ ($\angle HAD$) would be less than $\pi/2$ and would have a functional dependence on the length of AD, which I designate by q. He then demonstrated that by choosing a suitable unit of length, one has

$$\tan \frac{\phi(q)}{2} = e^{-q}. \tag{8.1}$$

(In the English version, this demonstration required 28 pages of nontrivial arguments.) Three pages later at the end of his treatise, Lobachevsky arrived at what can

J. Snygg, *A New Approach to Differential Geometry using Clifford's Geometric Algebra*, DOI 10.1007/978-0-8176-8283-5_8,

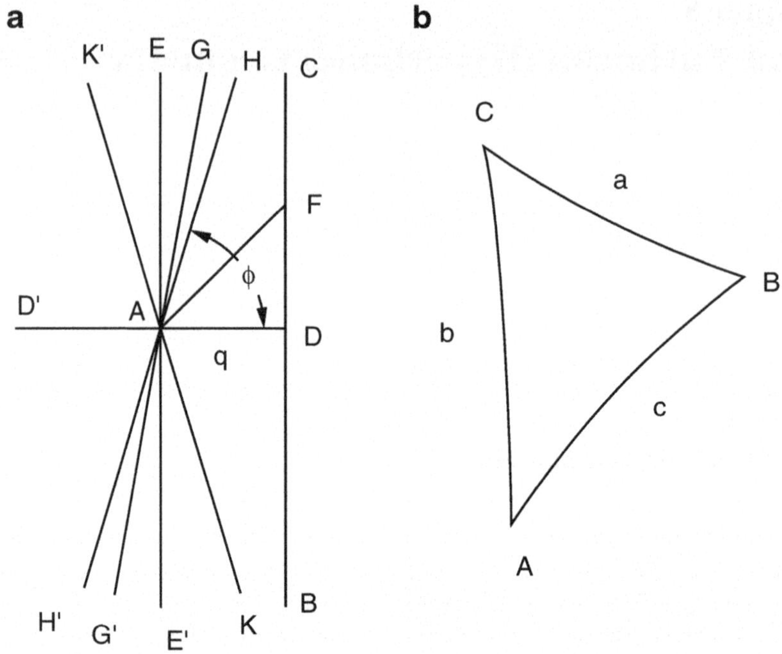

Fig. 8.1 (**a**) A diagram adapted from a work by Lobachevsky. (**b**) A triangle bounded by straight lines in the non-Euclidean geometry of Bolyai and Lobachevsky

now be identified as hyperbolic trigonometric formulas. Referring to Fig. 8.1b, two of these equations were

$$\sin A \tan\phi(a) = \sin B \tan\phi(b), \text{ and} \tag{8.2}$$

$$\cos A \cos\phi(b)\cos\phi(c) + \frac{\sin\phi(b)\sin\phi(c)}{\sin\phi(a)} = 1. \tag{8.3}$$

In the same year that Lobachevsky published his German treatise, Ferdinand Minding (Minding 1840, p. 324) investigating the nature of geodesic triangles on two-dimensional surfaces of constant negative curvature K, arrived at the formula:

$$\cos\left(a\sqrt{K}\right) = \cos\left(b\sqrt{K}\right)\cos\left(c\sqrt{K}\right) + \sin\left(b\sqrt{K}\right)\sin\left(c\sqrt{K}\right)\cos A. \tag{8.4}$$

This is identical to the law of cosines for the surface of a sphere that you hopefully derived in Prob. 58, when R is replaced by $1/\sqrt{K}$. Minding's achievement was to show that this equation remains valid for the case that K is negative and $\sqrt{K} = i\sqrt{|K|}$. In that case,

$$\cos\left(q\sqrt{K}\right) = \cosh\left(q\sqrt{|K|}\right), \text{ and} \tag{8.5}$$

$$\sin\left(q\sqrt{K}\right) = i\sinh\left(q\sqrt{|K|}\right). \tag{8.6}$$

(See Prob. 196.)

With these substitutions, (8.4) becomes

$$\cosh\left(a\sqrt{|K|}\right) = \cosh\left(b\sqrt{|K|}\right)\cosh\left(c\sqrt{|K|}\right) - \sinh\left(b\sqrt{|K|}\right)\sinh\left(c\sqrt{|K|}\right)\cos A. \tag{8.7}$$

It turns out that (8.7) is equivalent to (8.3) – the one derived by Lobachevsky. From (8.1), it can be shown that

$$e^{i\phi(q)} = \tanh q + \frac{i}{\cosh q}, \tag{8.8}$$

$$\sin\phi(q) = \frac{1}{\cosh q}, \tag{8.9}$$

$$\cos\phi(q) = \tanh q, \text{ and} \tag{8.10}$$

$$\tan\phi(q) = \frac{1}{\sinh q}. \tag{8.11}$$

(See Prob. 197.) With these substitutions, (8.2) and (8.3) become

$$\frac{\sin A}{\sinh a} = \frac{\sin B}{\sinh b} \text{ and} \tag{8.12}$$

$$\cosh a = \cosh b\cosh c - \sinh b\sinh c\cos A. \tag{8.13}$$

Equations (8.12) and (8.13) are, respectively, the Law of Sines and Law of Cosines for a surface of constant negative Gaussian curvature with $K = -1$. The fact that the constant $\sqrt{|K|}$ or some comparable constant does not appear in (8.12) and (8.13) stems from the fact that Lobachevsky's choice of unit length corresponds to selecting a system of measurement for which $K = -1$.

We see from a comparison of the work of Lobachevsky and Minding that the "straight lines" in the non-Euclidean geometry of Lobachevsky and Bolyai correspond to the geodesics on the surface of constant negative Gaussian curvature. However, this was not recognized for another 28 years. Robert Osserman has suggested that if Lobachevsky and Minding had read one another's papers, they might have recognized that they were writing about equivalent structures (Osserman 1995, p. 67 and pp. 186–187).

Why did not someone else see the coincidence and its consequence? It is true that Lobachevsky did not express his results in the form of (8.12) and (8.13). However, he did point out that if you replaced a, b, and c, respectively, by ia, ib, and ic then (8.2) and (8.3) would become

$$\frac{\sin A}{\sin a} = \frac{\sin B}{\sin b}, \text{ and} \tag{8.14}$$

$$\cos a = \cos b \cos c + \sin b \sin c \cos A. \tag{8.15}$$

Presumably for this reason, Lobachevsky used the term "Imaginary Geometry" to describe his work. He not only made this point in his short book in 1840 in German, but he also made the same point in Russian several years earlier. As a matter of fact in 1837, he had made the same point in a paper written in French entitled *Géométrie imaginaire* (Lobachevsky 1837, pp. 295–320). The 1837 paper by Lobachevsky and the 1840 paper of Minding were three years apart and written in two different languages (French and German) but they appeared in different volumes of the *same* journal!

The chief reason that this coincidence and its significance was not observed earlier was probably due to the fact that Bolyai and Lobachevsky had carried out their presentations in three dimensions and the nature of their arguments were totally unlike those of differential geometry. Few familiarized themselves with the work of Bolyai and Lobachevsky and those who did viewed their work as curiosities that were probably logically flawed. After Gauss died in 1855, mathematicians examined his correspondence and unpublished papers. It then became evident that Gauss had taken hyperbolic geometry seriously. As a result, the work of Bolyai and Lobachevsky was reexamined but even then the relation between their work and that of differential geometry was not self-evident to the mathematical community.

It was only when Eugenio Beltrami (1835–1900) attacked the problem of mapping geodesics from a curved 2-dimensional surface to straight lines in a plane, that the critical insight was made (Beltrami 1868, pp. 284–312). Even Beltrami did not recognize the full logical consequences of the identification of the "straight lines" of Bolyai and Lobachevsky with the geodesics on a surface of constant negative Gaussian curvature. From Beltrami's point of view, Bolyai and Lobachevsky had not introduced anything revolutionary. They had simply described the theory of geodesics on surfaces of negative curvatures using the term "straight line" for an entity that was not really "straight." Two years later in 1870, Guillaume Jules Hoüel (1823–1886) translated both Beltrami's paper and some work by Lobachevsky into French. In that publication, Hoüel pointed out that indeed Beltrami's observation demonstrated the independence of the parallel postulate (O'Connor and Robertson 2000).

Problem 196.

$$\sinh x = \frac{\exp(x) - \exp(-x)}{2} \quad \text{and} \quad \cosh x = \frac{\exp(x) + \exp(-x)}{2}.$$

(a) Use the definitions above to show that

$$\sin ix = i \sinh x \quad \text{and} \quad \cos ix = \cosh x.$$

(b) Determine the appropriate formulas for $\sinh(x \pm y)$, $\cosh(x \pm y)$, and then $\tanh(x \pm y)$. Note!

$$\tanh x = \frac{\sinh x}{\cosh x}.$$

Problem 197. (a) Use (8.1) to derive (8.8). Suggestion: Let $z = e^{i\phi/2}$ in (8.1) and then solve for z.

(b) Use the result of Part (a) to derive (8.9), (8.10), and (8.11).

8.2 *The Poincaré Model and Reflections

For many purposes, it is easier to use the formulation of Riemann to study hyperbolic geometry rather than the formulations of Bolyai, Lobachevsky, or Minding. Henri Poincaré (1854–1912) worked out the details of this approach. As a result, this approach is known as the *Poincaré model.* It is this approach that we will present in this chapter.

For 2-dimensions, the metric introduced by Riemann in his 1854 habilitationsvortrag address was

$$ds = \frac{\sqrt{(dx)^2 + (dy)^2}}{1 + \frac{\alpha}{4} r^2} \tag{8.16}$$

If you did Prob. 136, you know that the Gaussian curvature for this metric is

$$K = R^{12}{}_{12} = \alpha. \tag{8.17}$$

If K is negative, the domain of this space is limited to the region

$$x^2 + y^2 = r^2 < 4/|K|.$$

The equations are somewhat simpler if we use the "*unit disk*". That is the region for which $r < 1$. If we choose α to be -4, then we will get our desired unit disk. However, the results we would then obtain from this space would be limited to those for which the Gaussian curvature $K = -4$. We can remedy this situation by multiplying the right-hand side of (8.16) by some suitable constant β. This has the effect of multiplying the g_{ij}'s by β^2 and the g^{ij}'s by $1/\beta^2$. We note that

$$R^{\lambda}{}_{\eta\alpha\beta} = \frac{\partial}{\partial u^{\alpha}} \Gamma^{\lambda}_{\eta\beta} - \frac{\partial}{\partial u^{\beta}} \Gamma^{\lambda}_{\eta\alpha} + \Gamma^{\nu}_{\eta\beta} \Gamma^{\lambda}_{\nu\alpha} - \Gamma^{\nu}_{\eta\alpha} \Gamma^{\lambda}_{\nu\beta}, \text{ where}$$

$$\Gamma^{\lambda}_{\alpha\beta} = \frac{g^{\lambda\nu}}{2} \left[\frac{\partial g_{\lambda\alpha}}{\partial u^{\beta}} + \frac{\partial g_{\lambda\beta}}{\partial u^{\alpha}} - \frac{\partial g_{\alpha\beta}}{\partial u^{\nu}} \right].$$

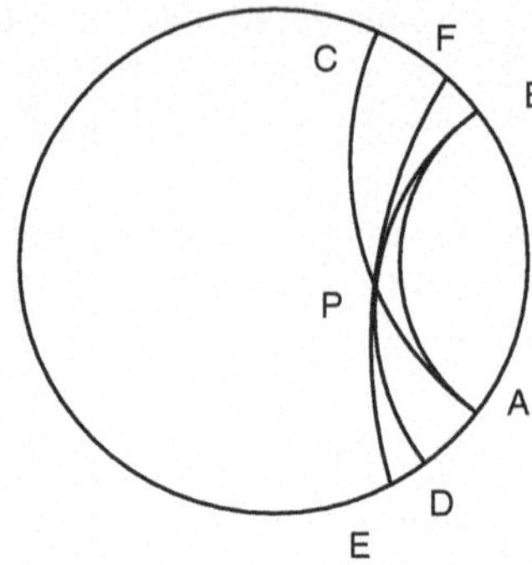

Fig. 8.2 Geodesics for the Poincaré model

Thus, $R^{\lambda}{}_{\eta\alpha\beta}$ is independent of β. However,

$$K = R^{12}{}_{12} = g^{2\alpha} R^{1}{}_{\alpha 12}.$$

Thus, multiplying the right-hand side of (8.16) by β results in a Gaussian curvature of $-4/\beta^2$. If we wish this to be K, then we must choose β to be $2/\sqrt{|K|}$. Thus are desired metric will be

$$ds = \frac{2}{\sqrt{|K|}} \frac{\sqrt{(dx)^2 + (dy)^2}}{1 - r^2}, \text{ where}$$
$$r^2 = x^2 + y^2. \tag{8.18}$$

With some difficulty, it can be shown that the geodesics for this space are either diameters or circles that intersect the boundary ($x^2 + y^2 = 1$) at a right angle. (See Prob. 199.) It should be noted that the angle at which two curves intersect is defined to be the angle between the tangents at the point of intersection.

The nature of "*parallel lines*" in this context is quite different than what we encounter in Euclidean geometry. (See Fig. 8.2.) The "line" AC and "line" EB are each said to be parallel to "line" AB but in different directions. Lines AC and EB intersect AB but only at infinity. Line DF does not intersect AB even at infinity. With this thought in mind, DF is said to be *ultra parallel* to AB.

On a surface of constant Gaussian curvature, one can slide a polygon around without stretching. The same is true for an observer using a Riemannian ruler on the Poincaré disk. Quite clearly such translations and rotations in the Poincaré disk would look quite differently to an observer looking at the disk from a Euclidean point of view. The Euclidean observer would agree that the angles at each vertex would not change. However, the Euclidean observer would see the lengths of the geodesics connecting adjacent vertices change while the Riemannian observer would see that the lengths of the "straight lines" joining adjacent vertices remain the same. How would these motions be described by a Euclidean observer?

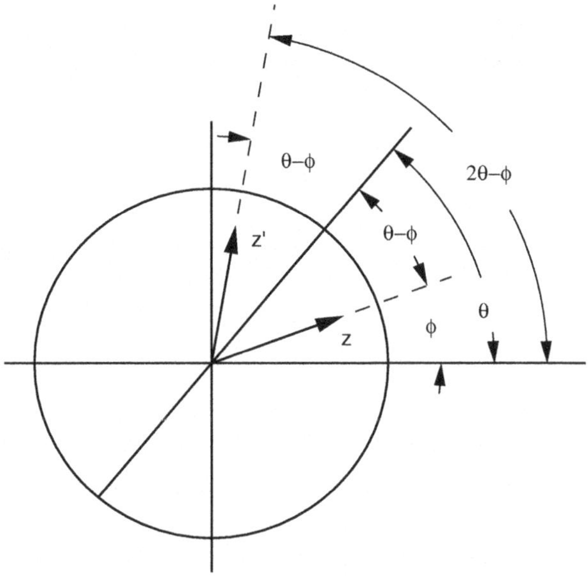

Fig. 8.3 z' is the reflection of z with respect to a diameter

In Euclidean geometry, distance preserving motions are composed of reflections, translations, and rotations. These motions all have their analogs in the Poincaré model. With that thought in mind, we will start with a discussion of reflections. In two-dimensional Euclidean geometry, one reflects a point with respect to a straight line. In the Poincaré model, one reflects a point with respect to a geodesic.

First consider the easy case where the geodesic is a diameter. (See Fig. 8.3.) Suppose the point $z = re^{i\phi}$ and we wish to reflect it with respect to the diameter $y = x \tan\theta$. The angle between the diameter and the vector associated with z is $\theta - \phi$. To get the reflected version z' of z, we need to increase the argument of z by $2(\theta - \phi)$. Thus, the new argument will be $\phi + 2(\theta - \phi) = 2\theta - \phi$. Therefore

$$z' = re^{2i\theta - i\phi} = re^{2i\theta}e^{-i\phi}.$$

However, $re^{-i\phi}$ is the complex conjugate of $re^{i\phi}$, so this last equation becomes

$$z' = e^{2i\theta}z^*, \text{ where } z^* \text{ is the complex conjugate of } z. \tag{8.19}$$

Now suppose the geodesic is a circle. In the context of circles, the analog of a reflection is not so obvious. In the context of circles, the analog of a reflection is known as an *inversion*(See Fig. 8.4.). To determine the inverse of a point z with respect to a circle centered at c, consider the straight line passing through c and z. The *inverse* z' of z lies on this straight line with the requirement that z and z' lie on the same side of c and $|z - c|\,|z' - c| = r^2$, where r is the radius of the circle. Observe that if z is inside the circle, then $|z - c| < r$ and $|z' - c| > r$, so z' is outside the circle. Similarly, if z is outside the circle, then its inverse z' lies inside the circle.

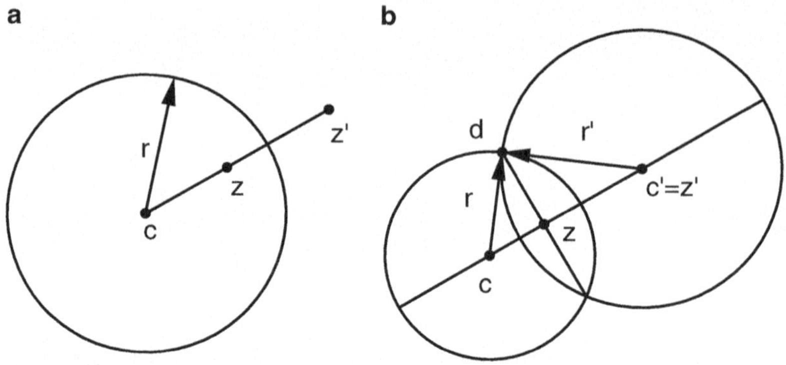

Fig. 8.4 In both figures, z' is the inverse of z

If z lies inside the circle, a compass and ruler construction for z' is illustrated in Fig. 8.4b. Point d is located by constructing a line perpendicular to the line cz passing through point z. Once line cd is drawn, c' is located by constructing a line perpendicular to cd passing through d. Since every triangle in sight is similar to one another,

$$\frac{|z-c|}{r} = \frac{r}{|c'-c|} \quad \text{and thus}$$
$$|z-c|\,|c'-c| = r^2.$$

Therefore, c' is the inverse of z with respect to the circle centered at point c. By a similar argument, one sees that point c is the inverse of point z with respect to the circle centered at point c'. One can also determine the inverse of z by compass and ruler construction if z lies outside its reference circle. (See Prob. 200.)

As one would expect with a "reflection", when the operation of inversion is applied a second time, a point is returned to its initial location.

In addition, if the center of the circle is at $(0,0)$, then

$$z' = r^2 z/\left|z^2\right| = r^2 z/zz^* = r^2/z^*. \tag{8.20}$$

An important feature of inverse points is described in the following theorem:

Theorem 198. *If A and A' are inverse points with respect to some reference circle, then any circle passing through both A and A' intersects the reference circle at right angles. Furthermore if a circle passing through A intersects the reference circle at right angles, then that circle must also pass through A'.*

Proof. Consider Fig. 8.5. To prove the first statement, suppose A and A' are inverse points and some circle passing through these two points intersects the reference circle at point B. Since A and A' are inverse points:

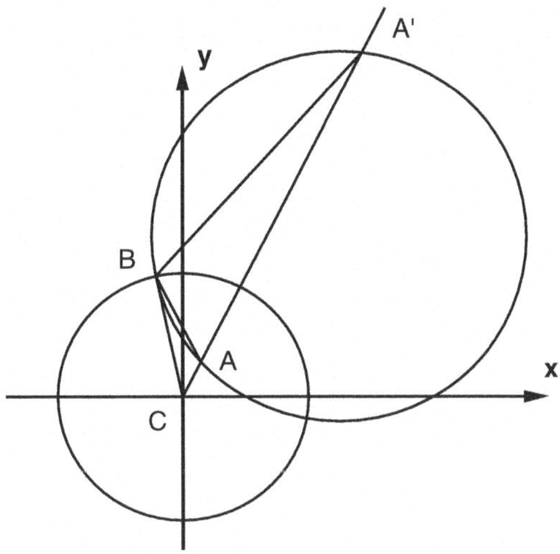

Fig. 8.5 $\angle ABC = \angle BA'C$. Why?

$$AC \cdot A'C = (BC)^2 \text{ and thus}$$

$$\frac{AC}{BC} = \frac{BC}{A'C} \tag{8.21}$$

This implies that $\triangle ACB$ is similar to $\triangle BCA'$ since they both share $\angle C$ and the sides adjacent to $\angle C$ are in proportion. Thus, $\angle CA'B = \angle CBA$. Since $\angle CA'B = (1/2)\text{arc}AB$, it follows that $\angle CBA$ is also equal to $(1/2)\text{arc}AB$. But that implies that BC must be tangent to the circle passing through A and A'. This, in turn, implies the two circles intersect at right angles.

To prove the second part, consider the case where A lies inside the reference circle. Then suppose the circle passing through A intersects the reference circle at right angles at point B. We wish to show that A' lies on this same circle. The same two triangles are similar for the reasons stated above. $\angle CBA = (1/2)\text{arc}AB$ because BC is tangent to the circle passing through A. Therefore, $\angle CA'B$ must also equal $(1/2)\text{arc}AB$ which implies A' lies on the circle.

Finally, suppose A lies outside the reference circle. To avoid drawing another figure, call this point A' and its inverse A. Consider a circle passing through A' that intersects the reference circle at right angles. Let D designate the point inside the reference circle where the circle passing through A' intersects $A'C$. We wish to show that D, not shown in the diagram, coincides with A. The same triangles mentioned above are again similar so $\angle CBA = \angle CA'B$. $\angle CBD$ and $\angle CA'B$ are both equal to $(1/2)arcAB$. Thus $\angle CBD = \angle CBA$. Since both A and D lie on line $A'C$, these two angles cannot be equal unless D coincides with A. □

How can a reflection with respect to a geodesic circle be carried out algebraically? Consider Fig. 8.6. Suppose $\text{arc}AB$ is a geodesic in the unit Poincaré disk centered at $(0, 0)$. Let us see how the reflection process works out. Suppose the complex number a is located at the center of the geodesic circle. The translation

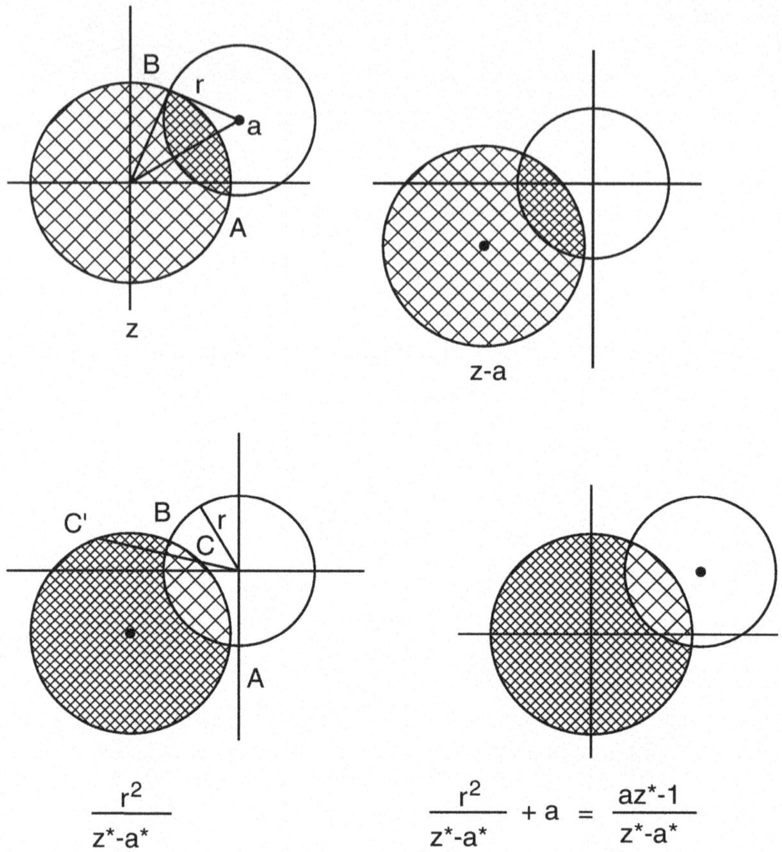

Fig. 8.6 Reflection with respect to a geodesic

$z \to z - a$ will translate the center of the geodesic circle to the origin. If we now carry out the inversion with respect to the geodesic circle, we have

$$z \to z - a \to \frac{r^2}{z^* - a^*}.$$

From Theorem 198, this step maps point C onto C' and C' onto C. More generally, arcACB is mapped onto $AC'B$ and vice versa. Furthermore, it is clear that the coarsely cross hatched region is mapped onto the finely cross hatched region and vice versa.

If we now translate this last configuration by adding a, we will return the center of the Poincaré disk to $(0, 0)$. Summarizing, we now have

$$z \to z - a \to \frac{r^2}{z^* - a^*} \to \frac{r^2}{z^* - a^*} + a = \frac{az^* - aa^* + r^2}{z^* - a^*}$$

From Fig. 8.6, it is clear from the Pythagorean Theorem that

$$aa^* - r^2 = |a|^2 - r^2 = 1.$$

Thus, the reflection with respect to the geodesic arc centered at a is

$$\alpha(z) = \frac{az^* - 1}{z^* - a^*}. \tag{8.22}$$

This formula is not unique. If you multiply both the denominator and the numerator by the same complex number, the image points will remain the same. However for the form shown with the coefficient of z^* in the denominator equal to 1, one can read off the coordinates of the center of the geodesic circle. We should note that the choice of a is not completely arbitrary. Since the center of a geodesic circle must lie outside the unit disk, we require that $|a| > 1$.

Problem 199. (Arduous)

(a) Show that the geodesics on Poincaré's disk are the arcs of circles that intersect the boundary at right angles as shown in Fig. 8.2. Suggestion: From (8.18),

$$\begin{aligned}
d\mathbf{s} &= \frac{2}{\sqrt{|K|}(1-r^2)}(\mathbf{e}_1 dx + \mathbf{e}_2 dy) \text{ and thus} \\
\mathbf{t}(s) &= \frac{d\mathbf{s}}{ds} = \frac{2}{\sqrt{|K|}(1-r^2)}(\mathbf{e}_1\dot{x} + \mathbf{e}_2\dot{y}) = \boldsymbol{\gamma}_1\dot{x} + \boldsymbol{\gamma}_2\dot{y} \text{ and} \\
\mathbf{n}(s) &= \mathbf{t}(s)\mathbf{e}_{12} = \frac{2}{\sqrt{|K|}(1-r^2)}(-\mathbf{e}_1\dot{y} + \mathbf{e}_2\dot{x}), \text{ where} \\
\dot{x} &= \frac{dx}{ds}, \text{ and } \dot{y} = \frac{dy}{ds}.
\end{aligned} \tag{8.23}$$

Here, s represents arc length for someone using a Riemannian measuring stick corresponding to Riemann's metric tensor for the disk. For someone using a Euclidean metric stick and using u to represent arc length:

$$\begin{aligned}
\mathbf{t}(s) &= \mathbf{e}_1\frac{dx}{du} + \mathbf{e}_2\frac{dy}{du} \text{ and} \\
\mathbf{n}(s) &= -\mathbf{e}_1\frac{dy}{du} + \mathbf{e}_2\frac{dx}{du}.
\end{aligned} \tag{8.24}$$

Comparing (8.23) and (8.24), it is clear that

$$\frac{ds}{du} = \frac{2}{\sqrt{|K|}(1-r^2)}.$$

Now since

$$\frac{d\mathbf{t}(u)}{du} = k(u)\mathbf{n}(u),$$

we can compute the curvature for an arbitrary curve. Using (5.50) in Prob. 79, you can determine the differential equations for the geodesics and use them to eliminate $\ddot{x}$ and $\ddot{y}$ in the formula you obtained for $k(u)$. You should then have

$$k = \frac{4(y\dot{x} - xy)}{\sqrt{|K|}(1 - r^2)^2}. \tag{8.25}$$

To show k is constant, show $dk/du = 0$ or $dk/ds = 0$. To do this, you will need to use the differential equations for geodesics again.

(b) From (8.25), you should be able to show that for an observer looking at the disk from an Euclidean point of view, the geodesic circular arcs strike the rim of the disk at right angles. Note! A tangent to the disk at the point of intersection has direction $(y, -x)$. (For an observer looking at the disk from a Riemannian point of view, the rim of the disk is at infinity.)

Problem 200. Suppose point z lies outside a circle centered at point c. Show how the inverse of z may be located by a compass and ruler construction. Hint: If one side of a triangle inscribed in a circle is a diameter, the angle opposite the diameter is a right angle. Why?

Problem 201. Suppose

$$\alpha(z) = \frac{az^* - 1}{z^* - a^*} \text{ and } \beta(z) = \frac{bz^* - 1}{z^* - b^*}.$$

Express $\alpha \circ \beta(z) = a(\beta(z))$ as a simple fraction.

8.3 *Direct Non-Euclidean Transformations

In 2-dimensional Euclidean geometry, the composition of two reflections is a rotation. What happens when we take the composition of two reflections in the Poincaré model? Suppose

$$\alpha(z) = \frac{az^* - 1}{z^* - a^*} \text{ and } \beta(z) = \frac{bz^* - 1}{z^* - b^*}.$$

If we compute

$$w(z) = \alpha(\beta(z)), \text{ we get}$$

$$w(z) = \frac{(ab^* - 1)z + (b - a)}{(b^* - a^*)z + (a^*b - 1)} \tag{8.26}$$

(If you have not done Prob. 201, you should do so now.)

Equation (8.26) is not the most general composition of two reflections. It does not include the rotation about the origin resulting from two reflections about diameters. It turns out that one needs to adjust the form of (8.26) only slightly to get the most general composition of two reflections. Namely

$$w(z) = \frac{cz + d}{d^*z + c^*}, \tag{8.27}$$

where c and d are arbitrary complex constants with one condition. The format of (8.27) is restrictive enough to guarantee that the boundary of the unit disk is mapped onto itself. (See Prob. 205.) However, it is not restrictive enough to guarantee that the interior of the unit disk is not mapped into the exterior and vice versa. The added restriction is that the image of (0, 0) must lie inside the unit disk. We note that

$$w(0) = \frac{d}{c^*}, \text{ so we require that}$$

$$|d| < |c|. \tag{8.28}$$

To demonstrate that (8.27) always represents the composition of two reflections, we will prove the following theorem:

Theorem 202. *If*

$$w(z) = \frac{cz + d}{d^*z + c^*}, \text{ where}$$

$$|d| < |c|,$$

then $w(z)$ is the composition of two reflections, at least one of which can be chosen to be a reflection with respect to a diameter.

Proof. Case 1: $d = 0$. In this case,

$$w(z) = \frac{c}{c^*}z.$$

If $c = re^{i\theta}$, then our equation becomes

$$w(z) = e^{2i\theta}z.$$

This is a rotation of angle 2θ about the origin. It results from two reflections with respect to diameters, which intersect with an angle of θ between them. (You may wish to review Fig. 2.2 in Sect. 1 of Chap. 2.)

Case 2: $d \neq 0$. For this case, we will demonstrate that $w(z)$ can be written as the composition of two successive reflections β and then α such that β is a reflection with respect to a diameter. If

$$w(z) = \alpha(\beta(z)), \text{ then}$$

$$\beta(z) = \alpha^{-1}(w(z)). \tag{8.29}$$

If α is a reflection, then $\alpha^{-1} = \alpha$. If α is a reflection, it can be written in the form;

$$\alpha(z) = \frac{az^* - 1}{z^* - a^*}.$$

With these observations, (8.29) becomes

$$\beta(z) = \frac{aw^* - 1}{w^* - a^*} = \frac{a\frac{c^*z^*+d^*}{dz^*+c} - 1}{\frac{c^*z^*+d^*}{dz^*+c} - a^*} = \frac{ac^*z^* + ad^* - dz^* - c}{c^*z^* + d^* - a^*dz^* - a^*c}$$

$$= \frac{(ac^* - d)z^* + (ad^* - c)}{(c^* - a^*d)z^* + (d^* - a^*c)}. \tag{8.30}$$

Since $d \neq 0$, we can set

$$a = c/d^*.$$

(It is required that $|a| > 1$. Why is that condition satisfied?) With this value for a, (8.30) becomes

$$\beta(z) = \frac{\left(\frac{cc^*}{d^*} - d\right) z^*}{\left(d^* - \frac{cc^*}{d}\right)} = \frac{\frac{1}{d^*}(cc^* - dd^*)\, z^*}{\frac{1}{d}(dd^* - cc^*)} = -\frac{d}{d^*} z^*. \tag{8.31}$$

Note! $(cc^* - dd^*) \neq 0$. Why?)

If $d = re^{i\theta}$, (8.31) becomes

$$\beta(z) = e^{i(2\theta + \pi)} z^*.$$

A review of the discussion preceding (8.19) reveals that this is a reflection with respect to the diagonal defined by the equation

$$y = x \tan\left(\theta + \frac{\pi}{2}\right).$$

To summarize: If

$$w(z) = \frac{cz + d}{d^*z + c^*}, \text{ with } d \neq 0, \text{ then}$$

$$w(z) = \alpha(\beta(z)), \text{ where}$$

$$\alpha(z) = \frac{(c/d^*)z^* - 1}{z^* - (c^*/d)}, \text{ and}$$

$$\beta(z) = -\frac{d}{d^*} z^*.$$

□

The composition of an even number of reflections is defined to be a *direct non-Euclidean transformation*, while the composition of an odd number of reflections is defined to be an *indirect non-Euclidean transformation*.

Theorem 203. *For a 2-dimensional Poincaré disk, a direct non-Euclidean transformation can be decomposed into the product of two reflections.*

Proof. We only need to show that the composition of four reflections can be reformulated as the composition of two reflections. That is if

$$w(z) = \frac{az+b}{b^*z+a^*} \text{ and } g(z) = \frac{cz+d}{d^*z+c^*}, \text{ where}$$
$$|b| < |a| \text{ and } |d| < |c|,$$

then

$$g(w(z)) = \frac{pz+q}{q^*z+p^*}, \text{ where } |q| < |p|.$$

Carrying out the obvious calculations, we get

$$g(w(z)) = \frac{(ac+b^*d)z + (a^*d+bc)}{(ad^*+b^*c^*)z + (a^*c^*+bd^*)}.$$

We note that this has the required form where $p = ac + b^*d$ and $q = a^*d + bc$. To show that $|p| > |q|$, we note that

$$\begin{aligned}|p|^2 - |q|^2 = pp^* - qq^* &= (ac+b^*d)(a^*c^*+bd^*) - (a^*d+bc)(ad^*+b^*c^*)\\ &= aa^*cc^* + abcd^* + a^*b^*c^*d + bb^*dd^* - aa^*dd^* - a^*b^*c^*d\\ &\quad - abcd^* - bb^*cc^*\\ &= aa^*cc^* + bb^*dd^* - aa^*dd^* - bb^*cc^*\\ &= (aa^* - bb^*)(cc^* - dd^*) > 0.\end{aligned}$$

□

Corollary 204. *For the 2-dimensional Poincaré disk, an indirect non-Euclidean transformation is either a single reflection or the composition of three reflections.*

Problem 205. Suppose

$$w(z) = \frac{az+b}{b^*z+a^*}, \text{ where}$$

$aa^* - bb^* \neq 0$. Show that $ww^* = 1$ implies $zz^* = 1$

Problem 206. The proof of Theorem 202 demonstrates that if

$$w(z) = \frac{cz+d}{d^*z+c^*}, \text{ where } 0 < |d| < |c|$$

then

$$w(z) = \alpha(\beta(z)), \text{ where } \beta(z) = -\frac{d}{d^*}z^*.$$

Determine the reflection operator $\bar{\alpha}(z)$ that has the property that

$$w(z) = \beta(\bar{\alpha}(z)), \text{ where again } \beta(z) = -\frac{d}{d*}z^*.$$

Problem 207. A direct non-Euclidean transformation results from the composition of two reflections. What is the nature of the fixed points when the reflections are done with respect to two geodesics which are

(a) Ultra parallel?
(b) Parallel?
(c) Intersecting?

In each case, given the two geodesics, describe how you can locate the fixed point or fixed points. Do this when both of the geodesics are arcs of circles and also when one of the geodesics is a diameter.

8.4 *Möbius Transformations

The group of direct non-Euclidean transformations is a subgroup of a wider class of mappings known as *Möbius transformations*. A Möbius transformation is a mapping $w(z)$ such that

$$w(z) = \frac{az + b}{cz + d}, \text{ where} \tag{8.32}$$

$a, b, c,$ and d are arbitrary complex constants with the condition that $ad - bc \neq 0$. From Prob. 210, the group of Möbius transformations corresponds to the group of 2×2 matrices with nonzero determinant. In this context, the group of Möbius transformations is usually identified as the group known as the *general linear transformations* for 2 dimensions – the group of 2x2 nonsingular matrices acting on the two-dimensional vector space for the field of complex numbers. (Technically, this is not quite true. If we multiply the numerator and denominator of the right-hand side of (8.32) by some nonzero complex constant, we do not change the image points of the function $w(z)$. This means that we could adjust $a, b, c,$ and d so that

$$\det\begin{bmatrix} a & b \\ c & d \end{bmatrix} = 1.$$

Therefore, the group of Möbius transformations is actually isomorphic to the group of *special linear transformations* – the general linear transformations whose determinant is 1.)

An important property of Möbius transformations is that they are *conformal*. That is, they preserve angles. This is a consequence of the following theorem:

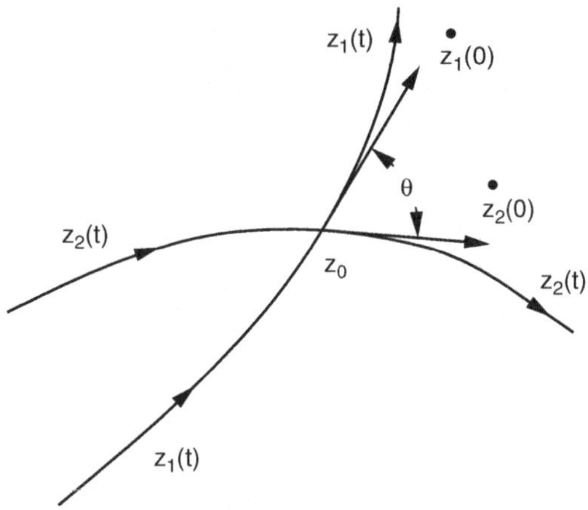

Fig. 8.7 The angle θ is preserved under a general linear transformation

Theorem 208. *Suppose the function $w(z)$ is such that dw/dz is well defined in the neighborhood of z_0 and*

$$\left.\frac{dw}{dz}\right|_{z=z_0} \neq 0, \text{ then}$$

any angle with vertex at $z = z_0$ is preserved by the mapping $w(z)$.

Proof. Consider two curves $z_1(t)$ and $z_2(t)$ that are both parameterized so that $z_1(0) = z_2(0) = z_0$. (See Fig. 8.7.) If $\dot{z}_k(t) = dz_k/dt$, then $\dot{z}_k(0)$ is tangent to $z_k(t)$ at $t = 0$ for $k = 1$ or 2. To obtain the angle θ between the two curves, we note that

$$\dot{z}_k(t) = |\dot{z}_k(t)| \exp(i\alpha_k(t)), \text{ where}$$

α_k is said to be the *argument* or *amplitude* of $\dot{z}_k(t)$. Clearly, $\theta = \alpha_1(0) - \alpha_2(0)$, so that

$$\theta = \arg\left(\frac{\dot{z}_1(0)}{\dot{z}_2(0)}\right).$$

The corresponding angle in the w plane is

$$\arg\left(\frac{\dot{w}(z_1(0))}{\dot{w}(z_2(0))}\right).$$

But

$$\dot{w}(z_k(t)) = \frac{dw}{dz_k}\frac{dz_k}{dt}, \text{ so}$$

$$\arg(\dot{w}(z_k(0))) = \arg\left(\left.\frac{dw}{dz}\right|_{z=z_0}\right) + \arg(\dot{z}_k(0))$$

$$= \arg\left(\left.\frac{dw}{dz}\right|_{z=z_0}\right) + \alpha_k(0).$$

Thus,

$$\arg\left(\frac{\dot{w}(z_1(0))}{\dot{w}(z_2(0))}\right) = \left[\arg\left(\left.\frac{dw}{dz}\right|_{z=z_0}\right) + \alpha_1(0)\right] - \left[\arg\left(\left.\frac{dw}{dz}\right|_{z=z_0}\right) + \alpha_2(0)\right]$$
$$= \alpha_1(0) - \alpha_2(0) = \theta.$$

This is what we set out to prove. □

You should note that the proof of the theorem breaks down if

$$\left.\frac{dw}{dz}\right|_{z=z_0} = 0, \text{ since}$$

in that circumstance

$$\arg\left(\left.\frac{dw}{dz}\right|_{z=z_0}\right) \text{ becomes meaningless.}$$

To show that the theorem applies to general linear transformations, you should check that $dw/dz \neq 0$, when

$$w(z) = \frac{az+b}{cz+d}.$$

(See Prob. 211.)

An obvious corollary is that direct non-Euclidean transformations are conformal.

If $\alpha(z)$ is an indirect non-Euclidean transformation, then $d\alpha/dz$ is not well defined. However, an indirect non-Euclidean transformation is the composition of a Möbius transformation for which the derivative with respect to z is well defined followed by the operation of complex conjugation. The operation of complex conjugation changes the signs of the arguments of all complex numbers and thereby reverses the orientation for the angles between curves at their point of intersection. Thus, an indirect non-Euclidean transformation preserves the magnitude of angles but reverses their orientation.

Problem 209. In the opening lines of this section, we defined a Möbius transformation as a mapping of the form:

$$w(z) = \frac{az+b}{cz+d}, \text{ with}$$

the restriction that $ad - bc \neq 0$. Why the restriction?

Problem 210. Suppose

$$\zeta(z) = \frac{ew(z)+f}{gw(z)+h}, \text{ and } w(z) = \frac{az+b}{cz+d}.$$

(a) Show that

$$\zeta(w(z)) = \frac{\alpha z + \beta}{\gamma z + \delta}, \text{ where}$$

$$\begin{bmatrix} \alpha & \beta \\ \gamma & \delta \end{bmatrix} = \begin{bmatrix} e & f \\ g & h \end{bmatrix} \begin{bmatrix} a & b \\ c & d \end{bmatrix}$$

(b) If

$$w(z) = \frac{az + b}{cz + d}, \text{ where } ad - bc = 1, \text{ what is } z(w)?$$

Problem 211. Suppose

$$w(z) = \frac{az + b}{bz + d}.$$

Compute dw/dz to demonstrate that $dw/dz \neq 0$.

8.4.1 *The Cross Ratio

It will be useful to note that a general linear transformation has the property that circles or straight lines are mapped onto circles or straight lines. To demonstrate this property, it is useful to use the concept of *cross ratio*. According to Howard Ewes, this is a concept that was investigated by the ancient Greeks. In the nineteenth century interest in this topic was reinvigorated and August Ferdinand Möbius (1790–1868) worked out many details in the context of projective geometry. It is interesting to note that for a short time in 1813, Möbius studied some astronomy under Gauss at Göttingen. Like Gauss, Möbius spent his entire professional life in academia as an astronomer and is better known for his contributions to mathematics. In mainland Europe, various labels were used but it was William Kingdon Clifford who coined the English term "cross ratio" in 1878 (Ewes 1963, pp. 86–87).

The cross ratio is constructed from four points. Given four points z_1, z_2, z_3, and z_4 in the complex Argand plane then the *cross ratio* is

$$\frac{(z_3 - z_1)(z_4 - z_2)}{(z_3 - z_2)(z_4 - z_1)}.$$

It can be shown that the cross ratio is preserved under a general linear transformation. That is if

$$w_k = \frac{az_k + b}{cz_k + d} \text{ for } k = 1, 2, 3, \text{ and } 4, \text{ then}$$

$$\frac{(w_3 - w_1)(w_4 - w_2)}{(w_3 - w_2)(w_4 - w_1)} = \frac{(z_3 - z_1)(z_4 - z_2)}{(z_3 - z_2)(z_4 - z_1)}. \tag{8.33}$$

(See Prob. 215.)

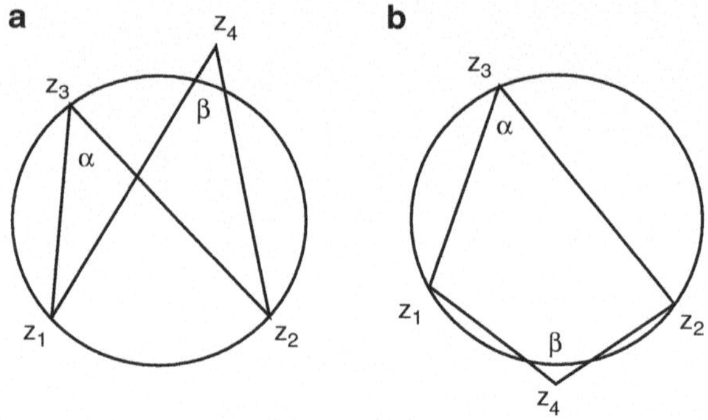

Fig. 8.8 (**a**) z_4 lies on circle $\iff \alpha = \beta$. (**b**) z_4 lies on circle $\iff \alpha + \beta = \pi$.

A useful theorem for our purposes is as follows:

Theorem 212. *The cross ratio is real if and only if the four points on the Argand plane lie on a circle or on a straight line.*

Proof. Case (1) The points z_1, z_2, and z_3 are not colinear. In this case, there is a unique circle that passes through the three points. Referring to Fig. 8.8a,

$$\frac{z_3 - z_1}{z_3 - z_2} = \left|\frac{z_3 - z_1}{z_3 - z_2}\right| \exp(-i\alpha), \text{ and } \frac{z_4 - z_2}{z_4 - z_1} = \left|\frac{z_4 - z_2}{z_4 - z_1}\right| \exp(i\beta).$$

Thus,

$$\frac{(z_3 - z_1)(z_4 - z_2)}{(z_3 - z_2)(z_4 - z_1)} = \left|\frac{(z_3 - z_1)(z_4 - z_2)}{(z_3 - z_2)(z_4 - z_1)}\right| \exp(i\beta - i\alpha), \text{ so}$$

$$\frac{(z_3 - z_1)(z_4 - z_2)}{(z_3 - z_2)(z_4 - z_1)} \text{ is real}$$

if and only if $\alpha = \beta$ and z_4 lies on the circle.

For Fig. 8.8b,

$$\frac{z_3 - z_1}{z_3 - z_2} = \left|\frac{z_3 - z_1}{z_3 - z_2}\right| \exp(-i\alpha), \text{ and } \frac{z_4 - z_2}{z_4 - z_1} = \left|\frac{z_4 - z_2}{z_4 - z_1}\right| \exp(-i\beta).$$

Thus,

$$\frac{(z_3 - z_1)(z_4 - z_2)}{(z_3 - z_2)(z_4 - z_1)} = \left|\frac{(z_3 - z_1)(z_4 - z_2)}{(z_3 - z_2)(z_4 - z_1)}\right| \exp(-i\alpha - i\beta), \text{ so}$$

$$\frac{(z_3 - z_1)(z_4 - z_2)}{(z_3 - z_2)(z_4 - z_1)} \text{ is real}$$

if and only if $\alpha + \beta = \pi$ and z_4 lies on the circle.

Starting with z_1 and moving clockwise around the circle determined by z_1, z_2, and z_3, it is easy to see that there are six possible sequences to consider. However, you may be able to demonstrate that taking the mirror image of a given sequence results in taking the complex conjugate of the cross ratio. In that sense, there is only one more case to deal with. (See Prob. 216.)

Case (2) The points z_1, z_2, and z_3 lie on a straight line. I will let you deal with that case. You should note that if z_4 does not lie on the same line, you can construct a circle through z_2, z_3, and z_4. □

Corollary 213. *General linear transformations map circles or straight lines into circles or straight lines.*

Proof. z lies on the circle or straight line passing through z_1, z_2, and z_3 if and only if

$$\frac{(z_3 - z_1)(z - z_2)}{(z_3 - z_2)(z - z_1)} = \text{some real number } \alpha.$$

From Prob. 215, if $w(z)$ is a general linear transformation, then

$$\frac{(w(z_3) - w(z_1))(w(z) - w(z_2))}{(w(z_3) - w(z_2))(w(z) - w(z_1))} = \alpha.$$

Thus, any point on the circle or straight line passing through z_1, z_2, and z_3 is mapped onto the circle or straight line passing through $w(z_1)$, $w(z_2)$, and $w(z_3)$. Furthermore, it is clear that any point on the circle or straight line passing through $w(z_1)$, $w(z_2)$, and $w(z_3)$ has a preimage on the circle or straight line passing through z_1, z_2, and z_3. □

Corollary 214. *Direct non-Euclidean transformations map geodesics onto geodesics.*

Proof. From Corollary 213, a direct non-Euclidean transformation maps geodesics onto straight lines or circles. Since direct non-Euclidean transformations are conformal, the image of a geodesic must meet the boundary of the unit disk at a right angle. Thus, the image of a geodesic must be a geodesic. □

8.4.2 *Fixed Points

The nature of a particular direct non-Euclidean transformation depends substantially on its fixed points. If

$$w(z) = \frac{az + b}{b^*z + a^*}, \text{ then}$$

z is a fixed point if

$$z = \frac{az + b}{b^*z + a^*}.$$

This equation is equivalent to the quadratic equation;

$$b^* z^2 + (a^* - a)z - b = 0.$$

If the discriminant is not zero, we get two roots:

$$z_+ = \frac{(a - a^*) + \sqrt{(a - a^*)^2 + 4bb^*}}{2b^*} \text{ and } z_- = \frac{(a - a^*) - \sqrt{(a - a^*)^2 + 4bb^*}}{2b^*}.$$

If we take the product of the two roots, we get

$$z_+ z_- = \frac{(a - a^*)^2 - \left[(a - a^*)^2 + 4bb^*\right]}{4(b^*)^2} = \frac{-b}{b^*}.$$

Thus,

$$|z_+|\,|z_-| = |z_+ z_-| = \left|\frac{-b}{b^*}\right| = 1.$$

It then follows that either both of the fixed points lie on the boundary of the unit disk or one fixed point lies inside the boundary and the other lies outside.

If the discriminant $(a-a^*)^2+4bb^*$ is zero, we get a single fixed point. We should note that $a - a^*$ is pure imaginary so $(a - a^*)^2$ is a negative real number, whereas $4bb^*$ is a positive real number. Thus, if the discriminant is zero, $|a - a^*| = |2b|$. Thus, for a single fixed point

$$|z| = \left|\frac{a - a^*}{2b^*}\right| = 1.$$

Therefore, when one has a single fixed point, it lies on the boundary of the Poincaré disk.

The nature of the mapping corresponding to each type of fixed point configuration is shown in Fig. 8.9.

In all three diagrams, a geodesic represented by an unbroken arc of a circle or a straight line is mapped onto the neighbor labeled with the next higher integer. For example in each diagram, the geodesic labeled with -1 is mapped onto the geodesic labeled 0; the geodesic 0 is mapped onto the geodesic 1 and so forth. (The inverse mappings would move the indicated geodesics in the opposite direction.)

Using a Riemannian ruler, all geodesic segments bounded by adjacent grid points are equal in all three diagrams. (A boundary point should not be counted as a "grid point." If one end of a geodesic lies on the boundary, it has infinite length for someone using a Riemannian ruler.)

The mapping in diagram A is called a *translation*. The mapping in diagram B is called a *limit rotation*. And the mapping in diagram C is called a *rotation*.

Problem 215. Verify (8.33). That is, verify the fact that the cross ratio is preserved under a general linear transformation.

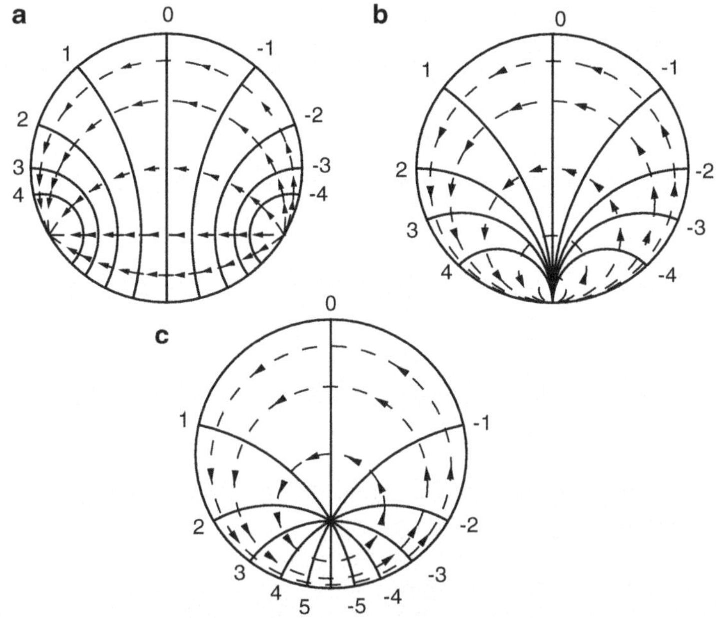

Fig. 8.9 (**a**) Translation. (**b**) Limit Rotation. (**c**) Rotation

Problem 216. Determine the missing sequence or sequences for case 1) in Theorem 212 and complete the proof for that case. Also prove the theorem for case 2).

Problem 217. Suppose we designate the cross ratio

$$\frac{(z_3 - z_1)(z_4 - z_2)}{(z_3 - z_2)(z_4 - z_1)} \quad \text{by} \quad (z_1 z_2 z_3 z_4).$$

It is not difficult to show that

$$\frac{(z_3 - z_1)(z_4 - z_2)}{(z_3 - z_2)(z_4 - z_1)} = \frac{(z_1 - z_3)(z_2 - z_4)}{(z_1 - z_4)(z_2 - z_3)}.$$

Thus, we could define

$$(z_1 z_2 z_3 z_4) = \frac{(z_1 - z_3)(z_2 - z_4)}{(z_1 - z_4)(z_2 - z_3)}.$$

I think this alternate definition will make the problems below slightly easier.

(a) $(z_1 z_2 z_3 z_4)$ is invariant under several permutations of z_1, z_2, z_3, and z_4. Determine those permutations of z_1, z_2, z_3, and z_4 that leave $(z_1 z_2 z_3 z_4)$ invariant. You can determine those permutations by filling in the appropriate values of x, y, and z in each case listed below:

$$\frac{(z_2 - y)(x - z)}{(z_2 - z)(x - y)} = (z_2 xyz) = (z_1 z_2 z_3 z_4);$$

$$\frac{(z_3 - y)(x - z)}{(z_3 - z)(x - y)} = (z_3 xyz) = (z_1 z_2 z_3 z_4); \text{ and}$$

$$\frac{(z_4 - y)(x - z)}{(z_4 - z)(x - y)} = (z_4 xyz) = (z_1 z_2 z_3 z_4).$$

(b) Show

$$1 - (z_1 z_2 z_3 z_4) = (z_4 z_2 z_3 z_1) = (z_1 z_3 z_2 z_4).$$

Note! Switching the first and last entries has the same result as switching the middle two entries.

(c) Show $(z_2 z_1 z_3 z_4) = (z_1 z_2 z_4 z_3) = 1/(z_1 z_2 z_3 z_4)$.
Note! Switching the first two entries has the same result as switching the last two entries.
Knowing the value for $(z_1 z_2 z_3 z_4)$, one can use the results of parts b) and c) to compute the value of any other permutation. For example, suppose $(z_1 z_2 z_3 z_4) = k$ and we wish to compute $(z_2 z_3 z_4 z_1)$. We observe that

$$(z_2 z_3 z_4 z_1) = \frac{1}{(z_3 z_2 z_4 z_1)} = \frac{1}{1 - (z_1 z_2 z_4 z_3)} = \frac{1}{1 - \frac{1}{(z_1 z_2 z_3 z_4)}}$$
$$= \frac{1}{1 - \frac{1}{k}} = \frac{k}{k - 1}.$$

(d) Suppose $(z_1 z_2 z_3 z_4) = k$. Determine the values for $(z_1 z_2 z_4 z_3)$, $(z_1 z_3 z_2 z_4)$, $(z_1 z_3 z_4 z_2)$, $(z_1 z_4 z_2 z_3)$, and $(z_1 z_4 z_3 z_2)$. (You should get $1 - k$, $1/k$, $k/(k - 1)$, $(k - 1/k$, and $1/(1 - k)$ – not necessarily in that order.)

(e) There are a total of 24 permutations for $(z_1 z_2 z_3 z_4)$. Without computing the values for all 24, demonstrate that given $(z_1 z_2 z_3 z_4) = k$, show that the possible values for permutations is exhausted by the list in part d).

8.5 *The Distance Function

As you might suspect, the Riemann metric for the Poincaré disk is invariant under direct non-Euclidean transformations. (See Prob. 218.) From Th. 198, it is clear that a geodesic through a point z in the Poincaré disk may be described either as the arc of a circle that passes through z and its inverse $1/z^*$ or the diameter that is a segment of the straight line that passes through the same two points. This means that the geodesic that connects z_1 and z_2 is the arc of the circle or a segment of the straight line that passes through z_1, z_2, $1/z_1^*$, and $1/z_2^*$.

Furthermore when the Poincaré disk is subjected to a direct non-Euclidean transformation, the image of the inverse of a point z is the inverse of the image of z. To see this, we note that if

$$w(z) = \frac{az+b}{b^*z+a^*}, \text{ then}$$

$$w(1/z^*) = \frac{a(1/z^*)+b}{b^*(1/z^*)+a^*} = \frac{a+bz^*}{b^*+a^*z^*} = \frac{1}{w^*}.$$

From what we know about cross ratios, it is now obvious that if $w(z)$ is a direct non-Euclidean transformation and

$$w_k = w(z_k) \text{ for } k = 1, 2, 3, \text{and } 4, \text{ then}$$

$$\frac{(w_2 - w_1)\left(\frac{1}{w_1^*} - \frac{1}{w_2^*}\right)}{\left(w_2 - \frac{1}{w_2^*}\right)\left(\frac{1}{w_1^*} - w_1\right)} = \frac{(z_2 - z_1)\left(\frac{1}{z_1^*} - \frac{1}{z_2^*}\right)}{\left(z_2 - \frac{1}{z_2^*}\right)\left(\frac{1}{z_1^*} - z_1\right)} \text{ or restated}$$

$$\frac{|w_2 - w_1|^2}{\left(1 - |w_2|^2\right)\left(1 - |w_1|^2\right)} = \frac{|z_2 - z_1|^2}{\left(1 - |z_2|^2\right)\left(1 - |z_1|^2\right)}. \tag{8.34}$$

We know that the Riemann metric for the Poincaré disk and therefore the Riemann distance between two points is also invariant under direct non-Euclidean mappings. It is too much to expect that our distance function is identical to the invariant indicated by (8.34). But it is plausible that some function of that invariant is the Riemann distance. If $m(z_1, z_2)$ is the Riemann distance between the points z_1 and z_2, then we may be able to find a function f such that

$$m(z_1, z_2) = f\left(\frac{|z_2 - z_1|}{\sqrt{(1 - |z_2|^2)(1 - |z_1|^2)}}\right). \tag{8.35}$$

Because of the invariance of the argument of f, we only need to investigate a convenient geodesic. To seek out our desired function, we will look at the x axis. From (8.18)

$$ds = \frac{2}{\sqrt{|K|}} \frac{dx}{1-x^2} = dm(0, x) = df\left(\frac{x}{\sqrt{1-x^2}}\right) dx$$

$$= f'\left(\frac{x}{\sqrt{1-x^2}}\right)\left(\frac{d}{dx}\frac{x}{\sqrt{1-x^2}}\right) dx = f'\left(\frac{x}{\sqrt{1-x^2}}\right) \frac{dx}{(1-x^2)^{3/2}}.$$

From this last equation, we have

$$f'\left(\frac{x}{\sqrt{1-x^2}}\right) = \frac{2}{\sqrt{|K|}}\frac{1}{\sqrt{1-x^2}}. \tag{8.36}$$

Substituting

$$u = \frac{x}{\sqrt{1-x^2}},$$

Equation (8.36) becomes

$$f'(u) = \frac{2}{\sqrt{|K|}\sqrt{1+u^2}}.$$

(You should fill in the steps.) Substituting $u = \sinh\phi$, we get

$$f'(u)du = \frac{2d\sinh\phi}{\sqrt{|K|}\sqrt{1+\sinh^2\phi}} = \frac{2\cosh\phi d\phi}{\sqrt{|K|}\cosh\phi} = \frac{2}{\sqrt{|K|}}d\phi.$$

Integrating, we have

$$f(u) = \frac{2}{\sqrt{|K|}}\phi + c = \frac{2}{\sqrt{|K|}}\sinh^{-1} u + c, \text{ where}$$

c is a constant that is to be determined. From (8.35), it is clear that $f(0) = 0$, so $c = 0$. Referring to (8.35) again, we now have our desired distance function. Namely;

$$m(z_1, z_2) = \frac{2}{\sqrt{|K|}}\sinh^{-1}\left(\frac{|z_2 - z_1|}{\sqrt{(1-|z_2|^2)(1-|z_1|^2)}}\right). \tag{8.37}$$

Problem 218. Suppose that

$$w(z) = \frac{az+b}{b^*z+a^*}.$$

Show that

$$\frac{dwdw^*}{(1-ww^*)^2} = \frac{dzdz^*}{(1-zz^*)^2}.$$

Also show that this equation is equivalent to the equation:

$$\frac{\sqrt{(du)^2+(dv)^2}}{1-u^2-v^2} = \frac{\sqrt{(dx)^2+(dy)^2}}{1-x^2-y^2}, \text{ where}$$

$$z = x + iy \text{ and } w = u + iv.$$

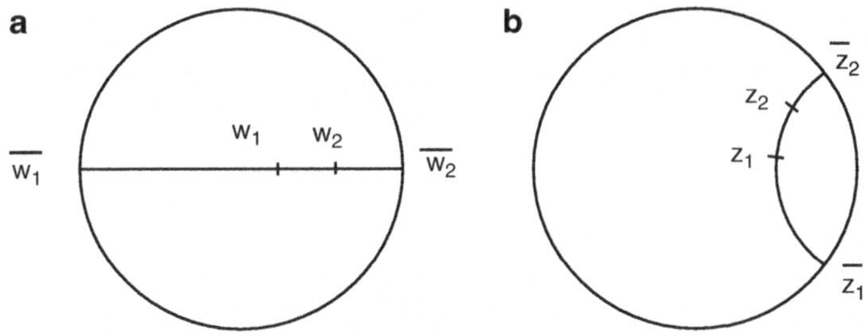

Fig. 8.10 The cross ratio $\frac{(\bar{w}_1 - w_2)(\bar{w}_2 - w_1)}{(\bar{w}_1 - w_1)(\bar{w}_2 - w_2)}$ in figure **a** equals the cross ratio $\frac{(\bar{z}_1 - z_2)(\bar{z}_2 - z_1)}{(\bar{z}_1 - z_1)(\bar{z}_2 - z_2)}$ in figure **b**

Problem 219. Refer to Fig. 8.10. From (8.18),

$$m(w_1, w_2) = \frac{2}{|K|} \int_{w_1}^{w_2} \frac{dx}{1 - x^2}.$$

Use the method of partial fractions to compute this integral and show that

$$m(w_1, w_2) = \frac{1}{\sqrt{|K|}} \ln \left(\frac{(w_2 - \bar{w}_1)(\bar{w}_2 - w_1)}{(w_1 - \bar{w}_1)(\bar{w}_2 - w_2)} \right) = \frac{1}{\sqrt{|K|}} \ln \left(\frac{(\bar{w}_1 - w_2)(\bar{w}_2 - w_1)}{(\bar{w}_1 - w_1)(\bar{w}_2 - w_2)} \right).$$

Using the fact that

$$\frac{(\bar{w}_1 - w_2)(\bar{w}_2 - w_1)}{(\bar{w}_1 - w_1)(\bar{w}_2 - w_2)} \text{ is a cross ratio,}$$

and referring to Fig. 8.10b, we see that in general:

$$m(z_1, z_2) = \frac{1}{\sqrt{|K|}} \ln \left(\frac{(\bar{z}_1 - z_2)(\bar{z}_2 - z_1)}{(\bar{z}_1 - z_1)(\bar{z}_2 - z_2)} \right). \tag{8.38}$$

This was the distance formula derived by Poincaré. A difficulty with this formula is that given z_1 and z_2, it is very difficult to compute $\bar{z}_1$ and $\bar{z}_2$. (At least for me.)

Problem 220. By permuting the four points $z_1, z_2, 1/z_1^*$, and $1/z_2^*$, we can construct six cross ratios. However, besides the one I have used in this section, there is only one other with $(z_2 - z_1)$ in the numerator. Namely:

$$\frac{(z_2 - z_1)\left(\frac{1}{z_2^*} - \frac{1}{z_1^*}\right)}{\left(z_2 - \frac{1}{z_1^*}\right)\left(\frac{1}{z_2^*} - z_1\right)} = \frac{(z_2 - z_1)(z_1^* - z_2^*)}{(z_2 z_1^* - 1)(1 - z_1 z_2^*)}$$

$$= \frac{(z_2 - z_1)(z_2^* - z_1^*)}{(1 - z_2 z_1^*)(1 - z_1 z_2^*)}.$$

Show that using this cross ratio, one can obtain the distance formula:

$$m(z_1, z_2) = \frac{2}{\sqrt{|K|}} \tanh^{-1}\left(\frac{|z_2 - z_1|}{|1 - z_1 z_2^*|}\right). \tag{8.39}$$

This is the version of the distance formula that appears in *Geometry* written by *David A. Brannan, Matthew F. Esplen, and Jeremy J. Gray (Brannan et al. 1999, p.285).*

Problem 221. If

$$y = \sinh^{-1} x, \text{ then } x = \sinh y = \frac{1}{2}\left(e^y - e^{-y}\right).$$

Use this equation to show that

$$e^y = \sqrt{x^2 + 1} + x.$$

Use this result to show that

$$\sinh^{-1} x = \tanh^{-1}\left(\frac{x}{\sqrt{x^2 + 1}}\right) = \ln\left(\sqrt{x^2 + 1} + x\right).$$

Finally, use this last result to obtain a formula for the cross ratio used by Poincaré in (8.38) in terms of z_1, z_2, z_1^*, and z_2^* . *(Do not ignore the fact that the coefficient that appears in front of the* $\sinh^{-1}$ term in (8.37) and in front of the $\tanh^{-1}$ term in (8.39) differs from the coefficient that appears in front of the log term in (8.38) by a factor of 2.

8.6 *The Law of Cosines and the Law of Sines

As noted in (8.7), Minding was able to show that for surfaces of constant negative Gaussian curvature, the Law of Cosines is

$$\cosh\left(c\sqrt{|K|}\right) = \cosh\left(a\sqrt{|K|}\right)\cosh\left(b\sqrt{|K|}\right) - \sinh\left(a\sqrt{|K|}\right)\sinh\left(b\sqrt{|K|}\right)\cos C,$$

where this time angle C is opposite the edge of the triangle with length c.

For the remainder of this section, I will use a, b, c, etc. for Euclidean lengths and $\hat{a}$, $\hat{b}$, $\hat{c}$ etc. for Riemannian lengths. With this convention, we can now prove the Law of Cosines for surfaces of constant negative curvature K.

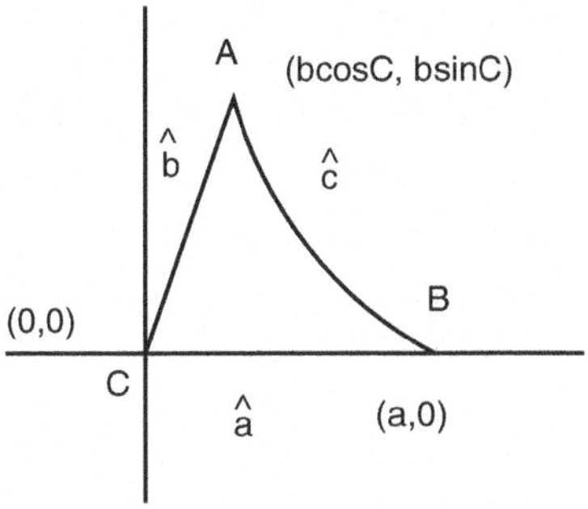

Fig. 8.11 A triangle bounded by three geodesics

Theorem 222. *(Law of Cosines for the surfaces of constant negative Gaussian curvature K.)*

$$\cosh\left(\sqrt{|K|}\hat{c}\right) = \cosh\left(\sqrt{|K|}\hat{a}\right)\cosh\left(\sqrt{|K|}\hat{b}\right) - \sinh\left(\sqrt{|K|}\hat{a}\right)\sinh\left(\sqrt{|K|}\hat{b}\right)\cos C,$$

where C is the angle opposite edge $\hat{c}$.

Proof. Referring to Fig. 8.11 and using (8.37), we have

$$\sinh\left(\frac{\sqrt{|K|}}{2}\hat{c}\right) = \frac{\sqrt{(b\cos C - a)^2 + b^2\sin^2 C}}{\sqrt{1-a^2}\sqrt{1-b^2}} = \frac{\sqrt{a^2+b^2-2ab\cos C}}{\sqrt{1-a^2}\sqrt{1-b^2}}.$$

Thus,

$$\begin{aligned}\cosh\left(\sqrt{|K|}\hat{c}\right) &= \cosh^2\left(\frac{\sqrt{|K|}}{2}\hat{c}\right) + \sinh^2\left(\frac{\sqrt{|K|}}{2}\hat{c}\right) = 1 + 2\sinh^2\left(\frac{\sqrt{|K|}}{2}\hat{c}\right)\\ &= \frac{\left(1-a^2\right)\left(1-b^2\right) + 2\left(a^2+b^2-2ab\cos C\right)}{\left(1-a^2\right)\left(1-b^2\right)}\\ &= \frac{\left(1+a^2\right)\left(1+b^2\right) - 4ab\cos C}{\left(1-a^2\right)\left(1-b^2\right)}. \qquad (8.40)\end{aligned}$$

Using (8.37) again, we have

$$\sinh\left(\frac{\sqrt{|K|}}{2}\hat{b}\right) = \frac{b}{\sqrt{1-b^2}}.$$

Since

$$\cosh^2\left(\frac{\sqrt{|K|}}{2}\hat{b}\right) = 1 + \sinh^2\left(\frac{\sqrt{|K|}}{2}\hat{b}\right) = \frac{1-b^2+b^2}{1-b^2} = \frac{1}{1-b^2},$$

it follows that

$$\cosh\left(\frac{\sqrt{|K|}}{2}\hat{b}\right) = \frac{1}{\sqrt{1-b^2}}.$$

(The hyperbolic cosine is always positive, so we do not need to consider the negative square root!)

We now have

$$\cosh\left(\sqrt{|K|}\hat{b}\right) = \cosh^2\left(\frac{\sqrt{|K|}}{2}\hat{b}\right) + \sinh^2\left(\frac{\sqrt{|K|}}{2}\hat{b}\right) = \frac{1+b^2}{1-b^2}, \text{ and} \tag{8.41}$$

$$\sinh\left(\sqrt{|K|}\hat{b}\right) = 2\sinh\left(\frac{\sqrt{|K|}}{2}\hat{b}\right)\cosh\left(\frac{\sqrt{|K|}}{2}\hat{b}\right) = \frac{2b}{1-b^2}. \tag{8.42}$$

A similar computation gives us

$$\cosh\left(\sqrt{|K|}\hat{a}\right) = \cosh^2\left(\frac{\sqrt{|K|}}{2}\hat{a}\right) + \sinh^2\left(\frac{\sqrt{|K|}}{2}\hat{a}\right) = \frac{1+a^2}{1-a^2}, \text{ and} \tag{8.43}$$

$$\sinh\left(\sqrt{|K|}\hat{a}\right) = 2\sinh\left(\frac{\sqrt{|K|}}{2}\hat{a}\right)\cosh\left(\frac{\sqrt{|K|}}{2}\hat{a}\right) = \frac{2a}{1-a^2}. \tag{8.44}$$

Using these last four equations, (8.40) becomes

$$\begin{aligned}\cosh\left(\sqrt{|K|}\hat{c}\right) &= \cosh\left(\sqrt{|K|}\hat{a}\right)\cosh\left(\sqrt{|K|}\hat{b}\right)\\ &\quad - \sinh\left(\sqrt{|K|}\hat{a}\right)\sinh\left(\sqrt{|K|}\hat{b}\right)\cos C.\end{aligned} \tag{8.45}$$

This is the Law of Cosines for surfaces of constant negative Gaussian curvature K that we set out to prove. □

We are now in a position to prove the following theorem:

Theorem 223. *(The Law of Sines for the surfaces of constant negative Gaussian curvature K.)*

$$\frac{\sin A}{\sinh\left(\sqrt{|K|}\hat{a}\right)} = \frac{\sin B}{\sinh\left(\sqrt{|K|}\hat{b}\right)} = \frac{\sin C}{\sinh\left(\sqrt{|K|}\hat{c}\right)}.$$

Proof.

$$\sinh^2\left(\sqrt{|K|}\hat{a}\right)\sinh^2\left(\sqrt{|K|}\hat{b}\right)\sin^2 C = X - Y, \text{ where} \tag{8.46}$$

$$X = \sinh^2\left(\sqrt{|K|}\hat{a}\right)\sinh^2\left(\sqrt{|K|}\hat{b}\right), \text{ and}$$

$$Y = \sinh^2\left(\sqrt{|K|}\hat{a}\right)\sinh^2\left(\sqrt{|K|}\hat{b}\right)\cos^2 C.$$

It is clear that

$$X = \left(\cosh^2\left(\sqrt{|K|}\hat{a}\right) - 1\right)\left(\cosh^2\left(\sqrt{|K|}\hat{b}\right) - 1\right).$$

From the Law of Cosines,

$$Y = \left[\cosh\left(\sqrt{|K|}\hat{a}\right)\cosh\left(\sqrt{|K|}\hat{b}\right) - \cosh\left(\sqrt{|K|}\hat{c}\right)\right]^2.$$

Carrying out the multiplications and organizing the resulting terms, we get

$$\begin{aligned} X - Y &= 1 + 2\cosh\left(\sqrt{|K|}\hat{a}\right)\cosh\left(\sqrt{|K|}\hat{b}\right)\cosh\left(\sqrt{|K|}\hat{c}\right) \\ &\quad - \cosh^2\left(\sqrt{|K|}\hat{a}\right) - \cosh^2\left(\sqrt{|K|}\hat{b}\right) - \cosh^2\left(\sqrt{|K|}\hat{c}\right). \end{aligned} \tag{8.47}$$

From 8.47, $X - Y$ is invariant under any permutations of the sides of the triangle ABC. For this reason, the left-hand side of 8.46 is invariant under the same permutations. Therefore

$$\begin{aligned} \sinh\left(\sqrt{|K|}\hat{a}\right)\sinh\left(\sqrt{|K|}\hat{b}\right)\sin C &= \sinh\left(\sqrt{|K|}\hat{c}\right)\sinh\left(\sqrt{|K|}\hat{a}\right)\sin B \\ &= \sinh\left(\sqrt{|K|}\hat{b}\right)\sinh\left(\sqrt{|K|}\hat{c}\right)\sin A. \end{aligned}$$

Dividing this last equation by $\sinh(\sqrt{|K|}\hat{a})\sinh(\sqrt{|K|}\hat{b})\sinh(\sqrt{|K|}\hat{c})$, we get

$$\frac{\sin A}{\sinh\left(\sqrt{|K|}\hat{a}\right)} = \frac{\sin B}{\sinh\left(\sqrt{|K|}\hat{b}\right)} = \frac{\sin C}{\sinh\left(\sqrt{|K|}\hat{c}\right)}.$$

Problem 224. Consider (8.45). Compute the first few terms of Maclaurin's series for the variable $\sqrt{|K|}$ and then demonstrate that when one takes the limit $\sqrt{|K|} \to 0$, one obtains the usual Law of Cosines for the Euclidean plane.

8.7 *Tessellations

In closing this section, it is fun to discuss *tessellations*. A method of covering a surface with congruent polygons is called a tessellation. On a surface of constant negative curvature, the interior angles of an equilateral triangle are each less than 60^0. This means that you cannot cover a surface of constant negative curvature with equilateral triangles if you insist on having six triangles share a common vertex. However by increasing the size of an equilateral triangle, you reduce each interior angle. Thus by adjusting the size of an equilateral

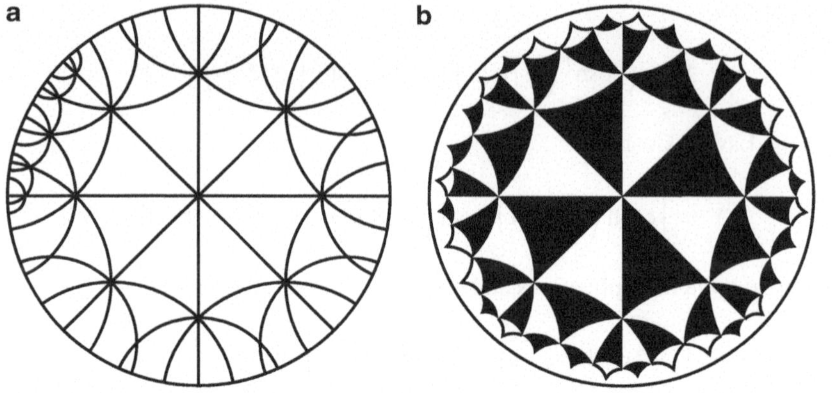

Fig. 8.12 The Gauss tessellation

triangle, you can have each interior angle $= 360^0/n$ for any integer $n \geq 7$. Figure 8.12 shows how the Riemann disk can be covered with equilateral triangles for $n = 8$. Figure 8.12a was found among Gauss' unpublished papers after he died (Gauss 1900, p.104). One might be inclined to conclude that it was inspired by Riemann's habilitationsvortrag address, but the diagram and the associated computations are undated. In private correspondence (April 4, 2004), John Stillwell indicated to me that, "– it is believed to date from well before Riemann's time. Probably in the early 1800s, when he (Gauss) also thought about tessellations related to the modular function."

Although the triangles in Fig. 8.12b appear to have an infinite number of sizes and shapes to an observer using a Euclidean metric, they are all congruent equilateral triangles for an observer using Riemann's metric.

Problem 225. Does it make sense either from the perspective of a Riemannian observer or a Euclidean observer to label a point of the Riemann disk as being a "center"?

Problem 226. Suppose that $\hat{r}$ designates the Riemann radius of a circle on a Poincaré disk. Show that the area A of the circle is $\frac{4}{|K|}\pi \sinh^2 \frac{\sqrt{|K|}\hat{r}}{2}$ and the circumference C is $\frac{2\pi}{\sqrt{|K|}} \sinh \sqrt{|K|}\hat{r}$. What are the corresponding formulas for circles on a sphere?

Problem 227. For the Poincaré disk, one can cover the surface with equilateral triangles where n triangles share a vertex if $n \geq 7$. What are the possibilities for covering a sphere with equilateral triangles. In each case, describe how many triangles would be required to cover the sphere? From the Gauss–Bonnet theorem, you can determine the area of each triangle and check this against the total area of the sphere. Is this question related to some of the regular solids?

In some limiting sense, is it possible to cover a sphere with two equilateral triangles? with one equilateral triangle?

Problem 228. Suppose $\hat{a}$, $\hat{b}$, and $\hat{c}$ are the magnitudes of the three geodesic sides of a triangle on a 2-dimensional surface of constant negative curvature. Use the Law of Cosines to show that

$$\left|\hat{a} - \hat{b}\right| < \hat{c} < \hat{a} + \hat{b}.$$

Problem 229. In view of (6.58), is there an upper bound for the area of a triangle (a geometric figure whose edges consist of three geodesics.) for the Poincaré disk? Make some intelligent comments about the lengths of the edges.

Chapter 9
*Ruled Surfaces Continued

9.1 *Lines of Stricture

Although a multitude of directrices can be used for a given ruled surface, there are particular directrices that are worth discussing.

For a cylinder in the n-dimensional Euclidean space E^n, it is natural to choose a directrix that lies in a $(n-1)$-dimensional plane orthogonal to the generators. Suppose $\mathbf{x}(u)$ is a directrix for the cylinder

$$\mathbf{y}(u,t) = \mathbf{x}(u) + t\mathbf{v}, \quad \text{where} \tag{9.1}$$

$\mathbf{v}$ is a constant unit vector. If you wish to construct an alternate representation of the same surface using a directrix that lies in the plane passing through the point $\mathbf{p}_0$ that is perpendicular to $\mathbf{v}$, you can write

$$\bar{\mathbf{y}}(u,\bar{t}) = \bar{\mathbf{x}}(u) + \bar{t}\mathbf{v}, \quad \text{where} \tag{9.2}$$

$$\bar{\mathbf{x}}(u) = \mathbf{x}(u) - \langle \mathbf{x}(u) - \mathbf{p}_0, \mathbf{v}\rangle\, \mathbf{v}, \quad \text{and} \tag{9.3}$$

$$t = t + \langle \mathbf{x}(u) - \mathbf{p}_0, \mathbf{v}\rangle. \tag{9.4}$$

We note that $\langle \bar{\mathbf{x}}(u) - \mathbf{p}_0, \mathbf{v}\rangle = 0$ and this means that $\bar{\mathbf{x}}(u)$ lies in the plane passing through $\mathbf{p}_0$ that is perpendicular to $\mathbf{v}$. Furthermore,

$$\bar{\mathbf{y}}(u,\bar{t}) = \mathbf{x}(u) - \langle \mathbf{x}(u) - \mathbf{p}_0, \mathbf{v}\rangle\, \mathbf{v} + (t + \langle \mathbf{x}(\mathbf{u}) - \mathbf{p}_0, \mathbf{v}\rangle)\, \mathbf{v} = \mathbf{y}(u,t),$$

so $\bar{\mathbf{y}}(u,\bar{t})$ represents the same surface as $\mathbf{y}(u,t)$.

For cones, we have already observed that it is most natural to use the vertex in place of any directrix curve.

For the remaining ruled surfaces, we can pick out a special directrix (known as the *line of stricture*). A line of stricture is the curve that passes through the *central point of each generator*. What is a central point? Consider a given generator and

J. Snygg, *A New Approach to Differential Geometry using Clifford's Geometric Algebra*, DOI 10.1007/978-0-8176-8283-5_9,

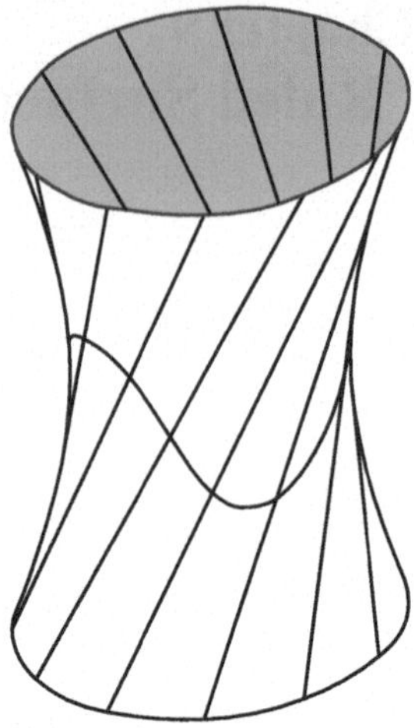

Fig. 9.1 One of the lines of stricture for the hyperboloid $\frac{x^2}{a^2} + \frac{y^2}{b^2} - \frac{z^2}{c^2} = 1$ when $a \neq b$

a second "nearby" generator. If the surface is not a cylinder, we know that if the two generators are sufficiently close, they will not be parallel. This means either they will be skew or they will intersect. In either case, each generator will have a point that is closest to the other generator. In the case of skew generators, these two points can be connected by a straight line segment, which is perpendicular to both generators. (See Fig. 9.2a.) If we now consider a sequence of generators that approach the given generator, then the corresponding closest points on the given generator will approach a limit point. It is that limit point that is the central point on the given generator. As indicated above, the curve passing through the central point on each generator is the line of stricture.

In general, a doubly ruled surface has two lines of stricture – one for each set of generators. An exception to this rule is a special class of one sheet hyperboloids. If $a = b$ in (7.91), the same curve is a line of stricture for both sets of generators.

Now let us consider the computation of a central point. Suppose $\mathbf{v}(u)$ is a unit vector indicating the direction of a generator. Then we can specify a point on that generator by the formula,

$$\mathbf{y}(u,t) = \mathbf{x}(u) + t\mathbf{v}(u), \tag{9.5}$$

where $\mathbf{x}(u)$ is the directrix. Similarly, we can specify a point on a nearby generator by the formula,

$$\mathbf{y}(u',t') = \mathbf{x}(u') + t'\mathbf{v}(u').$$

Our first task is to determine the values of t and t' that will minimize the length of the vector

$$\mathbf{y}(u',t') - \mathbf{y}(u,t) = \mathbf{x}(u') - \mathbf{x}(u) + t'\mathbf{v}(u') - t\mathbf{v}(u). \tag{9.6}$$

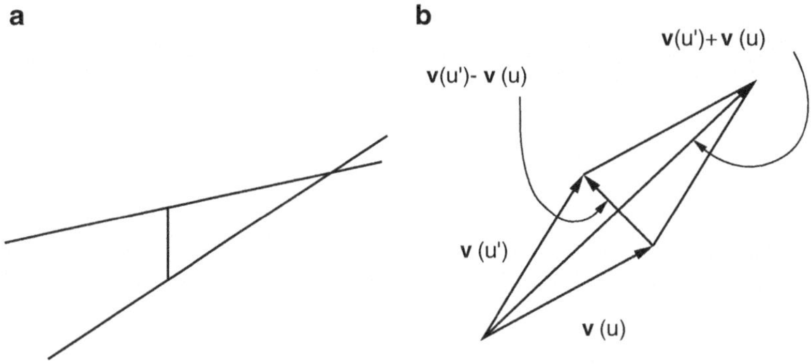

Fig. 9.2 (**a**) Two skew lines joined by a mutually orthogonal line segment. (**b**) Rhombus with diagonals that are mutually orthogonal

It will be slightly advantageous to make use of the fact that $(\mathbf{v}(u') + \mathbf{v}(u))$ is orthogonal to $(\mathbf{v}(u')-\mathbf{v}(u))$. This fact follows from the fact that since $\mathbf{v}(u')$ and $\mathbf{v}(u)$ both have unit length, they can be used to form two adjacent edges of a rhombus. In that situation, the vectors $(\mathbf{v}(u')+\mathbf{v}(u))$ and $(\mathbf{v}(u')-\mathbf{v}(u))$ can be interpreted as the mutually orthogonal diagonals of the rhombus. (See Fig. 9.2b.)

With this thought in mind, we will rewrite (9.6) in the form

$$\mathbf{y}(u',t') - \mathbf{y}(u,t) = \mathbf{x}(u') - \mathbf{x}(u) + t'\left[\frac{\mathbf{v}(u')+\mathbf{v}(u)}{2} + \frac{\mathbf{v}(u')-\mathbf{v}(u)}{2}\right]$$
$$- t\left[\frac{\mathbf{v}(u')+\mathbf{v}(u)}{2} - \frac{\mathbf{v}(u')-\mathbf{v}(u)}{2}\right].$$

Or

$$\mathbf{y}(u',t') - \mathbf{y}(u,t) = \Delta\mathbf{x} + (t'-t)\left[\frac{\mathbf{v}(u')+\mathbf{v}(u)}{2}\right] + \frac{(t'+t)}{2}\left[\Delta\mathbf{v}\right], \tag{9.7}$$

where

$$\Delta\mathbf{x} = \mathbf{x}(u') - \mathbf{x}(u) \quad \text{and} \quad \Delta\mathbf{v} = \mathbf{v}(u') - \mathbf{v}(u). \tag{9.8}$$

Now we wish to minimize

$$f(t,t') = \big\langle \mathbf{y}(u',t') - \mathbf{y}(u,t), \mathbf{y}(u',t') - \mathbf{y}(u,t) \big\rangle$$
$$= \langle \Delta\mathbf{x}, \Delta\mathbf{x}\rangle + (t'-t)\big\langle \Delta\mathbf{x}, \mathbf{v}(u') + \mathbf{v}(u)\big\rangle + (t'+t)\,\langle \Delta\mathbf{x}, \Delta\mathbf{v}\rangle$$
$$+ \frac{(t'-t)^2}{4}\big\langle \mathbf{v}(u') + \mathbf{v}(u), \mathbf{v}(u') + \mathbf{v}(u)\big\rangle + \frac{(t'+t)^2}{4}\,\langle \Delta\mathbf{v}, \Delta\mathbf{v}\rangle.$$

At the minimum, we have

$$\frac{\partial f}{\partial t} = \frac{\partial f}{\partial t'} = 0.$$

Therefore,

$$\frac{\partial f}{\partial t} = 0 = -\left\langle \Delta\mathbf{x}, \mathbf{v}(u') + \mathbf{v}(u) \right\rangle + \langle \Delta\mathbf{x},\Delta\mathbf{v}\rangle$$
$$- \frac{(t'-t)}{2}\left\langle \mathbf{v}(u') + \mathbf{v}(u), \mathbf{v}(u') + \mathbf{v}(u) \right\rangle + \frac{(t'+t)}{2}\langle \Delta\mathbf{v},\Delta\mathbf{v}\rangle. \tag{9.9}$$

And

$$\frac{\partial f}{\partial t'} = 0 = \left\langle \Delta\mathbf{x}, \mathbf{v}(u') + \mathbf{v}(u) \right\rangle + \langle \Delta\mathbf{x},\Delta\mathbf{v}\rangle$$
$$+ \frac{(t'-t)}{2}\left\langle \mathbf{v}(u') + \mathbf{v}(u), \mathbf{v}(u') + \mathbf{v}(u) \right\rangle + \frac{(t'+t)}{2}\langle \Delta\mathbf{v},\Delta\mathbf{v}\rangle. \tag{9.10}$$

The straightforward method is to eliminate t' from (9.9) and (9.10) and obtain a formula for $t = t(u,u')$. At that point, we can obtain the value of t (call it τ) corresponding to the central point by computing the limit

$$\tau = \lim_{u' \to u} t(u,u'). \tag{9.11}$$

On the other hand, the computation is not so messy if we are slightly devious. It should be clear that not only is (9.11) true, but it can also be said that

$$\tau = \lim_{u' \to u} t'(u,u'). \tag{9.12}$$

As a consequence, we can add (9.9) and (9.10) to get

$$2\,\langle \Delta\mathbf{x}, \Delta\mathbf{v}\rangle + (t'+t)\,\langle \Delta\mathbf{v}, \Delta\mathbf{v}\rangle = 0$$

or

$$\frac{t'+t}{2} = -\frac{\langle \Delta\mathbf{x}, \Delta\mathbf{v}\rangle}{\langle \Delta\mathbf{v}, \Delta\mathbf{v}\rangle} = -\frac{\langle \Delta\mathbf{x}/\Delta u, \Delta\mathbf{v}/\Delta u\rangle}{\langle \Delta\mathbf{v}/\Delta u, \Delta\mathbf{v}/\Delta u\rangle}, \tag{9.13}$$

where

$$\Delta u = u' - u.$$

Thus,

$$\tau = \lim_{\acute{u} \to u} \frac{t'+t}{2} = -\frac{\langle \dot{\mathbf{x}}(u), \dot{\mathbf{v}}(\mathbf{u})\rangle}{\langle \dot{\mathbf{v}}(u), \dot{\mathbf{v}}(u)\rangle}, \tag{9.14}$$

where $\dot{\mathbf{x}}(u) = \mathrm{d}\mathbf{x}(u)/\mathrm{d}u$ and $\dot{\mathbf{v}}(u) = \mathrm{d}\mathbf{v}(u)/\mathrm{d}u$.

Combining (9.14) and (9.5), we see that the formula for the central point $\mathbf{z}(u)$ for the generator passing through $\mathbf{x}(u)$ is

$$\mathbf{z}(u) = \mathbf{x}(u) - \frac{\langle \dot{\mathbf{v}}(u), \dot{\mathbf{x}}(u)\rangle}{\langle \dot{\mathbf{v}}(u), \dot{\mathbf{v}}(u)\rangle}\mathbf{v}(u). \tag{9.15}$$

Note! It should be noted that in (9.15), u is not necessarily an arc length parameter for **x**, **z** or any other curve.

It should also be noted that it is possible for the right-hand side of (9.15) to be a constant. In that case, we have the vertex of a cone instead of a line of stricture.

Having computed the central points, it is now possible to characterize them in a second way. We note that

$$\dot{\mathbf{z}}(u) = \dot{\mathbf{x}}(u) - \frac{\langle \dot{\mathbf{v}}(u), \dot{\mathbf{x}}(u) \rangle}{\langle \dot{\mathbf{v}}(u), \dot{\mathbf{v}}(u) \rangle} \dot{\mathbf{v}}(u) - \frac{\mathrm{d}}{\mathrm{d}u} \left(\frac{\langle \dot{\mathbf{v}}(u), \dot{\mathbf{x}}(u) \rangle}{\langle \dot{\mathbf{v}}(u), \dot{\mathbf{v}}(u) \rangle} \right) \mathbf{v}(u). \tag{9.16}$$

Since $\langle \mathbf{v}, \mathbf{v} \rangle = 1$, we know that $\langle \mathbf{v}, \dot{\mathbf{v}} \rangle = 0$. Using that fact, it is not too difficult to see that from (9.16),

$$\langle \dot{\mathbf{z}}(u), \dot{\mathbf{v}}(u) \rangle = 0. \tag{9.17}$$

What is worth observing is that $\mathbf{z}(u)$ is the only directrix that satisfies (9.17). That is, we could use (9.17) to define a central point or a line of stricture, although I would have difficulty understanding the geometric significance of such a definition. To justify this claim, we note that given a directrix passing through the point $\mathbf{x}(u)$, any alternative directrix $\mathbf{w}(u)$ must cross the generator passing through $\mathbf{x}(u)$. This means that

$$\mathbf{w}(u) = \mathbf{x}(u) + f(u)\mathbf{v}(u).$$

If we require that $\langle \dot{\mathbf{w}}(u), \dot{\mathbf{v}}(u) \rangle = 0$, then

$$\left\langle \dot{\mathbf{x}}(u) + f(u)\dot{\mathbf{v}}(u) + \dot{f}(u)\mathbf{v}(u), \dot{\mathbf{v}}(u) \right\rangle = 0, \text{ or simplified,}$$

$$\langle \dot{\mathbf{x}}(u) + f(u)\dot{\mathbf{v}}(u), \dot{\mathbf{v}}(u) \rangle = 0.$$

But this implies that

$$f(u) = -\frac{\langle \dot{\mathbf{v}}(u), \dot{\mathbf{x}}(u) \rangle}{\langle \dot{\mathbf{v}}(u), \dot{\mathbf{v}}(u) \rangle},$$

which agrees with (9.15). Thus (9.17) is a necessary and sufficient condition for $\mathbf{z}(u)$ to be a line of stricture.

It is interesting to examine the nature of the moving frame for a ruled surface when we let $\mathbf{E}_{\bar{1}} = \mathbf{v}(u)$. If

$$\mathbf{y}(t, u) = \mathbf{z}(u) + t\mathbf{v}(u),$$

then

$$\gamma_1 = \frac{\partial \mathbf{y}(t, u)}{\partial t} = \mathbf{v}(u) \quad \text{and} \quad \gamma_2 = \frac{\partial \mathbf{y}(t, u)}{\partial u} = \dot{\mathbf{z}}(u) + t\dot{\mathbf{v}}(u). \tag{9.18}$$

If we are not dealing with a cylinder, $\dot{\mathbf{v}}(u) \neq 0$. Since both $\langle \dot{\mathbf{z}}(u), \dot{\mathbf{v}}(u) \rangle$ and $\langle \mathbf{v}(u), \dot{\mathbf{v}}(u) \rangle$ equal zero, $\dot{\mathbf{z}}$ and $\mathbf{v}$ lie in the plane orthogonal to $\dot{\mathbf{v}}$. Thus, there are two cases:

Case 1: $\dot{\mathbf{z}}(u)$ is a scalar multiple of $\mathbf{v}(u)$.
Case 2: $\dot{\mathbf{z}}(u)$, $\dot{\mathbf{v}}(u)$, and $\mathbf{v}(u)$ are linearly independent.

For Case 1, $\boldsymbol{\gamma}_1$ and $\boldsymbol{\gamma}_2$ span the same space as $\mathbf{v}$ and $\dot{\mathbf{v}}$. Thus, we can let our "moving frame" be

$$\mathbf{E}_{\bar{1}}(t,u) = \mathbf{v}(u) \quad \text{and} \quad \mathbf{E}_{\bar{2}}(t,u) = \dot{\mathbf{v}}(u) / \left|\dot{\mathbf{v}}(u)\right|.$$

Thus for Case 1, not only does $\mathbf{E}_{\bar{1}}$ remain constant along any given generator but $\mathbf{E}_{\bar{2}}$ also remains constant along any given generator.

One might think that by constructing different surfaces for Case 2, one could have $\mathbf{E}_{\bar{2}}$ move in almost any way as long as it remained orthogonal to $\mathbf{E}_{\bar{1}}$. But that is not the case. For Case 2, we again let $\mathbf{E}_{\bar{1}} = \mathbf{v}(u)$. But this time

$$\mathbf{E}_{\bar{2}}\mathbf{E}_{\bar{1}} = \mathbf{E}_{\bar{2}}\mathbf{v} = \frac{\boldsymbol{\gamma}_2 \wedge \mathbf{v}}{|\boldsymbol{\gamma}_2 \wedge \mathbf{v}|}, \quad \text{and thus}$$

$$\mathbf{E}_{\bar{2}} = \frac{\boldsymbol{\gamma}_2 \wedge \mathbf{v}}{|\boldsymbol{\gamma}_2 \wedge \mathbf{v}|}\mathbf{v}. \tag{9.19}$$

Note that since $(\gamma_2 - \langle \gamma_2, \mathbf{v} \rangle \mathbf{v})$ and $\mathbf{v}$ are orthogonal to one another,

$$\boldsymbol{\gamma}_2 \wedge \mathbf{v} = (\boldsymbol{\gamma}_2 - \langle \boldsymbol{\gamma}_2, \mathbf{v} \rangle \mathbf{v}) \wedge \mathbf{v} = (\boldsymbol{\gamma}_2 - \langle \boldsymbol{\gamma}_2, \mathbf{v} \rangle \mathbf{v})\, \mathbf{v}. \tag{9.20}$$

Furthermore,

$$|\boldsymbol{\gamma}_2 \wedge \mathbf{v}| = |\boldsymbol{\gamma}_2 - \langle \boldsymbol{\gamma}_2, \mathbf{v} \rangle \mathbf{v}|\, |\mathbf{v}| = |\boldsymbol{\gamma}_2 - \langle \boldsymbol{\gamma}_2, \mathbf{v} \rangle \mathbf{v}|. \tag{9.21}$$

Now from (9.18),

$$\boldsymbol{\gamma}_2 - \langle \boldsymbol{\gamma}_2, \mathbf{v} \rangle \mathbf{v} = \dot{\mathbf{z}} + t\dot{\mathbf{v}} - \langle \dot{\mathbf{z}} + t\dot{\mathbf{v}}, \mathbf{v} \rangle \mathbf{v}$$

Since $\langle \dot{\mathbf{v}}, \mathbf{v} \rangle = 0$, we see that

$$\boldsymbol{\gamma}_2 - \langle \boldsymbol{\gamma}_2, \mathbf{v} \rangle \mathbf{v} = \dot{\mathbf{z}} + t\dot{\mathbf{v}} - \langle \dot{\mathbf{z}}, \mathbf{v} \rangle \mathbf{v}. \tag{9.22}$$

From (9.17), $\langle \dot{\mathbf{z}}, \dot{\mathbf{v}} \rangle = 0$. Using that fact along with (9.21) and (9.22), we get

$$|\boldsymbol{\gamma}_2 \wedge \mathbf{v}| = \left[\langle \dot{\mathbf{z}}, \dot{\mathbf{z}} \rangle - \langle \dot{\mathbf{z}}, \mathbf{v} \rangle^2 + (t)^2 \langle \dot{\mathbf{v}}, \dot{\mathbf{v}} \rangle\right]^{1/2}. \tag{9.23}$$

Combining (9.20), (9.22), and (9.23), we have

$$\mathbf{E}_{\bar{2}}(t,u) = \frac{(\dot{\mathbf{z}} + t\dot{\mathbf{v}} - \langle \dot{\mathbf{z}}, \mathbf{v} \rangle \mathbf{v})\, \mathbf{v}}{\left[\langle \dot{\mathbf{z}}, \dot{\mathbf{z}} \rangle - \langle \dot{\mathbf{z}}, \mathbf{v} \rangle^2 + (t)^2 \langle \dot{\mathbf{v}}, \dot{\mathbf{v}} \rangle\right]^{1/2}}\mathbf{v} = \frac{\dot{\mathbf{z}} + t\dot{\mathbf{v}} - \langle \dot{\mathbf{z}}, \mathbf{v} \rangle \mathbf{v}}{\left[\langle \dot{\mathbf{z}}, \dot{\mathbf{z}} \rangle - \langle \dot{\mathbf{z}}, \mathbf{v} \rangle^2 + (t)^2 \langle \dot{\mathbf{v}}, \dot{\mathbf{v}} \rangle\right]^{1/2}}. \tag{9.24}$$

We are now in a position to see how $\mathbf{E}_{\bar{2}}$ behaves when it is moved along a generator for Case 2. From (9.24),

$$\lim_{t\to-\infty} \mathbf{E}_{\bar{2}}(t,u) = -\frac{\dot{\mathbf{v}}}{|\dot{\mathbf{v}}|}, \quad \mathbf{E}_{\bar{2}}(0,u) = \frac{\dot{\mathbf{z}} - \langle \dot{\mathbf{z}}, \mathbf{v}\rangle \mathbf{v}}{\sqrt{\langle \dot{\mathbf{z}}, \dot{\mathbf{z}}\rangle - \langle \dot{\mathbf{z}}, \mathbf{v}\rangle^2}}, \quad \text{and}$$

$$\lim_{t\to+\infty} \mathbf{E}_{\bar{2}}(t,u) = \frac{\dot{\mathbf{v}}}{|\dot{\mathbf{v}}|}.$$

Since $\mathbf{E}_{\bar{2}}(0,u)$ is orthogonal to $\dot{\mathbf{v}}$, we can say that for Case 2, as $\mathbf{E}_{\bar{2}}$ moves along a generator, it rotates 90° as t goes from $-\infty$ to 0, which corresponds to the central point on the generator. Then $\mathbf{E}_{\bar{2}}$ rotates another 90° as t goes from 0 to $+\infty$.

If this 2-dimensional ruled surface is embedded in the 3-dimensional Euclidean space E^3, then a vector normal to the surface undergoes a similar rotation. According to Struik (1988, p. 194), the nature of this rotation is the motivation for the label "central point."

Returning to Case 1, if $\dot{\mathbf{z}}(u)$ is a scalar multiple of $\mathbf{v}(u)$, then the ruled surface is swept out by the tangent vector of the line of stricture. In this situation, the line of stricture is said to be an *edge of regression*. An example of such a surface is the developable helicoid of Prob. 167 and Fig. 7.6a.

The surfaces that are generated by tangent lines of a given curve are very special. Along with cylinders and cones, they are the only 2-dimensional ruled surfaces that have zero Gaussian curvature. That will be demonstrated in the next section.

Problem 230. (a) Determine the line of stricture for the one sheet hyperboloid:

$$\frac{x^2}{a^2} + \frac{y^2}{b^2} - \frac{z^2}{c^2} = 1, \tag{9.25}$$

when $a = b$.

(b) (More difficult!) Determine the line of stricture for one set of generators when $a \neq b$. Then show that the line of stricture lies in the surface:

$$\frac{y^2z^2}{b^2c^2}\left(\frac{1}{b^2} + \frac{1}{c^2}\right)^2 + \frac{z^2x^2}{c^2a^2}\left(\frac{1}{c^2} + \frac{1}{a^2}\right)^2 - \frac{x^2y^2}{a^2b^2}\left(\frac{1}{a^2} - \frac{1}{b^2}\right)^2 = 0.$$

From the symmetry of this equation, one can see that if it is valid for one set of generators, it is valid for the other set of generators. Nonetheless, this equation is consistent with the claim that there are two lines of stricture for the one sheet hyperboloid when $a \neq b$. Why?

(c) Describe the surface of (9.25) when $a = b$. When $a \neq b$, do the two lines of strictures cross? If so, where do they cross?

Problem 231. Determine both lines of stricture for the saddle surface:

$$z = xy \quad (\text{or } x^3 = x^1 x^2).$$

Problem 232. Determine the line of stricture for the helicoid:

$$\mathbf{y}(s,t) = \mathbf{x}(s) + t\mathbf{v}(s), \quad \text{where}$$

$$\mathbf{x}(s) = \mathbf{e}_1 b \cos \frac{s}{\sqrt{a^2+b^2}} + \mathbf{e}_2 b \sin \frac{s}{\sqrt{a^2+b^2}} + \mathbf{e}_3 \frac{as}{\sqrt{a^2+b^2}}, \quad \text{and}$$

$$\mathbf{v}(s) = \mathbf{E}_{\bar{2}}(s). \text{ (not } \mathbf{E}_{\bar{1}}).$$

(The helicoid should not be confused with the developable helicoid!)

9.2 *Gaussian Curvature of Ruled Surfaces

In this section, I will derive a formula for the Gaussian curvature of ruled surfaces. As a consequence, it will be shown that the Gaussian curvature of a *developable surface* is zero. What is a developable surface? A developable surface is a special kind of ruled surface. In particular, it is either a tangential developable, a cone, or a cylinder. Here is the theorem.

Theorem 233. *For the ruled surface:*

$$\mathbf{y}(u^1,u^2) = \mathbf{z}(u^2) + u^1\mathbf{v}(u^2), \tag{9.26}$$

we can consider two cases.

Case (1) $\dot{\mathbf{v}} = d\mathbf{v}/du^2 = 0$. For this case, we have a cylindrical surface for which the Gaussian curvature is zero. (See Prob. 166.)

Case (2) $\langle \dot{\mathbf{v}}, \dot{\mathbf{v}} \rangle \neq 0$. For this case, we can assume $\mathbf{z}(u^2)$ is a line of stricture and $\mathbf{v}(u^2)$ is a vector of unit length. Using the notation that $\dot{\mathbf{z}} = d\mathbf{z}/du^2$ and $\dot{\mathbf{v}} = d\mathbf{v}/du^2$, the Gaussian curvature is

$$K = R_{12}{}^{12} = \frac{-f(u^2)}{[f(u^2) + (u^1)^2]^2}, \quad \text{where} \tag{9.27}$$

$$f(u^2) = \frac{|\dot{\mathbf{z}} - \langle \dot{\mathbf{z}}, \mathbf{v} \rangle \mathbf{v}|^2}{\langle \dot{\mathbf{v}}, \dot{\mathbf{v}} \rangle} = \frac{\langle \dot{\mathbf{z}}, \dot{\mathbf{z}} \rangle - \langle \dot{\mathbf{z}}, \mathbf{v} \rangle^2}{\langle \dot{\mathbf{v}}, \dot{\mathbf{v}} \rangle}. \tag{9.28}$$

Proof. From (9.24), we have

$$\begin{aligned} \mathbf{E}_{\bar{2}}(u^1,u^2) &= \frac{(\dot{\mathbf{z}} - \langle \dot{\mathbf{z}}, \mathbf{v} \rangle \mathbf{v}) + u^1 \dot{\mathbf{v}}}{\sqrt{\langle \dot{\mathbf{z}}, \dot{\mathbf{z}} \rangle - \langle \dot{\mathbf{z}}, \mathbf{v} \rangle^2 + (u^1)^2 \langle \dot{\mathbf{v}}, \dot{\mathbf{v}} \rangle}} \\ &= \frac{(\dot{\mathbf{z}} - \langle \dot{\mathbf{z}}, \mathbf{v} \rangle \mathbf{v}) + u^1 \dot{\mathbf{v}}}{\langle \dot{\mathbf{v}}, \dot{\mathbf{v}} \rangle^{1/2} \left[f(u^2) + (u^1)^2 \right]^{1/2}}. \end{aligned} \tag{9.29}$$

Hopefully, from Sect. 5.6, you recall that

$$\nabla_j \mathbf{E}_{\bar{k}} = -c_{\bar{k}\bar{m}}(\boldsymbol{\gamma}_j)\mathbf{E}^{\bar{m}}, \quad \text{where}$$

$\nabla_j \mathbf{E}_{\bar{k}}$ is the projection of $\partial \mathbf{E}_{\bar{k}}/\partial u^j$ on the space spanned by $\mathbf{E}^{\bar{1}}$ and $\mathbf{E}^{\bar{2}}$ (or $\mathbf{E}_{\bar{1}}$ and $\mathbf{E}_{\bar{2}}$).

Since

$$\frac{\partial \mathbf{E}_{\bar{1}}}{\partial u^1} = \frac{\partial \mathbf{v}(u^2)}{\partial u^1} = 0, \text{ it follows that}$$

$$\nabla_1 \mathbf{E}_{\bar{1}} = 0 = -c_{\bar{1}\bar{2}}(\boldsymbol{\gamma}_1)\mathbf{E}^{\bar{2}}. \text{ Thus}$$

$$c_{\bar{1}\bar{2}}(\boldsymbol{\gamma}_1) = 0. \tag{9.30}$$

On the other hand,

$$\frac{\partial \mathbf{E}_{\bar{1}}}{\partial u^2} = \dot{\mathbf{v}} = -\, c_{\bar{1}\bar{2}}(\boldsymbol{\gamma}_2)\mathbf{E}^{\bar{2}} + \text{ terms orthogonal to } \mathbf{E}_{\bar{1}} \text{ and } \mathbf{E}_{\bar{2}}. \tag{9.31}$$

Thus, using (9.31) and (9.29), we have

$$c_{\bar{1}\bar{2}}(\boldsymbol{\gamma}_2) = -\left\langle \dot{\mathbf{v}}, \bar{\mathbf{E}}_{\bar{2}} \right\rangle = \frac{-u^1 \left\langle \dot{\mathbf{v}}, \dot{\mathbf{v}} \right\rangle}{\left\langle \dot{\mathbf{v}}, \dot{\mathbf{v}} \right\rangle^{1/2} [f(u^2) + (u^1)^2]^{1/2}}$$

$$= \frac{-u^1 \left\langle \dot{\mathbf{v}}, \dot{\mathbf{v}} \right\rangle^{1/2}}{[(u^1)^2 + f(u^2)]^{1/2}}.$$

Now since

$$\boldsymbol{\Gamma}_\alpha = \frac{1}{4} c_{\bar{j}\bar{k}}(\boldsymbol{\gamma}_\alpha)\mathbf{E}^{\bar{j}}\mathbf{E}^{\bar{k}} = \frac{1}{2} c_{\bar{1}\bar{2}}(\boldsymbol{\gamma}_\alpha)\mathbf{E}^{\bar{1}}\mathbf{E}^{\bar{2}}, \text{ we have}$$

$$\boldsymbol{\Gamma}_1 = 0 \quad \text{and} \quad \boldsymbol{\Gamma}_2 = \frac{-1}{2} \frac{u^1 \left\langle \dot{\mathbf{v}}, \dot{\mathbf{v}} \right\rangle^{1/2}}{[(u^1)^2 + f(u^2)]^{1/2}} \mathbf{E}^{\bar{1}}\mathbf{E}^{\bar{2}}. \tag{9.32}$$

For 2-dimensional surfaces,

$$\frac{1}{2}\mathbf{R}_{12} = \partial_1 \boldsymbol{\Gamma}_2 - \partial_2 \boldsymbol{\Gamma}_1 = \frac{-1}{2} \frac{\left\langle \dot{\mathbf{v}}, \dot{\mathbf{v}} \right\rangle^{1/2} f(u^2)}{[f(u^2) + (u^1)^2]^{3/2}} \bar{\mathbf{E}}^{\bar{1}}\bar{\mathbf{E}}^{\bar{2}}, \text{ or}$$

$$\mathbf{R}_{12} = \frac{-\left\langle \dot{\mathbf{v}}, \dot{\mathbf{v}} \right\rangle^{1/2} f(u^2)}{[(u^1)^2 + f(u^2)]^{3/2}} \frac{\boldsymbol{\gamma}_{12}}{|\boldsymbol{\gamma}_1 \wedge \boldsymbol{\gamma}_2|} \tag{9.33}$$

From (9.23),

$$|\boldsymbol{\gamma}_1 \wedge \boldsymbol{\gamma}_2| = |\boldsymbol{\gamma}_2 \wedge \mathbf{v}| = \left[\left\langle \dot{\mathbf{z}}, \dot{\mathbf{z}} \right\rangle - \left\langle \dot{\mathbf{z}}, \mathbf{v} \right\rangle^2 + (u^1)^2 \left\langle \dot{\mathbf{v}}, \dot{\mathbf{v}} \right\rangle \right]^{1/2}$$

$$= \left\langle \dot{\mathbf{v}}, \dot{\mathbf{v}} \right\rangle^{1/2} \left[f(u^2) + (u^1)^2 \right]^{1/2}.$$

Thus

$$\mathbf{R}_{12} = \frac{1}{2} R^{jk}_{12} \boldsymbol{\gamma}_{jk} = R^{12}_{12} \boldsymbol{\gamma}_{12} = \frac{-f(u^2)}{[(u^1)^2 + f(u^2)]^2} \boldsymbol{\gamma}_{12}.$$

So finally

$$K = R^{12}_{12} = \frac{-f(u^2)}{[(u^1)^2 + f(u^2)]^2}, \quad \text{where}$$

$$f(u^2) = \frac{|\dot{\mathbf{z}} - \langle \dot{\mathbf{z}}, \mathbf{v} \rangle \mathbf{v}|^2}{\langle \dot{\mathbf{v}}, \dot{\mathbf{v}} \rangle} = \frac{\langle \dot{\mathbf{z}}, \dot{\mathbf{z}} \rangle - \langle \dot{\mathbf{z}}, \mathbf{v} \rangle^2}{\langle \dot{\mathbf{v}}, \dot{\mathbf{v}} \rangle}.$$

□

Corollary 234. *For a tangential developable, where*

$$\mathbf{y}(u^1, u^2) = \mathbf{z}(u^2) + u^1 \mathbf{v}(u^2) \quad \textit{and} \quad \mathbf{v} = \mathbf{E}_1 = \dot{\mathbf{z}} / |\dot{\mathbf{z}}| \quad \textit{or}$$

for a cone, where

$$\mathbf{y}(u^1, u^2) = \mathbf{z} + u^1 \mathbf{v}(u^2) \quad \textit{and} \quad \dot{\mathbf{z}} = 0,$$

the Gaussian curvature $K = 0$.

Corollary 235. *For ruled surfaces that are not developable (not cylinders, not cones, and not tangential developables), the Gaussian curvature is negative.*

Corollary 236. *For ruled surfaces that are not developable, the maximum magnitude for the Gaussian curvature along any generator occurs at the central point. Furthermore, the magnitude of the Gaussian curvature decreases as you move along any given generator away from the central point in either direction. In addition, the Gaussian curvature at any point on a generator is equal to the Gaussian curvature at the point on the generator that is on the opposite side of the central point at the same distance from the central point.*

9.3 *The Cusp at the Edge of Regression

From Fig. 7.6a, we see that the tangential developable for a circular helix has a cusp at the edge of regression. This kind of behavior is characteristic of all tangential developables. Unless the edge of regression lies in a 2-dimensional plane, the tangential ray in the forward direction will sweep out a different surface than that swept out by the tangential ray in the trailing direction. The two surfaces then form a sharp edge at the edge of regression. This sharp edge is sometimes referred to as the *cuspidal edge* (Struik 1988, p. 68). What happens when the edge of regression lies in a 2-dimensional plane? (You should be able to figure that out.)

In this section, I will examine the nature of this cusp. I will do this by considering the curve that lies in the intersection of the tangential developable and the plane that is orthogonal to the edge of regression at the point $\mathbf{x}(s_0)$. (If the edge of regression lies in the n-dimensional Euclidean space E^n, then the plane would be $(n-1)$-dimensional.) Suppose the equation for the tangential developable is

$$\mathbf{y}(t,s) = \mathbf{x}(s) + t\mathbf{E}_{\bar{1}}(s) \quad \text{and}$$

$$\frac{\mathrm{d}\mathbf{x}(s)}{\mathrm{d}s} = \mathbf{E}_{\bar{1}}(s).$$

(That is to say, s is an arc length parameter for the edge of regression $\mathbf{x}(s)$.)

Since our curve lies in the tangential developable, it will have the form:

$$\mathbf{u}(s) = \mathbf{x}(s) + t(s)\mathbf{E}_{\bar{1}}(s), \quad \text{where} \tag{9.34}$$

$t(s)$ is to be determined. Actually, I just lied. I will not determine $t(s)$. What I will do is determine

$$\left.\frac{d^k t(s)}{\mathrm{d}s^k}\right|_{s=s_0} \quad \text{for } k = 0, 1, 2, \text{ and } 3,$$

so that the approximating curve

$$\bar{\mathbf{u}}(s) = \sum_{k=0}^{3} \frac{(s-s_0)^k}{k!} \left.\frac{d^k \mathbf{u}(s)}{ds^k}\right|_{s=s_0} \tag{9.35}$$

lies in the plane perpendicular to $\mathbf{E}_1(s)$ at $s = s_0$.

Our strategy is to begin with (9.34) and show that

$$\left.\frac{\mathrm{d}\mathbf{u}(s)}{\mathrm{d}s}\right|_{s=s_0} = a_1\mathbf{E}_1(s_0) + b_1\mathbf{E}_2(s_0),$$

$$\left.\frac{d^2\mathbf{u}(s)}{\mathrm{d}s^2}\right|_{s=s_0} = a_2\mathbf{E}_1(s_0) + b_2\mathbf{E}_2(s_0) + c_2\mathbf{E}_3(s_0), \quad \text{and}$$

$$\left.\frac{d^3\mathbf{u}(s)}{\mathrm{d}s^3}\right|_{s=s_0} = a_3\mathbf{E}_1(s_0) + b_3\mathbf{E}_2(s_0) + c_3\mathbf{E}_3(s_0) + d_3\mathbf{E}_4(s_0).$$

Since $\mathbf{u}(s)$ lies in a plane orthogonal to $\mathbf{E}_1(s_0)$, we know that $a_1 = a_2 = a_3 = 0$. Using this requirement along with the condition that $\mathbf{u}(s_0) = \mathbf{x}(s_0)$, we can determine

$$\left.\frac{d^k t(s)}{\mathrm{d}s^k}\right|_{s=s_0} \quad \text{for } k = 0, 1, 2, \text{ and } 3.$$

In turn, these results determine the values of the b's, c's, and d_3.

From (9.34), we have

$$\mathbf{u}(s) = \mathbf{x}(s) + t(s)\mathbf{E}_1(s), \tag{9.36}$$

$$\frac{\mathrm{d}\mathbf{u}}{\mathrm{d}s} = \left(1 + \frac{\mathrm{d}t}{\mathrm{d}s}\right)\mathbf{E}_1 + k_1 t\mathbf{E}_2, \tag{9.37}$$

$$\frac{d^2\mathbf{u}}{\mathrm{d}s^2} = \left(\frac{\mathrm{d}}{\mathrm{d}s}\left(1 + \frac{\mathrm{d}t}{\mathrm{d}s}\right)\right)\mathbf{E}_1 + \left(1 + \frac{\mathrm{d}t}{\mathrm{d}s}\right)\frac{\mathrm{d}}{\mathrm{d}s}\mathbf{E}_1 + \left(\frac{\mathrm{d}}{\mathrm{d}s}(k_1 t)\right)\mathbf{E}_2 + k_1 t\frac{\mathrm{d}}{\mathrm{d}s}\mathbf{E}_2, \text{ and} \tag{9.38}$$

$$\begin{aligned}\frac{d^3\mathbf{u}}{\mathrm{d}s^3} &= \left(\frac{d^2}{\mathrm{d}s^2}\left(1 + \frac{\mathrm{d}t}{\mathrm{d}s}\right)\right)\mathbf{E}_1 + 2\left(\frac{\mathrm{d}}{\mathrm{d}s}\left(1 + \frac{\mathrm{d}t}{\mathrm{d}s}\right)\right)\frac{\mathrm{d}}{\mathrm{d}s}\mathbf{E}_1 + \left(1 + \frac{\mathrm{d}t}{\mathrm{d}s}\right)\frac{d^2}{\mathrm{d}s^2}\mathbf{E}_1 \\ &\quad + \left(\frac{d^2}{\mathrm{d}s^2}(k_1 t)\right)\mathbf{E}_2 + 2\left(\frac{\mathrm{d}}{\mathrm{d}s}(k_1 t)\right)\frac{\mathrm{d}}{\mathrm{d}s}\mathbf{E}_2 + k_1 t\frac{d^2}{\mathrm{d}s^2}\mathbf{E}_2.\end{aligned} \tag{9.39}$$

Since $\mathbf{u}(s_0) = \mathbf{x}(s_0)$, (9.36) implies $t(s_0) = 0$. Equation (9.37) implies that

$$\left.\left(1 + \frac{dt}{ds}\right)\right|_{s=s_0} = 0, \quad \text{so } \left.\frac{\mathrm{d}t}{\mathrm{d}s}\right|_{s=s_0} = -1, \quad \text{and } \left.\frac{\mathrm{d}\mathbf{u}}{\mathrm{d}s}\right|_{s=s_0} = 0.$$

From (9.38), we get

$$\left.\left(\frac{\mathrm{d}}{\mathrm{d}s}\left(1 + \frac{\mathrm{d}t}{\mathrm{d}s}\right)\right)\right|_{s=s_0} = 0 \quad \text{or} \quad \left.\frac{d^2 t}{\mathrm{d}s^2}\right|_{s=s_0} = 0.$$

Furthermore, since

$$\frac{\mathrm{d}}{\mathrm{d}s}(k_1 t) = t\frac{\mathrm{d}}{\mathrm{d}s}k_1 + \frac{\mathrm{d}t}{\mathrm{d}s}k_1, \quad \text{and thus, } \left.\frac{\mathrm{d}}{\mathrm{d}s}(k_1 t)\right|_{s=s_0} = -k_1.$$

It then follows that

$$\left.\frac{d^2\mathbf{u}}{\mathrm{d}s^2}\right|_{s=s_0} = -k_1(s_0)\mathbf{E}_2(s_0).$$

Similar considerations give us

$$\begin{aligned}\left.\frac{d^3\mathbf{u}}{\mathrm{d}s^3}\right|_{s=s_0} &= -2\left(\left.\frac{\mathrm{d}k_1}{\mathrm{d}s}\right|_{s=s_0}\right)\mathbf{E}_2(s_0) - 2k_1(s_0)k_2(s_0)\mathbf{E}_3(s_0) \\ &= -2\dot{k}_1(s_0)\mathbf{E}_2(s_0) - 2k_1(s_0)k_2(s_0)\mathbf{E}_3(s_0).\end{aligned} \tag{9.40}$$

(See Prob. 237.)

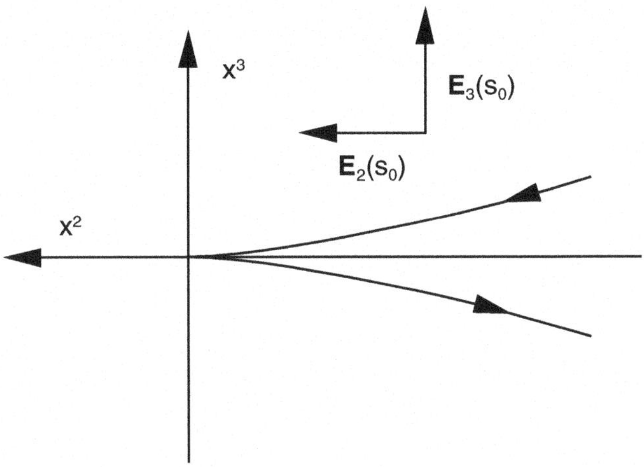

Fig. 9.3 Cusp at the edge of regression

Summarizing, we now have

$$\bar{\mathbf{u}}(s) = \mathbf{x}(s_0) - \frac{(s-s_0)^2}{2} k_1(s_0)\mathbf{E}_2(s_0) - \frac{(s-s_0)^3}{3}\left(\dot{k}_1(s_0)\mathbf{E}_2(s_0) + k_1(s_0)k_2(s_0)\mathbf{E}_3(s_0)\right). \tag{9.41}$$

If we let x^2 be the coefficient of $\mathbf{E}_2$ and x^3 be the coefficient of $\mathbf{E}_3$, then

$$x^2 = \frac{-(s-s_0)^2}{2} k_1(s_0) + \text{ higher powers of } (s-s_0) \text{ and}$$

$$x^3 = \frac{-(s-s_0)^3}{3} k_1(s_0)k_2(s_0) + \text{ higher powers of } (s-s_0),$$

or

$$(x^3)^2 \approx \frac{-8(k_2(s_0))^2}{9k_1(s_0)}(x^2)^3. \tag{9.42}$$

(See Fig. 9.3.)

Problem 237. Verify (9.40).

Problem 238. From (9.39), show that

$$\left.\frac{d^3 t}{ds^3}\right|_{s=s_0} = -2(k_1)^2.$$

Chapter 10
*Lines of Curvature

10.1 *Computing Lines of Curvature

10.1.1 *Intrinsic Projection Operators for Real Symmetric Linear Operators

In this section, I will discuss lines of curvature. A curve $\mathbf{x}(t)$ is said to be a *line of curvature* if

$$\frac{d\mathbf{x}(t)}{dt} = \mathbf{v}(t), \text{ where} \tag{10.1}$$

$\mathbf{v}(t)$ is a vector having a principal direction. (It is not necessary that $\mathbf{v}(t)$ be normalized unless you insist that t be an arc length parameter.)

In Sect. 7.5.3, I discussed how to compute the principal curvatures but the matter of computing the principal directions or eigenvectors of the shape operator was left as unfinished business.

Given a square matrix A, there is some positive integer p such that the set of matrices $\{I, A, A^2, \ldots, A^{p-1}\}$ are linearly independent while the set $\{I, A, A^2, \ldots, A^p\}$ are linearly dependent. This implies $\exists$ a polynomial $P(\lambda)$ s.t.

$$P(A) = \prod_{j=1}^{p}(A - \lambda_j I) = 0. \tag{10.2}$$

This polynomial is known as the *minimal polynomial* for the matrix A.

Given a coordinate system, there is an isomorphism between A and its matrix representation. Therefore, (10.2) is valid for any linear operator A that maps E^n into E^n. In this context,

$$A^2(\mathbf{v}) = A(A(\mathbf{v})),\ A^3(\mathbf{v}) = A(A(A(\mathbf{v}))), \text{ and}$$

higher powers of A are defined in a similar manner. Furthermore to say some polynomial $P(A) = 0$ is equivalent to saying that $P(A(\mathbf{v})) = 0$ for any $\mathbf{v}$ in E^n.

J. Snygg, *A New Approach to Differential Geometry using Clifford's Geometric Algebra*, DOI 10.1007/978-0-8176-8283-5_10, © Springer Science+Business Media, LLC 2012

For real symmetric linear operators in a space with a positive definite metric, a stronger version of (10.2) holds. Namely

Theorem 239. *If A is a real symmetric linear operator in a space with a positive definite metric, each root of the minimal polynomial occurs only once in the factorization. That is*

$$P(A) = \prod_{j=1}^{p} (A - \lambda_j I) = 0, \text{ where} \tag{10.3}$$

$\lambda_j \neq \lambda_k$ *if* $j \neq k$.

Proof. The proof is by contradiction. Suppose

$$P(A) = \prod_{j=1}^{q} (A - \lambda_j I)^{k_j}, \text{ where} \tag{10.4}$$

at least one of the k_j's is greater than one. Without loss of generality, assume $k_1 > 1$. Since $P(\lambda)$ is the minimal polynomial

$$\prod_{j=1}^{p} (A - \lambda_j I)^{k_j} = (A - \lambda_1 I)^{k_1} \prod_{j=2}^{p} (A - \lambda_j I)^{k_j} = 0, \text{ and} \tag{10.5}$$

$$B = (A - \lambda_1 I)^{k_1 - 1} \prod_{j=2}^{p} (A - \lambda_j I)^{k_j} \neq 0. \tag{10.6}$$

Since $B \neq 0$, $\exists \mathbf{v}$ such that

$$\mathbf{u} = B(\mathbf{v}) \neq \mathbf{0}.$$

To get our contradiction, I will show that $\mathbf{u} = \mathbf{0}$. Since A is symmetric, you can show that $A - \lambda_1 I$ is symmetric. This means that

$$\langle \mathbf{u}, \mathbf{u} \rangle = \langle B(\mathbf{v}), B(\mathbf{v}) \rangle = \langle C(\mathbf{v}), [(A - \lambda_1 I)B](\mathbf{v}) \rangle = \langle C(\mathbf{v}), P(A)(\mathbf{v}) \rangle = 0, \text{ where}$$

$$C = (A - \lambda_1 I)^{k_1 - 2} \prod_{j=2}^{p} (A - \lambda_j I)^{k_j}.$$

From (10.3), we can construct some useful projection operators. Namely

$$P_k = \prod_{j \neq k}^{p} \left(\frac{A - \lambda_j I}{\lambda_k - \lambda_j} \right) \text{ for } k = 1, 2, \ldots, p. \tag{10.7}$$

Remark 240. Equation (10.7) is not meaningful if the roots of the minimal polynomial are not distinct. If the linear operator A is not real and symmetric and acts on a space with a positive definite metric, things may be more complex. It is possible that the some of the roots of the minimal polynomial may be repeated. In this case, the theory is more complicated. See (Sobczyk 1997, 2001) or (Snygg 2002).

We note that from (10.7)

$$P_k(A - \lambda_k I) = 0 \text{ or } P_k A = \lambda_k P_k. \tag{10.8}$$

This means

$$(P_k)^2 = P_k \prod_{j \neq k}^{p} \left(\frac{A - \lambda_j I}{\lambda_k - \lambda_j} \right) = P_k \prod_{j \neq k}^{p} \left(\frac{\lambda_k - \lambda_j}{\lambda_k - \lambda_j} \right) I.$$

And therefore,

$$(P_k)^2 = P_k. \tag{10.9}$$

Furthermore if $j \neq k$, one of the factors of P_j is $(A - \lambda_k I)$. This means

$$P_k P_j = 0 \text{ if } j \neq k. \tag{10.10}$$

In general any set of operators that satisfy (10.9) and (10.10) are said to be *projection operators*. It is easy to show that projection operators are linearly independent. In particular, suppose

$$\sum_{j=1}^{p} \alpha_j P_j = 0.$$

If we multiply this equation by P_k, then we have

$$\begin{aligned} \alpha_k P_k &= 0 \text{ and thus} \\ \alpha_k &= 0 \text{ for } k = 1, 2, \ldots, p. \end{aligned}$$

Thus, the projection operators are linearly independent.

We will use these projection operators to project out our desired eigenvectors. However, before doing that we need to establish a few more simple relations.

Theorem 241. *The projection operators defined by (10.7) have the properties*

(a)

$$\sum_{j=1}^{p} P_j = I, \textit{ and} \tag{10.11}$$

(b)

$$A = \sum_{j=1}^{p} \lambda_j P_j. \tag{10.12}$$

(Rememeber! These relations apply only if the minimal polynomial has no repeated roots. However, shape operators are real and symmetric. So all of the roots for the minimal polynomial of a shape operator are distinct.)

Proof. To prove (a), we note that before I constructed the projection operators, we assumed that the set $\{I, A, A^2, \ldots, A^{p-1}\}$ were linearly independent and thus span a space of p dimensions. On the other hand, the projection operators are formed from linear combinations of members of the same set and they span a space of p dimensions. Thus, the space spanned by $\{I, A, A^2, \ldots, A^{p-1}\}$ is the same space spanned by the projection operators. As a result, it can be said that any member of $\{I, A, A^2, \ldots, A^{p-1}\}$ can be written as a linear combination of the projection operators. Thus,

$$I = \sum_{j=1}^{p} \alpha_j P_j. \tag{10.13}$$

To determine the α_j's, we multiply both sides of (10.13) by P_k and get

$$P_k = \sum_{j=1}^{p} \alpha_j P_k P_j = \alpha_k P_k.$$

Thus, $\alpha_k = 1$ for $k = 1, 2, \ldots, p$. And (10.13) becomes

$$I = \sum_{j=1}^{p} P_j. \tag{10.14}$$

To get (b), multiply both sides of (10.14) by A and then use (10.8) to get

$$A = \sum_{j=1}^{p} P_j A = \sum_{j=1}^{p} \lambda_j P_j. \tag{10.15}$$

□

You may note that these projection operators can be computed in a straight forward manner if you know the roots of the minimal polynomial. However how do we determine the minimal polynomial? Playing around with linear combinations of powers of some operator A does not seem to be a promising approach. In the general theory of linear operators, it is known that the minimal polynomial is a divisor of the characteristic polynomial for a matrix representation. This is helpful but for real symmetric linear operators, there is a stronger and more useful result. Namely

Theorem 242. *If A is a real symmetric linear operator, then the minimal polynomial for A is*

$$P(A) = \prod_{j=1}^{p} (A - \lambda_j I), \text{ where} \tag{10.16}$$

the λ_j's are eigenvalues of A. Furthermore, each eigenvalue appears once and only once in the product on the right-hand side of (10.16)

Proof. Repeating (10.8), we have

$$P_k A = \lambda_k P_k.$$

This means that each row of the projection operator P_k is a row eigenvector of A with eigenvalue λ_k. Thus, it is clear that every root of the minimal polynomial for A is an eigenvalue of A. From Theorem 239, we know that each root of the minimal polynomial appears only once in it s factorization. What is left to show is that every eigenvalue is a root of the minimal polynomial. We can show this by contradiction.

Suppose λ is an eigenvalue that is not a root of the minimal polynomial. Then there is a nonzero eigenvector $\mathbf{v}$ of A such that

$$A\mathbf{v} = \lambda\mathbf{v}.$$

Thus,

$$\begin{aligned}
P(A)\mathbf{v} &= (A - \lambda_1)\cdots(A - \lambda_{p-1})(A - \lambda_p)\mathbf{v} \\
&= (A - \lambda_1)\cdots(A - \lambda_{p-1})(\lambda - \lambda_p)\mathbf{v} \\
&= (A - \lambda_1)\cdots(A - \lambda_{p-2})(\lambda - \lambda_{p-1})(\lambda - \lambda_p)\mathbf{v} \\
&= \prod_{j=1}^{p}(\lambda - \lambda_j)\mathbf{v} \neq 0.
\end{aligned}$$

However, $P(A) = 0$, so that $P(A)\mathbf{v} = 0$. Thus, we have our contradiction. □

The next theorem is essentially a useful observation.

Theorem 243. *The eigenvectors of a real symmetric linear operator A span the domain of A.*

Proof. From (10.14), we have

$$I = \sum_{j=1}^{p} P_j.$$

From this equation, we can infer that the columns(rows) of the projection operators span the space of column(row) vectors. (See Problem 250.) Since every column(row) is a column(row) eigenvector of A, we can conclude that the column(row) eigenvectors span the space of column(row) vectors under consideration. □

Remark 244. The eigenvectors of a real symmetric linear operator can be orthogonalized. If every column(row) of some chosen projection operator is a scalar multiple of a single selected column(row) in that particular projection operator then there is essentially only one eigenvector associated with that particular projection operator. Since the eigenvectors associated with distinct eigenvalues are orthogonal, this particular eigenvector will be orthogonal to any other eigenvector.

If the columns(rows) of a projection operator span a space of more than one dimension, you can apply the Gram–Schmidt process to obtain on orthogonal set of eigenvectors associated with the eigenvalue identified with its projection operator. These eigenvectors will not only be orthogonal to one another but will also be orthogonal to any eigenvector associated with a different eigenvalue.

I am now in a position to prove a version II of Euler's Theorem for Normal Curvature. In version I, we decomposed our space into eigenvectors and then used them to compute the normal curvature for any given vector. In that version, there was a certain arbitrariness in the choice of some of the eigenvectors. In this version, we decompose our space into a direct sum eigenspaces each of which can be identified with a distinct eigenvalue or projection operator.

Theorem 245. *Euler's Theorem for Normal Curvature. (Version II) Suppose*

$$P_k = \prod_{j \neq k}^{p} \left(\frac{S - \lambda_j I}{\lambda_k - \lambda_j} \right) \text{ for } k = 1, 2, \ldots, p, \text{ where}$$

S is the shape operator. Suppose $\mathbf{v}$ is a unit vector and $\mathbf{v}(k) = P_k(\mathbf{v}) / |P_k(\mathbf{v})|$ if $P_k(\mathbf{v}) \neq 0$ and $\mathbf{v}(k) = 0$ if $P_k(\mathbf{v}) = 0$. Then the normal curvature for the unit vector $\mathbf{v}$ is

$$k_N = \sum_{k=1}^{p} \lambda_k \cos^2 \theta(k), \text{ where}$$

$$\cos \theta(k) = \langle \mathbf{v}, \mathbf{v}(k) \rangle .$$

Proof. We first note that

$$\langle \mathbf{v}, \mathbf{v}(k) \rangle = \langle \mathbf{v}, P_k(\mathbf{v}) \rangle / |P_k(\mathbf{v})| = \sum_{j=1}^{p} \left\langle P_j(\mathbf{v}), P_k(\mathbf{v}) \right\rangle / |P_k(\mathbf{v})| \tag{10.17}$$

From Problem 249, $\left\langle P_j(\mathbf{v}), P_k(\mathbf{v}) \right\rangle = 0$, unless $j = k$. Thus (10.17) becomes

$$\cos \theta(k) = \langle \mathbf{v}, \mathbf{v}(k) \rangle = \langle P_k(\mathbf{v}), P_k(\mathbf{v}) \rangle / |P_k(\mathbf{v})| = |P_k(\mathbf{v})| .$$

Furthermore,

$$\mathbf{v} = \sum_{k=1}^{p} P_k(\mathbf{v}), \text{ so from Prob. 249,}$$

$$S(\mathbf{v}) = \sum_{k=1}^{p} S(P_k(\mathbf{v})) = \sum_{k=1}^{p} \lambda_k P_k(\mathbf{v}).$$

Therefore,

$$k_N = \langle \mathbf{v}, S(\mathbf{v}) \rangle = \sum_{k=1}^{p} \lambda_k \langle \mathbf{v}, P_k(\mathbf{v}) \rangle$$

$$= \sum_{k=1}^{p} \lambda_k \sum_{j=1}^{p} \left\langle P_j(\mathbf{v}), P_k(\mathbf{v}) \right\rangle.$$

Since $\left\langle P_j(\mathbf{v}), P_k(\mathbf{w}) \right\rangle = 0$, whenever $j \neq k$, we now have

$$k_N = \sum_{k=1}^{p} \lambda_k \langle P_k(\mathbf{v}), P_k(\mathbf{v}) \rangle, \text{ or}$$

$$k_N = \sum_{k=1}^{p} \lambda_k \cos^2 \theta(k).$$

We will now consider the computation of lines of curvature. □

10.1.2 *The Computation of Principal Directions from Intrinsic Projection Operators

Example 246. The saddle surface continued.

From Example 172, for the saddle surface, we have

$$A = \begin{bmatrix} h_1^1 & h_2^1 \\ h_1^2 & h_2^2 \end{bmatrix} = \begin{bmatrix} \dfrac{-u^1 u^2}{w^{3/2}} & \frac{1+(u^1)^2}{w^{3/2}} \\ \dfrac{1+(u^2)^2}{w^{3/2}} & \frac{-u^1 u^2}{w^{3/2}} \end{bmatrix}, \quad \text{where}$$

$$w = 1 + (u^1)^2 + (u^2)^2.$$

From the same example, we also have

$$\lambda = \frac{-u^1 u^2 \pm \sqrt{(1+(u^1)^2)(1+(u^2)^2}}{w^{3/2}}.$$

From (10.7)

$$P_+ = \frac{A - \lambda_- I}{\lambda_+ - \lambda_-}, \text{ and } P_- = \frac{A - \lambda_+ I}{\lambda_- - \lambda_+}.$$

Carrying out the calculations, you should get

$$P_{+}=\frac{1}{2\sqrt{(1+(u^1)^2)(1+(u^2)^2)}}\begin{bmatrix}\sqrt{(1+(u^1)^2)(1+(u^2)^2} & 1+(u^1)^2\\ 1+(u^2)^2 & \sqrt{(1+(u^1)^2)(1+(u^2)^2}\end{bmatrix}$$

and

$$P_{-}=\frac{1}{2\sqrt{(1+(u^1)^2)(1+(u^2)^2)}}\begin{bmatrix}\sqrt{(1+(u^1)^2)(1+(u^2)^2} & -1-(u^1)^2\\ -1-(u^2)^2 & \sqrt{(1+(u^1)^2)(1+(u^2)^2}\end{bmatrix}.$$

To get the eigenvector or eigenvectors associated with λ_+, the usual method is to solve the equation

$$S(\mathbf{v}) = \lambda_+\mathbf{v} \text{ for } \mathbf{v}, \text{ or}$$

$$\begin{bmatrix} h_1^1 & h_2^1 \\ h_1^2 & h_2^2 \end{bmatrix}\begin{bmatrix} v_+^1 \\ v_+^2 \end{bmatrix} = \lambda_+\begin{bmatrix} v_+^1 \\ v_+^2 \end{bmatrix}.$$

For our example, this means solving the equation,

$$\begin{bmatrix} \frac{-u^1u^2}{w^{3/2}} & \frac{1+(u^1)^2}{w^{3/2}} \\ \frac{1+(u^2)^2}{w^{3/2}} & \frac{-u^1u^2}{w^{3/2}} \end{bmatrix}\begin{bmatrix} x \\ y \end{bmatrix} = \frac{-u^1u^2+\sqrt{(1+(u^1)^2)(1+(u^2)^2}}{w^{3/2}}\begin{bmatrix} x \\ y \end{bmatrix},$$

for x and y. However, since $SP_+ = \lambda_+P_+$, the columns of P_+ are column eigenvvectors of S. (If we use the convention that the contravariant indices are row indices, the column eigenvectors will give us the contravariant components of the eigenvectors.) Examining these columns, we note that the first column is

$$\frac{1}{2\sqrt{1+(u^1)^2}}\begin{bmatrix}\sqrt{1+(u^1)^2}\\ \sqrt{1+(u^2)^2}\end{bmatrix},$$

and the second column is

$$\frac{1}{2\sqrt{1+(u^2)^2}}\begin{bmatrix}\sqrt{1+(u^1)^2}\\ \sqrt{1+(u^2)^2}\end{bmatrix}.$$

Thus all column eigenvectors associated with λ_+ are scalar multiples of one vector. Thus the pair of contravariant components of the eigenvector associated with λ_+,

$$\begin{bmatrix} v_+^1 \\ v_+^2 \end{bmatrix}, \text{ is a scalar multiple of } \begin{bmatrix}\sqrt{1+(u^1)^2}\\ \sqrt{1+(u^2)^2}\end{bmatrix}. \tag{10.18}$$

To get the covariant components of the same vector $\mathbf{v}$, we read off the rows of P_+

$$\begin{bmatrix} x & y \end{bmatrix} P_+.$$

But this would result in a linear combination of the rows of P_+ and then note that any pair of covariant components of the eigenvector associated with λ_+

$$\left(v_1^+, v_2^+\right) \text{ has to be a scalar multiple of } \left(\sqrt{1+(u^2)^2}, \sqrt{1+(u^1)^2}\right).$$

For the eigenvectors associated with λ_-, the pair of contravariant components

$$\begin{bmatrix} v_-^1 \\ v_-^2 \end{bmatrix} \text{ is a scalar multiple of } \begin{bmatrix} \sqrt{1+(u^1)^2} \\ \\ -\sqrt{1+(u^2)^2} \end{bmatrix} \tag{10.19}$$

We have now determined the principal directions for the lines of curvature on the saddle surface. This sets the stage for us to determine the lines of curvature.

10.1.3 *The Computation of Lines of Curvature

Example 247. The saddle surface continued some more.

If $\mathbf{x}(t^+)$ represents a line of curvature associated with λ_+ for the saddle surface, we can write,

$$\mathbf{x}(t^+) = \mathbf{e}_1 u^1(t^+) + \mathbf{e}_2 u^2(t^+) + \mathbf{e}_3 u^1(t^+) u^2(t^+). \tag{10.20}$$

And thus

$$\begin{aligned} \frac{\mathrm{d}\mathbf{x}(t^+)}{\mathrm{d}t+} &= \mathbf{v}(t^+) = \gamma_1 v_+^1(t^+) + \gamma_2 v_+^2(t^+) \\ &= \frac{\partial \mathbf{x}}{\partial u^1}\frac{\mathrm{d}u^1}{\mathrm{d}t+} + \frac{\partial \mathbf{x}}{\partial u^2}\frac{\mathrm{d}u^2}{\mathrm{d}t+} = \gamma_1 \frac{\mathrm{d}u^1}{\mathrm{d}t+} + \gamma_2 \frac{\mathrm{d}u^2}{\mathrm{d}t+}. \end{aligned} \tag{10.21}$$

Combining (10.21) with (10.18), we have

$$\frac{\mathrm{d}u^1}{\mathrm{d}t+} = \sqrt{1+(u^1)^2}, \text{ and } \frac{\mathrm{d}u^2}{\mathrm{d}t+} = \sqrt{1+(u^2)^2}.$$

Therefore,

$$\frac{\mathrm{d}u^1}{\sqrt{1+(u^1)^2}} = \mathrm{d}t^+. \tag{10.22}$$

Substituting

$$u^1 = \sinh(\phi + \phi_0),$$

Equation (10.22) becomes

$$\begin{aligned} d\phi &= dt^+, \text{ which implies that} \\ \phi + \phi_0 &= t^+ + a^+, \text{ and thus} \\ u^1(t^+) &= \sinh(t^+ + a^+), \text{ where} \end{aligned}$$

a^+ is an arbitrary constant. Similarly

$$u^2(t^+) = \sinh(t^+ + b^+).$$

Thus for one family of lines of curvature, we have

$$\mathbf{x}(t^+) = \mathbf{e}_1 \sinh(t^+ + a^+) + \mathbf{e}_2 \sinh(t^+ + b^+) + \mathbf{e}_3 \sinh(t^+ + a^+) \sinh(t^+ + b^+). \tag{10.23}$$

Starting with (10.19), we get an equation for the other family of lines of curvature. In particular,

$$\begin{aligned} \mathbf{x}(t^-) &= \mathbf{e}_1 \sinh(t^- + a^-) - \mathbf{e}_2 \sinh(t^- + b^-) - \mathbf{e}_3 \sinh(t^- + a^-) \sinh(t^- + b^-) \\ &= \mathbf{e}_1 \sinh(a^- + t^-) + \mathbf{e}_2 \sinh(-b^- - t^-) + \mathbf{e}_3 \sinh(t^- + a^-) \sinh(-b^- - t^-) \end{aligned} \tag{10.24}$$

Equations (10.23) and (10.24) can be combined to get a parameterization of the saddle surface for which the coordinate curves are lines of curvature. Namely

$$\mathbf{x}(t^+, t^-) = \mathbf{e}_1 \sinh(t^+ + t^-) + \mathbf{e}_2 \sinh(t^+ - t^-) + \mathbf{e}_3 \sinh(t^+ + t^-) \sinh(t^+ - t^-). \tag{10.25}$$

(Clearly, the coordinate curves for the above parameterization are lines of curvature. You should check the parameterization to see that any point on the saddle surface is defined uniquely by some pair (t^+, t^-).)

Remark 248. When I set about plotting the points for Fig. 10.1, I discovered that my elderly version of Maple did not deal with the hyperbolic functions in a stable manner. Thus, I let

$$u = \sinh\left(t^+\right) \text{ and } v = \sinh\left(t^-\right). \tag{10.26}$$

Since

$$\begin{aligned} \sinh(t^+ + t^-) &= \sinh\left(t^+\right) \cosh\left(t^-\right) + \sinh\left(t^-\right) \cosh\left(t^+\right) \\ &= \sinh\left(t^+\right) \sqrt{1 + \sinh^2\left(t^-\right)} + \sinh\left(t^-\right) \sqrt{1 + \sinh^2\left(t^+\right)}. \end{aligned}$$

It follows that

$$\sinh(t^+ + t^-) = u\sqrt{1 + (v)^2} + v\sqrt{1 + (u)^2}.$$

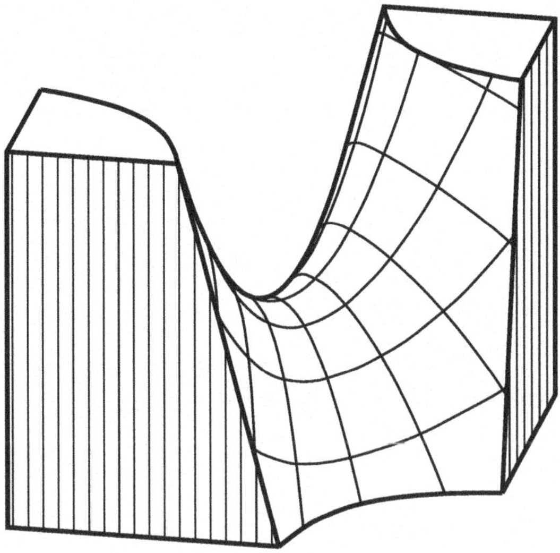

Fig. 10.1 Lines of curvature for the saddle surface

Carrying out similar computations, I got

$$\mathbf{x}(u,v) = \mathbf{e}_1\left(u\sqrt{1+(v)^2} + v\sqrt{1+(u)^2}\right) + \mathbf{e}_2\left(u\sqrt{1+(v)^2} - v\sqrt{1+(u)^2}\right) + \mathbf{e}_3\left(u^2 - v^2\right). \tag{10.27}$$

Seeing the difficulties of getting an explicit representation for the lines of curvature on the saddle surface, it should be clear that for many surfaces, it is impossible to get explicit formulas for lines of curvature.

In the next section, I will discuss a pair of theorems that can enable one to identify lines of curvature for a substantial set of surfaces.

Problem 249. (a) Suppose P_j is a projection operator associated with the real symmetric linear operator A and $\mathbf{v}$ is a vector such that $P_j(\mathbf{v}) \neq 0$. Show that $P_j(\mathbf{v})$ is an eigenvector of A.

(b) Using the fact that P_j and P_k are the projection operators of a real symmetric linear operator A, show that $\langle P_j(\mathbf{v}), P_k(\mathbf{w})\rangle = 0$ if $j \neq k$. (Note! The projection operators are real, symmetric, and linear. Why?)

Problem 250. Using the fact that

$$I = \sum_{j=1}^{p} P_j \text{ show that}$$

the columns(rows) of the P_k's span the space of column(row) vectors.

Problem 251. From (10.14) and (10.15),

$$I = \sum_{j=1}^{p} P_j, \text{ and}$$

$$A = \sum_{j=1}^{p} \lambda_j P_j.$$

Generalize these equations, first for A^n and then for more general functions of a real symmetric matrix A.

Problem 252. Consider the matrix

$$A = \begin{bmatrix} -1 & 3 \\ 3 & -1 \end{bmatrix}.$$

(a) Determine the characteristic and minimal polynomials for A.
(b) Determine the projection operators.
(c) Determine a formula for each of the four components of A^{100}.

Problem 253. Consider the matrix

$$A = \begin{bmatrix} -1 & 0 & 2 \\ 0 & 1 & 0 \\ 2 & 0 & -1 \end{bmatrix}.$$

(a) Determine the characteristic and minimal polynomials.
(b) Determine the projection operators.
(c) What is the dimension of the eigenspace associated with each projection operator? How is this reflected in the columns or rows of each projection operator?
(d) Determine a formula for each component of A^{100}.

Problem 254. HELICOID cont.

In Problem 179, we saw that one formula for the helicoid is:

$$\mathbf{y}(u, v) = \mathbf{e}_3 au + a \sinh(v) \left[\mathbf{e}_1 \cos(u) + \mathbf{e}_2 \sin(u)\right].$$

Changing the variables again with,

$$u = s - t \text{ and } v = s + t, \text{ we have}$$

$$\mathbf{x}(s, t) = \mathbf{e}_3(s - t) + a \sinh(s + t) \left[\mathbf{e}_1 \cos(s - t) + \mathbf{e}_2 \sin(s - t)\right].$$

Show that for this parameterization, the lines of curvature are the same as the coordinate curves.

Problem 255. One parameterization for the saddle surface is

$$\mathbf{x}(u^1, u^2) = \mathbf{e}_1 u^1 + \mathbf{e}_2 u^2 + \mathbf{e}_3 u^1 u^2.$$

However, the coordinate curves are not lines of curvature. Why are the coordinate curves for the parameterization of (10.27) also lines of curvature?

10.2 *Two Useful Theorems for Lines of Curvature

In this section, I will present two theorems that can be used to identify lines of curvature for a wide variety of surfaces. Our first theorem relates lines of curvature to the 2-dimensional ruled surface swept out by a line orthogonal with the hypersurface as it moves along a line of curvature.

Theorem 256. $\mathbf{x}(s)$ *is a line of curvature on a given surface.* $\Leftrightarrow$ *The ruled surface* $\mathbf{y}(s,t)$ *is a developable, where* $\mathbf{y}(s,t) = \mathbf{x}(s) + t\mathbf{N}(s)$*. (*$\mathbf{N}(s)$ *is the unit vector orthogonal to the given surface at* $\mathbf{x}(s)$*.)*

Proof. If $\mathbf{N}(s)$ is constant, $\mathbf{x}(s)$ is a line of curvature and $\mathbf{y}(s,t)$ is a cylinder and thus a developable.

If $\dot{\mathbf{N}}(s) = \mathrm{d}\mathbf{N}/\mathrm{d}s \neq 0$, then we can replace the directrix $\mathbf{x}(s)$ by $\mathbf{z}(s)$ representing either the vertex of a cone or a line of striction. In particular, from (9.15), we can write

$$\mathbf{y}(s,\bar{t}) = \mathbf{z}(s) + \bar{t}\mathbf{N}(s), \text{ where} \tag{10.28}$$

$$\mathbf{z}(s) = \mathbf{x}(s) - \frac{\langle \dot{\mathbf{N}}, \dot{\mathbf{x}} \rangle}{\langle \dot{\mathbf{N}}, \dot{\mathbf{N}} \rangle}\mathbf{N}(s) \text{ and} \tag{10.29}$$

$$\bar{t} = t + \frac{\langle \dot{\mathbf{N}}, \dot{\mathbf{x}} \rangle}{\langle \dot{\mathbf{N}}, \dot{\mathbf{N}} \rangle}.$$

If s is an arc length parameter for $\mathbf{x}(s)$ and $\mathbf{x}(s)$ is a line of curvature, then applying the shape operator, we have

$$S(\mathbf{E}_1(s)) = -\dot{\mathbf{N}}(s) = \lambda \mathbf{E}_1(s) = \lambda \dot{\mathbf{x}}(s)\,.$$

Since

$$\frac{\langle \dot{\mathbf{N}}, \dot{\mathbf{x}} \rangle}{\langle \dot{\mathbf{N}}, \dot{\mathbf{N}} \rangle} = \frac{\langle -\lambda \dot{\mathbf{x}}, \dot{\mathbf{x}} \rangle}{\langle -\lambda \dot{\mathbf{x}}, -\lambda \dot{\mathbf{x}} \rangle} = \frac{-1}{\lambda},$$

Equation (10.29) becomes

$$\mathbf{z}(s) = \mathbf{x}(s) + \frac{1}{\lambda}\mathbf{N}(s) \text{ and}$$

$$\dot{\mathbf{z}}(s) = \mathbf{E}_1(s) - \mathbf{E}_1(s) + \left[\frac{\mathrm{d}}{\mathrm{d}s}\left(\frac{1}{\lambda}\right)\right]\mathbf{N}(s). \text{ Or}$$

$$\dot{\mathbf{z}}(s) = \left[\frac{\mathrm{d}}{\mathrm{d}s}\left(\frac{1}{\lambda}\right)\right]\mathbf{N}(s).$$

If $\mathrm{d}\lambda/\mathrm{d}s = 0$, $\dot{\mathbf{z}}(s) = 0$ and $\mathbf{z}(s)$ in (10.28) is the vertex of a cone. If $\mathrm{d}\lambda/\mathrm{d}s \neq 0$, $\mathbf{N}(s)$ is tangent to $\mathbf{z}(s)$ and $\mathbf{z}(s)$ is the edge of regression for a tangential developable. In either case, $\mathbf{y}(s,\bar{t})$ or $\mathbf{y}(s,t)$ is a developable. We have now demonstrated that if $\mathbf{x}(s)$ is a line of curvature, $\mathbf{y}(s,t)$ is a developable.

To prove the converse, we assume that $\mathbf{z}(s)$ is either a vertex of a cone or the edge of regression for a tangential developable. We note that

$$\mathbf{z}(s) = \mathbf{x}(s) - \frac{\langle \dot{\mathbf{N}}, \dot{\mathbf{x}}\rangle}{\langle \dot{\mathbf{N}}, \dot{\mathbf{N}}\rangle}\mathbf{N}(s), \text{ so}$$

$$\dot{\mathbf{z}}(s) = \left[\mathbf{E}_1(s) - \frac{\langle \dot{\mathbf{N}}, \dot{\mathbf{x}}\rangle}{\langle \dot{\mathbf{N}}, \dot{\mathbf{N}}\rangle}\dot{\mathbf{N}}(s)\right] - \left(\frac{\mathrm{d}}{\mathrm{d}s}\frac{\langle \dot{\mathbf{N}}, \dot{\mathbf{x}}\rangle}{\langle \dot{\mathbf{N}}, \dot{\mathbf{N}}\rangle}\right)\mathbf{N}(s).$$

If $\mathbf{z}(s)$ is the vertex of a cone, then $\dot{\mathbf{z}}(s) = 0$. If $\mathbf{z}(s)$ is an edge of regression for a tangential developable, then $\dot{\mathbf{z}}(s)$ is the scalar multiple of $\mathbf{N}(s)$. In either case

$$S(\mathbf{E}_1(s)) = -\dot{\mathbf{N}}(s) = -\frac{\langle \dot{\mathbf{N}}, \dot{\mathbf{N}}\rangle}{\langle \dot{\mathbf{N}}, \dot{\mathbf{x}}\rangle}\mathbf{E}_1(s) \text{ and}$$

$\mathbf{x}(s)$ is a line of curvature. □

Corollary 257. *Any curve on an n-dimensional hypersphere is a line of curvature since the corresponding ruled surface is a cone with its vertex at the center of the hypersphere.*

Corollary 258. *For an n-dimensional hypersurface of revolution that is parameterized in the manner of Example 173, the coordinate curves are all lines of curvature. Note! I previously proved this result, when I first defined* line of curvature *(Definition 174). Nonetheless, I feel the following discussion is enlightening. From Example 173, we had*

$$\mathbf{x}(u^1, u^2, \ldots, u^n) = r(u^n)\mathbf{w}(u^1, u^2, \ldots, u^{n-1}) + \mathbf{e}_{n+1}u^n, \text{ where}$$

$$\mathbf{w}(u^1, u^2, \ldots, u^{n-1}) = \mathbf{e}_1 \cos u^1 \cos u^2 \cos u^3 \cdots \cos u^{n-1}$$

$$+ \mathbf{e}_2 \sin u^1 \cos u^2 \cos u^3 \cdots \cos u^{n-1}$$
$$+ \mathbf{e}_3 \sin u^2 \cos u^3 \cdots \cos u^{n-1}$$
$$+ \cdots + \mathbf{e}_{n-1} \sin u^{n-2} \cos u^{n-1}$$
$$+ \mathbf{e}_n \sin u^{n-1}, \text{ and}$$
$$\mathbf{N} = \frac{\mathbf{w} - \dot{r}\mathbf{e}_{n+1}}{\sqrt{1 + (\dot{r})^2}}.$$

The coordinate curve for u^n is

$$\mathbf{x}(u^n) = r(u^n)\mathbf{w} + \mathbf{e}_{n+1}u^n, \text{ where}$$

$\mathbf{w}$ *is a constant vector. For the 2-dimensional surface of revolution, this curve would be a meridian. The corresponding ruled surface of Theorem 256 would be*

$$\mathbf{y}(u^n, t) = r(u^n)\mathbf{w} + \mathbf{e}_{n+1}u^n + t\frac{\mathbf{w} - \dot{r}\mathbf{e}_{n+1}}{\sqrt{1 + (\dot{r})^2}}. \tag{10.30}$$

Clearly, this surface is a 2-dimensional plane spanned by the constant vectors $\mathbf{w}$ *and* $\mathbf{e}_{n+1}$. *Note! An anomaly may occur when* $\ddot{r}(u^n) = 0$. *(See Problem 261.) Nonetheless, except at isolated points, the condition for Theorem 256 is satisfied and* $\mathbf{x}(u^n)$ *is a line of curvature.*

For the 2-dimensional surface of revolution, the other coordinate curves would be parallels. For the n-dimensional surface of revolution, the situation is quite similar to what you would expect for the 2-dimensional case. For the coordinate curves other than those for u^n, the ruled surface of Theorem 256 has the form

$$\mathbf{y}(u^k, t) = r(u^n)\mathbf{w}(u^1, u^2, \ldots, u^k, \ldots, u^{n-1}) + \mathbf{e}_{n+1}u^n + t\mathbf{N}, \text{ where} \tag{10.31}$$

all parameters are held constant except u^k and t. If you reformulate this surface in terms of the line of stricture, you get

$$\mathbf{y}(u^k, t) = \mathbf{e}_{n+1}\left[u^n + r(u^n)\dot{r}(u^n)\right] + t\mathbf{N}. \tag{10.32}$$

Thus, the line of stricture reduces to a single point on the $\mathbf{e}_{n+1}$ *axis. Therefore,* $\mathbf{y}(u^k, t)$ *is a cone and Theorem 256 implies that the corresponding coordinate curve is a line of curvature.*

Actually in Example 173, all the principal curvatures except the one associated with u^n were identical. As a consequence for higher dimensions, any curve on the hypersurface for a surface of revolution whose tangent remains orthogonal to $\mathbf{e}_{n+1}$ *is a line of curvature.*

Corollary 259. *TORUS*

In Problem 98, we parameterized the surface of a torus by

$$\mathbf{x}(\theta,\phi) = \mathbf{e}_1(R + a\cos\theta)\cos\phi + \mathbf{e}_2(R + a\cos\theta)\sin\phi + \mathbf{e}_3 a\sin\theta.$$

I will leave it to you to describe the developables.

The second theorem in this section arises from orthogonal coordinate systems. For many coordinate systems that are nice to use, points are located by the intersection of n orthogonal coordinate surfaces. For example in spherical coordinates, points are located by the intersection of a sphere, cone, and a plane, where the three surfaces are mutually orthogonal. Also coordinate curves can be identified with lines of curvature. In particular for spherical coordinates,

$$\mathbf{x}(r,\theta,\phi) = \mathbf{e}_1 r\sin\theta\cos\phi + \mathbf{e}_2 r\sin\theta\sin\phi + \mathbf{e}_3 r\cos\theta.$$

If we set $\theta = \theta_0$ and $\phi = \phi_0$, we have the coordinate curve

$$\mathbf{x}(r) = \mathbf{e}_1 r\sin\theta_0\cos\phi_0 + \mathbf{e}_2 r\sin\theta_0\sin\phi_0 + \mathbf{e}_3 r\cos\theta_0.$$

This coordinate curve is a line of curvature on the cone defined by $\theta = \theta_0$ and on the plane defined by $\phi = \phi_0$. Similarly, if we hold r and θ constant, we get a coordinate curve that is a line of curvature on a sphere and on a cone.

This situation generalizes to other orthogonal coordinate systems both in three and higher dimensions. Suppose one has an orthogonal coordinate system in the Euclidean space E^{n+1}. That is

$$\mathbf{x}(u^1,u^2,\ldots,u^{n+1}) = \mathbf{e}_j x^j(u^1,u^2,\ldots,u^{n+1}) \text{ and}$$

$$\langle \boldsymbol{\gamma}_j, \boldsymbol{\gamma}_k\rangle = 0 \text{ if } j \neq k, \text{ where}$$

$$\boldsymbol{\gamma}_j = \frac{\partial \mathbf{x}}{\partial u^j}.$$

In this circumstance, you can obtain a family of n-dimensional surfaces for each value of j by allowing u^j to assume any value in a family of constants. It can then be said that the family of surfaces corresponding to one value of j is orthogonal to the family of surfaces corresponding to another value of j in the sense that any member of one family is orthogonal to any member of the other family.

In this situation, it can be proven that a coordinate curve on any of these n-dimensional surfaces is a line of curvature.

Stated in the form of a theorem, we have

Theorem 260. *Suppose we have an orthogonal coordinate system in the Euclidean space E^{n+1}. That is*

$$\mathbf{x}(u^1,u^2,\ldots,u^{n+1}) = \mathbf{e}_j x^j(u^1,u^2,\ldots,u^{n+1}) \textit{ and}$$

$$\langle \boldsymbol{\gamma}_j, \boldsymbol{\gamma}_k\rangle = 0 \textit{ if } j \neq k, \textit{ where}$$

$$\boldsymbol{\gamma}_j = \frac{\partial \mathbf{x}}{\partial u^j}.$$

Then the coordinate curve

$$\mathbf{x}(u^k) = \mathbf{x}(c^1, c^2, \ldots, c^{k-1}, u^k, c^{k+1}, \ldots, c^{n+1})$$

is a line of curvature on any surface on which points are located by the function

$$\mathbf{x}(u^1, u^2, \ldots, u^{j-1}, c^j, u^{j+1}, \ldots, u^{n+1}) \textit{ where } j \neq k.$$

(It is understood that the c^m's are constants.)

Proof. Without loss of generality, consider the n-dimensional surface

$$\mathbf{x}(u^1, u^2, \ldots, u^n, c^{n+1}) = \mathbf{e}_j x^j(u^1, u^2, \ldots, u^n, c^{n+1}).$$

The normal to this surface is

$$\mathbf{N} = \frac{\boldsymbol{\gamma}_{n+1}}{|\boldsymbol{\gamma}_{n+1}|}.$$

We wish to show that

$$\frac{\partial \mathbf{N}}{\partial u^k} = -\lambda_k \frac{\boldsymbol{\gamma}_k}{|\boldsymbol{\gamma}_k|} \quad \text{for any } k \neq n+1, \text{ where}$$

λ_k is some scalar. With that goal in mind, I will first prove that

$$\left\langle \frac{\partial \boldsymbol{\gamma}_{n+1}}{\partial u^k}, \boldsymbol{\gamma}_j \right\rangle = 0 \text{ where } j \neq k \text{ and}$$

neither j nor k is equal to $n+1$. Observe

$$\begin{aligned}
\left\langle \frac{\partial \boldsymbol{\gamma}_{n+1}}{\partial u^k}, \boldsymbol{\gamma}_j \right\rangle &= \frac{\partial}{\partial u^k}\langle \boldsymbol{\gamma}_{n+1}, \boldsymbol{\gamma}_j \rangle - \left\langle \boldsymbol{\gamma}_{n+1}, \frac{\partial \boldsymbol{\gamma}_j}{\partial u^k} \right\rangle \\
&= -\left\langle \boldsymbol{\gamma}_{n+1}, \frac{\partial \boldsymbol{\gamma}_k}{\partial u^j} \right\rangle = \left\langle \frac{\partial \boldsymbol{\gamma}_{n+1}}{\partial u^j}, \boldsymbol{\gamma}_k \right\rangle \\
&= \left\langle \frac{\partial \boldsymbol{\gamma}_j}{\partial u^{n+1}}, \boldsymbol{\gamma}_k \right\rangle = -\left\langle \boldsymbol{\gamma}_j, \frac{\partial \boldsymbol{\gamma}_k}{\partial u^{n+1}} \right\rangle \\
&= -\left\langle \boldsymbol{\gamma}_j, \frac{\partial \boldsymbol{\gamma}_{n+1}}{\partial u^k} \right\rangle = -\left\langle \frac{\partial \boldsymbol{\gamma}_{n+1}}{\partial u^k}, \boldsymbol{\gamma}_j \right\rangle.
\end{aligned}$$

Since, we have shown that

$$\left\langle \frac{\partial \boldsymbol{\gamma}_{n+1}}{\partial u^k}, \boldsymbol{\gamma}_j \right\rangle = -\left\langle \frac{\partial \boldsymbol{\gamma}_{n+1}}{\partial u^k}, \boldsymbol{\gamma}_j \right\rangle, \text{ it follows that}$$

$$2\left\langle \frac{\partial \boldsymbol{\gamma}_{n+1}}{\partial u^k}, \boldsymbol{\gamma}_j \right\rangle = 0 \text{ and thus}$$

$$\left\langle \frac{\partial \boldsymbol{\gamma}_{n+1}}{\partial u^k}, \boldsymbol{\gamma}_j \right\rangle = 0, \text{ for } j \neq k.$$

From this last equation, it follows that

$$\frac{\partial \boldsymbol{\gamma}_{n+1}}{\partial u^k} = \alpha \boldsymbol{\gamma}_k + \beta \boldsymbol{\gamma}_{n+1}.$$

for some pair of scalars α and β. It now follows that

$$\begin{aligned} \frac{\partial \mathbf{N}}{\partial u^k} &= \frac{\partial}{\partial u^k} \frac{\boldsymbol{\gamma}_{n+1}}{|\boldsymbol{\gamma}_{n+1}|} \\ &= \left(\frac{\partial}{\partial u^k} \frac{1}{|\boldsymbol{\gamma}_{n+1}|} \right) \boldsymbol{\gamma}_{n+1} + \frac{1}{|\boldsymbol{\gamma}_{n+1}|} \left(\alpha \boldsymbol{\gamma}_k + \beta \boldsymbol{\gamma}_{n+1} \right). \end{aligned} \tag{10.33}$$

But $\langle \mathbf{N}, \mathbf{N} \rangle = 1$, so $\langle \mathbf{N}, \partial \mathbf{N}/\partial u^k \rangle = 0$ and thus $\langle \boldsymbol{\gamma}_{n+1}, \partial \mathbf{N}/\partial u^k \rangle = 0$. Combining this result with (10.33), we have

$$\frac{\partial \mathbf{N}}{\partial u^k} = \frac{\alpha}{|\boldsymbol{\gamma}_{n+1}|} \boldsymbol{\gamma}_k = \frac{\alpha \, |\boldsymbol{\gamma}_k|}{|\boldsymbol{\gamma}_{n+1}|} \frac{\boldsymbol{\gamma}_k}{|\boldsymbol{\gamma}_k|} = -\lambda_k \frac{\boldsymbol{\gamma}_k}{|\boldsymbol{\gamma}_k|}.$$

In the next section, I will show how this theorem can be applied to a non-trivial situation. □

Problem 261. (a) Determine the line of stricture for the ruled surface defined by (10.30). You should see that an anomaly occurs when $\ddot{r}(u^n) = 0$. Discuss the nature of this anomaly when this occurs on an interval and when this occurs at an isolated point.

(b) Use (10.31) to compute the line of stricture and thereby confirm (10.32).

Problem 262. After Theorem 256, I mentioned three corollaries. Can Theorem 260 be used to prove any of those corollaries or related results? If so, which ones?

10.3 *Confocal Coordinates in n-Dimensions

When coordinate systems are orthogonal, it is usually fairly obvious that they are orthogonal. One coordinate system that is orthogonal but not obviously orthogonal is the system obtained by the intersection of n families of n-dimensional confocal surfaces. In particular,

$$\frac{(x^1)^2}{(a^1)^2 - u^1} + \frac{(x^2)^2}{(a^2)^2 - u^1} + \frac{(x^3)^2}{(a^3)^2 - u^1} + \cdots + \frac{(x^{n+1})^2}{(a^{n+1})^2 - u^1} = 1,$$

$$\frac{-(x^1)^2}{u^2 - (a^1)^2} + \frac{(x^2)^2}{(a^2)^2 - u^2} + \frac{(x^3)^2}{(a^3)^2 - u^2} + \cdots + \frac{(x^{n+1})^2}{(a^{n+1})^2 - u^2} = 1,$$

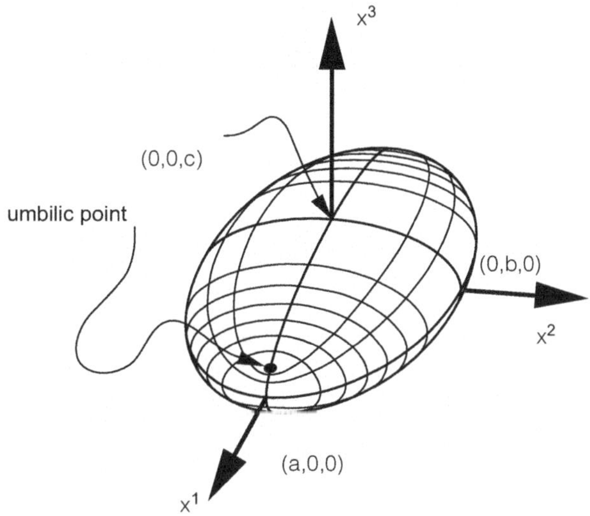

Fig. 10.2 Lines of curvature for an asymmetric ellipsoid

$$\frac{-(x^1)^2}{u^3-(a^1)^2}+\frac{-(x^2)^2}{u^3-(a^2)^2}+\frac{(x^3)^2}{(a^3)^2-u^3}+\cdots+\frac{(x^{n+1})^2}{(a^{n+1})^2-u^3}=1,$$

$$\cdots=1, \text{ and}$$

$$\frac{-(x^1)^2}{u^{n+1}-(a^1)^2}+\frac{-(x^2)^2}{u^{n+1}-(a^2)^2}+\cdots+\frac{-(x^n)^2}{u^{n+1}-(a^3)^2}+\frac{(x^{n+1})^2}{(a^{n+1})^2-u^{n+1}}=1.$$

(It is understood here that $-\infty < u^1 < (a^1)^2 < u^2 < (a^2)^2 < \cdots < u^{n+1} < (a^{n+1})^2$.)

When $n = 1$, the first equation represents a family of confocal ellipses and the second equation represents a family of confocal hyperbolas. (Hence, the name "confocal.") For $n > 1$, there are no focal points to be shared, but the resulting coordinate system is still said to be "confocal." For $n = 2$, the first equation represents a family of asymmetric ellipsoids. The second equation represents a family of one-sheeted hyperboloids and the third equation represents a family of two-sheeted hyperboloids. The fourth equation represents nothing since it does not exist.

What we wish to do is to solve the system of equations for each x^k as a function of $u^1, u^2, \ldots, u^{n+1}$ and then show that the resulting coordinate system is orthogonal. We will then not only have a plausibly useful coordinate system but also a parameterization for the lines of curvature on each of the surfaces. It was this parameterization that enabled me to construct Fig. 10.2.

Solving the system of equations for x^k in a straightforward way is a formidable task even for $n = 2$. However, a brilliant trick appears in Margenau and Murphy's *The Mathematics of Physics and Chemistry* (1956, p. 179).

Each equation in the system is of the form

$$\frac{(x^1)^2}{(a^1)^2-q}+\frac{(x^2)^2}{(a^2)^2-q}+\frac{(x^3)^2}{(a^3)^2-q}+\cdots+\frac{(x^{n+1})^2}{(a^{n+1})^2-q}=1.$$

Getting rid of the denominators, we have

$$\sum_{k=1}^{n+1}\left[(x^k)^2\prod_{j\neq k}\left((a^j)^2-q\right)\right]-\prod_{j=1}^{n+1}\left((a^j)^2-q\right)=0. \tag{10.34}$$

The left-hand side of (10.34) is a polynomial in q of order $n+1$. We know that the roots of the polynomial are $u^1, u^2, u^3, \ldots, u^{n+1}$. From the last term on the left-hand side of (10.34), we also know that the highest order term is

$$-\prod_{j=1}^{n+1}(-q)=(-1)^n q^{n+1}.$$

With this knowledge, we can infer that

$$\sum_{k=1}^{n+1}\left[(x^k)^2\prod_{j\neq k}\left((a^j)^2-q\right)\right]-\prod_{j=1}^{n+1}\left[(a^j)^2-q\right]=-\prod_{j=1}^{n+1}\left[u^j-q\right]. \tag{10.35}$$

Since (10.35) is an identity, it remains valid for any number that we assign to q. If we let $q=(a^m)^2$, most terms on the left-hand side of (10.35) are zero. What we then have is

$$(x^m)^2\prod_{j\neq m}\left[(a^j)^2-(a^m)^2\right]=-\prod_{j=1}^{n+1}\left[u^j-(a^m)^2\right]. \tag{10.36}$$

$$\text{That is,}\quad (x^m)^2=\frac{-\prod_{j=1}^{n+1}\left[u^j-(a^m)^2\right]}{\prod_{j\neq m}\left[(a^j)^2-(a^m)^2\right]}. \tag{10.37}$$

(You should confirm for yourself that the right-hand side of (10.36) is positive.)

When I wrote down the original system of equations, u^j was not allowed to assume the value of $(a^j)^2$, because one of the denominators in the jth equation would then be zero. None of the surfaces in the original system pass through the origin except in some limiting sense. However in (10.36), we can allow

$$u^j=(a^j)^2 \text{ for } j=1,2,\ldots,n+1.$$

Thus, (10.36) can be used to include the origin, where $x^1=x^2=\cdots=x^{n+1}=0$. Another feature of (10.36) is that x^m is squared. This means that corresponding to one set of values for $\{u^1, u^2, \ldots, u^{n+1}\}$, one has 2^{n+1} points in E^{n+1}. (One for each quadrant in E^2, one for each octant in E^3, and one for each whatever for E^{n+1}.) This complication creates no difficulty for local computations and few difficulties for global computations.

Our next step is to compute $\boldsymbol{\gamma}_k = \boldsymbol{\gamma}_{u^k}$ and then show that

$$\langle \boldsymbol{\gamma}_j, \boldsymbol{\gamma}_k \rangle = 0 \text{ if } j \neq k.$$

Since

$$\mathbf{x}(u^1, u^2, \ldots, u^{n+1}) = \mathbf{e}_j x^j (u^1, u^2, \ldots, u^{n+1}),$$

$$\boldsymbol{\gamma}_k = \frac{\partial \mathbf{x}}{\partial u^k} = \mathbf{e}_m \frac{\partial x^m}{\partial u^k}. \tag{10.38}$$

From (10.36),

$$2 \ln x^m = \ln \left[\pm (u^k - (a^m)^2 \right] + \text{ terms not involving } u^k.$$

For either sign,

$$\frac{2}{x^m} \frac{\partial x^m}{\partial u^k} = \frac{1}{u^k - (a^m)^2}. \tag{10.39}$$

Combining this result with (10.38), we have

$$\boldsymbol{\gamma}_k = \sum_{m=1}^{n+1} \mathbf{e}_m \frac{x^m}{2\left(u^k - (a^m)^2\right)}. \tag{10.40}$$

And

$$\langle \boldsymbol{\gamma}_j, \boldsymbol{\gamma}_k \rangle = \sum_{m=1}^{n+1} \frac{(x^m)^2}{4\left((a^m)^2 - u^j\right)(a^m)^2 - u^k)}. \tag{10.41}$$

Using the method of partial fractions,

$$\frac{1}{(w - u^j)\left(w - u^k\right)} = \frac{1}{\left(u^j - u^k\right)} \left[\frac{1}{w - u^j} - \frac{1}{w - u^k} \right]. \tag{10.42}$$

Using this result, (10.41) becomes

$$\langle \boldsymbol{\gamma}_j, \boldsymbol{\gamma}_k \rangle = \frac{1}{4\left(u^j - u^k\right)} \sum_{m=1}^{n+1} \left[\frac{(x^m)^2}{\left((a^m)^2 - u^j\right)} - \frac{(x^m)^2}{\left((a^m)^2 - u^k\right)} \right]. \tag{10.43}$$

If we reexamine the system of equations at the beginning of this section, we see that

$$\sum_{m=1}^{n+1} \frac{(x^m)^2}{\left((a^m)^2 - u^r\right)} = 1 \text{ for any value of } r \text{ between 1 and } n+1.$$

Thus,

$$\langle \boldsymbol{\gamma}_j, \boldsymbol{\gamma}_k \rangle = 0 \text{ if } j \neq k. \tag{10.44}$$

We now know that we have an orthogonal system of coordinates. From Theorem 260, this implies that every coordinate curve is a line of curvature in each of the n distinct surfaces in which it lies.

To obtain the principal curvatures for the kth surface on the list, we need to compute

$$S\left(\frac{\boldsymbol{\gamma}_j}{|\boldsymbol{\gamma}_j|}\right) = -\frac{\partial}{\partial u^j}\mathbf{N}_k = \lambda_j \frac{\boldsymbol{\gamma}_j}{|\boldsymbol{\gamma}_j|}, \text{ where}$$

$\mathbf{N}_k$ is the normal of unit length to the kth surface on the list of surfaces at the beginning of this section.

What is an appropriate formula for $\mathbf{N}_k$? In the new coordinate system, the kth surface

$$\sum_{j=1}^{n+1} \frac{(x^j)^2}{(a^j)^2 - c^k} = 1, \text{ may}$$

be represented in the form

$$\mathbf{x} = \mathbf{e}_m x^m(u^1, u^2, \ldots, u^{k-1}, c^k, u^{k+1}, \ldots, u^{n+1}).$$

This implies that

$$\frac{\partial \mathbf{x}}{\partial u^j} = \mathbf{e}_m \frac{\partial x^m}{\partial u^j} = \boldsymbol{\gamma}_j \text{ is}$$

tangent to the kth surface along as $j \neq k$. Since

$$\langle \boldsymbol{\gamma}_j, \boldsymbol{\gamma}_k \rangle = 0, \text{ if } j \neq k, \text{ it follows}$$

that $\boldsymbol{\gamma}_k$ is orthogonal to the kth surface. Therefore,

$$\mathbf{N}_k = \pm \frac{\boldsymbol{\gamma}_k}{|\boldsymbol{\gamma}_k|}. \tag{10.45}$$

From (10.40),

$$\boldsymbol{\gamma}_k = \sum_{m=1}^{n+1} \mathbf{e}_m \frac{x^m}{2\left(u^k - (a^m)^2\right)}. \tag{10.46}$$

Using (10.39), choosing the plus sign in (10.45), and noting that $j \neq k$, we have

$$\frac{\partial \boldsymbol{\gamma}_k}{\partial u^j} = \sum_{m=1}^{n+1} \mathbf{e}_m \frac{1}{2\left(u^k - (a^m)^2\right)} \frac{\partial x^m}{\partial u^j} = \sum_{m=1}^{n+1} \mathbf{e}_m \frac{1}{2\left(u^k - (a^m)^2\right)} \left(\frac{x^m}{2(u^j - (a^m)^2)}\right).$$

Using (10.42) and (10.46), we have

$$\frac{\partial \boldsymbol{\gamma}_k}{\partial u^j} = \frac{-1}{2(u^k - u^j)}\left[\boldsymbol{\gamma}_k - \boldsymbol{\gamma}_j\right].$$

Now, using this result

$$S\left(\frac{\boldsymbol{\gamma}_j}{|\boldsymbol{\gamma}_j|}\right) = -\frac{\partial}{\partial u^j}\mathbf{N}_k = \frac{\partial}{\partial u^j}\frac{\boldsymbol{\gamma}_k}{|\boldsymbol{\gamma}_k|} = \frac{-1}{|\boldsymbol{\gamma}_k|}\frac{\partial}{\partial u^j}\boldsymbol{\gamma}_k - \left(\frac{\partial}{\partial u^j}\frac{1}{|\boldsymbol{\gamma}_k|}\right)\boldsymbol{\gamma}_k$$

$$= \frac{1}{2(u^k - u^j)\,|\boldsymbol{\gamma}_k|}\left[\boldsymbol{\gamma}_k - \boldsymbol{\gamma}_j\right] - \left(\frac{\partial}{\partial u^j}\frac{1}{|\boldsymbol{\gamma}_k|}\right)\boldsymbol{\gamma}_k. \tag{10.47}$$

Since $\langle \mathbf{N}_k, \mathbf{N}_k\rangle = 1$,

$$\frac{1}{|\boldsymbol{\gamma}_k|}\left\langle \boldsymbol{\gamma}_k, \frac{\partial \mathbf{N}_k}{\partial u^j}\right\rangle = \left\langle \mathbf{N}_k, \frac{\partial \mathbf{N}_k}{\partial u^j}\right\rangle = 0.$$

This means that the terms involving $\boldsymbol{\gamma}_k$ on the right-hand side of (10.47) must cancel out. What remains is

$$S\left(\frac{\boldsymbol{\gamma}_j}{|\boldsymbol{\gamma}_j|}\right) = \frac{-1}{2(u^k - u^j)\,|\boldsymbol{\gamma}_k|}\boldsymbol{\gamma}_j = \frac{-|\boldsymbol{\gamma}_j|}{2(u^k - u^j)\,|\boldsymbol{\gamma}_k|}\frac{\boldsymbol{\gamma}_j}{|\boldsymbol{\gamma}_j|} = \lambda_j\frac{\boldsymbol{\gamma}_j}{|\boldsymbol{\gamma}_j|}. \tag{10.48}$$

Thus the principal curvature for the line of curvature associated with u^j on the kth surface is

$$\lambda_j = \frac{-|\boldsymbol{\gamma}_j|}{2(u^k - u^j)\,|\boldsymbol{\gamma}_k|}. \tag{10.49}$$

To finish our task, we need to obtain a formula for $|\boldsymbol{\gamma}_\alpha|$ or $\langle \boldsymbol{\gamma}_\alpha, \boldsymbol{\gamma}_\alpha\rangle$.

Theorem 263. *For the confocal coordinate system defined by (10.36),*

$$\langle \boldsymbol{\gamma}_\alpha, \boldsymbol{\gamma}_\alpha\rangle = \frac{1}{4}\frac{\prod_{j\neq\alpha}\left(u^j - u^\alpha\right)}{\prod_j\left[(a^j)^2 - u^\alpha\right]}. \tag{10.50}$$

Proof. From (10.46)

$$\boldsymbol{\gamma}_\alpha = \sum_{m=1}^{n+1}\mathbf{e}_m\frac{x^m}{2\left(u^\alpha - (a^m)^2\right)}, \quad \text{so}$$

$$\langle \boldsymbol{\gamma}_\alpha, \boldsymbol{\gamma}_\alpha\rangle = \frac{1}{4}\sum_{m=1}^{n+1}\frac{(x^m)^2}{\left(u^\alpha - (a^m)^2\right)^2}. \tag{10.51}$$

From (10.36),

$$(x^m)^2 = \frac{-\prod_{j=1}^{n+1}\left[u^j - (a^m)^2\right]}{\prod_{j\neq m}\left[(a^j)^2 - (a^m)^2\right]}.$$

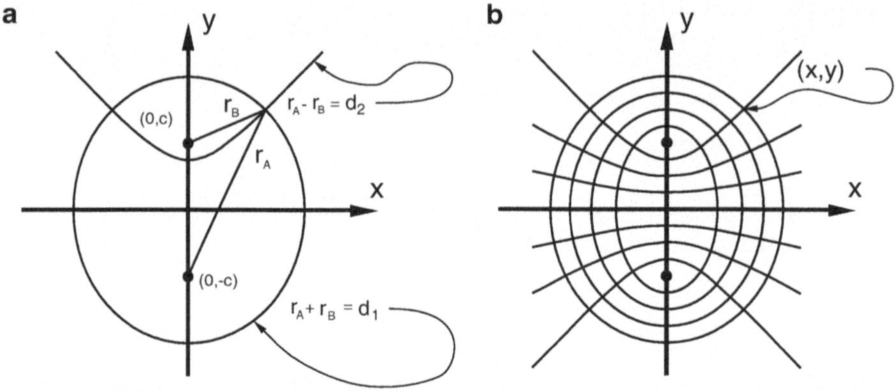

Fig. 10.3 Confocal ellipses and hyperbolas

So

$$\langle \boldsymbol{\gamma}_\alpha, \boldsymbol{\gamma}_\alpha \rangle = \frac{1}{4} \sum_{m=1}^{n+1} \frac{\prod_{j \neq \alpha} \left[u^j - (a^m)^2 \right]}{\prod_{j \neq m} ((a^j)^2 - (a^m)^2)^2} \frac{1}{(a^m)^2 - u^\alpha}. \tag{10.52}$$

For $n = 2$, the terms in the sum of the right-hand side of (10.52) can be combined by brute force to obtain (10.50). Margenau and Murphy (1956, pp. 178–180) point out that even in that case, it is easier to prove the equality by taking the right-hand side of (10.50) and expanding it in partial fractions. From the 3-dimensional case worked out by Margenau and Murphy, it is easy to guess at (10.50) for higher dimensions. I will leave it to you to expand the right hand side of (10.50) in partial fractions to show that (10.50) is equivalent to (10.52). See Problem 265. □

Margenau and Murphy (1956, p. 180) state that for the case $n = 2$, "The confocal ellipsoidal coordinate system has proved useful in problems of mechanics, potential theory, electrodynamics and hydrodynamics." They also cite several sources.

Problem 264. For $n = 1$, there are two equations in the list at the beginning of the section. Namely

$$\frac{(x^1)^2}{(a^1)^2 - u^1} + \frac{(x^2)^2}{(a^2)^2 - u^1} = 1, \text{ and}$$

$$\frac{-(x^1)^2}{u^2 - (a^1)^2} + \frac{(x^2)^2}{(a^2)^2 - u^2} = 1, \text{ where}$$

$$-\infty < u^1 < (a^1)^2 < u^2 < (a^2)^2.$$

Demonstrate that these equations represents families of confocal ellipses and hyperbolas. In particular, determine the location of the focal points common to both families. You may wish to refer to Fig. 10.3.

Problem 265. Suppose

$$f(u^\alpha) = \frac{\prod_{j\neq\alpha}(u^j - u^\alpha)}{4\prod_j ((a^j)^2 - u^\alpha)}.$$

Expand this function in partial fractions to show

$$f(u^\alpha) = \frac{1}{4}\sum_{m=1}^{n+1} \frac{A_m}{(a^m)^2 - u^\alpha}, \text{ where}$$

$$A_m = \frac{\prod_{j\neq\alpha}\left[u^j - (a^m)^2\right]}{\prod_{j\neq m}\left[(a^j)^2 - (a^m)^2\right]}.$$

10.4 *The Nondeformability of a Hypersurface

For a 2-dimensional surface embedded in E^3, the intrinsic observer cannot determine the principal curvatures. The intrinsic observer can determine the product of the principal curvatures at any point but not the two separate curvatures. Usually, this means that the surface can be distorted without stretching. (Stretching would cause a change in the metric tensor for the 2-dimensional surface that would be detectable to the intrinsic observer.) There are exceptions. For example, a complete sphere is rigid. However, if some portion of the sphere is cut away, it is no longer rigid.

In higher dimensions, the situation is different. For higher dimensions, we have the following theorem:

Theorem 266. *If three or more of the principal curvatures for an n-dimensional hypersurface are nonzero, the hypersurface cannot be distorted without stretching or contracting. (Stretching or contracting would change the distance between points on the surface and modify the metric tensor.)*

Proof. To prove this theorem, we show that both the principal curvatures and the principal directions can be determined from the Riemann tensor. Since the Riemann tensor is computed from the metric tensor, this result demonstrates that both the principal curvatures and the principal directions are determined by the metric tensor.

From (7.157), we have

$$R^{\alpha k}{}_{ij} = h_i^\alpha h_j^k - h_j^\alpha h_i^k. \tag{10.53}$$

If we define

$$\mathbf{H}_k = h_k^\alpha \boldsymbol{\gamma}_\alpha, \text{ then} \tag{10.54}$$

$$\mathbf{R}_{ij} = \frac{1}{2} R^{\alpha k}{}_{ij} \boldsymbol{\gamma}_\alpha \boldsymbol{\gamma}_k = \frac{1}{2} \left(h_i^\alpha h_j^k - h_j^\alpha h_i^k \right) \boldsymbol{\gamma}_\alpha \boldsymbol{\gamma}_k$$
$$= \frac{1}{2} \left(\mathbf{H}_i \mathbf{H}_j - \mathbf{H}_j \mathbf{H}_i \right)$$

or restated,

$$\mathbf{R}_{ij} = \mathbf{H}_i \wedge \mathbf{H}_j . \tag{10.55}$$

From (10.54), knowing $\mathbf{H}_j$ is equivalent to knowing the components of the jth column of the matrix

$$\left[h^i{}_j \right].$$

(Remember, I used the convention that the upper index is the row index and the lower index is the column index.)

The curvature 2-form is an intrinsic entity that can be determined from the metric tensor for the n-dimensional surface. Therefore if we can solve the system of equations represented by (10.55) for the $\mathbf{H}_k$'s, we know all the columns of $[h^i{}_j]$. If we know all the components of $[h^i{}_j]$, we can determine all the principal curvatures and the lines of curvature. Therefore to prove this theorem, it is sufficient to show that the system of equations represented by (10.55) can be solved for the $\mathbf{H}_k$'s.

If three of the principal curvatures are nonzero, the rank of the matrix $[h^i{}_j]$ must be at least three so at least three of the $\mathbf{H}_k$'s must be linearly independent. Without loss of generality, we can assume that $\mathbf{H}_1$, $\mathbf{H}_2$, and $\mathbf{H}_3$ are linearly independent. (If they are not, pick out three columns that are linearly independent and relabel the coordinates.) We are now faced with solving the system

$$\mathbf{H}_2 \wedge \mathbf{H}_3 = \mathbf{R}_{23}, \tag{10.56}$$

$$\mathbf{H}_3 \wedge \mathbf{H}_1 = \mathbf{R}_{31}, \text{ and} \tag{10.57}$$

$$\mathbf{H}_1 \wedge \mathbf{H}_2 = \mathbf{R}_{12}. \tag{10.58}$$

From (10.58), we know that $\mathbf{R}_{12}$ is the exterior product of two independent vectors in the 2-dimensional plane spanned by $\mathbf{H}_1$ and $\mathbf{H}_2$. You may be able to factor $\mathbf{R}_{12}$ into the exterior product of two such vectors but this may be difficult in practice.

Another technique to get two vectors that span the same space spanned by $\mathbf{H}_1$ and $\mathbf{H}_2$ is to return to (10.53). From this equation,

$$R^{\alpha k}{}_{12} \boldsymbol{\gamma}_\alpha = h_2^k \mathbf{H}_1 - h_1^k \mathbf{H}_2. \tag{10.59}$$

If we allow k to run through the values from 1 to n in (10.59), we will get a list of n vectors. At least two of them must be independent. (Otherwise, $h_2^k = \alpha h_1^k$ for all values of k and the second column of $[h^i{}_j]$ would be a scalar multiple of the first column.)

Suppose two values of k that result in independent vectors are p and q. This implies that the 2 dimensional space spanned by $R^{\alpha p}{}_{12} \boldsymbol{\gamma}_\alpha$ and $R^{\alpha q}{}_{12} \boldsymbol{\gamma}_\alpha$ is the same as that spanned by $\mathbf{H}_1$ and $\mathbf{H}_2$. That is in general

$$AR^{\alpha p}{}_{12}\boldsymbol{\gamma}_\alpha + BR^{\alpha q}{}_{12}\boldsymbol{\gamma}_\alpha = \lambda\mathbf{H}_1 + \mu\mathbf{H}_2.$$

To determine a linear combination of $R^{\alpha p}{}_{12}\boldsymbol{\gamma}_\alpha$ and $R^{\alpha q}{}_{12}\boldsymbol{\gamma}_\alpha$ that results in a scalar multiple of $\mathbf{H}_1$, we note that

$$\mathbf{R}_{31} = \mathbf{H}_3 \wedge \mathbf{H}_1.$$

So

$$\mathbf{R}_{31} \wedge \left(AR^{\alpha p}{}_{12}\boldsymbol{\gamma}_\alpha + BR^{\alpha q}{}_{12}\boldsymbol{\gamma}_\alpha\right) = \mu\mathbf{H}_3 \wedge \mathbf{H}_1 \wedge \mathbf{H}_2.$$

Therefore,

$$\begin{aligned} \mathbf{R}_{31} \wedge \left(AR^{\alpha p}{}_{12}\boldsymbol{\gamma}_\alpha + BR^{\alpha q}{}_{12}\boldsymbol{\gamma}_\alpha\right) = 0 \Leftrightarrow & \\ AR^{\alpha p}{}_{12}\boldsymbol{\gamma}_\alpha + BR^{\alpha q}{}_{12}\boldsymbol{\gamma}_\alpha = \lambda\mathbf{H}_1. & \end{aligned} \tag{10.60}$$

Thus, finding nonzero values of A and B that satisfy (10.60) is equivalent to determining a scalar multiple of $\mathbf{H}_1$. In a similar fashion, we can determine scalar multiples of $\mathbf{H}_2$ and $\mathbf{H}_3$.

Having determined the directions of $\mathbf{H}_1$, $\mathbf{H}_2$, and $\mathbf{H}_3$; we can determine normalized versions of each. That is

$$\hat{\mathbf{H}}_k = \frac{\mathbf{H}_k}{|\mathbf{H}_k|}, \text{ for } k = 1, 2, \text{and } 3.$$

What remains is the task of computing the magnitudes of $\mathbf{H}_1$, $\mathbf{H}_2$, and $\mathbf{H}_3$. Suppose $\mathbf{H}_1 = \lambda_1\hat{\mathbf{H}}_1$, $\mathbf{H}_2 = \lambda_2\hat{\mathbf{H}}_2$, and $\mathbf{H}_3 = \lambda_3\hat{\mathbf{H}}_3$. Then

$$\begin{aligned} \lambda_2\lambda_3\hat{\mathbf{H}}_2 \wedge \hat{\mathbf{H}}_3 &= \mathbf{R}_{23}, \\ \lambda_3\lambda_1\hat{\mathbf{H}}_3 \wedge \hat{\mathbf{H}}_1 &= \mathbf{R}_{31}, \text{ and} \\ \lambda_1\lambda_2\hat{\mathbf{H}}_1 \wedge \hat{\mathbf{H}}_2 &= \mathbf{R}_{12}. \end{aligned}$$

From these three equations, we can determine the products $\lambda_2\lambda_3$, $\lambda_3\lambda_1$, and $\lambda_1\lambda_2$. To get λ_1, we note that

$$(\lambda_1)^2 = \frac{(\lambda_3\lambda_1)(\lambda_1\lambda_2)}{(\lambda_2\lambda_3)}. \tag{10.61}$$

Solving (10.61), we encounter a sign ambiguity for λ_1. This is the same sign ambiguity, we ran into when we had to choose a direction for the normal vector $\mathbf{N}$. Once the sign is chosen for λ_1, the signs for λ_2 and λ_3 are determined since

$$\lambda_2 = \frac{(\lambda_1\lambda_2)}{\lambda_1} \text{ and } \lambda_3 = \frac{(\lambda_3\lambda_1)}{\lambda_1}.$$

Having determined $\mathbf{H}_1$, $\mathbf{H}_2$, and $\mathbf{H}_3$, I will leave it to you to figure out how to compute any of the other $\mathbf{H}_k$'s. □

Chapter 11
*Minimal Surfaces

11.1 *Why are Minimal Surfaces Said to be Minimal?

The term *minimal surface* is applied to any 2-dimensional surface embedded in E^3 with zero mean curvature. It is understood that *mean curvature* is the average of the two principal curvatures.

Surprisingly (at least to me), this concept can be generalized in a very elegant manner to 2-dimensional surfaces embedded in higher dimensional Euclidean spaces (Osserman 2002). However, I will limit the presentation in this book to 2-dimensional surfaces embedded in E^3.

What makes this field of differential geometry attractive to some is that it connects pictures of soap films stretched across various frames with the theory of complex variables. Both the theory of complex variables and pictures of soap films are aesthetically pleasing.

If a surface bounded by some closed loop has the least possible area, then that surface has zero mean curvature at each of its interior points. It is for this reason that any surface with mean zero curvature is said to be "minimal." However, it should be noted that "surface with least area" is not synonymous with "minimal surface." This point is illustrated by *Enneper's surface* defined by the equation:

$$\mathbf{x}(u,v) = \left(u - \frac{u^3}{3} + uv^2, -v + \frac{v^3}{3} - u^2 v, u^2 - v^2 \right). \tag{11.1}$$

(See Fig. 11.1a where the portion of Enneper's surface bounded by the loop $u^2 + v^2 = (1.5)^2$ is shown.)

If we consider the portion of Enneper's surface bounded by the loop $u^2 + v^2 = (1.5)^2$, we have a surface that is "minimal" in the sense that its mean curvature is zero. However, Enneper's surface for the given loop does not have the least area. By numerical integration, it can be shown that the cylindrical surface bounded by the same loop shown in Fig. 11.1b has less area.

J. Snygg, *A New Approach to Differential Geometry using Clifford's Geometric Algebra*, DOI 10.1007/978-0-8176-8283-5_11,

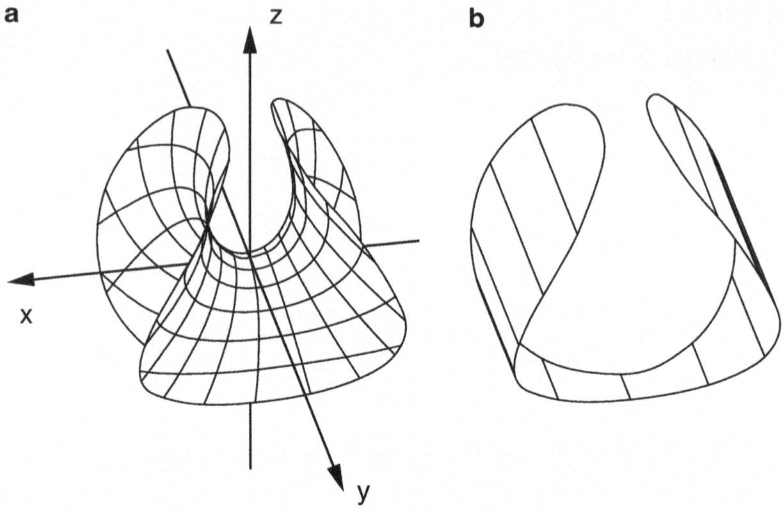

Fig. 11.1 (**a**) Portion of Enneper's surface bounded by a loop. (**b**) Cylindrical surface bounded by the same loop

The theory is developed well enough so that it is known that the surface with least area has zero mean curvature but an explicit formula for this particular problem is unknown. The most important theorems on this topic are discussed in John Oprea's text *Differential Geometry And Its Applications* (Oprea 1997, pp. 219–253).

Perhaps, the most important theorem is as follows:

Theorem 267. *A necessary condition for a surface bounded by a regular closed curve to have the least possible area is that the mean curvature of the surface at all interior points be zero.*

Proof. We first note that

$$\mathrm{d}\mathbf{A} = \boldsymbol{\gamma}_1 \mathrm{d}u^1 \wedge \boldsymbol{\gamma}_2 \mathrm{d}u^2 \text{ so}$$

$$A = \int\int_R |\boldsymbol{\gamma}_{12}|\, \mathrm{d}u^1 \mathrm{d}u^2, \text{ where} \tag{11.2}$$

R is the portion of the given surface bounded by the given regular closed curve. Since

$$|\boldsymbol{\gamma}_{12}|^2 = \boldsymbol{\gamma}_{12}\boldsymbol{\gamma}_{21} = g_{11}g_{22} - g_{12}g_{21} = g, \text{ we have}$$

$$A = \int\int_R \sqrt{g}\mathrm{d}u^1 \mathrm{d}u^2. \tag{11.3}$$

(The following argument was lifted from Manfredo do Carmo's *Differential Geometry of Curves and Surfaces* (do Carmo 1976, pp. 197–199).)

If $\mathbf{x}(u^1, u^2)$ is the surface that has the least area for the given boundary, then an alternate surface corresponding to a *normal variation* would have a greater area. A family of alternate surfaces for a normal variation can be represented in the form:

$$\bar{\mathbf{x}}(u^1, u^2) = \mathbf{x}(u^1, u^2) + tw(u^1, u^2)\mathbf{N}(u^1, u^2), \text{ where} \tag{11.4}$$

$w(u^1, u^2)$ is any differentiable function that is zero on the boundary and $\mathbf{N}(u^1, u^2)$ is the unit normal for the surface $\mathbf{x}(u^1, u^2)$. (It is understood that each member of the family corresponds to a particular value of t.)

From (11.4), it can be said that for each member of the family,

$$\begin{aligned}\bar{\boldsymbol{\gamma}}_j &= \frac{\partial \bar{\mathbf{x}}}{\partial u^j} \\ &= \frac{\partial \mathbf{x}}{\partial u^j} + tw\frac{\partial \mathbf{N}}{\partial u^j} + t\frac{\partial w}{\partial u^j}\mathbf{N} \\ &= \boldsymbol{\gamma}_j - twh_j^\alpha \boldsymbol{\gamma}_\alpha + t\frac{\partial w}{\partial u^j}\mathbf{N}.\end{aligned}$$

Therefore using the fact that $\langle \boldsymbol{\gamma}_k, \mathbf{N}\rangle = 0$,

$$\bar{g}_{jk} = \langle \bar{\boldsymbol{\gamma}}_j, \bar{\boldsymbol{\gamma}}_k\rangle = \langle \boldsymbol{\gamma}_j, \boldsymbol{\gamma}_k\rangle - 2twh_j^\alpha \langle \boldsymbol{\gamma}_\alpha, \boldsymbol{\gamma}_k\rangle + t^2 A_{jk}.,$$

or

$$\bar{g}_{jk} = g_{jk} - 2twh_{jk} + t^2 A_{jk}. \tag{11.5}$$

(We do not need to compute A_{jk}.)

Using (11.5) and ignoring powers of t higher than one, we have

$$\begin{aligned}\bar{g} &= \bar{g}_{11}\bar{g}_{22} - \bar{g}_{12}\bar{g}_{21} \\ &= (g_{11} - 2twh_{11})(g_{22} - 2twh_{22}) - (g_{12} - 2twh_{12})(g_{21} - 2twh_{21}) \\ &= g_{11}g_{22} - g_{12}g_{21} - 2tw(g_{11}h_{22} + g_{22}h_{11} - g_{12}h_{21} - g_{21}h_{12}) \\ &= g - 2tw(g_{11}g_{2\alpha}h_2^\alpha + g_{22}g_{1\alpha}h_1^\alpha - g_{12}g_{2\alpha}h_1^\alpha - g_{21}g_{1\alpha}h_2^\alpha).\end{aligned} \tag{11.6}$$

Summing over α on the R.H.S. of (11.6), we find some terms cancel out and we get

$$\begin{aligned}\bar{g} &= g - 2tw(g_{11}g_{22} - g_{12}g_{21})(h_1^1 + h_2^2) \\ &= g\left[1 - 2tw(h_1^1 + h_2^2)\right],\end{aligned}$$

or more precisely:

$$\bar{g} = g\left[1 - 2tw(h_1^1 + h_2^2) + t^2 P(t)\right], \text{ where}$$

$P(t)$ is a quadratic polynomial in the variable "t."

If we designate the mean curvature by H, then

$$H = \frac{h_1^1 + h_2^2}{2}. \tag{11.7}$$

(See Problem 268.)

It is now clear that

$$\begin{aligned}
\bar{g} &= g\left[1 - 4twH + t^2P(t)\right], \text{ so} \\
\bar{A}(t) &= \int\int_R \sqrt{\bar{g}}\mathrm{d}u^1\mathrm{d}u^2 \\
&= \int\int_R \sqrt{g}\sqrt{1 - 4twH + t^2P(t)}\mathrm{d}u^1\mathrm{d}u^2
\end{aligned}$$

If $\mathbf{x}(u^1, u^2)$ is the surface with least area, then

$$\left.\frac{\mathrm{d}\bar{A}(t)}{\mathrm{d}t}\right|_{t=0} = 0 = -2\int\int_R w(u^1, u^2)H\sqrt{g}\mathrm{d}u^1\mathrm{d}u^2. \tag{11.8}$$

This condition must hold for an arbitrary differentiable function $w(u^1, u^2)$ that is zero on the boundary. For reasons that you should fill in, this implies that $H = 0$. (See Problem 269.) □

Problem 268. Consider the quadratic equation for the curvatures of a 2-dimensional surface embedded in E^3. Namely:

$$\det\begin{bmatrix} h_1^1 - \lambda & h_2^1 \\ h_1^2 & h_2^2 - \lambda \end{bmatrix} = 0.$$

Use this equation to confirm (11.7).

Problem 269. Demonstrate that (11.8) implies that $H = 0$. Suggestion: It may be useful to consider the function

$$w(u^1, u^2) = \begin{cases} \exp\left(\dfrac{1}{(u^1 - a)^2 + (u^2 - b)^2 - r^2}\right) & \text{for } (u^1 - a)^2 + (u^2 - b)^2 < r^2 \\ 0 \text{ for } (u^1 - a)^2 + (u^2 - b)^2 \geq r^2. \end{cases}$$

(Is this function differentiable?)

Problem 270. a. Use (11.1) to show that for Enneper's surface:

$$\begin{aligned}
\gamma_{uv} &= (1 + u^2 + v^2)\left[\mathbf{e}_{23}2u + \mathbf{e}_{31}2v + \mathbf{e}_{12}(-1 + u^2 + v^2)\right] \text{ and} \\
|\gamma_{uv}| &= (1 + u^2 + v^2)^2.
\end{aligned}$$

b. Let $u = r\cos\theta$ and $v = r\sin\theta$ to compute the area of Enneper's surface bounded by the loop $u^2 + v^2 = (1.5)^2$.
c. Use MAPLE or MATHEMATICA to obtain a numerical approximation for the area of the cylindrical surface shown in Fig. 11.1b. Compare your answer for this part with your answer for part (b).

Problem 271. a. Using (11.1), determine the symmetries of Enneper's surface.
b. If Enneper's surface is extended beyond the portion shown in Fig. 11.1a, it intersects itself in the y-z plane for $z > 0$. Where else does it intersect itself? Determine equations for the two curves of intersection.

Problem 272. a. Determine the principal curvatures for Enneper's surface.
b. Show that the coordinate curves arising from (11.1) for Enneper's surface are also lines of curvature.

11.2 *Minimal Surfaces and Harmonic Functions

If the metric tensor is a scalar multiple of the identity matrix, the corresponding coordinates are said to be *isothermic*. (I find this terminology rather strange but it has been used for a long time. Dirk Struik traced it back to a paper written by Gabriel Lamé (1795–1870) in 1833 (Struik 1988, p. 175).)

In theory, any 2-dimensional surface can be parameterized by isothermal coordinates. A proof by Robert Osserman for minimal surfaces appears in Oprea's text (Oprea 1997, p. 222). In addition, if one uses isothermal coordinates for a minimal surface, then each component of the surface is *harmonic*. That is if the surface is represented by the equation:

$$\mathbf{x}(u^1, u^2) = \mathbf{e}_k x^k(u^1, u^2), \text{ then}$$

$$\Delta\mathbf{x}(u^1, u^2) = \mathbf{e}_k \Delta x^k(u^1, u^2) = \mathbf{e}_k \left(\frac{\partial^2 x^k}{(\partial u^1)^2} + \frac{\partial^2 x^k}{(\partial u^2)^2} \right) = 0. \tag{11.9}$$

The easiest way to study harmonic functions is in the context of the theory of complex variables. If $z = x + iy$ where x and y are real then z^* designates the *complex conjugate of* z, where

$$z^* = x - iy. \tag{11.10}$$

Suppose

$$z = x + iy \text{ and } w = u + iv, \text{ then}$$

it is not too difficult to show that

$$(z+w)^* = z^* + w^*, \tag{11.11}$$

$$(zw)^* = z^* w^*, \text{ and} \tag{11.12}$$

$$(z^n)^* = (z^*)^n, \text{ where} \tag{11.13}$$

n is an integer. (See Problem 273.) It then follows that if

$$f(z) = \sum_{k=0}^{\infty} a_k (z - z_0)^k, \text{ then } (f(z))^* = \sum_{k=0}^{\infty} a_k^* (z^* - z_0^*)^k = f^*(z^*).$$

As you might expect, the derivative of $f(z)$ at $z = z_0$ is defined by the limit

$$f'(z) = \lim_{\Delta z \to 0} \frac{f(z + \Delta z) - f(z)}{\Delta z}. \tag{11.14}$$

The consequence of this definition is quite different than it is for real variables. The limit in (11.14) is meaningful only if it is independent of the direction that Δz approaches zero in the x-y plane. As a result, it can be shown that if the first derivative exists in some open neighborhood of z_0, then the derivative of any order exists in that same neighborhood. Furthermore if $f'(z)$ exists when

$$|z - z_0| = \sqrt{(x - x_0)^2 + (y - y_0)^2} < R, \text{ then}$$

$$f(z) = \sum_{k=0}^{\infty} \frac{f^{(k)}(z_0)}{k!} (z - z_0)^k \text{ for } |z - z_0| < R. \tag{11.15}$$

Due to this situation, if $f'(z)$ exists, $f(z)$ is not described as being merely differentiable – it is said to be *analytic*.

It turns out that a function $f(z)$ that is analytic in a region $|z - z_0| < R$ can be characterized by the fact that there exists an infinite (or possibly finite) series $\sum_{k=0}^{\infty} a_k (z - z_0)^k$ such that

$$f(z) = \sum_{k=0}^{\infty} a_k (z - z_0)^k \text{ for } |z - z_0| < R \text{ where the } a_k\text{'s are possibly complex.}$$

(We will use this definition of an analytic function in the next section.)

What does all this have to do with harmonic functions? Suppose

$$f(z) = p(x, y) + i q(x, y), \text{ where}$$

$$z = x + iy, \text{ and}$$

both p and q are real. Since the limit in (11.14) is independent of the direction that Δz approaches zero,

$$\lim_{\Delta x \to 0} \lim_{\Delta y \to 0} \frac{f(z+\Delta z) - f(z)}{\Delta z} = \lim_{\Delta y \to 0} \lim_{\Delta x \to 0} \frac{f(z+\Delta z) - f(z)}{\Delta z}.$$

That is

$$\lim_{\Delta x \to 0} \frac{p(x+\Delta x, y) + iq(x+\Delta x, y) - p(x,y) - iq(x,y)}{\Delta x}$$
$$= \lim_{\Delta y \to 0} \frac{p(x, y+\Delta y) + iq(x, y+\Delta y) - p(x,y) - iq(x,y)}{i\Delta y}.$$

This implies that

$$\frac{\partial p}{\partial x} + i\frac{\partial q}{\partial x} = -i\frac{\partial p}{\partial y} + \frac{\partial q}{\partial y}. \tag{11.16}$$

Equating the real and imaginary parts of (11.16), we have

$$\frac{\partial p}{\partial x} = \frac{\partial q}{\partial y} \text{ and } \frac{\partial p}{\partial y} = -\frac{\partial q}{\partial x}. \tag{11.17}$$

This pair of equations is known as the *Cauchy–Riemann equations*. An immediate consequence is

$$\frac{\partial^2 p}{(\partial x)^2} = \frac{\partial^2 q}{\partial x \partial y} = -\frac{\partial^2 p}{(\partial y)^2} \text{ and thus}$$

$p(x, y)$ is harmonic. That is

$$\frac{\partial^2 p}{(\partial x)^2} + \frac{\partial^2 p}{(\partial y)^2} = 0.$$

(I leave it to you to demonstrate that $q(x, y)$ is also harmonic.)

If

$$f^k(z) = p^k(x,y) + iq^k(x,y), \text{ and}$$
$$f^{k*}(z^*) = p^k(x,y) - iq^k(x,y), \text{ then}$$

it is clear that $f^k(z) + f^{k*}(z^*)$ is a real harmonic function. Assuming that I have not lied to you at the beginning of this section, we can represent a minimal surface $\mathbf{x}(u^1, u^2)$ in the form:

$$\mathbf{x}(u^1, u^2) = \mathbf{F}(z) + \mathbf{F}^*(z^*) = \mathbf{e}_k \left(f^k(z) + f^{k*}(z^*)\right), \text{ where} \tag{11.18}$$

this time $z = u^1 + iu^2$.

As we shall see below, this representation does not guarantee that the coordinates are isothermal unless we also require that

$$\left\langle \frac{\partial \mathbf{F}}{\partial z}, \frac{\partial \mathbf{F}}{\partial z} \right\rangle = 0. \tag{11.19}$$

You should note that this also implies that

$$\left\langle \frac{\partial \mathbf{F}^*}{\partial z^*}, \frac{\partial \mathbf{F}^*}{\partial z^*} \right\rangle = 0.$$

It will be shown below that (11.18) and (11.19) are sufficient for the surface to be minimal. To show that a minimal surface can be parameterized by isothermal coordinates with harmonic components is considerably more difficult. If you wish to pursue this further, I recommend (Osserman 2002) or (Oprea 1997). In this short presentation, I am limiting my discussion to 2-dimensional surfaces embedded in E^3. Osserman discusses the generalization to 2-dimensional surfaces embedded in E^k, where $k \geq 3$.

Now from (11.18),

$$\boldsymbol{\gamma}_1 = \frac{\partial \mathbf{x}}{\partial u^1} = \frac{\partial \mathbf{F}}{\partial z}\frac{\partial z}{\partial u^1} + \frac{\partial \mathbf{F}^*}{\partial z^*}\frac{\partial z^*}{\partial u^1} = \frac{\partial \mathbf{F}}{\partial z} + \frac{\partial \mathbf{F}^*}{\partial z^*}, \text{ and} \tag{11.20}$$

$$\boldsymbol{\gamma}_2 = \frac{\partial \mathbf{x}}{\partial u^2} = \frac{\partial \mathbf{F}}{\partial z}\frac{\partial z}{\partial u^2} + \frac{\partial \mathbf{F}^*}{\partial z^*}\frac{\partial z^*}{\partial u^2} = i\frac{\partial \mathbf{F}}{\partial z} - i\frac{\partial \mathbf{F}^*}{\partial z^*}. \tag{11.21}$$

This implies:

$$\begin{aligned} g_{11} = \langle \boldsymbol{\gamma}_1, \boldsymbol{\gamma}_1 \rangle &= \left\langle \frac{\partial \mathbf{F}}{\partial z} + \frac{\partial \mathbf{F}^*}{\partial z^*}, \frac{\partial \mathbf{F}}{\partial z} + \frac{\partial \mathbf{F}^*}{\partial z^*} \right\rangle \\ &= \left\langle \frac{\partial \mathbf{F}}{\partial z}, \frac{\partial \mathbf{F}}{\partial z} \right\rangle + \left\langle \frac{\partial \mathbf{F}^*}{\partial z^*}, \frac{\partial \mathbf{F}^*}{\partial z^*} \right\rangle + 2\left\langle \frac{\partial \mathbf{F}}{\partial z}, \frac{\partial \mathbf{F}^*}{\partial z^*} \right\rangle. \end{aligned}$$

Similarly,

$$g_{22} = -\left\langle \frac{\partial \mathbf{F}}{\partial z}, \frac{\partial \mathbf{F}}{\partial z} \right\rangle - \left\langle \frac{\partial \mathbf{F}^*}{\partial z^*}, \frac{\partial \mathbf{F}^*}{\partial z^*} \right\rangle + 2\left\langle \frac{\partial \mathbf{F}}{\partial z}, \frac{\partial \mathbf{F}^*}{\partial z^*} \right\rangle, \text{ and}$$

$$g_{12} = g_{21} = i\left\langle \frac{\partial \mathbf{F}}{\partial z}, \frac{\partial \mathbf{F}}{\partial z} \right\rangle - i\left\langle \frac{\partial \mathbf{F}^*}{\partial z^*}, \frac{\partial \mathbf{F}^*}{\partial z^*} \right\rangle.$$

For isothermal coordinates, $g_{11} = g_{22}$. This implies that

$$\left\langle \frac{\partial \mathbf{F}}{\partial z}, \frac{\partial \mathbf{F}}{\partial z} \right\rangle + \left\langle \frac{\partial \mathbf{F}^*}{\partial z^*}, \frac{\partial \mathbf{F}^*}{\partial z^*} \right\rangle = 0.$$

For isothermal coordinates, we also require that $g_{12} = g_{21} = 0$. Therefore,

$$\left\langle \frac{\partial \mathbf{F}}{\partial z}, \frac{\partial \mathbf{F}}{\partial z} \right\rangle - \left\langle \frac{\partial \mathbf{F}^*}{\partial z^*}, \frac{\partial \mathbf{F}^*}{\partial z^*} \right\rangle = 0.$$

From these last two equations, we see that for isothermal coordinates:

$$\left\langle \frac{\partial \mathbf{F}}{\partial z}, \frac{\partial \mathbf{F}}{\partial z} \right\rangle = \left\langle \frac{\partial \mathbf{F}^*}{\partial z^*}, \frac{\partial \mathbf{F}^*}{\partial z^*} \right\rangle = 0. \tag{11.22}$$

It should now be obvious that

$$\begin{bmatrix} g_{11} & g_{12} \\ g_{21} & g_{22} \end{bmatrix} = \begin{bmatrix} 2\left\langle \frac{\partial \mathbf{F}}{\partial z}, \frac{\partial \mathbf{F}^*}{\partial z^*} \right\rangle & 0 \\ 0 & 2\left\langle \frac{\partial \mathbf{F}}{\partial z}, \frac{\partial \mathbf{F}^*}{\partial z^*} \right\rangle \end{bmatrix} \tag{11.23}$$

and

$$\begin{bmatrix} g^{11} & g^{12} \\ g^{21} & g^{22} \end{bmatrix} = \begin{bmatrix} \frac{1}{2\left\langle \frac{\partial \mathbf{F}}{\partial z}, \frac{\partial \mathbf{F}^*}{\partial z^*} \right\rangle} & 0 \\ 0 & \frac{1}{2\left\langle \frac{\partial \mathbf{F}}{\partial z}, \frac{\partial \mathbf{F}^*}{\partial z^*} \right\rangle} \end{bmatrix}. \tag{11.24}$$

To compute the normal vector $\mathbf{N}$, we note that

$$\frac{\boldsymbol{\gamma}_{12}\mathbf{N}}{|\boldsymbol{\gamma}_{12}|} = \mathbf{e}_{123} \text{ and thus}$$

$$\mathbf{N} = \frac{\boldsymbol{\gamma}_{21}}{|\boldsymbol{\gamma}_{12}|}\mathbf{e}_{123}, \text{ where} \tag{11.25}$$

$$\boldsymbol{\gamma}_{12} = \frac{\partial \mathbf{x}}{\partial u^1} \wedge \frac{\partial \mathbf{x}}{\partial u^2} = \left(\frac{\partial \mathbf{F}}{\partial z} + \frac{\partial \mathbf{F}^*}{\partial z^*} \right) \wedge i \left(\frac{\partial \mathbf{F}}{\partial z} - \frac{\partial \mathbf{F}^*}{\partial z^*} \right). \tag{11.26}$$

Since

$$\frac{\partial \mathbf{F}}{\partial z} \wedge \frac{\partial \mathbf{F}}{\partial z} = \frac{\partial \mathbf{F}^*}{\partial z^*} \wedge \frac{\partial \mathbf{F}^*}{\partial z^*} = 0,$$

Equation (11.26) becomes

$$\boldsymbol{\gamma}_{12} = -2i \frac{\partial \mathbf{F}}{\partial z} \wedge \frac{\partial \mathbf{F}^*}{\partial z^*}, \text{ and thus}$$

Equation (11.25) becomes

$$\mathbf{N} = \frac{i\,\frac{\partial \mathbf{F}}{\partial z} \wedge \frac{\partial \mathbf{F}^*}{\partial z^*}}{\left| i\,\frac{\partial \mathbf{F}}{\partial z} \wedge \frac{\partial \mathbf{F}^*}{\partial z^*} \right|} \mathbf{e}_{123}. \tag{11.27}$$

To get the components of the second fundamental form, we note that

$$\frac{\partial \mathbf{N}}{\partial u^j} = -h_j^\alpha \frac{\partial \mathbf{x}}{\partial u^\alpha} \quad \text{or}$$

$$\left\langle \frac{\partial \mathbf{N}}{\partial u^j}, \frac{\partial \mathbf{x}}{\partial u^\beta} \right\rangle = -h_j^\alpha \left\langle \frac{\partial \mathbf{x}}{\partial u^\alpha}, \frac{\partial \mathbf{x}}{\partial u^\beta} \right\rangle = -h_j^\alpha g_{\alpha\beta} = -h_{j\beta}. \tag{11.28}$$

However,

$$\left\langle \frac{\partial \mathbf{N}}{\partial u^j}, \frac{\partial \mathbf{x}}{\partial u^\beta} \right\rangle = \frac{\partial}{\partial u^j} \left\langle \mathbf{N}, \frac{\partial \mathbf{x}}{\partial u^\beta} \right\rangle - \left\langle \mathbf{N}, \frac{\partial^2 \mathbf{x}}{\partial u^j \partial u^\beta} \right\rangle \quad \text{or}$$

$$h_{j\beta} = \left\langle \mathbf{N}, \frac{\partial^2 \mathbf{x}}{\partial u^j \partial u^\beta} \right\rangle. \tag{11.29}$$

Since

$$\frac{\partial \mathbf{x}}{\partial u^1} = \frac{\partial \mathbf{F}}{\partial z} + \frac{\partial \mathbf{F}^*}{\partial z^*},$$

$$\frac{\partial^2 \mathbf{x}}{(\partial u^1)^2} = \frac{\partial^2 \mathbf{F}}{(\partial z)^2} \frac{\partial z}{\partial u^1} + \frac{\partial^2 \mathbf{F}^*}{(\partial z^*)^2} \frac{\partial z^*}{\partial u^1} = \frac{\partial^2 \mathbf{F}}{(\partial z)^2} + \frac{\partial^2 \mathbf{F}}{(\partial z^*)^2}.$$

Similarly,

$$\frac{\partial^2 \mathbf{x}}{\partial u^2 \partial u^1} = i\,\frac{\partial^2 \mathbf{F}}{(\partial z)^2} - i\,\frac{\partial^2 \mathbf{F}^*}{(\partial z^*)^2}, \quad \text{and}$$

$$\frac{\partial^2 \mathbf{x}}{(\partial u^2)^2} = -\frac{\partial^2 \mathbf{F}}{(\partial z)^2} - \frac{\partial^2 \mathbf{F}^*}{(\partial z^*)^2}.$$

Combining these results with (11.29), we have

$$\begin{bmatrix} h_{11} & h_{12} \\ h_{21} & h_{22} \end{bmatrix} = \begin{bmatrix} \left\langle \mathbf{N}, \frac{\partial^2 \mathbf{F}}{(\partial z)^2} + \frac{\partial^2 \mathbf{F}^*}{(\partial z^*)^2} \right\rangle & \left\langle \mathbf{N}, i\,\frac{\partial^2 \mathbf{F}}{(\partial z)^2} - i\,\frac{\partial^2 \mathbf{F}^*}{(\partial z^*)^2} \right\rangle \\ \left\langle \mathbf{N}, i\,\frac{\partial^2 \mathbf{F}}{(\partial z)^2} - i\,\frac{\partial^2 \mathbf{F}^*}{(\partial z^*)^2} \right\rangle & -\left\langle \mathbf{N}, \frac{\partial^2 \mathbf{F}}{(\partial z)^2} + \frac{\partial^2 \mathbf{F}^*}{(\partial z^*)^2} \right\rangle \end{bmatrix}$$

Since $h_i^j = h_{i\alpha} g^{\alpha j}$, it is not too difficult to discover that

$$\begin{bmatrix} h_1^1 & h_2^1 \\ h_1^2 & h_2^2 \end{bmatrix} = \frac{1}{2\left\langle \frac{\partial \mathbf{F}}{\partial z}, \frac{\partial \mathbf{F}^*}{\partial z^*} \right\rangle} \begin{bmatrix} \left\langle \mathbf{N}, \frac{\partial^2 \mathbf{F}}{(\partial z)^2} + \frac{\partial^2 \mathbf{F}^*}{(\partial z^*)^2} \right\rangle & \left\langle \mathbf{N}, i\frac{\partial^2 \mathbf{F}}{(\partial z)^2} - i\frac{\partial^2 \mathbf{F}^*}{(\partial z^*)^2} \right\rangle \\ \left\langle \mathbf{N}, i\frac{\partial^2 \mathbf{F}}{(\partial z)^2} - i\frac{\partial^2 \mathbf{F}^*}{(\partial z^*)^2} \right\rangle & -\left\langle \mathbf{N}, \frac{\partial^2 \mathbf{F}}{(\partial z)^2} + \frac{\partial^2 \mathbf{F}^*}{(\partial z^*)^2} \right\rangle \end{bmatrix}. \tag{11.30}$$

To obtain the principal curvatures, we must solve the equation:

$$\det \begin{bmatrix} h_1^1 - \lambda & h_2^1 \\ h_1^2 & h_2^2 - \lambda \end{bmatrix} = 0.$$

Doing this, we get

$$\lambda = \frac{\pm\sqrt{\left\langle \mathbf{N}, \frac{\partial^2 \mathbf{F}}{(\partial z)^2} + \frac{\partial^2 \mathbf{F}^*}{(\partial z^*)^2} \right\rangle^2 + \left\langle \mathbf{N}, i\frac{\partial^2 \mathbf{F}}{(\partial z)^2} - i\frac{\partial^2 \mathbf{F}^*}{(\partial z^*)^2} \right\rangle^2}}{2\left\langle \frac{\partial \mathbf{F}}{\partial z}, \frac{\partial \mathbf{F}^*}{\partial z^*} \right\rangle}, \text{ where} \tag{11.31}$$

$$\mathbf{N} = \frac{i\frac{\partial \mathbf{F}}{\partial z} \wedge \frac{\partial \mathbf{F}^*}{\partial z^*}}{\left| i\frac{\partial \mathbf{F}}{\partial z} \wedge \frac{\partial \mathbf{F}^*}{\partial z^*} \right|} \mathbf{e}_{123}. \tag{11.32}$$

From (11.31), it is clear that the mean curvature is zero as was predicted. Repeating ourselves, we note that to construct a minimal surface, we require that

$$\left\langle \frac{\partial \mathbf{F}}{\partial z}, \frac{\partial \mathbf{F}}{\partial z} \right\rangle = 0.$$

An example of a vector function **F** that satisfies this condition is

$$\frac{\partial \mathbf{F}}{\partial z} = \frac{a}{2}(-\cos z, -\sin z, -i).$$

In this case,

$$\mathbf{F} = \frac{a}{2}(-\sin z, \cos z, -iz) \text{ plus} \tag{11.33}$$

a possible constant. This corresponds to a catenoid. (See Problem 274.)

Once one has a vector function $\mathbf{F}(z)$ to represent a minimal surface, one can construct an entire family of minimal surfaces that can be deformed into one another without stretching. (This can be said to be a family of *isometric* surfaces.) To do this, simply replace $\mathbf{F}(z)$ by $\mathbf{F}(z)e^{i\beta}$, where β is a constant. Each member of the family has the same metric tensor since

$$g_{11} = g_{22} = 2\left\langle \frac{\partial\left(\mathbf{F}e^{i\beta}\right)}{\partial z}, \frac{\partial\left(\mathbf{F}^{*}e^{-i\beta}\right)}{\partial z^{*}} \right\rangle = 2\left\langle \frac{\partial \mathbf{F}}{\partial z}, \frac{\partial \mathbf{F}^{*}}{\partial z^{*}} \right\rangle.$$

Also

$$g_{12} = g_{21} = 0. \text{ Why?}$$

Using essentially the same argument and (11.32), it becomes clear that for any two members of the family, points with the same coordinates have the same normal vector.

Surprisingly (at least to me), even the curvatures at points with the same coordinates are identical. To see this, first consider the case for which $\beta = 0$. If we let

$$\frac{\left\langle \mathbf{N}, \dfrac{\partial^2 \mathbf{F}}{(\partial z)^2} + \dfrac{\partial^2 \mathbf{F}^{*}}{(\partial z^{*})^2} \right\rangle}{2\left\langle \dfrac{\partial \mathbf{F}}{\partial z}, \dfrac{\partial \mathbf{F}^{*}}{\partial z^{*}} \right\rangle} = A\cos\theta, \text{ and}$$

$$\frac{\left\langle \mathbf{N}, i\dfrac{\partial^2 \mathbf{F}}{(\partial z)^2} - i\dfrac{\partial^2 \mathbf{F}^{*}}{(\partial z^{*})^2} \right\rangle}{2\left\langle \dfrac{\partial \mathbf{F}}{\partial z}, \dfrac{\partial \mathbf{F}^{*}}{\partial z^{*}} \right\rangle} = A\sin\theta, \text{ where}$$

$$A = \frac{\sqrt{\left\langle \mathbf{N}, \dfrac{\partial^2 \mathbf{F}}{(\partial z)^2} + \dfrac{\partial^2 \bar{\mathbf{F}}}{(\partial \bar{z})^2} \right\rangle^2 + \left\langle \mathbf{N}, i\dfrac{\partial^2 \mathbf{F}}{(\partial z)^2} - i\dfrac{\partial^2 \mathbf{F}^{*}}{(\partial z^{*})^2} \right\rangle^2}}{2\left\langle \dfrac{\partial \mathbf{F}}{\partial z}, \dfrac{\partial \mathbf{F}^{*}}{\partial z^{*}} \right\rangle}, \text{ then}$$

Equation (11.30) becomes

$$\begin{bmatrix} h_1^1 & h_2^1 \\ h_1^2 & h_2^2 \end{bmatrix} = \begin{bmatrix} A\cos\theta & A\sin\theta \\ A\sin\theta & -A\cos\theta \end{bmatrix}, \text{ and}$$

Equation (11.31) becomes

$$\lambda = \pm\sqrt{A^2\cos^2\theta + A^2\sin^2\theta} = \pm A.$$

If $\beta \neq 0$, then

$$h_1^1 = -h_2^2 = \frac{\left\langle \mathbf{N}, \dfrac{\partial^2\left(\mathbf{F}e^{i\beta}\right)}{(\partial z)^2} + \dfrac{\partial^2\left(\mathbf{F}^* e^{-i\beta}\right)}{(\partial z^*)^2} \right\rangle}{2\left\langle \dfrac{\partial\left(\mathbf{F}e^{i\beta}\right)}{\partial z}, \dfrac{\partial\left(\mathbf{F}^* e^{-i\beta}\right)}{\partial z^*} \right\rangle}$$

$$= \frac{\left\langle \mathbf{N}, \dfrac{\partial^2\mathbf{F}}{(\partial z)^2} + \dfrac{\partial^2\mathbf{F}^*}{(\partial z^*)^2} \right\rangle}{2\left\langle \dfrac{\partial\mathbf{F}}{\partial z}, \dfrac{\partial\mathbf{F}^*}{\partial z^*} \right\rangle}\cos\beta + \frac{\left\langle \mathbf{N}, i\dfrac{\partial^2\mathbf{F}}{(\partial z)^2} - i\dfrac{\partial^2\mathbf{F}^*}{(\partial z^*)^2} \right\rangle}{2\left\langle \dfrac{\partial\mathbf{F}}{\partial z}, \dfrac{\partial\mathbf{F}^*}{\partial z^*} \right\rangle}\sin\beta$$

$$= A\cos\theta\cos\beta + A\sin\theta\sin\beta = A\cos(\theta - \beta).$$

Furthermore

$$h_2^1 = h_1^2 = \frac{\left\langle \mathbf{N}, i\dfrac{\partial^2\left(\mathbf{F}e^{i\beta}\right)}{(\partial z)^2} - i\dfrac{\partial^2\left(\mathbf{F}^* e^{-i\beta}\right)}{(\partial z^*)^2} \right\rangle}{2\left\langle \dfrac{\partial\left(\mathbf{F}e^{i\beta}\right)}{\partial z}, \dfrac{\partial\left(\mathbf{F}^* e^{-i\beta}\right)}{\partial z^*} \right\rangle}$$

$$= \frac{\left\langle \mathbf{N}, i\dfrac{\partial^2\mathbf{F}}{(\partial z)^2} - i\dfrac{\partial^2\mathbf{F}^*}{(\partial z^*)^2} \right\rangle}{2\left\langle \dfrac{\partial\mathbf{F}}{\partial z}, \dfrac{\partial\mathbf{F}^*}{\partial z^*} \right\rangle}\cos\beta - \frac{\left\langle \mathbf{N}, \dfrac{\partial^2\mathbf{F}}{(\partial z)^2} + \dfrac{\partial^2\mathbf{F}^*}{(\partial z^*)^2} \right\rangle}{2\left\langle \dfrac{\partial\mathbf{F}}{\partial z}, \dfrac{\partial\mathbf{F}^*}{\partial z^*} \right\rangle}\sin\beta$$

$$= A\sin\theta\cos\beta - A\cos\theta\sin\beta = A\sin(\theta - \beta).$$

To get the principal curvatures, we solve the equation:

$$\det\begin{bmatrix} h_1^1 - \lambda & h_2^1 \\ h_1^2 & h_2^2 - \lambda \end{bmatrix} = \begin{bmatrix} A\cos(\theta-\beta) - \lambda & A\sin(\theta-\beta) \\ A\sin(\theta-\beta) & -A\cos(\theta-\beta) - \lambda \end{bmatrix} = 0.$$

This gives us

$$\lambda = \sqrt{A^2\cos^2(\theta-\beta) + A^2\sin^2(\theta-\beta)} = \pm A.$$

Thus when we consider two members of the family, we can say that when we pair the points with the same coordinates, the metric tensor, the normal vector $\mathbf{N}$, and the principal curvatures are the same. Yet, there is a very distinct difference between different members of the same family.

As β increases, the principal directions rotate with respect to the coordinate lines. To see this, we note that the equation for the direction of the principal curvature corresponding to the principal curvature λ is

$$\begin{bmatrix} h_1^1 & h_2^1 \\ h_1^2 & h_2^2 \end{bmatrix} \begin{bmatrix} v^1 \\ v^2 \end{bmatrix} = \lambda \begin{bmatrix} v^1 \\ v^2 \end{bmatrix}.$$

For our problem, this becomes

$$\begin{bmatrix} A\cos(\theta-\beta) & A\sin(\theta-\beta) \\ A\sin(\theta-\beta) & -A\cos(\theta-\beta) \end{bmatrix} \begin{bmatrix} v^1 \\ v^2 \end{bmatrix} = \lambda \begin{bmatrix} v^1 \\ v^2 \end{bmatrix}.$$

For $\lambda = A$, we find

$$\begin{bmatrix} v^1 \\ v^2 \end{bmatrix} = \begin{bmatrix} \cos\left(\frac{\theta-\beta}{2}\right) \\ \sin\left(\frac{\theta-\beta}{2}\right) \end{bmatrix}. \tag{11.34}$$

For $\lambda = -A$, we find

$$\begin{bmatrix} v^1 \\ v^2 \end{bmatrix} = \begin{bmatrix} -\sin\left(\frac{\theta-\beta}{2}\right) \\ \cos\left(\frac{\theta-\beta}{2}\right) \end{bmatrix}. \tag{11.35}$$

For the family of the catenoid, we have

$$\mathbf{x}(u,v) = \mathbf{e}_1 x^1(u,v) + \mathbf{e}_2 x^2(u,v) + \mathbf{e}_3 x^3(u,v), \text{ where} \tag{11.36}$$

$$x^1(u,v) = a(-\cos\beta \sin u \cosh v + \sin\beta \cos u \sinh v), \tag{11.37}$$

$$x^2(u,v) = a(\cos\beta \cos u \cosh v + \sin\beta \sin u \sinh v), \text{ and} \tag{11.38}$$

$$x^3(u,v) = a(v\cos\beta + u\sin\beta). \tag{11.39}$$

For $\beta = 0$, we get a catenoid and for $\beta = \pi/2$, we get a helicoid. (See Figs. 11.2 and 11.3.)

Problem 273. Verify (11.11)–(11.13).

Problem 274. Suppose

$$\mathbf{F}(z) = \frac{a}{2}\exp(i\beta)(-\sin z, \cos z, -iz), \text{ where} \tag{11.40}$$

$$z = u + iv.$$

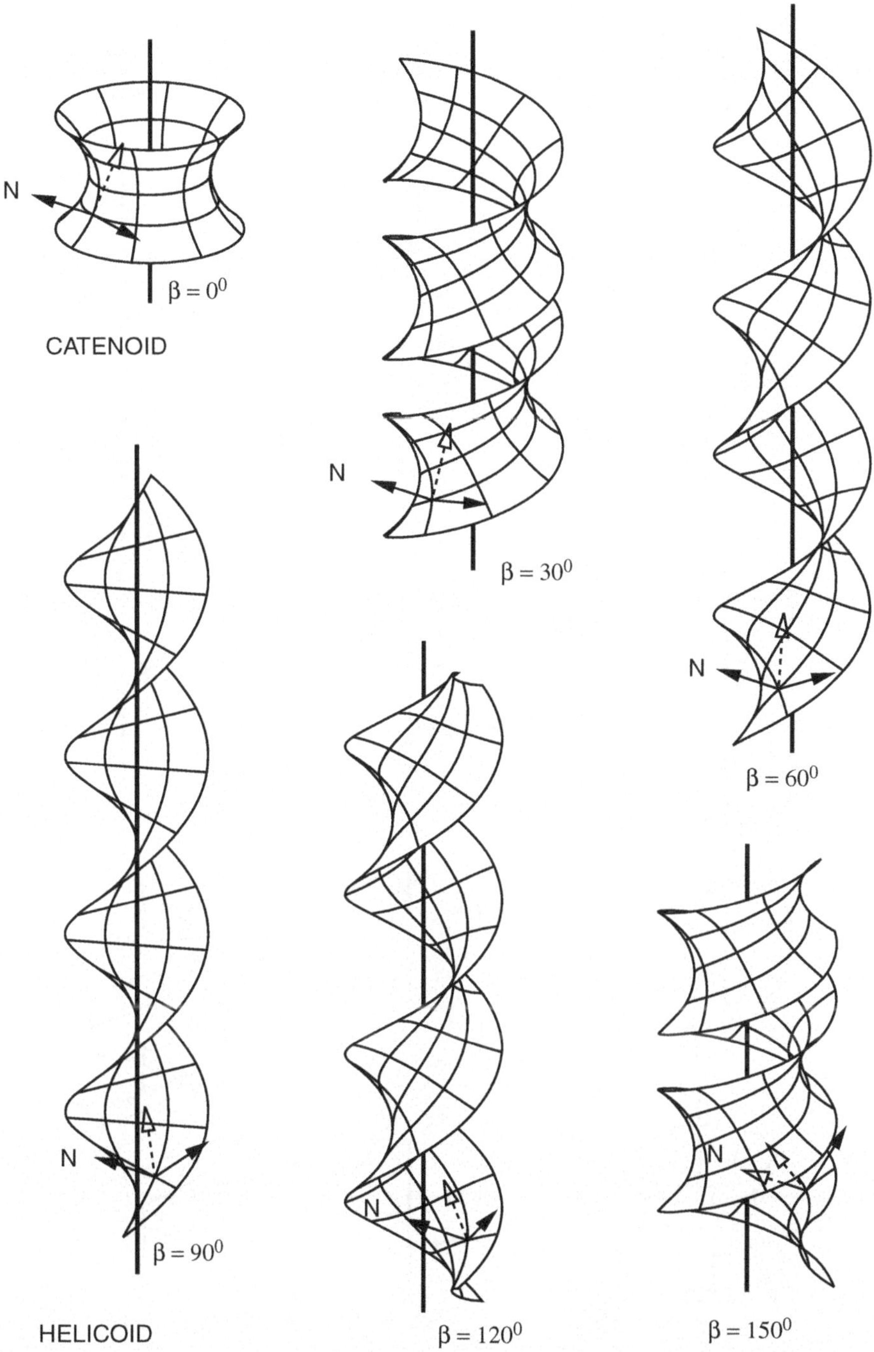

Fig. 11.2 A family of surfaces with zero mean curvature

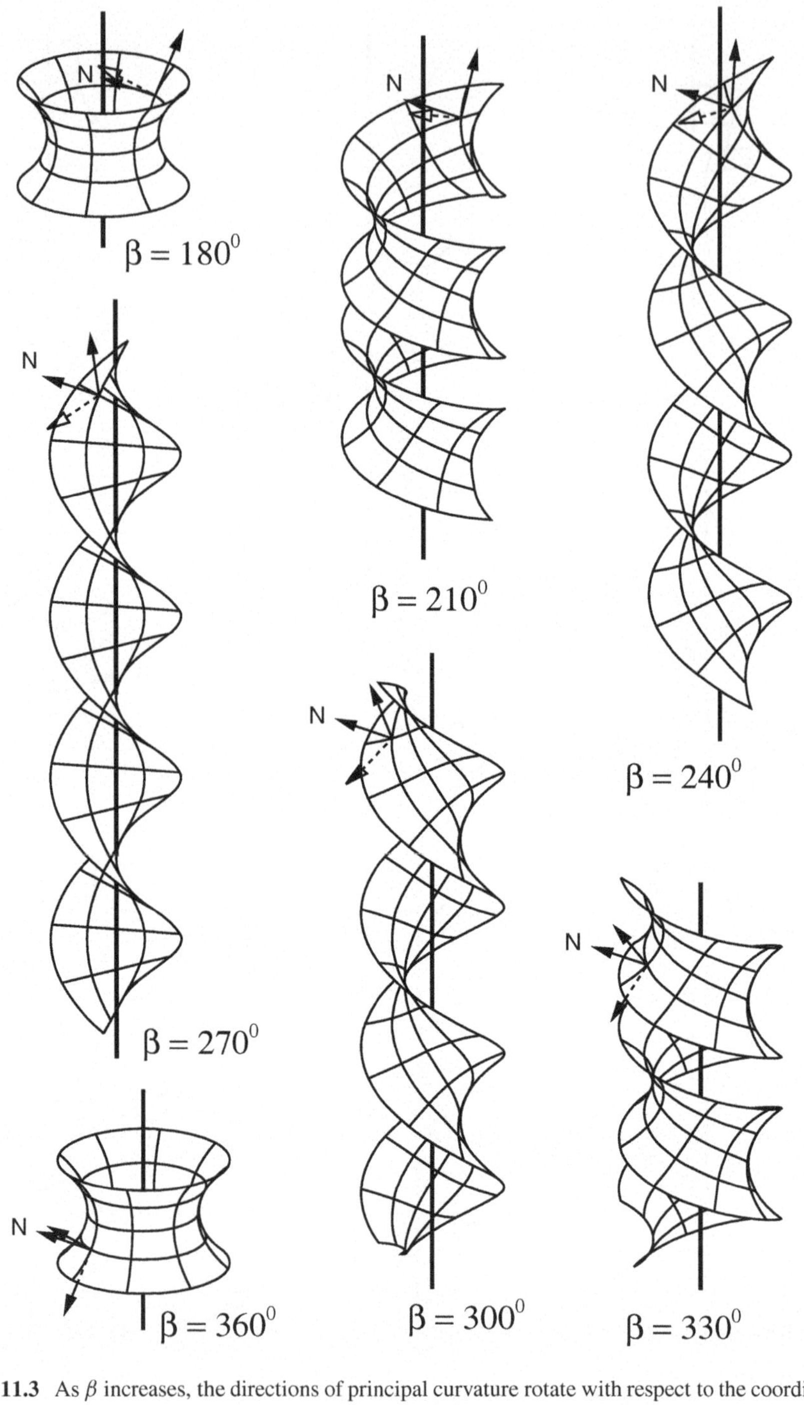

Fig. 11.3 As β increases, the directions of principal curvature rotate with respect to the coordinate lines

(a) Use (11.18) to show that this is equivalent to (11.37), (11.38), and (11.39). Note! The formulas for trigonometric functions and hyperbolic functions with complex arguments are pretty much what you should expect. For example:

$$\sin z = \sin(u + iv) = \sin u \cos(iv) + \cos u \sin(iv).$$

$$\cos(iv) = \frac{\exp(i(iv) + \exp(-i(iv)}{2} = \frac{\exp(-v) + \exp(v)}{2} = \cosh v, \text{ and}$$

$$\sin(iv) = \frac{\exp(i(iv)) - \exp(-i(iv))}{2i} = i\frac{\exp(v) - \exp(-v)}{2} = i \sinh v.$$

(b) Convince yourself that (11.40) corresponds to a catenoid when $\beta = 0$ and a helicoid when $\beta = \pi/2$. (You may wish to refer to (7.148) and (7.149).)
(c) Do the results of this section imply that a segment of the catenoid surface can be deformed into a segment of the helicoid surface without stretching?

Problem 275. Verify (11.34) and (11.35).

11.3 *The Enneper–Weierstrass Representations

The theory of how to construct minimal surfaces that can be extended indefinitely was worked out by Albert Enneper (1830–1885) and Karl Weierstrass (1815–1897). Enneper published his results first and his work is applicable to general coordinate systems (Enneper 1864). Weierstrass' results are restricted to isothermal coordinates and harmonic functions (Weierstrass 1866). However, Weierstrass also showed that it is hard to conceive of a circumstance in which you would not want to use isothermal coordinates.

From the material covered in the last section, you should have reason to believe that it is easy to construct formulas representing minimal surfaces. Indeed, this is true but there are difficulties that you need to be aware of. If you have

$$\frac{\partial \mathbf{F}}{\partial z} = (f(z), g(z), i\sqrt{(f(z)^2 + (g(z)^2))}, \text{ where}$$

$$z = u + iv, \text{ then}$$

$$\left\langle \frac{\partial \mathbf{F}}{\partial z}, \frac{\partial \mathbf{F}}{\partial z} \right\rangle = 0, \text{ but}$$

you may have a problem in the neighborhood of any point, where $(f(z)^2 + g(z)^2) = 0$. You are faced with the problem of choosing a plus or minus sign for the square root. If the point in question is an interior point of some region you wish to consider, there is no way to consistently choose a sign.

For example if we let $z = z_0 + r\exp(i\theta)$, then

$$\sqrt{z - z_0} = \pm r^{\frac{1}{2}} \exp\left(i\frac{\theta}{2}\right) = \pm r^{\frac{1}{2}}\left[\cos\frac{\theta}{2} + i\sin\frac{\theta}{2}\right].$$

Assuming $r > 0$, and $r^{\frac{1}{2}} > 0$, we have two choices for the sign that we associate with $\sqrt{z-z_0}$. In either case, if θ increases by 2π, the quantity $\left[\cos\frac{\theta}{2} + i\sin\frac{\theta}{2}\right]$ will change sign. Thus, you cannot avoid the problem of dealing with both signs for $\sqrt{z-z_0}$. You should note that the same problem will occur if

$$\sqrt{(f(z)^2 + (g(z)^2)} = (z - z_0)^{\frac{2n+1}{2}} \sum_{k=0}^{\infty} a_k (z - z_0)^k \quad \text{where } n \text{ is a positive integer.}$$

Such troublesome points are said to be *branch points*. Riemann introduced the notion of what are now described as *Riemann surfaces* to deal with branch points in the context of complex variables. However, a Riemann surface is not what we usually think of when we study surfaces in the context of differential geometry.

In certain circumstances, the square root problem disappears if you restrict yourself to certain regions. For example, if you restrict yourself to a bounded region that is simply connected (no holes), then you can choose either square root if none of the interior points are branch points. However if you wish to study minimal surfaces that can be indefinitely extended than you need to avoid branch points altogether.

One way of avoiding branch points suggested by Weierstrass is to require that

$$\frac{\partial \mathbf{F}}{\partial z} = \frac{1}{2}\left((G(z))^2 - (H(z))^2, i(G(z))^2 + i(H(z))^2, 2G(z)H(z)\right). \tag{11.41}$$

(You should check that this requirement implies that $\left\langle \frac{\partial \mathbf{F}}{\partial z}, \frac{\partial \mathbf{F}}{\partial z} \right\rangle = 0$.) If both $G(z)$ and $H(z)$ are analytic functions, then $\mathbf{F}$ will represent a minimal surface that can be extended indefinitely. However, this does not exhaust the possibilities. If

$$G(z) = (z - z_0)^{\frac{2m+1}{2}} \sum_{j=0}^{\infty} a_j (z - z_0)^j \quad \text{and}$$

$$H(z) = (z - z_0)^{\frac{2n+1}{2}} \sum_{k=1}^{\infty} b_k (z - z_0)^k, \quad \text{where}$$

both m and n are positive integers and both series converge for all values of z, then all three components of $\partial \mathbf{F}/\partial z$ are analytic even though neither G nor H is analytic.

A somewhat superior formulation also introduced by Enneper and Weierstrass is to require that

$$\frac{\partial \mathbf{F}}{\partial z} = \frac{1}{2}\left((G(z))^2\left(1 - \frac{(H(z))^2}{(G(z))^2}\right), i(G(z))^2\left(1 + \frac{(H(z))^2}{(G(z))^2}\right), 2(G(z))^2\frac{H(z)}{G(z)}\right)$$

or restated,

$$\frac{\partial \mathbf{F}}{\partial z} = \frac{1}{2}\left(f(z)(1-(g(z))^2), if(z)(1+(g(z))^2), 2f(z)g(z)\right), \text{ where}$$

$$f(z) = (G(z))^2 \text{ and } g(z) = \frac{H(z)}{G(z)}. \tag{11.42}$$

This formulation eliminates the branch point problem. (Why?) However, it introduces another problem. If $G(z_0) = 0$, then unless $H(z_0)$ is also zero with at least the same multiplicity, $g(z)$ will have a representation of the form

$$g(z) = \frac{b_{-n}}{(z-z_0)^n} + \frac{b_{-n+1}}{(z-z_0)^{n-1}} + \cdots + \sum_{k=0}^{\infty} b_k(z-z_0)^k = \sum_{k=-n}^{\infty} b_k(z-z_0)^k.$$

Such a function is said to have a *pole* of order n. Since we need all three components of $\partial \mathbf{F}/\partial z$ to be analytic, this imposes a restriction on $f(z)$. In particular since the first and second components of $\partial \mathbf{F}/\partial z$ have to be analytic, $f(z)$ must have a representation of the form

$$f(z) = \sum_{j=2n}^{\infty} a_j(z-z_0)^j, \text{ where}$$

n is the order of the pole possessed by $g(z)$ at $z = z_0$. To compute curvatures and any other features of a surface, it is generally easier to deal with poles than branch points. Therefore to derive general formulas, it is usually more useful to use (11.42) rather than (11.41).

In this context, Enneper's surface is perhaps the most simple minimal surface beyond a plane. In that case

$$f(z) = 1 \text{ and } g(z) = z.$$

(See Problem 277.)

Problem 276. Even to the casual observer, it is obvious that (11.42) is a sufficient condition for $\mathbf{F}(z)$ to represent a minimal surface that can be extended indefinitely. However, it is also a necessary condition. Suppose

$$\frac{\partial \mathbf{F}}{\partial z} = \frac{1}{2}(\phi_1, \phi_2, \phi_3), \text{ where}$$

ϕ_1, ϕ_2, and ϕ_3 are each analytic and

$$\left\langle \frac{\partial \mathbf{F}}{\partial z}, \frac{\partial \mathbf{F}}{\partial z} \right\rangle = \frac{1}{4}\left[(\phi_1)^2 + (\phi_2)^2 + (\phi_3)^2\right] = 0.$$

Show that given ϕ_1, ϕ_2, and ϕ_3 satisfying the above conditions, there exist functions $f(z)$ and $g(z)$ such that

$$\begin{aligned} f(z)\left(1 - (g(z))^2\right) &= \phi_1(z), \\ if(z)\left(1 + (g(z))^2\right) &= \phi_2(z), \text{ and} \\ 2f(z)g(z) &= \phi_3(z). \end{aligned}$$

(Actually, there appears to be an exception. The exception occurs when $\phi_3(z) = 0$, and $\phi_1(z) = i\phi_2(z) \neq 0$. In this case, $\mathbf{F}(z)$ represents a plane parallel to the x-y plane.) You should show that you can still choose a pair of functions $f(z)$ and $g(z)$ so that the corresponding $\mathbf{F}(z)$ represents a plane parallel to the x-y plane.

Problem 277. Verify that if $f(z) = 1$ and $g(z) = z$, then (11.42) is a representation of Enneper's surface.

Problem 278. Use (11.42) to invent your own minimal surface. Then use MAPLE, MATHEMATICA, or some other computer program to plot it.

Problem 279. Show that the normal vector $\mathbf{N}$ for a minimal surface is independent of the choice of of $f(z)$ in the formulation of (11.42).

Chapter 12
Some General Relativity

12.1 Einstein's Theory

It was in 1905 that Einstein wrote his Ph.D. thesis and three papers each of which would individually receive nominations for the Nobel Prize. He would soon be working on his General Theory of Relativity. The Special Theory of Relativity clearly predicts how measurements for the same phenomena will differ when the observers and objects move at constant speeds relative to one another. But what happens when these relative speeds are not constant?

In 1922, in an address delivered in Kyoto, Japan, Einstein recalled the moment in 1907 that would give direction to his efforts to answer this question.

> I was sitting in a chair in the patent office at Bern when all of a sudden a thought occurred to me, 'If a person falls freely he will not feel his own weight.' I was startled. This simple thought made a deep impression on me. It impelled me toward a theory of gravitation.

At another time, he described this as the "happiest thought of my life." (Pais 1983, pp. 178–179) The year 1907 was also the same year, Einstein was initially denied a part-time position at the University of Berne. However, it may have been fortuitous that Einstein was still working in the patent office. The general relativity theorist, Engelbert Schücking, has pointed out that about this time the application for the Otis elevator patent would have arrived at the Berne patent office. The Otis elevator made it feasible to construct buildings above five stories. It was designed so that no matter what fails, passengers do not go into free fall (unless they step into an open shaft). Prof. Schücking suggests that the Otis patent may have been "Einstein's apple."

The relevant records in the Berne patent office have been destroyed, so it cannot be determined whether Einstein was assigned to review the Otis application. One piece of evidence that supports Schücking's theory is that in popular presentations, Einstein would describe the limitations of an observer in a room-sized box pulled by a rope attached to its lid (Einstein 1961, pp. 66–70).

J. Snygg, *A New Approach to Differential Geometry using Clifford's Geometric Algebra*, DOI 10.1007/978-0-8176-8283-5_12,

Whether this room-sized box was the cage of an Otis elevator , the idea that came to Einstein became known as the *equivalence principle*. What was his equivalence principle? In this day and age, it seems more appropriate to discuss its consequences for an observer inside a space ship rather than for an observer in a large box pulled by a rope.

Suppose one is shut up in a windowless space ship with a noiseless, vibrationless engine. If the space ship was accelerating, one would feel an apparent force. But could one determine whether the force experienced was due to an acceleration of the space ship in outer space or due to the force of gravity from some planet on which the space ship happened to be parked?

If the planet was small, an occupant might be able to observe that while parked on that planet, two objects dropped would converge slightly since each object would move toward the center of the planet. The idea that the occupant of the space ship could not determine the difference of the two situations is known as the equivalence principle. Clearly, it is flawed but it was a good starting point for Einstein. Furthermore, if he could figure out how to deal with accelerations, he would have a theory of gravity.

None of Einstein's groundbreaking papers in 1905 required mathematics beyond the high school level. It was probably for this reason that Einstein underestimated the amount of mathematics he would need to deal with his new problem. In 1911, while in Prague, he thought he was making some progress. But by the time he returned to Switzerland in the summer of 1912, he realized that he was encountering some geometric questions that he was not equipped to deal with (Pais 1983, p. 22).

Seeking help from his longtime friend Marcel Grossman, he was pointed in the direction of Riemannian geometry. I do not pretend to know what thoughts ran thorough Einstein's mind. Generally, when theoretical physicists seek out a law, they are happiest when they can find some entity that when maximized or minimized results in the things that actual happen. For example, consider Fig. 12.1.

Given points A and B, there are many conceivable paths from A to the mirror and then to B. However, the shortest of these paths is that for which the angle of incidence is equal to the angle of reflection. (Note! The length of the path AC_1B is the same as the length of the path AC_1B_1, where B_1 is the mirror image of B. The shortest path from A to B_1 is a straight line. Thus it is clear that the point C, which corresponds to the shortest path, is also the point where the angle of incidence θ_i is equal to the angle of reflection θ_r.)

As you see, this light ray problem can be dealt with using very simple math. However, when the problem is finding a function that will minimize or maximize an entity even calculus is insufficient to deal with the situation. To deal with this kind of problem, one needs to use the *calculus of variations*. The first problem solved by the calculus of variations was the *brachistochrone problem*.

In 1696, Johann Bernoulli (1667–1748) challenged the mathematical community with this problem (Bernoulli, Johann 1696). Given points A and B, what curve should join the two points so that a point mass sliding down such a curve would get from A to B in the shortest time? (Assuming no friction.) Johann Bernoulli already knew the answer and during the following year, the problem was solved

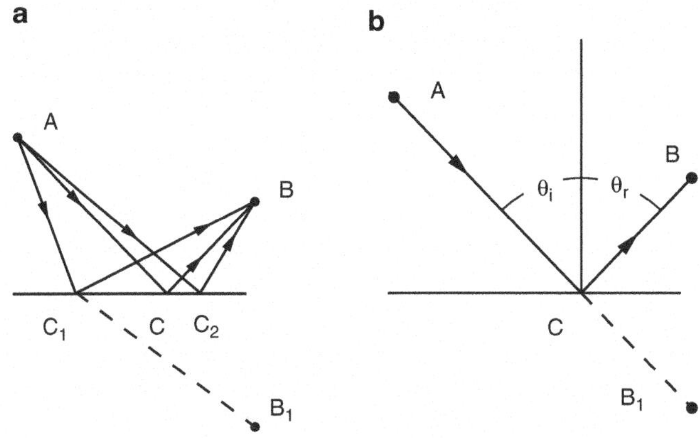

Fig. 12.1 (**a**) Possible paths from point A to the mirror and then from the mirror to point B. (**b**) For the shortest path, the angle of incidence θ_i is equal to the angle of reflection θ_r

independently by l'Hôpital, Leibniz, and Newton. However, Johann's older brother Jacob came up with a solution that became the first major step in the development of the calculus of variations (Bernoulli, Jacob 1697).

Leonard Euler was able to show that the approach used by Jacob Bernoulli for the brachistochrone problem could be generalized to solve virtually any problem in mechanics that does not include friction. It appeared that essentially all of Newtonian physics could be derived from a rule that Maupertius called the *Principle of Least Action*.

For many in the Age of Enlightenment, the Principle of Least Action had religious implications. Maupertuis, Voltaire, Euler, and others viewed the Principle of Least Action as evidence for the existence of God. Some viewed the principle as evidence of a God who after creating the universe would not interfere in the daily affairs of men (or women). Thus, all events were predetermined. Voltaire went so far as to conclude that man had no free will. He argued that if all things – planets, stars, falling objects, everything in nature – act according to certain laws, why, he asked should "a little animal five feet high – act as he pleased." (Muir 1961, pp. 171–172)

As an observant Calvinist, Euler could not accept Voltaire's conclusion on free will. However, in the context of the available evidence of the times, he was hard pressed to construct a convincing counter argument. Things have changed since that time. The calculus of variations no longer has the religious impact that it once did. General relativity and various areas of quantum mechanics each requires its own variational principle. Because there does not seem to be any single overriding variational principle, the discovery of these principles seems to reveal more about the cleverness of physicists than they do about the nature of a Creator. Furthermore, quantum mechanics does not have the determinism of Newton's laws. This fact undermines the thought that all events were determined at the moment of a creation.

Going back to the seventeenth and the eighteenth centuries, the success of Newton's theory, with or without the Principle of Least Action, undercut the authority of the European monarchy. European monarchs claimed that they had the *Divine Right to Rule*, that they were acting as God's agents. Defiance of a king was defiance of God. This approach worked well for King Louis XIV of France at the beginning of the eighteenth century and he was able to rule as an absolute monarch. The fact that, for eternity, the motion of the planets seemed to be ruled by the laws of Newton suggested that God did not monitor the motion of the planets and would not interfere in the daily lives of men (or women). The doctrine of the Divine Right to Rule became more difficult to sell at the end of the eighteenth century. The idea that kings had a special pipeline to God had lost credibility. In 1793, King Louis XVI was executed by people who had no fear that they might offend God by killing someone who might be acting under God's instructions.

With this historical background, Einstein was presumably looking for a law that could be stated in terms of the calculus of variations. One such law (or problem?) that arises in differential geometry is that of determining the shortest path between two given points in a space with some given metric. (In General Relativity, it is the problem of determining the path of maximum length.) If somehow, Einstein could find the right metric, the paths of the planets would be geodesics. Actually, he was looking for a more general law – that would not be restricted to the solar system.

Many times, Einstein arrived at a theory that he would later reject. At last in November of 1915 during World War I, he achieved his goal. As he was putting the final touches on his theory, Einstein presented a series of four lectures on each Thursday of that month at the Prussian Academy of Sciences. On the last Thursday (November 25), he presented his final form. This final form is

$$R_{\alpha\beta} - \frac{1}{2} g_{\alpha\beta} R = k T_{\alpha\beta}, \text{ where} \tag{12.1}$$

$$R_{\alpha\beta} = R^{\lambda}{}_{\alpha\beta\lambda} \text{ is the \textit{Ricci tensor},} \tag{12.2}$$

$$R = g^{\alpha\beta} R_{\alpha\beta} \text{ is the \textit{scalar curvature}, and} \tag{12.3}$$

$$T_{\alpha\beta} \text{ is the \textit{energy-momentum tensor}.}$$

(Sign conventions for both the Riemann curvature tensor and the Ricci tensor vary!)

For mass free regions, $T_{\alpha\beta} = 0$. In that case,

$$g^{\alpha\beta}\left(R_{\alpha\beta} - \frac{1}{2} g_{\alpha\beta} R\right) = R - \frac{1}{2}\delta^{\alpha}_{\alpha} R = R - 2R = -R = 0.$$

Note! Einstein was dealing with a 4-dimensional space so $\delta^{\alpha}_{\alpha} = 1 + 1 + 1 + 1 = 4$.

Thus for mass free regions, $R = 0$ and (12.1) becomes

$$R_{\alpha\beta} = 0. \tag{12.4}$$

By the time, Einstein made his presentation to the Prussian Academy of Science, he had already derived an approximate solution to (12.4) that would be valid for the solar system outside the radius of the sun. Less than 2 months after Einstein's presentation, Einstein read a paper on behalf of Karl Schwarzschild, a German astronomer who was serving in the army at that time on the Alsatian front (not the Russian front as is usually stated) (Schwarzschild 1916, pp. 189–196). In that paper, Schwarzschild presented an exact solution, which is valid for a small test particle (a planet) in the presence of a massive sphere (the sun).

12.2 *Karl Schwarzschild 1873–1916

The figure of a WWI German soldier conjures up an image of a young draftee. Actually, Karl Schwarzschild had had a long distinguished career as an astronomer before he mailed his famous solution to Einstein. After his death in 1916, his friend Otto Blumenthal constructed a list of 112 publications he could attribute to Schwarzschild (Blumenthal 1917, pp. 70–75).

Because of his age, he would not have been drafted – at least during the early stages of the war. At the age of 42, 6 years older than Einstein, Schwarzschild volunteered when war broke out. His wife, Else, later revealed that as a prominent Jew, Karl thought that his display of patriotism would neutralize some of the antisemitism that would later become virulent under the Nazis (Schwarzschild 1992, Vol. I, p. 23).

Karl Schwarzschild was born on October 9, 1873. He demonstrated a talent for astronomy and mathematics at an early age and he published his first paper in *Astronomische Nachrichten* in 1890 while he was only 16. He published a second paper in the same journal in the same year (Schwarzschild 1890a, 1890b).

In 1901, he became director of the Göttingen Observatory. Like Gauss, 100 years earlier at the same institution, he was an astronomer who also taught advanced courses in mathematics. According to Hans-Heinrich Voigt;

> Schwarzschild liked experimenting and was often brilliant at it. Whenever he needed some accessory, it was immediately taken from another temporarily unused instrument; if a diaphragm was needed immediately, a hole was soon made in a lens cap. The Observatory's inventory list contains numerous entries made by his successor in red ink, where such occurrences were deplored in the strongest terms (Schwarzschild 1992, Vol. I, p. 16).

In 1909, Schwarzschild left the Göttingen Observatory to become director of the Astrophysical Observatory in Potsdam, the most prestigious observatory in Germany. Schwarzschild was very receptive to Einstein's efforts to devise a general theory of relativity. In 1913, he tried without success to measure a shift in wave length of light emitted by the Sun – a shift predicted by both Einstein's preliminary and final theories of general relativity (Schwarzschild 1992, Vol. I, p. 23).

In August 1914, at the outbreak of World War I, Schwarzschild volunteered to join the Army. During his short 19 months of active duty, Karl Schwarzschild was awarded the iron cross (second class) and was promoted from the rank of a noncommissioned officer to the rank of Lieutenant. He served at various locations in Belgium, France, and Russia.

On March 6, 1916, his active service ended when he sent a telegram to his wife from Brussels announcing he had a 14-day pass to return home (Schwarzschild 1975, Sect. 7, Reel 13). Sometime in March, he presented his last paper, "Zur Quantenhypothese" to the Academy in Berlin. He had completed this paper in eight short days after he received some results from Sommerfeld (Blumenthal 1917, p. 69). However, he was soon hospitalized in nearby Potsdam with pemphigus, a rare and painful autoimmune skin disease. In 1916, doctors had no way of dealing with this disease and for Schwarzschild, pemphigus would prove to be fatal. It started with a few blisters near his mouth and slowly spread until the blisters covered his entire body (Blumenthal 1917, p. 70). A few days before his death, he read the proofs of his last paper, which appeared in print on the last day of his life, May 11, 1916.

At the time of his death, several obituaries indicated that he had acquired the disease on some battlefront. A few years ago, I asked my personal physician, George Lombardi, about the nature of pemphigus. He informed me that it was likely that Schwarzschild's pemphigus was set off by exposure to poison gas. It is indeed plausible that because of Schwarzschild's efforts to make himself useful to his fatherland, he became exposed to poison gas.

Many of Germany's most outstanding scientists were engaged in developing ever more effective poison gas concoctions and ever more effective delivery methods. The leading figure was Fritz Haber, who would won the Nobel Prize in chemistry for 1918. He won the prize for the synthesis of ammonia using atmospheric nitrogen. His method of synthesis is known as the Haber–Bosch process. Ammonia is the basis for fertilizers which revolutionized agriculture in the twentieth century. Daniel Charles, a biographer of Fritz Haber, has written;

> According to one careful estimate, about one third of all the people on earth, about two billion souls, could not survive in the absence of the Haber-Bosch process. Left to its own devices, Earth simply could not grow enough food to feed all six billion of us our accustomed diet (Charles 2005, p. 103).

Because of his activities during World War I, Fritz Haber is also known as the father of chemical warfare. He made a point of being present with his team of fellow scientists on the occasion of the first significant use of poison gas on the Western Front. Chlorine gas was released from canisters at the Second Battle of Ypres in Belgium against Algerian, French, and Canadian troops on April 22, 1915. Casualty figures for this event are unreliable. The Germans did not want to be portrayed as war criminals so they downplayed the numbers. For the same reason, the Allies tended to inflate the figures. The Germans claimed that on April 22 their hospitals dealt with a mere 200 gas victims, of whom twelve died later. The Allies, on the other hand, reported 15,000 and 5,000 killed (Haber 1986, p. 39). Author Simon Jones suggests that between 800 and 1400 were killed and another 2000–3000 were injured (Jones 2007, p. 6).

To the disappointment of Fritz Haber, German troops did not fully exploit gaps opened up in the Allied lines. Nonetheless, the results seemed to show that poison gas used against troops unprotected by gas masks could break the stalemate associated with trench warfare.

Fritz Haber recruited a number of future Nobel Prize winners for his team. In particular, his team included James Franck (Physics 1925), Gustav Hertz (Physics 1925), and Otto Hahn (Chemistry 1944). In addition, Walther Nernst (Chemistry 1920) led the development of munitions that enabled the German Army to propel gas into Allied lines via mortars (Haber 1986, p. 30). Richard Willstater (Chemistry 1915) and Heinrich Wieland (Chemistry 1927) also endeavored to advance the effectiveness of chemical warfare (Hahn 1970, p. 124).

For the first 10 months of his service in the Army, Karl Schwarzschild served as director of a field weather station in Namur, Belgium. At his request, he was assigned to an artillery regiment attached to the Fifth Army, which put him on the front lines in the Argonne forest. On June 20, 1915, the day of his arrival in Grandpré, the Fifth Army was engaged in an attack, which involved one of the earliest use of poison gas delivered by howitzer shells. On that day up to 25,000 of those shells rained down on French troops. Ten days later, in a similar attack, the Germans captured 3,000 prisoners and forced the French to abandon their positions (Jones 2007, p. 12 and Mosier 2001, p. 159).

While still at Namur, Schwarzschild had begun work on a paper entitled, *Über den Einfluss von Wind und Luftdichte auf die Geschussbahn* (The Effect of Wind Velocity and Air Density on the Trajectory of Artillery shells) (Schwarzschild 1992, p. 23). The usual artillery tactic was to fire a first shot and then use observers to adjust successive shots to close in on the target. It was Schwarzschild's hope that his computations would enable his countrymen to benefit from a lethal surprise on the first salvo before enemy troops could take cover.

The introduction of poison gas would have opened up many more opportunities for Schwarzschild to contribute to his adopted cause. Even at this early date, it had become clear that using poison gas was good for disabling enemy troops but did not necessarily result in territorial gains. There were many questions that could be posed to a mathematician. How many shells should be fired into what size area to have the desired effect? What time of day should the poison be delivered so that rising air would not mitigate the intended effect? From battlefield observation, Fritz Haber had already determined that if the product of the concentration and time of exposure was sufficiently high, one would get the desired lethal effect.

In August of 1915, Karl Schwarzschild along with his artillery brigade was assigned to the Tenth Army on the Russian Front at Kovoso in present day Lithuania. The Russian troops provided a tempting target for chemical warfare because their troops had no gas masks until close to the end of the war. Some historians estimate that Russia was the country that suffered the greatest number of chemically induced casualties, but others say that the figures are so unreliable that it is impossible to make any estimate with confidence.

The earliest German gas attacks were not on the Western Front but against the Russians. However, the results for the Germans were not completely satisfactory. On January 31, 1914, the Germans attempted to inflict a compound of bromine on the Russian troops at Bolimov. Because of cold weather, the liquid failed to vaporize and the Russians failed to notice they were the intended victims of a gas attack (Jones 2007, p. 3).

A later attack, on May 31, 1915, caught the Russians during a time of troop replacement when trenches were crowded and somewhat disorganized. Without gas masks, the Russians may have suffered as many as 5,000 casualties but a shift in the wind resulted in 56 German gas casualties (Jones 2007, pp. 11–12). When Otto Hahn was present at a subsequent attack on June 12, the wind reversed direction again causing 350 German casualties. The last of this series of attacks occurred on July 6. This time things were much worse for the Germans. At least 1,450 Germans were gassed of whom 130 were killed (Jones 2007, p. 12). One of the victims was Gustav Hertz who took several months to recover (Hahn 1970, pp. 120–121).

During this time, Erich Ludendorff was the German General responsible for the Russian Front. Later he wrote in his memoirs:

> In accordance with the instructions of General Headquarters the Ninth Army was now to attack at Skierniewice. We had received a supply of gas and anticipated great tactical result from its use, as the Russians were not yet fully protected against gas. We also had reason to expect local successes from an attack by the Tenth Army, east of Suwalki, and instructions were issued accordingly.
>
> The gas attacks by the Ninth Army, which took place on May 2, were not a success. The wind was favorable, but the troops had not been properly instructed. The gas was emitted as intended, but the troops imagined that the enemy ought not to be able to move at all. As the latter were still firing in places and our own artillery did not cooperate as it should have done, the infantry did not attack. It assumed that the gas had no effect. The Ninth Army was unlucky with gas. When it repeated the gas attack at the same place later, but not in connection with these operations, the wind veered round. We suffered severe losses by gassing. The troops were not fond of gas: the installation took too long and both officers and men disliked waiting with full gas-containers in the trenches for the wind (Ludendorff 1919, p. 167).

We do not know the detailed activities of Schwarzschild while on the scene a few months later but we do know that during his short stay on the Russian Front he was provided a report on a weather instrument, which was designed to give short-term forecasts of wind velocity and wind direction with the explicit intent of making the use of gas-filled howitzer shells a more viable tactic (Schwarzschild 1975, Reel 8, Sect. 4). It was during his time at the Russian Front that Schwarzschild received his promotion to the rank of Lieutenant.

According to most biographies of Einstein, Schwarzschild sent his famous solution to Einstein from the Russian Front in a letter dated December 22, 1915. However, correspondence with his wife shows that by the end of September he had been relocated to Mulhouse in Alsace. He was then relocated again to someplace else but by December 1, he was back in Mulhouse. Historian Tillman Sauer has drawn my attention to a letter Schwarzschild wrote to Arnold Sommerfeld on the same date that Schwarzschild mailed his solution to Einstein (Schwarzschild 1915). In the letter to Sommerfeld, Schwarzschild describes hearing cannon fire from Hartmannweilerkopf, which is about 10 km from Mulhouse. (Because of its commanding elevation, control of Hartmannweilerkopf was considered to be a necessary condition for control of the surrounding region. In 1915 alone, the hilltop changed hands four times, each time at great cost of life. Today, tourists can observe a landscape still scarred by the conflict and a crypt containing the bones of 12,000 unknown soldiers.)

Was Schwarzschild exposed to poison gas? With his "hands on" approach, it seems likely. Anyone who dedicated himself to the advancement of chemical warfare was vulnerable. We have already mentioned the incident with Gustav Hertz on the Russian Front. On April 3, 1915, about 3 weeks before the Germans inflicted chlorine on Allied troops at Ypres, Fritz Haber, and an army officer were riding horseback behind a cloud of chlorine gas during a field test. They got too close and nearly suffocated. Haber became sick but recovered after a few days (Charles 2005, p. 161). Otto Hahn described splashing some phosgene into one of his eyes, an incident that required medical attention to avoid loss of vision (Hahn 1970, pp. 123–124). Otto Hahn was also exposed to low concentrations of phosgene several times and at one point it took him a month of rest to recover. He wrote, "I came to no harm but Dr. Günther, a chemist from Luverkusen, was fatally poisoned, and Professor Freundlich, from Haber's laboratory, exposed himself to such an extent that for some time he was in danger of losing his life (Hahn 1970, pp. 126–128).

In June of 2009, I attended a college reunion where I encountered Jerry Ostriker. Professor Ostriker served in the Astrophysics Department of Princeton University for many years with Martin Scharzschild. Martin Schwarzschild was a son of Karl Schwarzschild who also became a world renowned astronomer. I mentioned the speculation of my physician that Karl Schwarzschild's pemphigus was induced by poison gas. Jerry responded by saying that he was under the impression that Karl Schwarzschild had died from exposure to poison gas. He told me that he did not remember how he got that impression but that the only logical explanation was that Martin had told him.

Regardless of the circumstance of Schwarzschild's death, his premature death was a loss to the scientific community. On June 29, 1916, Einstein delivered a eulogy to the Berlin Academy:

> – What is specially astonishing about Schwarzschild's theoretical work was his easy command of mathematical methods and the almost casual way in which he could penetrate to the essence of astronomical or physical questions. Rarely has so much mathematical erudition been adapted to reasoning about physical reality. And so it was, that he grappled with many problems from which others shrank on account of mathematical difficulties. The mainsprings of Schwarzschild's motivations in his restless theoretical quests seem less from a curiosity to learn to deeper inner relationships among the different aspects of Nature than from an artist's delight in discerning delicate mathematical patterns. – And in the very last month of his life, much weakened by a skin disease, he yet succeeded in making some profound contributions to quantum theory. – (Schwarzschild 1992, pp. 34–35).

12.3 The Schwarzschild Metric

I will now present a derivation of Schwarzschild's solution. Using various symmetry arguments, Schwarzschild hypothesized that the line element corresponding to a spherically symmetric solution of the equation $R_{\alpha\beta} = 0$ would have the form

$$(\mathrm{d}s)^2 = f(r)c^2(\mathrm{d}t)^2 - h(r)(\mathrm{d}r)^2 - r^2(\mathrm{d}\theta)^2 - r^2\sin^2\theta(\mathrm{d}\phi)^2. \tag{12.5}$$

(Here, it is understood that c represents the speed of light.)

In this curved space–time, you cannot use the same coordinate to designate the distance from the center and the circumference of a great circle about the origin divided by 2π. For the Schwarzschild metric, we note that on the surface of a sphere centered at the origin, $\mathrm{d}r = 0$, so

$$(\mathrm{d}s)^2 = -r^2\left[(\mathrm{d}\theta)^2 + \sin^2\theta(\mathrm{d}\phi)^2\right].$$

On the circumference of the equatorial circle, $\theta = \pi/2$ and $(\mathrm{d}s)^2 = -r^2(\mathrm{d}\phi)^2$. Thus, the circumference is

$$\int_0^{2\pi} r\mathrm{d}\phi = 2\pi r. \tag{12.6}$$

On the other hand, the radial distance between two points would be

$$\int_{r_1}^{r_2} (h(r))^{1/2}\mathrm{d}r \neq r_2 - r_1. \tag{12.7}$$

Now let us consider the problem of deriving the Schwarzschild metric. From (12.5),

$$[g_{\alpha\beta}] = \begin{bmatrix} c^2 f(r) & 0 & 0 & 0 \\ 0 & -h(r) & 0 & 0 \\ 0 & 0 & -r^2 & 0 \\ 0 & 0 & 0 & -r^2\sin^2\theta \end{bmatrix} \tag{12.8}$$

and

$$[g^{\alpha\beta}] = \begin{bmatrix} 1/(c^2 f(r)) & 0 & 0 & 0 \\ 0 & -1/h(r) & 0 & 0 \\ 0 & 0 & -1/r^2 & 0 \\ 0 & 0 & 0 & -1/(r^2\sin^2\theta) \end{bmatrix}. \tag{12.9}$$

From these two equations, it is clear how to construct a moving orthonormal frame

$$\gamma^t = \frac{1}{cf^{\frac{1}{2}}}\mathbf{E}^0 = \frac{1}{cf^{\frac{1}{2}}}\mathbf{E}_0 = \frac{1}{c^2 f}\gamma_t, \tag{12.10}$$

$$\gamma^r = \frac{1}{h^{\frac{1}{2}}}\mathbf{E}^1 = \frac{-1}{h^{\frac{1}{2}}}\mathbf{E}_1 = \frac{-1}{h}\gamma_r, \tag{12.11}$$

$$\gamma^{\theta} = \frac{1}{r}\mathbf{E}^2 = -\frac{1}{r}\mathbf{E}_2 = -\frac{1}{r^2}\gamma_{\theta}, \text{ and} \tag{12.12}$$

$$\gamma^{\phi} = \frac{1}{r\sin\theta}\mathbf{E}^3 = \frac{-1}{r\sin\theta}\mathbf{E}_3 = \frac{-1}{r^2\sin^2\theta}\gamma_{\phi}. \tag{12.13}$$

Since the metric is diagonal, we can use (5.96). That is

$$\mathbf{\Gamma}_{\alpha} = \frac{1}{4}\gamma^{\alpha\eta}\frac{\partial g_{\alpha\alpha}}{\partial u^{\eta}}, \text{ where} \tag{12.14}$$

the η index is summed but the α index is not. Thus,

$$\mathbf{\Gamma}_t = \frac{c^2}{4}\frac{\mathrm{d}f}{\mathrm{d}r}\gamma^{tr} = \frac{c}{4(fh)^{\frac{1}{2}}}\frac{\mathrm{d}f}{\mathrm{d}r}\mathbf{E}^{01}, \tag{12.15}$$

$$\mathbf{\Gamma}_r = 0, \tag{12.16}$$

$$\mathbf{\Gamma}_{\theta} = -\frac{r}{2}\gamma^{\theta r} = \frac{1}{2h^{\frac{1}{2}}}\mathbf{E}^{12}, \text{ and} \tag{12.17}$$

$$\mathbf{\Gamma}_{\phi} = -\frac{r\sin^2\theta}{2}\gamma^{\phi r} - \frac{r^2\sin\theta\cos\theta}{2}\gamma^{\phi\theta} = \frac{-\sin\theta}{2h^{\frac{1}{2}}}\mathbf{E}^{31} + \frac{\cos\theta}{2}\mathbf{E}^{23}. \tag{12.18}$$

To obtain the curvature 2-forms, we use (5.84). Namely,

$$\frac{1}{2}R_{\alpha\beta} = \partial_{\alpha}\mathbf{\Gamma}_{\beta} - \partial_{\beta}\mathbf{\Gamma}_{\alpha} + \mathbf{\Gamma}_{\alpha}\mathbf{\Gamma}_{\beta} - \mathbf{\Gamma}_{\beta}\mathbf{\Gamma}_{\alpha}. \tag{12.19}$$

From this formula, we get

$$\begin{aligned}\mathbf{R}_{tr} &= \left[\frac{-c}{2(fh)^{\frac{1}{2}}}\frac{\mathrm{d}^2 f}{\mathrm{d}r^2} + \frac{c}{4(fh)^{\frac{3}{2}}}\frac{\mathrm{d}f}{\mathrm{d}r}\frac{\mathrm{d}(fh)}{\mathrm{d}r}\right]\mathbf{E}^{01}\\ &= \left[\frac{-c^2}{2}\frac{\mathrm{d}^2 f}{\mathrm{d}r^2} + \frac{c^2}{4(fh)}\frac{\mathrm{d}f}{\mathrm{d}r}\frac{\mathrm{d}(fh)}{\mathrm{d}r}\right]\gamma^{tr},\end{aligned} \tag{12.20}$$

$$\mathbf{R}_{t\theta} = \frac{-c}{2f^{\frac{1}{2}}h}\frac{\mathrm{d}f}{\mathrm{d}r}\mathbf{E}^{02} = -\frac{c^2 r}{2h}\frac{\mathrm{d}f}{\mathrm{d}r}\gamma^{t\theta}, \tag{12.21}$$

$$\mathbf{R}_{t\phi} = \frac{-c\sin\theta}{2f^{\frac{1}{2}}h}\frac{\mathrm{d}f}{\mathrm{d}r}\mathbf{E}^{03} = -\frac{c^2 r\sin^2\theta}{2h}\frac{\mathrm{d}f}{\mathrm{d}r}\gamma^{t\phi}, \tag{12.22}$$

$$\mathbf{R}_{\theta\phi} = -\frac{\sin\theta(h-1)}{h}\mathbf{E}^{23} = -\frac{r^2\sin^2\theta(h-1)}{h}\gamma^{\theta\phi}, \tag{12.23}$$

$$\mathbf{R}_{\phi r} = -\frac{\sin\theta}{2h^{\frac{3}{2}}}\frac{\mathrm{d}h}{\mathrm{d}r}\mathbf{E}^{31} = -\frac{r\sin^2\theta}{2h}\frac{\mathrm{d}h}{\mathrm{d}r}\gamma^{\phi r}, \text{ and} \tag{12.24}$$

$$\mathbf{R}_{r\theta} = \frac{-1}{2h^{\frac{3}{2}}}\frac{\mathrm{d}h}{\mathrm{d}r}\mathbf{E}^{12} = \frac{-r}{2h}\frac{\mathrm{d}h}{\mathrm{d}r}\gamma^{r\theta}. \tag{12.25}$$

To extract the components of the Ricci tensor, it is useful to use the formula

$$\mathbf{R}_{\alpha\beta}\boldsymbol{\gamma}^{\beta} = \frac{1}{2}R_{\alpha\beta\eta\nu}\boldsymbol{\gamma}^{\eta}\boldsymbol{\gamma}^{\nu}\boldsymbol{\gamma}^{\beta} = -R_{\alpha\beta}\boldsymbol{\gamma}^{\beta}. \tag{12.26}$$

(See Problem 284.)

Using this last formula, we have

$$\begin{aligned}\mathbf{R}_{t\lambda}\boldsymbol{\gamma}^{\lambda} &= \mathbf{R}_{tr}\boldsymbol{\gamma}^{r} + \mathbf{R}_{t\theta}\boldsymbol{\gamma}^{\theta} + \mathbf{R}_{t\phi}\boldsymbol{\gamma}^{\phi}\\ &= \left[\frac{c^2}{2h}\frac{\mathrm{d}^2 f}{\mathrm{d}r^2} - \frac{c^2}{4f(h)^2}\frac{\mathrm{d}f}{\mathrm{d}r}\frac{\mathrm{d}(fh)}{\mathrm{d}r} + \frac{c^2}{rh}\frac{\mathrm{d}f}{\mathrm{d}r}\right]\boldsymbol{\gamma}^{t}.\end{aligned}$$

That is,

$$R_{t\lambda}\boldsymbol{\gamma}^{\lambda} = -\left[\frac{c^2}{2h}\frac{\mathrm{d}^2 f}{\mathrm{d}r^2} - \frac{c^2}{4f(h)^2}\frac{\mathrm{d}f}{\mathrm{d}r}\frac{\mathrm{d}(fh)}{\mathrm{d}r} + \frac{c^2}{rh}\frac{\mathrm{d}f}{\mathrm{d}r}\right]\boldsymbol{\gamma}^{t}. \tag{12.27}$$

Similarly,

$$R_{r\lambda}\boldsymbol{\gamma}^{\lambda} = -\left[\frac{-1}{2f}\frac{\mathrm{d}^2 f}{\mathrm{d}r^2} + \frac{1}{4(f)^2 h}\frac{\mathrm{d}f}{\mathrm{d}r}\frac{\mathrm{d}(fh)}{\mathrm{d}r} + \frac{1}{rh}\frac{\mathrm{d}h}{\mathrm{d}r}\right]\boldsymbol{\gamma}^{r}, \tag{12.28}$$

$$R_{\theta\lambda}\boldsymbol{\gamma}^{\lambda} = -\left[\frac{-r}{2(fh)}\frac{\mathrm{d}f}{\mathrm{d}r} + \frac{r}{2(h)^2}\frac{\mathrm{d}h}{\mathrm{d}r} + \frac{h-1}{h}\right]\boldsymbol{\gamma}^{\theta}, \text{ and} \tag{12.29}$$

$$R_{\phi\lambda}\boldsymbol{\gamma}^{\lambda} = -\sin^2\theta\left[\frac{-r}{2(fh)}\frac{\mathrm{d}f}{\mathrm{d}r} + \frac{r}{2(h)^2}\frac{\mathrm{d}h}{\mathrm{d}r} + \frac{h-1}{h}\right]\boldsymbol{\gamma}^{\phi}. \tag{12.30}$$

Thus, we see that most of the components of the Ricci tensor are already zero because of Schwarzschild's hypothesized form of the line element. Einstein's field equations now become

$$R_{tt} = \frac{c^2 f}{h}\left[\frac{-1}{2f}\frac{\mathrm{d}^2 f}{\mathrm{d}r^2} + \frac{1}{4(f)^2 h}\frac{\mathrm{d}f}{\mathrm{d}r}\frac{\mathrm{d}(fh)}{\mathrm{d}r} - \frac{1}{rf}\frac{\mathrm{d}f}{\mathrm{d}r}\right] = 0, \tag{12.31}$$

$$R_{rr} = \frac{1}{2f}\frac{\mathrm{d}^2 f}{\mathrm{d}r^2} - \frac{1}{4(f)^2 h}\frac{\mathrm{d}f}{\mathrm{d}r}\frac{\mathrm{d}(fh)}{\mathrm{d}r} - \frac{1}{rh}\frac{\mathrm{d}h}{\mathrm{d}r} = 0, \text{ and} \tag{12.32}$$

$$R_{\phi\phi} = \sin^2\theta R_{\theta\theta} = \sin^2\theta\left[\frac{r}{2(fh)}\frac{\mathrm{d}f}{\mathrm{d}r} - \frac{r}{2(h)^2}\frac{\mathrm{d}h}{\mathrm{d}r} - \frac{h-1}{h}\right] = 0. \tag{12.33}$$

Taking an obvious linear combination of (12.31) and (12.32), we have

$$\frac{1}{f}\frac{\mathrm{d}f}{\mathrm{d}r} + \frac{1}{h}\frac{\mathrm{d}h}{\mathrm{d}r} = 0. \tag{12.34}$$

Integrating, we have

$$\ln f + \ln h = \ln(fh) = \ln k, \text{ or}$$
$$fh = k, \text{ where } k \text{ is some constant.}$$

At long distances from the sun, the metric should approach the flat space–time Lorentz metric used for special relativity, so

$$\lim_{r\to\infty} f(r) = \lim_{r\to\infty} h(r) = 1.$$

Thus, we can conclude that our constant k is 1 and we then have

$$f(r)h(r) = 1. \tag{12.35}$$

Using (12.34), (12.33) becomes

$$\frac{r}{(h)^2}\frac{\mathrm{d}h}{\mathrm{d}r} + \frac{h-1}{h} = 0 \text{ or equivalently, } \frac{1}{h(h-1)}\frac{\mathrm{d}h}{\mathrm{d}r} + \frac{1}{r} = 0.$$

From a partial fraction expansion, this becomes

$$\frac{-\mathrm{d}h}{h} + \frac{\mathrm{d}h}{h-1} + \frac{\mathrm{d}r}{r} = 0.$$

Integrating this equation gives us

$$-\ln h + \ln(h-1) + \ln r = \ln C \text{ or } \frac{r(h-1)}{h} = C, \text{ where}$$

C is some constant. Solving for h, we get

$$h(r) = \left(1 - \frac{C}{r}\right)^{-1}.$$

From (12.35),

$$f(r) = \left(1 - \frac{C}{r}\right).$$

Traditionally, the constant C is designated by $2m$, so the line element becomes

$$(\mathrm{d}s)^2 = \left(1 - \frac{2m}{r}\right)c^2(\mathrm{d}t)^2 - \left(1 - \frac{2m}{r}\right)^{-1}(\mathrm{d}r)^2 - r^2(\mathrm{d}\theta)^2 - r^2\sin^2\theta(\mathrm{d}\phi)^2. \tag{12.36}$$

From the form of the Schwarzschild metric, we see that there is a singularity at $r = 2m$. This is known as the *Schwarzschild radius*. Mathematically, this singularity can be removed by another choice of coordinates like the singularities that occur at the North and South poles for spherical coordinates. To emphasize this point, current astronomers refer to the *horizon*. This is a term advanced by Wolfgang Rindler, a physicist at Cornell University in the 1950s.

Nonetheless, it is this peculiar aspect of the Schwarzschild metric that stimulated speculation that black holes might exist. The theory opened up the possibility of black holes but did not predict there existence. The Schwarzschild solution is only valid for mass free space. For the sun, the Schwarzschild radius is about 3 km, which is well within the surface of the sun, which is the boundary of the mass free region. (In view of (12.6) and (12.7), it makes more sense to say that the *critical circumference* for a body with the mass of the sun is 18.5 km.)

It took a stretch of imagination to suggest that their might be objects in outer space that were so dense that their boundaries might be inside their horizons. For a long time, astronomers were unable to construct a plausible scenario for the creation of a black hole. Furthermore, they were not highly motivated to do so. The existence of black holes seemed highly speculative and the two most prominent proponents of the general theory of relativity, Einstein and Eddington were convinced that black holes could not exist.

However as observational instruments became more sophisticated, it was discovered that two families of stars were extremely dense – white dwarfs and neutron stars. Faced with the task of developing theories for these objects, the existence of black holes became more plausible.

The term *black hole* was introduced late in the game. The Russians used the term *frozen star* and in the West, the favored term was *collapsed star.* John Wheeler is responsible for much of the theoretical groundwork on these super dense objects. Late in 1967, after much careful thought, Wheeler introduced the term black hole in a couple of lectures. Within months, this term became almost universally excepted. The exception being the French who resisted the term for several years because the literal translation *trou noir* has obscene connotations (Thorne 1994, p. 257).

The gravitational field of a black hole is so strong that nothing (including light) can escape from inside the horizon. The first black hole to be identified was Cygnus X-1, which was discovered in 1972. Its existence was inferred by X-rays emitted by gas particles that are pulled toward the black hole from all directions at tremendous rates of acceleration (Thorne 1994, pp. 24–25).

To obtain some insight into the long road required to achieve a comprehension of the nature of black holes, I recommend Kip Thorne's *Black Holes & Time Warps – Einstein's Outrageous Legacy*.

Problem 280. Snell's Law: Consider Fig. 12.2. The velocity of light in glass is slower than the velocity of light in air. Because of this the quickest path from A to B (or from B to A) is not a straight line. Use calculus or whatever method you can to show the quickest path occurs when

$$\frac{\sin\theta_a}{v_a} = \frac{\sin\theta_g}{v_g}, \text{ where}$$

v_a is the velocity of light in air and v_g is the velocity of light in glass. This is known as *Snell's Law*. Snell's Law is frequently stated in terms of the index of refraction. For glass, the index of refraction is $n_g = v_a/v_g$.

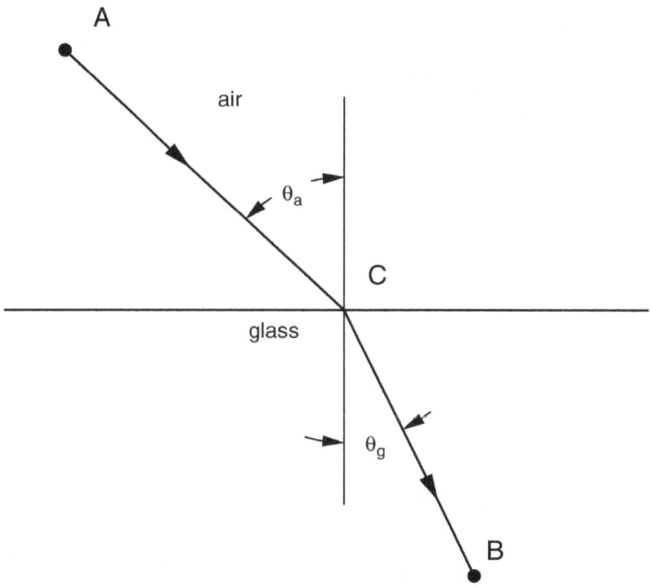

Fig. 12.2 The quickest path from A to B is not a straight line

Problem 281. Sometime before I graduated from high school, I obtained a job as a swmming instructor for a summer Y.M.C.A. "learn to swim" program. The swimming pool had a uniform depth of 4 ft (120 cm). Most of the students had a height of approximately 4 ft. I was supposed to tell the members of each incoming class to climb down a ladder and then grasp a rail that was attached to one of the walls of the pool. However, sometimes I would forget and a child would jump in and suddenly discover that the water was over his or her head. (If my memory is correct it was always a "his head" rather than a "her head.") I would then have to jump into the pool to rescue the child from my negligence. Fortunately, no more than one child ever did this at the same time. Because of refraction, the depth of the pool appeared to be more shallow than it actually was. (See Fig. 12.3.) The index of refraction for water varies slightly with temperature but it is essentially $4/3$. Determine the apparent depth of a 4 ft deep swimming pool looking straight down. Is this illusion sufficient to explain the behavior of my swimming students or should they have looked a little better before they jumped in?

Problem 282. (a) Consider Fig. 12.4. A parabola may be defined as the locus (set) of points that are equidistant from a straight line called a *directrix* and a point known as the *focal point*. Using that definition and an argument similar to the one I used in this section for the flat mirror show that the shortest path from the focal point at F to the parabola and then to point A is the one that is parallel to the axis of symmetry after hitting the parabola. (This shows that any light ray emitted from the focal point of a parabolic mirror is reflected into a path parallel

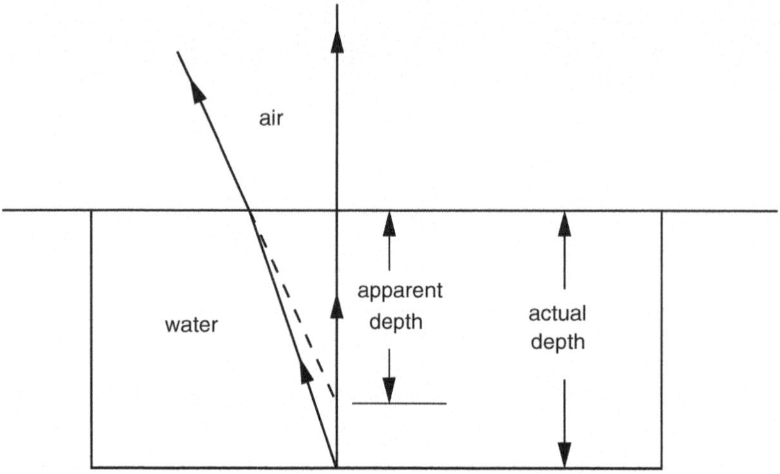

Fig. 12.3 Because of refraction, the apparent depth of a swimming pool is less than the actual depth. Diverging light rays originating from a point on the floor of the pool appear to originate from a point above the floor

Fig. 12.4 Possible light rays originating at the focal point F of a parabolic mirror and passing through the point A

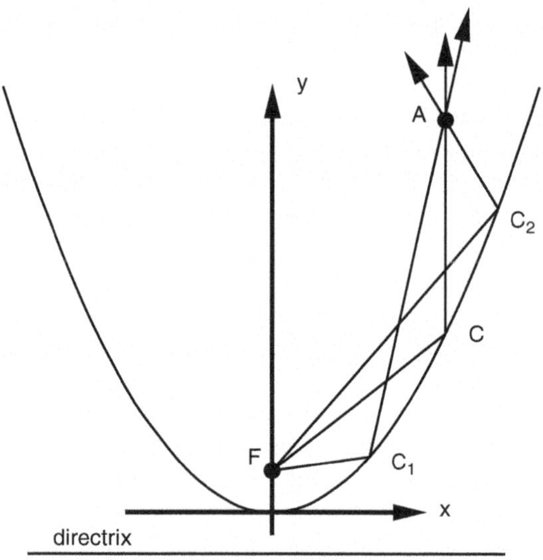

to the axis of symmetry.) Conversely, light ray directed toward the concave side of a parabolic mirror parallel to the axis of symmetry will be reflected toward the focal point. In this manner, a parabolic mirror directed toward the sun will have a very high temperature at the focal point.

(b) Show that at the point of reflection determined in part (a), the angle of incidence equals the angle of reflection.

Problem 283. Using the definition for the Ricci tensor,

$$R_{\alpha\beta} = R^{\lambda}{}_{\alpha\beta\lambda} \text{ show that}$$

$$R_{\alpha\beta} = R_{\beta\alpha}.$$

(You may wish to review (5.52)–(5.55).

Problem 284. Using the fact that

$$\boldsymbol{\gamma}^{\eta}\boldsymbol{\gamma}^{\nu}\boldsymbol{\gamma}^{\beta} = \boldsymbol{\gamma}^{\eta\nu\beta} + g^{\eta\nu}\boldsymbol{\gamma}^{\beta} - g^{\eta\beta}\boldsymbol{\gamma}^{\nu} + g^{\nu\beta}\boldsymbol{\gamma}^{\eta}, \text{ show}$$

$$\mathbf{R}_{\alpha\beta}\boldsymbol{\gamma}^{\beta} = \frac{1}{2}R_{\alpha\beta\eta\nu}\boldsymbol{\gamma}^{\eta}\boldsymbol{\gamma}^{\nu}\boldsymbol{\gamma}^{\beta} = -R_{\alpha\beta}\boldsymbol{\gamma}^{\beta}.$$

(You may wish to review the same equations suggested in Problem 283.)

Problem 285. The Schwarzschild radius has an interpretation even in Newtonian physics. According to Newtonian physics, an object in the gravitational field of a planet satisfies the equation,

$$\frac{1}{2}mv^2 - \frac{mMG}{r} = E, \text{ where}$$

m is the mass of the object, v is the velocity of the object, M is the mass of the planet, G is the universal gravitational constant, r is the distance between the object and the center of the planet, and E is a constant known as the total energy of the system.

Consider the problem of throwing an object from the surface of the planet into space. The constant E will be determined by the radius r_0 of the planet and the initial velocity v_0. As r increased, v decreases. If the object is not given sufficient velocity at the surface of the planet, the constant E will be negative and eventually r will attain a magnitude such that,

$$\frac{mMG}{r} + E = \frac{mMG}{r} - |E| = 0.$$

At that point, the velocity of the object will be zero and the object will begin a descent back to the surface of the planet. However if the initial velocity v_0 is sufficiently great, the object will continue to move into outer space. There exists a minimum value of v_0 that is sufficiently great so that the object will continue to move into outer space. This minimum value for v_0 is known as the *escape velocity*.

(a) Determine a formula for the escape velocity in terms of M, G, and the radius of the planet r_0.
(b) Suppose the speed of light is c. Determine a formula for the maximum radius of the planet with the property that an object given an initial velocity of c would not escape. On the basis of a similar calculation, John Mitchell predicted the existence of "dark stars" in 1783 (Thorne 1994, p. 122).

(c) To what extent would you have to shrink the earth without changing its mass so that the escape velocity would be equal to the speed of light. (The mass of earth $M = 6 \times 10^{24}$ kg, $G = 6.67 \times 10^{-11}$ m^3 (s^{-2} kg), and $c = 3.00 \times 10^8$ m s^{-1}).

Problem 286. In the discussion of Gauss's *Theorema Egregium*, it was shown that a necessary condition for a curved space to be a hypersurface is that there exists a tensor h^α_β such that

$$R^{\alpha\beta}{}_{\nu\eta} = h^\alpha_\nu h^\beta_\eta - h^\alpha_\eta h^\beta_\nu, \text{ which can be rewritten in the form,}$$

$$\mathbf{R}_{\nu\eta} = \frac{1}{2} R^{\alpha\beta}{}_{\nu\eta} \boldsymbol{\gamma}_\alpha \boldsymbol{\gamma}_\beta = \mathbf{H}_\nu \wedge \mathbf{H}_\eta, \text{ where}$$

$$\mathbf{H}_\alpha = h^\beta_\alpha \boldsymbol{\gamma}_\beta .$$

In view of this, is it plausible that the four-dimensional Schwarzschild space–time could be embedded in some flat five dimensional space?

12.4 The Precession of Mercury

From Problem 79, we know that the equations for geodesics can be written in the form,

$$\frac{\mathrm{d}}{\mathrm{d}s} \frac{\partial F}{\partial \acute{u}^\alpha} - \frac{\partial F}{\partial u^\alpha} = 0, \text{ where } \alpha = 1, 2, \ldots, n \text{ and} \tag{12.37}$$

$$F = g_{\alpha\beta} \acute{u}^\alpha \acute{u}^\beta . \tag{12.38}$$

This equation looks unnatural in the context of differential geometry. However, it arises naturally in the context of the calculus of variations. To determine geodesics, F is defined by (12.38). For other problems, F is defined differently. Equation (12.37) was introduced by Euler to generalize the approach that Jacob Bernoulli used to solve the brachistochrone problem to solve a wide variety of problems in physics. Later, Joseph Louis Lagrange (1736–1813) investigated some analytic aspects of the equation. For these reasons, (12.37) became known as the *Euler–Lagrange* equation.

For the Schwarzschild metric, we have

$$(\mathrm{d}s)^2 = \left(1 - \frac{2m}{r}\right)(c\mathrm{d}t)^2 - \left(1 - \frac{2m}{r}\right)^{-1} (\mathrm{d}r)^2 - (r\mathrm{d}\theta)^2 - (r \sin\theta \mathrm{d}\phi)^2. \tag{12.39}$$

so

$$F = \left(1 - \frac{2m}{r}\right)(c\acute{t})^2 - \left(1 - \frac{2m}{r}\right)^{-1} (\acute{r})^2 - (r\acute{\theta})^2 - (r \sin\theta \acute{\phi})^2, \text{ where} \tag{12.40}$$

$$\acute{t} = \frac{\mathrm{d}t}{\mathrm{d}s}, \ \acute{r} = \frac{\mathrm{d}r}{\mathrm{d}s}, \ \acute{\theta} = \frac{\mathrm{d}\theta}{\mathrm{d}s}, \text{ and } \acute{\phi} = \frac{\mathrm{d}\phi}{\mathrm{d}s}. \tag{12.41}$$

The Euler–Lagrange equations are

$$\frac{\mathrm{d}}{\mathrm{d}s}\frac{\partial F}{\partial \acute{t}} - \frac{\partial F}{\partial t} = 0, \tag{12.42}$$

$$\frac{\mathrm{d}}{\mathrm{d}s}\frac{\partial F}{\partial \acute{r}} - \frac{\partial F}{\partial r} = 0, \tag{12.43}$$

$$\frac{\mathrm{d}}{\mathrm{d}s}\frac{\partial F}{\partial \acute{\theta}} - \frac{\partial F}{\partial \theta} = 0, \text{ and} \tag{12.44}$$

$$\frac{\mathrm{d}}{\mathrm{d}s}\frac{\partial F}{\partial \acute{\phi}} - \frac{\partial F}{\partial \phi} = 0. \tag{12.45}$$

Equation (12.42) becomes

$$\frac{\mathrm{d}}{\mathrm{d}s}\left[2\left(1 - \frac{2m}{r}\right)c^2\acute{t}\right] = 0.$$

This implies that

$$\left(1 - \frac{2m}{r}\right)\acute{t} = \text{ a constant.} \tag{12.46}$$

Equations (12.44) and (12.45) become

$$\frac{\mathrm{d}}{\mathrm{d}s}(-2r^2\acute{\theta}) + 2r^2\sin\theta\cos\theta(\acute{\phi})^2 = 0, \text{ and}$$

$$\frac{\mathrm{d}}{\mathrm{d}s}\left[-2(r\sin\theta)^2\acute{\phi}\right] = 0.$$

If we seek out a solution in the x–y plane, then $\theta = \pi/2$ and $\acute{\theta} = 0$. This takes care of the first equation and the second equation then becomes

$$r^2\acute{\phi} = \frac{2\alpha}{c}, \text{ where} \tag{12.47}$$

α is a constant that I will discuss later.

If we write out (12.43), we will have a second-order differential equation for r. However, we can avoid some work if we use (12.39). From that equation, we have

$$1 = \left(1 - \frac{2m}{r}\right)c^2(\acute{t})^2 + \left(1 - \frac{2m}{r}\right)^{-1}(\acute{r})^2 - r^2(\acute{\theta})^2 - r^2\sin^2\theta(\acute{\phi})^2.$$

Substituting $\theta = \pi/2$, $\acute{\theta} = 0$, and $\acute{\phi} = 2\alpha/cr^2$ and reorganizing terms, this last equation becomes

$$(\acute{r})^2 = \left[1 - c^2\left(1 - \frac{2m}{r}\right)^2(\acute{t})^2\right] - \frac{2m}{r} - \frac{4\alpha^2}{c^2r^2} + \frac{8m\alpha^2}{c^2r^3}.$$

Since $(1 - \frac{2m}{r})(\acute{t})$ is a constant, this last equation can be written as

$$(\acute{r})^2 = \frac{-C}{c^2} - \frac{2m}{r} - \frac{4\alpha^2}{c^2r^2} + \frac{8m\alpha^2}{c^2r^3}.$$

The corresponding equation in Newtonian physics is

$$\left(\frac{\mathrm{d}r}{\mathrm{d}t}\right)^2 = -C - \frac{2MG}{r} - \frac{4\alpha^2}{r^2}, \text{ where} \tag{12.48}$$

M is the mass of the sun, G is the universal gravitational constant, and C is a constant proportional to the energy of the system.

To make the two equations easier to compare, I am going to substitute

$$s = c\tau \text{ and } m = MG/c^2. \tag{12.49}$$

Then $\acute{r} = \mathrm{d}r/c\mathrm{d}\tau$ and Einstein's equation for the geodesic path of Mercury becomes

$$\left(\frac{\mathrm{d}r}{\mathrm{d}\tau}\right)^2 = -C - \frac{2MG}{r} - \frac{4\alpha^2}{r^2} + \frac{8MG\alpha^2}{c^2r^3}. \tag{12.50}$$

In some sense, the Newtonian theory of gravitation may be considered the approximation of the Einstein theory made when the speed of light is treated as being infinite. We see that (12.48) and (12.50) do indeed become identical if we replace the speed of light by infinity. Furthermore, we should note that the additional Einstein term becomes most significant when r is small. Thus, the Einstein correction will have a greater impact on the orbit of Mercury than it will for any of the other planets.

We must make a distinction between the "τ" in (12.50) and the "t" that appears in the other Einstein relativity equations in this section. The variable "s" is the arc length parameter for a path through a 4-dimensional space. The variable "τ" is a scalar multiple of s. If two travelers separate and then come together again, they will agree that they began their trips at a common point in their 4-dimensional space and they ended their separate excursions at another common point in their 4-dimensional space. What may differ for the two travelers is the arc lengths of their trips. Each traveler can measure his (or her) arc length by carrying along a personal clock When they meet a second time, these clocks may disagree. This has been demonstrated for very accurate atomic clocks and in theory it should be true for less accurate biological clocks.

From (12.39), we have

$$c^2 = \left(1 - \frac{2m}{r}\right)\left(c\frac{\mathrm{d}t}{\mathrm{d}\tau}\right)^2 - \left(1 - \frac{2m}{r}\right)^{-1}\left(\frac{\mathrm{d}r}{\mathrm{d}\tau}\right)^2 - \left(r\frac{\mathrm{d}\theta}{\mathrm{d}\tau}\right)^2 - \left(r\sin\theta\frac{\mathrm{d}\phi}{\mathrm{d}\tau}\right)^2.$$

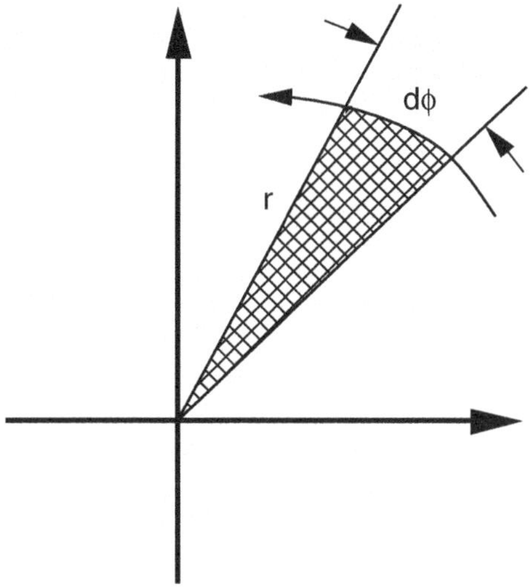

Fig. 12.5 Infinitesimal area swept out by a planet revolving around the sun

If the velocity is nonrelativistic ($\mathrm{d}r/\mathrm{d}\tau \ll c$, $r\,(\mathrm{d}\theta/\mathrm{d}\tau) \ll c$, and $r \sin\theta\,(\mathrm{d}\phi/\mathrm{d}\tau) \ll c$) and the gravitational field is weak ($m/r \approx 0$), then $dt/d\tau \approx 1$. In this circumstance, the change in the path parameter will approximate the change in the time parameter. That is

$$\Delta t = \int \mathrm{d}t \approx \Delta\tau = \int \mathrm{d}\tau.$$

However if a traveler moves around at relativistic speed during the traveler's excursion, then it is possible that $\Delta t \gg \Delta\tau$. Thus, it is theoretically possible that a high speed traveler could return home to discover that a twin brother (or sister) had aged more than he (or she) had. (For both twins, the Δt would be the same but for the stay at home twin $\Delta t \approx \Delta\tau$.)

In the Newtonian model, "t" is a path parameter in a 3-dimensional space. It is neither an arc length parameter nor a constant multiple of an arc length parameter.

Before attacking the problem of solving (12.50), I wish to tell you the significance of the constant "α" that appeared in (12.47). The constant α was chosen so that it can be interpreted as the *areal velocity*. From (12.47) and (12.49), we have

$$\alpha = \frac{1}{2} r^2 \frac{\mathrm{d}\phi}{\mathrm{d}\tau}.$$

We note that $(1/2)r^2\mathrm{d}\phi$ represent an infinitesimal area swept out by the radial vector from the sun to an orbiting planet. (See Fig. 12.5.) Thus, $\alpha = (1/2)r^2(\mathrm{d}\phi/\mathrm{d}\tau)$ is the areal velocity – that is the rate at which the area is swept out. The fact that the areal velocity is constant is known as *Kepler's second law*.

I will now return to the problem of solving (12.50). Namely

$$\left(\frac{\mathrm{d}r}{\mathrm{d}\tau}\right)^2 = -C - \frac{2MG}{r} - \frac{4\alpha^2}{r^2} + \frac{8MG\alpha^2}{c^2r^3}. \tag{12.51}$$

This equation is easier to solve if we substitute

$$r = \frac{1}{u} \text{ and}$$

then solve for u as a function of ϕ rather than as a function of τ. Since

$$\frac{\mathrm{d}r}{\mathrm{d}\tau} = \frac{-1}{u^2}\frac{\mathrm{d}u}{\mathrm{d}\tau} = \frac{-1}{u^2}\frac{\mathrm{d}u}{\mathrm{d}\phi}\frac{\mathrm{d}\phi}{\mathrm{d}\tau} = -r^2\frac{\mathrm{d}\phi}{\mathrm{d}\tau}\frac{\mathrm{d}u}{\mathrm{d}\phi} = -2\alpha\frac{\mathrm{d}u}{\mathrm{d}\phi}, \text{ (12.51) becomes} \tag{12.52}$$

$$4\left(\alpha\right)^2\left(\frac{\mathrm{d}u}{\mathrm{d}\phi}\right)^2 = -C - 2MGu - 4\alpha^2u^2 + \frac{8MG\alpha^2u^3}{c^2} \quad \text{or}$$

$$\left(\frac{\mathrm{d}u}{\mathrm{d}\phi}\right)^2 = \frac{-C}{4\alpha^2} - \frac{MGu}{2\alpha^2} - u^2 + \frac{2MGu^3}{c^2}. \tag{12.53}$$

The Newtonian version of this equation can be solved exactly and the relativistic term can be considered a small perturbation. With that thought in mind, I will present the solution of the Newtonian version first. That is, I will discuss the solution of the equation:

$$\left(\frac{\mathrm{d}u}{\mathrm{d}\phi}\right)^2 = -u^2 - \frac{MGu}{2\left(\alpha_N\right)^2} - \frac{C}{4\left(\alpha_N\right)^2} \tag{12.54}$$

(I have chosen to designate the areal velocity by α_N so that I can pair off a Newtonian orbit with an Einstein orbit later in this section.)

The point in the orbit that is closest to the sun is said to be the *perihelion*. At this position, r is a minimum so u is a maximum. If we use "R" to designate the minimum distance, then:

$$\frac{\mathrm{d}u}{\mathrm{d}\phi} = 0 \text{ when } u = \rho_P = \frac{1}{R}.$$

(See Fig. 12.6a.) From this observation and (12.54), we have

$$0 = -\left(\rho_P\right)^2 + \frac{MG}{2(\alpha_N)^2}\rho_P - \frac{C}{4(\alpha_N)^2}.$$

Subtracting this equation from (12.54) gives us

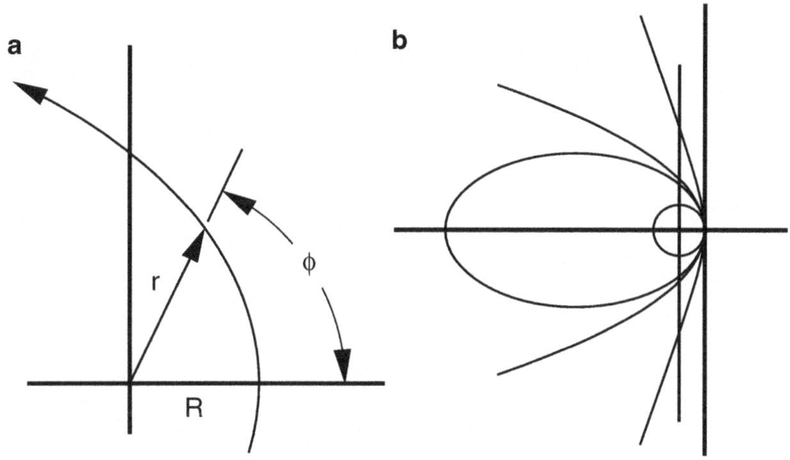

Fig. 12.6 (**a**) R is the minimum distance between a planet and the sun. (**b**) Possible orbits of objects passing near the sun

$$\left(\frac{\mathrm{d}u}{\mathrm{d}\phi}\right)^2 = (\rho_P)^2 - u^2 - \frac{MG}{2(\alpha_N)^2}(\rho_P - u) = (\rho_P - u)\left(\rho_P + u - \frac{MG}{2(\alpha_N)^2}\right), \text{ or}$$

$$\left(\frac{\mathrm{d}u}{\mathrm{d}\phi}\right)^2 = (\rho_P - u)(u - \rho_A), \text{ where} \tag{12.55}$$

$$\rho_A = \frac{MG}{2(\alpha_N)^2} - \rho_P. \tag{12.56}$$

If $\rho_A > 0$, then

$$\frac{1}{\rho_P} < r = \frac{1}{u} < \frac{1}{\rho_A}.$$

In this case, the orbit is bounded and the position where $u - \rho_A$ is said to be the *aphelion*. This is the position of the planet when it is farthest from the sun.

Expanding the right-hand side of (12.55) and completing the square gives us

$$\left(\frac{\mathrm{d}u}{\mathrm{d}\phi}\right)^2 = \left[\frac{\rho_P - \rho_A}{2}\right]^2 - \left[u - \frac{\rho_P + \rho_A}{2}\right]^2, \text{ or}$$

$$\frac{\mathrm{d}u}{\sqrt{\left[\frac{\rho_P-\rho_A}{2}\right]^2 - \left[u - \frac{\rho_P+\rho_A}{2}\right]^2}} = \pm \mathrm{d}\phi, \text{ which implies}$$

$$\arccos \frac{u - \frac{\rho_P+\rho_A}{2}}{\frac{\rho_P-\rho_A}{2}} = \mp\phi + \phi_0. \tag{12.57}$$

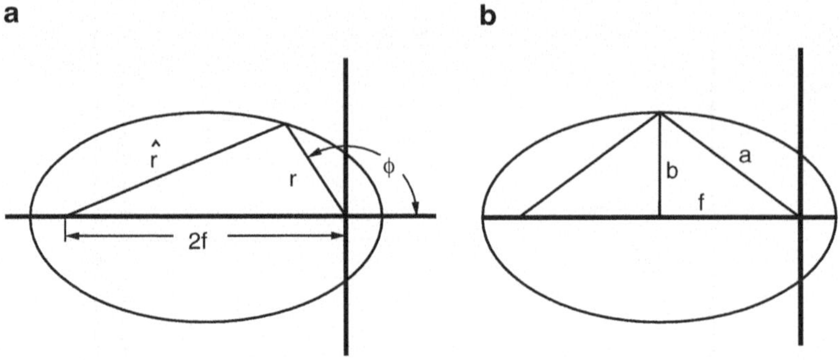

Fig. 12.7 (**a**) An ellipse is defined by the fact that $\hat{r} + r =$ the length of the major axis $2a$. $2f$ is the distance between the focal points. (**b**) $a^2 = b^2 + f^2$, where $2b$ is the length of the minor axis

If we use perihelion for our initial position, then $u = \rho_P$ when $\phi = 0$. Thus,

$$\arccos \frac{\rho_P - \frac{\rho_P + \rho_A}{2}}{\frac{\rho_P - \rho_A}{2}} = \arccos 1 = \phi_0.$$

Therefore, $\phi_0 = 0$. Taking the cosine of each side of (12.57), now gives us

$$u = \frac{1}{r} = \frac{\rho_P + \rho_A}{2} + \frac{\rho_P - \rho_A}{2} \cos\phi. \tag{12.58}$$

This is

a circle if $\rho_A = \rho_P$, or $\left(r^2 \frac{d\phi}{d\tau}\right)^2 = MGR$;

an ellipse if $0 < \rho_A < \rho_P$, or $MGR < \left(r^2 \frac{d\phi}{d\tau}\right)^2 < 2MGR$;

a parabola if $\rho_A = 0$, or $\left(r^2 \frac{d\phi}{d\tau}\right)^2 = 2MGR$;

(Since a parabola is not closed, ρ_A does not indicate an aphelion position.),

a hyperbola if $-\rho_P < \rho_A < 0$, or $2MGR < \left(r^2 \frac{d\phi}{d\tau}\right)^2$; and

a straight line if $\rho_A = -\rho_P$, or $M = 0$.

(If $\rho_A < -\rho_P$, then M would have to be negative.)

(See Fig. 12.6b.)

To examine the case of the ellipse more deeply, you should note that a point on an ellipse has the property that the sum of its distances from the focal points are constant. (See Fig. 12.7a.) From the law of cosines,

$$(\hat{r})^2 = (r)^2 + (2f)^2 + 4rf\cos\phi. \text{ Also,}$$

$$\hat{r} + r = 2a.$$

Therefore,

$$\begin{aligned}
(2a - r)^2 &= (r)^2 + (2f)^2 + 4rf\cos\phi, \\
4a^2 - 4ar &= 4f^2 + 4rf\cos\phi, \\
a^2 - f^2 &= r(a + f\cos\phi), \text{ or} \\
\frac{1}{r} &= \frac{a}{b^2} + \frac{f}{b^2}\cos\phi, \text{ where} \qquad (12.59) \\
b^2 &= a^2 - f^2. \qquad (12.60)
\end{aligned}$$

As you might expect from Fig. 12.7b, in Cartesian coordinates, this equation becomes

$$\frac{(x+f)^2}{a^2} + \frac{y^2}{b^2} = 1. \qquad (12.61)$$

(See Problem 287.) Thus, we see (or should see) that the semi major axis (half the long axis) is "a" and the semi minor axis is "b."

In passing, it is worthwhile to note that *Kepler's first law* is that the orbits of the planets are ellipses, where the sun is located at one of the focal points.

Now let us return to Einstein's equation for the orbit of Mercury. That is

$$\left(\frac{du}{d\phi}\right)^2 = \frac{-C}{4\alpha^2} - \frac{MGu}{2\alpha^2} - u^2 + \frac{2MGu^3}{c^2}. \qquad (12.62)$$

For a Newtonian orbit, the choice of the perihelion distance $R = 1/\rho_P$ and the magnitude of the areal velocity α determines the aphelion distance $1/\rho_A$. To compare an Einstein orbit with a Newton orbit, we wish to make the orbits as close as possible. With that thought in mind, I will choose the same perihelion distance and adjust α so that the aphelion distance is the same. The right-hand side of (12.62) can then be factored accordingly. Thus,

$$\left(\frac{du}{d\phi}\right)^2 = (\rho_P - u)(u - \rho_A)(A - Bu), \text{ where} \qquad (12.63)$$

A and B are to be determined. Multiplying out the right-hand side of (12.63), we have

$$\left(\frac{du}{d\phi}\right)^2 = Bu^3 - [A + B(\rho_P + \rho_A)]u^2 + [A(\rho_P + \rho_{2A}) + B(\rho_P\rho_A)]u - A\rho_P\rho_A. \qquad (12.64)$$

Matching the coefficients of u^3 and u^2 in this equation with those in (12.62), we have

$$B = \frac{2MG}{c^2} \text{ and } A = 1 - \frac{2MG}{c^2}(\rho_P + \rho_A). \qquad (12.65)$$

Thus, (12.64) becomes

$$\left(\frac{\mathrm{d}u}{\mathrm{d}\phi}\right)^2 = \frac{2MG}{c^2}u^3 - u^2 + (\rho_P + \rho_A)\left[1 - \frac{2MG}{c^2}\left(\rho_P + \rho_A - \frac{\rho_P\rho_A}{\rho_P + \rho_A}\right)\right]u - A\rho_P\rho_A.$$

From (12.56),

$$\rho_P + \rho_A = \frac{MG}{2(\alpha_N)^2}, \text{ so}$$

we now have

$$\left(\frac{\mathrm{d}u}{\mathrm{d}\phi}\right)^2 = \frac{2MG}{c^2}u^3 - u^2 + \frac{MG}{2(\alpha_N)^2}\left[1 - \frac{2MG}{c^2}\left(\rho_P + \rho_A - \frac{\rho_P\rho_A}{\rho_P + \rho_A}\right)\right]u - A\rho_P\rho_A. \tag{12.66}$$

Comparing the coefficient of u in this equation with the coefficient of u in (12.62), we see that they match if we equate

$$\alpha^2 = (\alpha_N)^2\left[1 - \frac{2MG}{c^2}\left(\rho_P + \rho_A - \frac{\rho_P\rho_A}{\rho_P + \rho_A}\right)\right]^{-1}.$$

It should be observed that if r_M is the "radius" of the orbit of Mercury, then

$$\rho_P \approx \rho_A \approx \frac{1}{r_M} \text{ and}$$

$$\frac{2MG}{c^2}\left(\rho_P + \rho_A - \frac{\rho_P\rho_A}{\rho_P + \rho_A}\right) = 2m\left(\rho_P + \rho_A - \frac{\rho_P\rho_A}{\rho_P + \rho_A}\right) \approx \frac{3m}{r_M}.$$

Since m is half the Schwarzschild radius for the sun and r_M is the radius of the orbit of Mercury, it is clear that the difference between α and α_N is infinitesimal.

As for the constant term

$$\frac{-C}{4\alpha^2} \text{ that}$$

occurs in the polynomial on the right-hand side of (12.62), the C must be adjusted so that ρ_P is a root of that polynomial. Doing that makes the constant terms on the right-hand sides of (12.66) and (12.62) agree.

Now from (12.65) and (12.65), we have

$$\begin{aligned}\left(\frac{\mathrm{d}u}{\mathrm{d}\phi}\right)^2 &= (\rho_P - u)(u - \rho_A)\left[1 - \frac{2MG}{c^2}(u + \rho_P + \rho_A)\right]\\ &= (\rho_P - u)(u - \rho_A)\left[1 - 2m(u + \rho_P + \rho_A)\right].\end{aligned}$$

Thus,

$$\frac{\mathrm{d}u}{\sqrt{(\rho_P - u)(u - \rho_A)\left[1 - 2m(u + \rho_P + \rho_A)\right]}} = \pm \mathrm{d}\phi \tag{12.67}$$

Using the first two terms in a Taylor series expansion in powers m, we have

$$\left[1 - 2m(u + \rho_P + \rho_A)\right]^{-\frac{1}{2}} \approx 1 + m(u + \rho_P + \rho_A).$$

From the Newtonian calculation, we know that

$$(\rho_P - u)(u - \rho_A) = \left[\frac{\rho_P - \rho_A}{2}\right]^2 - \left[u - \frac{\rho_P + \rho_A}{2}\right]^2. \tag{12.68}$$

Thus, (12.67) becomes

$$\frac{\left[1 + m(u + \rho_P + \rho_A)\right]\mathrm{d}u}{\sqrt{\left[\frac{\rho_P - \rho_A}{2}\right]^2 - \left[u - \frac{\rho_P + \rho_A}{2}\right]^2}} = \pm \mathrm{d}\phi.$$

Regrouping terms, so that things become integrable, we get

$$\frac{\left[1 + \frac{3m(\rho_P + \rho_A)}{2}\right]\mathrm{d}u}{\sqrt{\left[\frac{\rho_P - \rho_A}{2}\right]^2 - \left[u - \frac{\rho_P + \rho_A}{2}\right]^2}} + \frac{m\left[u - \frac{\rho_P + \rho_A}{2}\right]\mathrm{d}u}{\sqrt{\left[\frac{\rho_P - \rho_A}{2}\right]^2 - \left[u - \frac{\rho_P + \rho_A}{2}\right]^2}} = \pm \mathrm{d}\phi.$$

Integrating, we get

$$-\left[1 + \frac{3m(\rho_P + \rho_A)}{2}\right] \arccos \frac{u - \frac{\rho_P + \rho_A}{2}}{\frac{\rho_P - \rho_A}{2}} - m\sqrt{\left[\frac{\rho_P - \rho_A}{2}\right]^2 - \left[u - \frac{\rho_P + \rho_A}{2}\right]^2}$$
$$= \pm \phi + \phi_0. \tag{12.69}$$

Comparing (12.58) and (12.59), we find that

$$\frac{\rho_P + \rho_A}{2} = \frac{a}{b^2}.$$

Using this result and (12.69), we have

$$-\left(1 + \frac{3ma}{b^2}\right) \arccos \frac{u - \frac{\rho_P + \rho_A}{2}}{\frac{\rho_P - \rho_A}{2}} - m\sqrt{(\rho_P - u)(u - \rho_A)} = \pm \phi + \phi_0.$$

Using perihelion as our initial position as we did for our Newtonian solution, $\phi_0 = 0$. Near perihelion, ϕ should be an increasing function of r or a decreasing function of u. This forces us to choose the minus sign on the right-hand side of this last equation, so we finally have

$$\phi = \left(1 + \frac{3ma}{b^2}\right) \arccos \frac{u - \frac{\rho_P + \rho_A}{2}}{\frac{\rho_P - \rho_A}{2}} + m\sqrt{(\rho_P - u)(u - \rho_A)}. \tag{12.70}$$

Initially, when $u = \rho_P$,

$$\phi = \left(1 + \frac{3ma}{b^2}\right) \arccos 1 = 0.$$

After one revolution, when the planet returns to perihelion, we again have

$$\phi = \left(1 + \frac{3ma}{b^2}\right) \arccos 1.$$

But this time with a slight abuse of notation, we must interpret $\arccos 1 = 2\pi$. So, the new angle is

$$2\pi \left(1 + \frac{3ma}{b^2}\right).$$

Thus, after each revolution, the position of the perihelion advances by the amount

$$\Delta\phi = \frac{6\pi ma}{b^2}. \tag{12.71}$$

This rotation of the major axis of what is essentially an ellipse is known as *precession*. For the case of Mercury, (12.71) translates into 43.03 s of arc per century! Remember, a minute is 1/60 of a degree and a second is 1/60 of a minute. When Einstein discovered that this minuscule precession could be derived from his theory, he was elated. His biographer, Abraham Pais, who personally knew Einstein during his years in Princeton, wrote, "This discovery was, I believe, by far the strongest emotional experience in Einstein's scientific life, perhaps in all his life" (Pais 1983, p. 253).

Why was this result so significant to Einstein? In 1915, there were few known discrepancies in Newton's theory that could be explained by Einstein's General Theory of Relativity. In the past, any plausible discrepancy was eventually explained in terms of Newton's theory. For example, as already noted in Sect. 2 of Chap. 3, during the nineteenth century an anomalous behavior in the orbit of Uranus was observed. The British astronomer John Couch Adams and the French astronomer Urbain Jean Joseph Leverrier independently made very lengthy computations to predict the position of the then undiscovered planet Neptune. Leverrier sent his results to Johann Gottfried Galle of the Berlin Observatory. With this information, Galle was able to locate Neptune on September 23, 1846. This was a triumph for Newton's theory.

It is worth noting that Adams had completed his calculations 2 years earlier while still a student at Cambridge. However, being young and unknown, he was unable

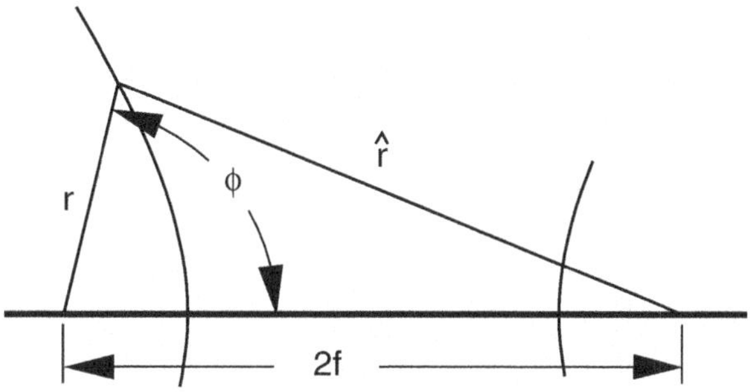

Fig. 12.8 A hyperbola can be defined by the relation $\hat{r} - r = 2a$, where $2a$ is the distances between the vertices of the two branches and $2f$ is the distance between the focal points

to persuade the authorities in charge of the Cambridge Observatory to initiate a systematic search for the planet in a timely manner.

When Leverrier turned his attention to the advance of the perihelion of Mercury, he found that he could account for most of the advance by the perturbations due to other planets. However, there was an unexplained residual advance. To account for this discrepancy, Leverrier hypothesized the existence of a planet between Mercury and the sun. This theoretical planet became known as "Vulcan" and many attempts were made to find it but all ended in failure. In 1898, the American Simon Newcomb published a paper in which he calculated the residual advance unaccounted by the known planets to be $41.24'' \pm 2.09''$ per century (Newcomb 1898). Einstein was convinced that there was no "Vulcan" and part of the reason he was dissatisfied with one of his earlier theories was that it failed to predict the correct precession for Mercury's orbit (Levenson 2004, p. 111).

More accurate estimates of this residual advance have been made in recent years, but they remain consistent with the value of $43.03''$ predicted by Einstein's theory.

Problem 287. Derive (12.61) from (12.59).

Problem 288. A hyperbola may be defined by the relation: $\hat{r} - r = 2a$, where $2a$ is the distance between the vertices of the two branches. (See Fig. 12.8.)

(a) Determine the equation for the branch pictured on the left in Fig. 12.8, in polar coordinates. Your equation should be expressed in terms of r, ϕ, a, f, and b, where $b^2 = f^2 - a^2$.
(b) Obtain an equation for both branches of the same hyperbola, in Cartesian coordinates.

Problem 289. Use the fact that α is the areal velocity and the fact that the area of an ellipse is πab to obtain a formula for the period T, where it is understood that the *period* is the time required for the planet to complete one revolution. Your final formula should involve only T, M, G, a, and b. Actually, the variable b will disappear when you eliminate α. Your formula should give you a Newtonian explanation for *Kepler's third law*. Namely, the ratio of the square of the period to the cube of the semi-major axis is the same for all planets in the solar system.

12.5 The Bending of Light

In 1905, Einstein had created a theory that was invariant under changes in coordinate systems that were related to one another by constant speeds. Now (in 1915), he had created a theory that was covariant – that is it was invariant under any change of coordinate systems.

By doing this, he not only removed some logical inconsistencies in Newton's theory but he was able to give a precise explanation for the precession of Mercury. For Einstein, the theory was too good not to be true.

However, the reasons that convinced Einstein that his theory was valid would not be enough to convince most of his fellow physicists. As a rule, physicists at that time were unaware of the fact that there were logical inconsistencies in Newton's theory. The fact that the hypothetical "Vulcan" had not been found was a point against Newton's theory. However, it was understood that it would be difficult to observe. Furthermore, for anyone who was not as sophisticated a student of differential geometry as a reader of this book, the mathematics looked hopelessly complicated. Perhaps, Einstein had somehow adjusted his theory so that it would fit the known precession of Mercury.

What would impress scientists and non-scientists alike would be a prediction of something that had never been measured before. A measurement that had not been made was the bending of light.

In 1913, Einstein had constructed a theory that also enabled him to predict the amount of bending for a light ray passing through a spherically symmetric gravitational field. This bending would be perceptible for a light ray from a distant star passing near the surface of the sun. Of course, usually one cannot see many stars during the daytime – particularly stars that would have any position in the sky near that of the sun. However, such stars are visible during a total solar eclipse.

A young German astronomer, Erwin Freundlich, was eager to test Einstein's prediction. A total eclipse would soon occur on August 21, 1914. Einstein and others raised funds to enable Freundlich and two other Germans to mount an expedition to Crimea in southern Russia, where the eclipse would occur. Freundlich and some others left Berlin on July 19, 1914 and arrived at their destination the following week (Levenson 2003, p. 45). Unfortunately for Freundlich, political events now unfolding in Europe made it impossible for him to fully execute his carefully laid plans.

Earlier that summer on June 28, Archduke Franz Ferdinand was murdered in Sarajevo. Visiting Sarajevo on June 28 was not a wise decision for the Archduke. Five hundred and twenty five years earlier, the Serbs had been disgraced in a battle they lost to the Turks at Kosovo. Although the Archduke of Austria was not a Turk, Austria had recently dictated the boundaries of Serbia in a way that the Serbs thought to be high-handed. Furthermore, Austria seemed poised to annex Serbia as a province of Austria.

June 28 was observed as a national anniversary of an old humiliation and the Archduke was viewed as a present-day humiliator. Thus, Ferdinand's decision to visit Sarajevo that day with the expectation that he be received with reverence was extremely naive and truly insensitive.

Soon after the assassination, Austria took the view that the Serbs should be punished. The Czar of Russia took the position that the Serbs should be respected. Kaiser Wilhelm II of Germany came to the support of Austria and demanded France remain neutral. France refused so Germany declared war on Russia on August 1, declared war on France on August 3, and sent troops into neutral Belgium 1 day later on August 4. (Going through Belgium to get to France instead of moving across the border Germany shared with France was considered good military strategy by the German high command.)

All of a sudden, Erwin Freundlich and his crew were in an enemy country with some of the world's most sophisticated camera equipment – equipment quite suitable for espionage. Naturally, the Russians were not inclined to behave as gracious hosts. They seized the camera equipment and arrested Freundlich and two of his friends. According to Abraham Pais, as war broke out, the crew was warned in time to avoid arrest and some did so (Pais 1982, p. 303). Freundlich's crew had made contact with a group of astronomers from Argentina who hoped to get a photo of the never to be discovered Vulcan. The fact that the Argentines planned on using the German equipment was not enough to inspire the Russian authorities to allow the international science project to proceed. To underline the futility of the venture, on the day of the eclipse, it was too cloudy to take any useful pictures.

Fortunately for Einstein's friends, they became subjects of one of the first prisoner exchanges of the war. They were swapped with some Russian officers and returned to Berlin sometime in September.

For Einstein, the failure of Freundlich's efforts turned out to be a lucky break. His 1913 prediction for the magnitude of the bending of light was wrong! It was 1/2 the correct value. Had the 1914 expedition to Crimea been successful, Einstein's 1915 prediction would not have been a prediction. It would have been merely a plausible explanation for something that was already known. This would have made it relatively easy to dismiss his theory. Few would have appreciated the fact that Einstein's 1915 theory removed some logical inconsistencies in Newton's theory – something that his 1913 theory did not.

Let us turn to the geometry of Einstein's 1915 prediction for the bending of light. As in the previous calculation,

$$(\mathrm{d}s)^2 = \left(1 - \frac{2m}{r}\right)(c\mathrm{d}t)^2 - \left(1 - \frac{2m}{r}\right)^{-1}(\mathrm{d}r)^2 - (r\mathrm{d}\theta)^2 - (r\sin\theta\mathrm{d}\phi)^2. \quad (12.72)$$

However for photons, $\mathrm{d}s = 0$. This implies that s cannot be used as a parameter. After introducing w as a parameter in place of s, most of the equations for the path of a light ray passing near the sun have the same form as those for a planetary orbit.

In particular,

$$\frac{\mathrm{d}}{\mathrm{d}w}\frac{\partial F}{\partial \dot{t}} - \frac{\partial F}{\partial t} = 0, \tag{12.73}$$

$$\frac{\mathrm{d}}{\mathrm{d}w}\frac{\partial F}{\partial \dot{\theta}} - \frac{\partial F}{\partial \theta} = 0, \text{ and} \tag{12.74}$$

$$\frac{\mathrm{d}}{\mathrm{d}w}\frac{\partial F}{\partial \dot{\phi}} - \frac{\partial F}{\partial \phi} = 0, \text{ where} \tag{12.75}$$

$$F = \left(1 - \frac{2m}{r}\right)(c\dot{t})^2 - \left(1 - \frac{2m}{r}\right)^{-1}(\dot{r})^2 - (r\dot{\theta})^2 - (r\sin\theta\dot{\phi})^2, \text{ and}$$

$$\dot{t} = \frac{\mathrm{d}t}{\mathrm{d}w}, \; \dot{r} = \frac{\mathrm{d}r}{\mathrm{d}w}, \; \dot{\theta} = \frac{\mathrm{d}\theta}{\mathrm{d}w}, \text{ and } \dot{\phi} = \frac{\mathrm{d}\phi}{\mathrm{d}w}.$$

Thus, we have as before

$$\theta = \frac{\pi}{2}, \; r^2\dot{\phi} = \frac{2\alpha}{c}, \text{ and } \left(1 - \frac{2m}{r}\right)\dot{t} = k.$$

The big difference is that from (12.72),

$$\left(\frac{\mathrm{d}s}{\mathrm{d}w}\right)^2 = 0 = \left(1 - \frac{2m}{r}\right)(c\dot{t})^2 - \left(1 - \frac{2m}{r}\right)^{-1}(\dot{r})^2 - (r\dot{\theta})^2 - (r\sin\theta\dot{\phi})^2, \text{ or}$$

$$(\dot{r})^2 = (ck)^2 - \frac{4\alpha^2}{c^2r^2} + \frac{8m\alpha^2}{c^2r^3}.$$

As before, let

$$r = \frac{1}{u} \text{ and } \dot{r} = \frac{\mathrm{d}r}{\mathrm{d}w} = -\frac{1}{u^2}\frac{\mathrm{d}u}{\mathrm{d}w} = -\frac{1}{u^2}\frac{\mathrm{d}u}{\mathrm{d}\phi}\dot{\phi} = -\frac{2\alpha}{c}\frac{\mathrm{d}u}{\mathrm{d}\phi}.$$

With this substitution, we have

$$\left(\frac{\mathrm{d}u}{\mathrm{d}\phi}\right)^2 = \frac{c^4k^2}{4\alpha^2} - u^2 + 2mu^3. \tag{12.76}$$

If R is the minimum distance of the light ray from the center of the sun, then $\mathrm{d}u/\mathrm{d}\phi = 0$, when $u = \rho = 1/R$. (Actually, R is the Schwarzschild version of the distance.) From (12.76),

$$0 = \frac{c^4k^2}{4\alpha^2} - \rho^2 + 2m\rho^3.$$

Subtracting this result from (12.76) gives us

$$\begin{aligned}\left(\frac{\mathrm{d}u}{\mathrm{d}\phi}\right)^2 &= \rho^2 - u^2 - 2m\left(\rho^3 - u^3\right)\\ &= (\rho - u)\left[\rho + u - 2m(\rho^2 + \rho u + u^2\right]\\ &= (\rho - u)\left[-2mu^2 + (1 - 2m\rho)\,u + \rho\,(1 - 2m\rho)\right]. \end{aligned} \tag{12.77}$$

To factor the quadratic term, we note that if

$$-2mu^2 + (1 - 2m\rho)\,u + \rho\,(1 - 2m\rho) = 0, \text{ then}$$

$$\begin{aligned} u &= \frac{-(1 - 2m\rho) + \sqrt{\left((1 - 2m\rho)^2 + 8m\rho\,(1 - 2m\rho)\right)}}{-4m}\\ &= \frac{-(1 - 2m\rho) \pm \sqrt{1 + 4m\rho - 12(m\rho)^2}}{-4m}. \end{aligned}$$

Using the first three terms for a Taylor's series expansion in powers of $m\rho$,

$$\sqrt{1 + 4m\rho - 12(m\rho)^2} \approx 1 + 2m\rho - 8(m\rho)^2.$$

Using this result,

$$u = \frac{-(1 - 2m\rho) \pm \left(1 + 2m\rho - 8(m\rho)^2\right)}{-4m}, \text{ or}$$

$$u = \begin{cases} \dfrac{4m\rho - 8\,(m\rho)^2}{-4m} & = -\rho + 2m\rho^2, \text{ or}\\ \dfrac{-2 + 8(m\rho)^2}{-4m} & = \frac{1}{2m} - 2m\rho^2. \end{cases}$$

Equation (12.77) now becomes

$$\begin{aligned}\left(\frac{\mathrm{d}u}{\mathrm{d}\phi}\right)^2 &= (\rho - u)\left(\rho + u - 2m\rho^2\right)\left(u - \frac{1}{2m} + 2m\rho^2\right)(-2m)\\ &= \left(\left[\rho - m\rho^2\right] - \left[u - m\rho^2\right]\right)\left(\left[\rho - m\rho^2\right] + \left[u - m\rho^2\right]\right)\left(1 - 2mu - 4(m\rho)^2\right). \end{aligned}$$

Thus,

$$\frac{\mathrm{d}u}{\sqrt{[\rho - m\rho^2]^2 - [u - m\rho^2]^2}\,\sqrt{1 - 2mu - 4(m\rho)^2}} = \pm\mathrm{d}\phi. \tag{12.78}$$

Using a Taylor's series expansion again

$$\left(1 - 2mu - 4(m\rho)^2\right)^{-1/2} \approx 1 + mu.$$

With this approximation, (12.78) becomes

$$\frac{(1+mu)\,\mathrm{d}u}{\sqrt{[\rho - m\rho^2]^2 - [u - m\rho^2]^2}} = \pm \mathrm{d}\phi.$$

Reorganizing terms, so that we can integrate everything, we have

$$\frac{\left(1+(m\rho)^2\right)\mathrm{d}u}{\sqrt{[\rho - m\rho^2]^2 - [u - m\rho^2]^2}} + \frac{m\left(u - m\rho^2\right)\mathrm{d}u}{\sqrt{[\rho - m\rho^2]^2 - [u - m\rho^2]^2}} = \pm \mathrm{d}\phi.$$

Carrying out the integration, we get

$$-\left(1+(m\rho)^2\right)\arccos\frac{u - m\rho^2}{\rho - m\rho^2} - m\left(\left[\rho - m\rho^2\right]^2 - \left[u - m\rho^2\right]^2\right)^{1/2} = \pm\phi + \phi_0.$$

Using previous arguments, $\phi_0 = 0$ and after the light ray has passed the sun, we should choose the minus sign for ϕ. We then have

$$\phi = \left(1+(m\rho)^2\right)\arccos\frac{u - m\rho^2}{\rho - m\rho^2} + m\left(\left[\rho - m\rho^2\right]^2 - \left[u - m\rho^2\right]^2\right)^{1/2}.$$

Taking the limit when $r \to \infty$ or $u \to 0$, we have

$$\phi = \left(1+(m\rho)^2\right)\arccos\frac{-m\rho}{1 - m\rho} + m\rho\left([1 - m\rho]^2 - [m\rho]^2\right)^{1/2}. \tag{12.79}$$

Taking the first two terms in yet another Taylor series expansion,

$$\arccos(-x) = \frac{\pi}{2} + x.$$

Using this fact and retaining only first-order terms, (12.79) becomes

$$\phi = \frac{\pi}{2} + 2m\rho.$$

This means that the total deflection is

$$\Delta\phi = 4m\rho = \frac{4m}{R}, \text{ where}$$

R is the closest distance between the light ray and the center of the sun. If R is the radius of the sun, and m is half the Schwarzschild radius for the sun, then

$$\Delta\phi = 1.75\,\text{s of arc}.$$

(See Fig. 12.9.)

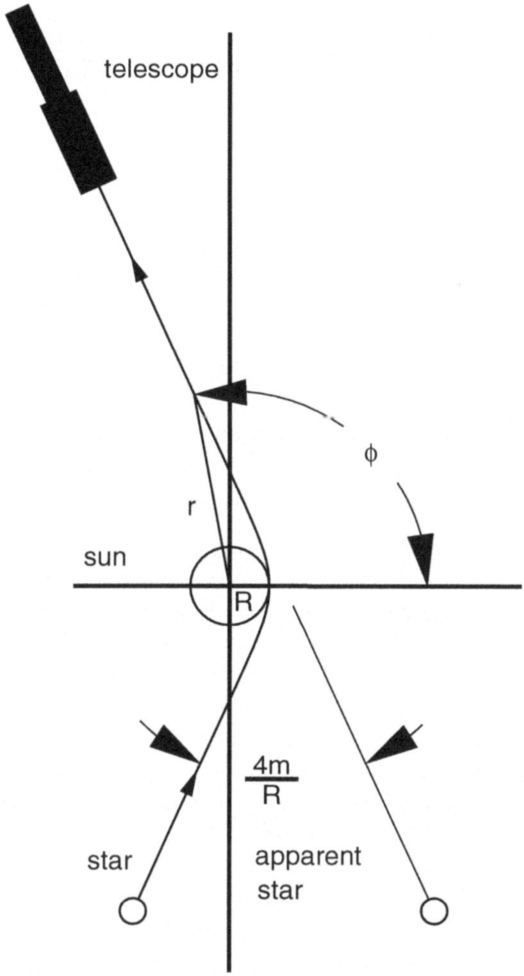

Fig. 12.9 The bending of light due to the curvature of space-time

This is twice the deflection predicted by Einstein's 1913 theory. When this prediction was confirmed by a group of English astronomers organized by Arthur Eddington during the eclipse of May 29, 1919, Einstein became an international celebrity – a status he never lost.

In recent years, it has been discovered that this bending of light can be used to extract information from the cosmos. It turns out that a massive entity such as a galaxy or black hole can act as a lens (a "gravitational lens") for light rays from a strong light source located at an exceptional distance. In 1979, a pair of quasars were spotted in the sky very close to one another. Because the spectrum of a quasar tends to be unique, it was soon decided that the "pair" of quasars were two images of one quasar. (See Fig. 12.10.)

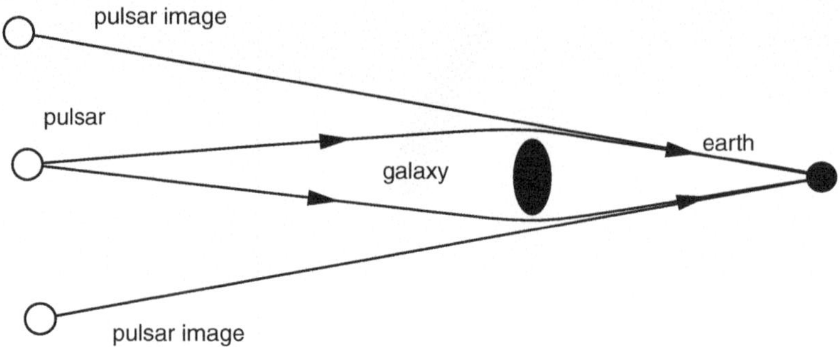

Fig. 12.10 A galaxy acting as a gravitational lens

Since 1979, using the Hubble telescope, many more examples of multiple images of quasars have been discovered and astronomers are now using them to extract information about the mass content of certain galaxies.

Problem 290. In 1801, Thomas Young showed that he could make light interfere with itself. This demonstrated that light had a wave behavior. For this reason, during the nineteenth century, no prominent physicist suggested that gravity would have any effect on a light path. During Newton's time, the situation had been different. On the one hand, Christian Huygens was able to show that one could construct a very plausible explanation for the refraction of light from a wave theory. On the other hand, it was clear that light did not bend around corners of buildings like sound does. Thus, Newton was inclined to believe that light rays were composed of streams of particles. In his book, *Einstein – The Passions of a Scientist* (2003, p. 206), Barry Parker writes,

"– as it turned out, Newton had also made a prediction that a light beam would be deflected by a gravitational field, in an appendix of his book *Opticks*. He had not made a numerical prediction, but Eddington showed that it would have been about –. Because of this, many people in England thought of the controversy as one between the Englishman, Newton, and the German, Einstein (Einstein was, of course, not a German citizen)."

It is interesting to note, that Einstein got the Nobel Prize not for his special or general relativity theories but for demonstrating that the result of an experiment by Philipp Lenard could best be explained if light not only has a wave behavior but also a particle behavior.

After all this chit chat, the problem I present to you is to determine the magnitude of the deflection of light that would have been predicted by Newton. Actually, photons do not behave quite like a Newtonian particle. A Newtonian particle passing near the sun would speed up as it got near the sun and then slow down as it moved away from the sun. To get a Newtonian prediction, ignore this fact and assume $2\alpha_N = r(r\dot{\phi}) = Rc = c/\rho_P$. Also, use the fact that $m = MG/c^2$. (You may wish to review the solution of the Newtonian equations for the orbit of Mercury.) How does the Newtonian prediction compare with Einstein's, 1913 prediction?

Appendix A
A Matrix Representation of a Clifford Algebra

To construct a matrix representation for a Clifford algebra corresponding to a Euclidean or pseudo-Euclidean space, it is not sufficient to construct a set $\{\mathbf{e}_1, \mathbf{e}_2, \ldots, \mathbf{e}_n\}$ such that

$$\mathbf{e}_j\mathbf{e}_k + \mathbf{e}_k\mathbf{e}_j = 2n_{jk}\mathbf{I}, \text{ where} \tag{A.1}$$

$$n_{kk} = \langle \mathbf{e}_k, \mathbf{e}_k \rangle = 1 \text{ for } k = 1, 2, \ldots, p,$$

$$n_{kk} = \langle \mathbf{e}_k, \mathbf{e}_k \rangle = -1 \text{ for } k = p+1, p+2, \ldots, p+q = n, \text{ and}$$

$$n_{jk} = \langle \mathbf{e}_j, \mathbf{e}_k \rangle = 0 \text{ for } j \neq k.$$

We must also consider all possible products of the form $\mathbf{M}_1\mathbf{M}_2 \ldots \mathbf{M}_n$ where $\mathbf{M}_k = \mathbf{e}_k$ or $\mathbf{I}$. These products form a set of 2^n matrices that should be linearly independent. For our purposes, this set should be linearly independent. One's first thought is that to check that these products are linearly independent would be a formidable task. However due to a theorem proven by Ian Porteus, there is a simple test to check for this required linear independence (Porteus 1981, pp. 243–245).

Theorem 291. *A set $\{\mathbf{e}_1, \mathbf{e}_2, \ldots, \mathbf{e}_n\}$ consisting of n matrices representing orthonormal 1-vectors generates a vector space (and therefore an algebra) of dimension 2^n unless the product $\mathbf{e}_1\mathbf{e}_2 \ldots \mathbf{e}_n = \mathbf{e}_{12\ldots n}$ is a scalar multiple of* $\mathbf{I}$.

Proof. With one exception, a product of one or more distinct orthonormal 1-vectors will anticommute with at least one of the 1-vectors. To see this, we note that a product of an even number of the 1-vectors will anticommute with any 1-vector appearing in the product. We also see that the product of an odd number of Dirac matrices will anticommute with any Dirac matrix that does not appear in the product. The one product that commutes with all the 1-vectors in the set is the product $\mathbf{e}_{12\ldots n}$ where n is odd. The proof now proceeds by self-contradiction. Suppose the products are not linearly independent. In that case, there exists a set of coefficients $A^{j_1 j_2 \ldots j_k}$ (not all zero) such that

DOI 10.1007/978-0-8176-8283-5,

$$\sum_{k=0}^{n} \sum_{j_1<j_2<\cdots<j_k} A^{j_1 j_2 \ldots j_k} \mathbf{e}_{j_1 j_2 \ldots j_k} = 0. \tag{A.2}$$

If the coefficient of $\mathbf{I}$ in (A.2) is not zero, divide (A.2) by that coefficient and obtain the equation

$$\mathbf{I} + \sum_{j_1<j_2<\cdots<j_k} B^{j_1 j_2 \ldots j_k} \mathbf{e}_{j_1 j_2 \ldots j_k} = 0, \text{ where} \tag{A.3}$$

the sum does not include the identity matrix. If the coefficient of $\mathbf{I}$ in (A.2) is zero, pick out a term in the equation with a nonzero coefficient (say $\mathbf{e}_{m_1 m_2 \ldots m_k}$) and multiply by $((1/A^{m_1 m_2 \ldots m_k})\,(\mathbf{e}_{m_1 m_2 \ldots m_k})^{-1})$ where $(\mathbf{e}_{m_1 m_2 \ldots m_k})^{-1} = \pm \mathbf{e}_{m_1 m_2 \ldots m_k}$. In this fashion, one can always obtain an equation with the form of (A.3) from (A.2).

If the $B^{j_1 j_2 \ldots j_k}$'s are all zero, we already have the desired contradiction. If the sum in (A.3) contains a product (say $\mathbf{e}_{m_1 m_2 \ldots m_k}$) that anti-commutes with $\mathbf{e}_m$, then one can multiply (A.3) on the left by $\mathbf{e}_m$ and on the right by $(\mathbf{e}_m)^{-1}$ and obtain

$$\mathbf{I} + \sum_{j_1<j_2<\cdots<j_k} B^{j_1 j_2 \ldots j_k} \mathbf{e}_m \mathbf{e}_{j_1 j_2 \ldots j_k} (\mathbf{e}_m)^{-1} = 0. \tag{A.4}$$

We note that

$$\mathbf{e}_m \mathbf{e}_{m_1 m_2 \ldots m_k} (\mathbf{e}_m)^{-1} = -\mathbf{e}_{m_1 m_2 \ldots m_k}.$$

Therefore, we can add (A.3) and (A.4) and thereby obtain a new equation that except for a factor of 2 is identical to (A.3) except at least one less term $(\mathbf{e}_{m_1 m_2 \ldots m_k})$ will now appear in the sum.

If n is even, this process can be continued until the sum reduces to zero and thus obtain the contradiction $\mathbf{I} = 0$. If n is odd, the process can be continued until we have

$$\mathbf{I} + \alpha \mathbf{e}_{12\ldots n} = 0. \tag{A.5}$$

Thus, we see that our desired contradiction occurs unless $\mathbf{e}_{12\ldots n}$ is a scalar multiple of $\mathbf{I}$ and so the theorem is proved. □

Before continuing, I should make a few remarks. Suppose a set of 1-vectors $\{\mathbf{e}_1, \mathbf{e}_2, \ldots, \mathbf{e}_n\}$ satisfies (A.1). If this set also generates an algebra of dimension 2^n, I have defined the resulting algebra to be a Clifford algebra. However, it is possible to have a set of 1-vectors $\{\mathbf{e}_1, \mathbf{e}_2, \ldots, \mathbf{e}_n\}$ that satisfies (A.1) but generates an algebra with dimension less than 2^n. What is to be said of such an algebra? From a purely algebraic point of view, such an algebra is also a Clifford algebra. Indeed, it can be shown that such an algebra is isomorphic to a Clifford algebra associated with a Euclidean or pseudo-Euclidean space of dimension $n - 1$.

For the study of geometry (at least the geometry presented in this book), such a Clifford algebra would be unsuitable to study an n-dimensional space. If the Dirac matrices satisfy (A.1) and generate an algebra of dimension 2^n, Ian Porteus describes the resulting algebra as an *universal Clifford algebra*. Universal Clifford algebras are the only kind of Clifford algebras used in this book. Other authors use more abstract definitions to avoid nonuniversal Clifford algebras.

Some physicists have written papers describing five-dimensional theories using a nonuniversal Clifford algebra. Since these papers do not mention the fact that they are using a nonuniversal Clifford algebra, it suggests that they have inadvertently made a mistake. This is understandable since in 1958, Marcel Riesz thought he had proven that nonuniversal Clifford algebras do not exist (1993, pp. 10–12).

Having proven Theorem 291, we are now in a much better position to construct a matrix representation for a Clifford algebra. Perhaps, the best way to construct explicit representations is by using the *Kronecker product* of matrices. Suppose A is an $n \times n$ matrix and B is an $m \times m$ matrix. In particular if $A = \left[a_{ij}\right]$, then the Kronecker product is defined by the partitioned matrix:

$$A \circ B = \begin{bmatrix} a_{11}B & a_{12}B & - & - & - & a_{1n}B \\ a_{21}B & a_{22}B & - & - & - & a_{2n}B \\ - & - & - & - & - & - \\ - & - & - & - & - & - \\ - & - & - & - & - & - \\ a_{n1}B & a_{n1}B & - & - & - & a_{nn}B \end{bmatrix} \tag{A.6}$$

For example if

$$A = \begin{bmatrix} a_{11} & a_{12} \\ a_{21} & a_{22} \end{bmatrix} \text{ and } B = \begin{bmatrix} b_{11} & b_{12} \\ b_{21} & b_{22} \end{bmatrix}, \text{ then}$$

$$A \circ B = \begin{bmatrix} a_{11}b_{11} & a_{11}b_{12} & a_{12}b_{11} & a_{12}b_{12} \\ a_{11}b_{21} & a_{11}b_{22} & a_{12}b_{21} & a_{12}b_{22} \\ a_{21}b_{11} & a_{21}b_{12} & a_{22}b_{11} & a_{22}b_{12} \\ a_{21}b_{21} & a_{21}b_{22} & a_{22}b_{21} & a_{22}b_{22} \end{bmatrix}.$$

The properties of the Kronecker product are thoroughly discussed in a book entitled *Kronecker Products and Matrix Calculus with Applications* by Alexander Graham (1981). Using a blackboard or large sheet of paper, it is not too difficult to convince oneself that

$$(A \circ B)(C \circ D) = AC \circ BD, \text{ where} \tag{A.7}$$

$(A \circ B)(C \circ D)$ represents the ordinary matrix product of $(A \circ B)$ with $(C \circ D)$. Similarly, AC and BD are, respectively, the ordinary matrix products of A with C and B with D.

Furthermore, it can be shown that

$$A \circ (B \circ C) = (A \circ B) \circ C. \tag{A.8}$$

From this result, it is not difficult to show that

$$(A_1 \circ A_2 \circ \ldots A_p)(B_1 \circ B_2 \circ \ldots \circ B_p)$$
$$= (A_1 B_1) \circ (A_2 \circ B_2) \circ \ldots \circ (A_n \circ B_p).$$

To see that our proposed matrix representations satisfy (A.1), it will be useful to note that

$$(A_1 \circ A_2 \circ \ldots \circ A_p)(B_1 \circ B_2 \circ \ldots \circ B_p) = -(B_1 \circ B_2 \circ \ldots \circ B_p)(A_1 \circ A_2 \circ \ldots \circ A_p) \text{ if}$$

$$A_j B_j = -B_j A_j \text{ for an odd number of } j\text{'s between 1 and } p \text{ and}$$
$$A_j B_j = B_j A_j \text{ for the remainder of the } j\text{'s between 1 and } p.$$

To carry out our construction, I will use the Pauli spin matrices:

$$\boldsymbol{\sigma}_1 = \begin{bmatrix} 0 & 1 \\ 1 & 0 \end{bmatrix}, \boldsymbol{\sigma}_2 = \begin{bmatrix} 0 & -i \\ i & 0 \end{bmatrix}, \text{ and } \boldsymbol{\sigma}_3 = \begin{bmatrix} 1 & 0 \\ 0 & -1 \end{bmatrix}. \tag{A.9}$$

For $\mathbf{R}_{2,0}$ (the Clifford algebra that corresponds to the Euclidean space E^2 (or $R^{2,0}$), we use $\mathbf{e}_1 = \boldsymbol{\sigma}_1$ and $\mathbf{e}_2 = \boldsymbol{\sigma}_2$. Note! $\mathbf{e}_1\mathbf{e}_2 = \boldsymbol{\sigma}_1\boldsymbol{\sigma}_2 = i\boldsymbol{\sigma}_3 \neq \pm\mathbf{I}$. For $\mathbf{R}_{4,0}$, we can let $\mathbf{e}_1 = \boldsymbol{\sigma}_1 \circ \boldsymbol{\sigma}_1$, $\mathbf{e}_2 = \boldsymbol{\sigma}_1 \circ \boldsymbol{\sigma}_2$, $\mathbf{e}_3 = \boldsymbol{\sigma}_1 \circ \boldsymbol{\sigma}_3$, and $\mathbf{e}_4 = \boldsymbol{\sigma}_2 \circ \mathbf{I}$. ($\mathbf{e}_1\mathbf{e}_2\mathbf{e}_3\mathbf{e}_4 =$?).

This process can be continued by induction. Suppose it is possible to construct a matrix representation of size $2^m \times 2^m$ for $\mathbf{R}_{2m,0}$. Suppose also that we designate the matrix representation of the Dirac matrices for this space by

$$\mathbf{e}_k(2m) \text{ for } k = 1, 2, \ldots, 2m.$$

We can then obtain a matrix representation for the Dirac matrices of $\mathbf{R}_{2m+2,0}$ as follows:

$$\mathbf{e}_k(2m+2) = \boldsymbol{\sigma}_1 \circ \mathbf{e}_k(2m) \text{ for } k = 1, 2, \ldots, 2m. \tag{A.10}$$
$$\mathbf{e}_{2m+1}(2m+2) = i^m \boldsymbol{\sigma}_1 \circ \mathbf{e}_{12\ldots 2m}(2m), \text{ and} \tag{A.11}$$
$$\mathbf{e}_{2m+2}(2m+2) = \boldsymbol{\sigma}_2 \circ I. \tag{A.12}$$

(The factor i^m that appears in (A.11) has been chosen to guarantee that $(\mathbf{e}_{2m+1}(2m+2))^2 = \mathbf{I}$.)

It is not difficult to show that the system of matrices just constructed is indeed an orthonormal system of 1-vectors, where

$$(\mathbf{e}_k)^2 = \mathbf{I} \text{ and}$$
$$\mathbf{e}_j\mathbf{e}_k + \mathbf{e}_k\mathbf{e}_j = 0 \text{ for } j \neq k.$$

From Theorem 291, this is all that is required to generate the universal Clifford algebra $\mathbf{R}_{2m+2,0}$ since $2m + 2$ is an even integer.

To construct the Dirac matrices for universal Clifford algebras of the type $\mathbf{R}_{p,q}$ where $p + q = 2m$ is now a simple matter. One can take the Dirac matrices constructed for $\mathbf{R}_{2m,0}$ and simply leave the first p matrices unchanged and then multiply each of the remaining q Dirac matrices by i.

To obtain a matrix representation for the Clifford algebras associated with odd dimensional vector spaces is now also an easy matter. Starting with the matrix representation constructed above for $\mathbf{R}_{2m,0}$, we can write

$$\mathbf{e}_k(2m+1) = \boldsymbol{\sigma}_3 \circ \mathbf{e}_k(2m) \text{ for } k = 1, 2, \ldots, 2m, \text{ and} \tag{A.13}$$

$$\mathbf{e}_{2m+1}(2m+1) = i^m \boldsymbol{\sigma}_3 \circ \mathbf{e}_{12\ldots 2m}(2m). \tag{A.14}$$

(We could use $\boldsymbol{\sigma}_1$ in place of $\boldsymbol{\sigma}_3$ in these last two equations but $\boldsymbol{\sigma}_3$ gives a more desirable form. In particular if we use $\boldsymbol{\sigma}_3$, then any Dirac matrix is of the form

$$\begin{bmatrix} A & 0 \\ 0 & -A \end{bmatrix},$$

where A is a matrix of size $2^m \times 2^m$.

Using (A.13) and (A.14), it is not difficult to show that (A.1) is satisfied. What is left to show is that

$$\mathbf{e}_1(2m+1)\mathbf{e}_2(2m+1)\ldots\mathbf{e}_{2m+1}(2m+1) \text{ is not a multiple of } \mathbf{I}.$$

But from (A.13) and (A.14), we have

$$\mathbf{e}_1(2m+1)\mathbf{e}_2(2m+1)\ldots\mathbf{e}_{2m+1}(2m+1) = (-i)^m \boldsymbol{\sigma}_3 \circ \mathbf{I}.$$

To construct matrix representations for universal Clifford algebras associated with odd dimensional vector spaces but with non-Euclidean signatures, we can use the same trick that was used for even dimensional vector spaces. That is leave the first p 1-vectors unchanged and multiply the remaining q 1-vectors by i.

Appendix B
Construction of Matrix Representations for Dirac Vectors

If we know how a space is embedded in a higher dimensional flat space, it may be a trivial matter to construct a set of coordinate Dirac vectors. However, one may simply be given the metric tensor. In that case, construction of a system of Dirac matrices is slightly more complicated. Nonetheless, we can make use of the fact that we already know how to construct a matrix representation for an orthonormal basis. In particular,

$$\mathbf{e}_j \mathbf{e}_k + \mathbf{e}_k \mathbf{e}_j = 2n_{jk}\mathbf{I}, \text{ where}$$

$$\begin{aligned} n_{kk} &= 1 \text{ for } 1 \leq k \leq p, \\ n_{kk} &= -1 \text{ for } p+1 \leq k \leq p+q = n, \text{ and} \\ n_{jk} &= 0 \text{ for } j \neq k. \end{aligned}$$

Suppose we denote the matrix corresponding to the matrix tensor $g_{\alpha\beta}$ by G. That is

$$G = \begin{bmatrix} g_{11} & g_{12} & - & - & - & g_{1n} \\ g_{21} & g_{22} & - & - & - & g_{2n} \\ - & - & - & - & - & - \\ - & - & - & - & - & - \\ - & - & - & - & - & - \\ g_{n1} & g_{n2} & - & - & - & g_{nn} \end{bmatrix}.$$

An arbitrary metric tensor is real and symmetric, that is $g_{\alpha\beta} = g_{\beta\alpha}$. From matrix theory, it is well known that a real symmetric matrix can be diagonalized by a real orthogonal matrix. Thus, we have

$$G = ODO^T, \text{ where} \tag{B.1}$$

O is an orthogonal matrix; O^T is the transpose of O, which is identical to O^{-1}; and D is a real diagonal matrix. Furthermore, the columns of O can be ordered so that

J. Snygg, *A New Approach to Differential Geometry using Clifford's Geometric Algebra*, DOI 10.1007/978-0-8176-8283-5,

all the positive components of D are identified with the first p indices. That is

$$D = \begin{bmatrix} (\alpha_1)^2 & 0 & & & & & & & \\ 0 & (\alpha_2)^2 & & & & & & & \\ & & - & & & & & & \\ & & & - & & & & & \\ & & & & - & & & & \\ & & & & & (\alpha_p)^2 & 0 & & \\ & & & & & 0 & -(\alpha_{p+1})^2 & & \\ & & & & & & & - & \\ & & & & & & & & - \\ & & & & & & & & & - \\ & & & & & & & & & & -(\alpha_{p+q})^2 \end{bmatrix},$$

where $p + q = n$. Clearly, D can be decomposed further into a product of the form

$$D = AnA, \text{ where} \tag{B.2}$$

$$n = \left[n_{ij} \right], \text{ and}$$

$$A = \begin{bmatrix} \alpha_1 & 0 & - & - & - & 0 \\ 0 & \alpha_2 & - & - & - & 0 \\ - & - & - & - & - & - \\ - & - & - & - & - & - \\ - & - & - & - & - & - \\ 0 & 0 & - & - & - & \alpha_n \end{bmatrix}.$$

From (B.1) and (B.2), we have

$$G = OAnA^T O^T = WnW^T \quad \text{where} \quad W = OA.$$

For 2-dimensions, this equation takes the form

$$\begin{bmatrix} g_{11} & g_{12} \\ g_{21} & g_{22} \end{bmatrix} = \begin{bmatrix} w_1^1 & w_1^2 \\ w_2^1 & w_2^2 \end{bmatrix} \begin{bmatrix} n_{11} & n_{12} \\ n_{21} & n_{22} \end{bmatrix} \begin{bmatrix} w_1^1 & w_2^1 \\ w_1^2 & w_2^2 \end{bmatrix}.$$

For arbitrary dimension, we can write

$$g_{\alpha\beta} = w_\alpha^j w_\beta^k n_{jk}. \tag{B.3}$$

If we now define

$$\boldsymbol{\gamma}_\alpha = w_\alpha^j \mathbf{e}_j, \text{ then} \tag{B.4}$$

$$\boldsymbol{\gamma}_\alpha \boldsymbol{\gamma}_\beta + \boldsymbol{\gamma}_\beta \boldsymbol{\gamma}_\alpha = w_\alpha^j w_\beta^k (\mathbf{e}_j \mathbf{e}_k + \mathbf{e}_k \mathbf{e}_j) = 2 w_\alpha^j w_\beta^k n_{jk} \mathbf{I}.$$

From (B.3), we now have our desired result:

$$\boldsymbol{\gamma}_\alpha \boldsymbol{\gamma}_\beta + \boldsymbol{\gamma}_\beta \boldsymbol{\gamma}_\alpha = 2 g_{\alpha\beta} \mathbf{I}.$$

Appendix C
A Few Terms of the Taylor's Series for the Urdī-Copernican Model for the Outer Planets

In Fig. C.1, $(d, 0)$ is the center of the eccentric deferent (not drawn), a is the radius of the deferent, ρ is the radius of the epicycle, and $\omega = 2\pi/T$, where T is the time required for the complete orbit of the planet in question. It is not too hard to see that if (x, y) are the coordinates of the planet, then

$$x = d + a\cos\omega t + \rho\cos(\pi - 2\omega t), \text{ and}$$
$$y = a\sin\omega t - \rho\sin(\pi - 2\omega t),$$

or slightly simplified:

$$x = a\cos\omega t + d - \rho\cos 2\omega t, \text{ and}$$
$$y = a\sin\omega t - \rho\sin 2\omega t. \tag{C.1}$$

It now follows that

$$\begin{aligned} r^2 &= x^2 + y^2 \\ &= a^2\cos^2\omega t + 2ad\cos\omega t - 2a\rho\cos\omega t\cos 2\omega t + d^2 - 2d\rho\cos 2\omega t \\ &\quad + \rho^2\cos^2 2\omega t + a^2\sin\omega t - 2a\rho\sin\omega t\sin 2\omega t + \rho^2\sin^2 2\omega t \\ &= a^2 + 2ad\cos\omega t - 2a\rho\cos\omega t + d^2 - 2d\rho\cos 2\omega t + \rho^2 \\ &= a^2\left[1 + \frac{2(d-\rho)}{a}\cos\omega t + \frac{d^2 - 2d\rho\cos 2\omega t + \rho^2}{a^2}\right]. \end{aligned} \tag{C.2}$$

From our knowledge of Taylor's series,

$$(1+x)^{1/2} = 1 + \frac{1}{2}x - \frac{1}{8}x^2 + \text{ higher powers of } x. \tag{C.3}$$

J. Snygg, *A New Approach to Differential Geometry using Clifford's Geometric Algebra*, DOI 10.1007/978-0-8176-8283-5, © Springer Science+Business Media, LLC 2012

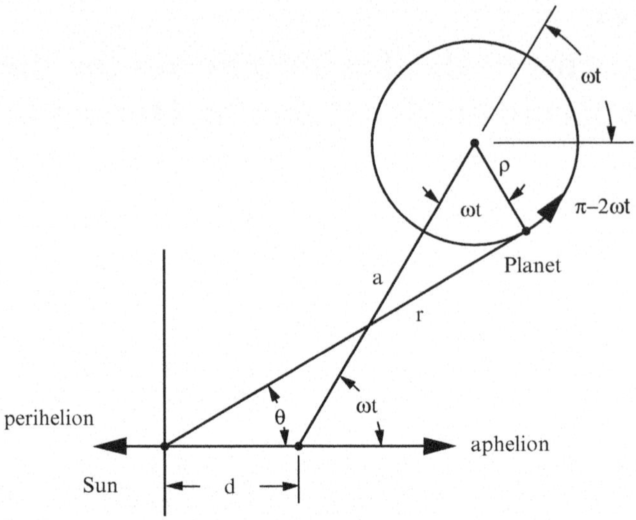

Fig. C.1 The Urdī-Copernican model for the outer planets

Combining (C.2) and (C.3) and retaining powers of d/a and ρ/a, which are less than three, we get:

$$\begin{aligned} r &= a\left[1+\frac{(d-\rho)}{a}\cos\omega t+\frac{d^2-2d\rho\cos 2\omega t+\rho^2}{2a^2}-\frac{(d-\rho)^2}{2a^2}\cos^2\omega t\right] \\ &= a\left[1+\frac{(d-\rho)}{a}\cos\omega t+\frac{\left(d^2+\rho^2\right)\left(1-\cos^2\omega t\right)-2d\rho(\cos 2\omega t-\cos^2\omega t)}{2a^2}\right] \\ &= a\left[1+\frac{(d-\rho)}{a}\cos\omega t+\frac{(d+\rho)^2\sin^2\omega t}{2a^2}+\text{higher powers of }\frac{d}{a}\text{ and }\frac{\rho}{a}.\right] \end{aligned} \tag{C.4}$$

From (C.1) and (C.4), we have

$$\sin\theta=\frac{y}{r}=\frac{a\sin\omega t-\rho\sin 2\omega t}{r}=\frac{a\sin\omega t\left[1-\frac{2\rho}{a}\cos\omega t\right]}{r}. \tag{C.5}$$

Since

$$(1+x)^{-1}=1-x+x^2+\text{higher powers of }x,$$

it follows from (C.4) that

$$\frac{1}{r} = \frac{1}{a}\left[1 - \frac{d-\rho}{a}\cos\omega t - \frac{(d+\rho)^2\sin^2\omega t}{2a^2} + \frac{(d-\rho)^2}{a^2}\cos^2\omega t\right]$$

So from (C.5), it follows that

$$\begin{aligned}\sin\theta &= \sin\omega t\left[1 - \frac{2\rho}{a}\cos\omega t\right]\left[1 - \frac{d-\rho}{a}\cos\omega t - \frac{(d+\rho)^2\sin^2\omega t}{2a^2}\right.\\ &\qquad\left. + \frac{(d-\rho)^2}{a^2}\cos^2\omega t\right]\\ &= \sin\omega t\left[1 - \left(\frac{d-\rho}{a} + \frac{2\rho}{a}\right)\cos\omega t - \frac{(d+\rho)^2}{2a^2}\sin^2\omega t\right.\\ &\qquad\left. + \left(\frac{(d-\rho)^2}{a^2} + \frac{2\rho(d-\rho)}{a^2}\right)\cos^2\omega t\right].\end{aligned}$$

That is

$$\begin{aligned}\sin\theta &= \sin\omega t\left[1 - \frac{d+\rho}{a}\cos\omega t - \frac{(d+\rho)^2}{2a^2}\sin^2\omega t + \frac{d^2-\rho^2}{a^2}\cos^2\omega t\right]\\ &\quad + \text{higher powers of } \frac{d}{a} \text{ and } \frac{\rho}{a}. \qquad \text{(C.6)}\end{aligned}$$

Appendix D
A Few Terms of the Taylor's Series for Kepler's Orbits

The coordinates for a Kepler orbit cannot be expressed in terms of elementary functions of time. However, they can be considered to be analytic functions of the eccentricity e, which means they can be expressed as an infinite power series of e. In this section, I will derive the first few terms. Namely up to e^2. It will be useful to use the notation $O(e^3)$ to indicate powers of three or higher for e to indicate the terms not computed.

We first note that the equation for an ellipse in polar coordinates with a horizontal major axis and origin at the left focal point is

$$r = \frac{b^2}{a(1 - e\cos\theta)}, \text{ where} \tag{D.1}$$

$$\begin{aligned} a &\text{ is the semimajor axis,} \\ b &\text{ is the semiminor axis, and} \\ c = ae &\text{ is the focal distance.} \end{aligned}$$

Since

$$b^2 = a^2 - c^2 = a^2(1 - e^2),$$

Equation (D.1) can be rewritten in the form

$$r = \frac{a(1 - e^2)}{(1 - e\cos\theta)} \tag{D.2}$$

There are several Taylor series, which will be helpful. In particular:

$$(1 + x)^{1/2} = 1 + \frac{x}{2} - \frac{x^2}{8} + O(x^3), \tag{D.3}$$

$$\frac{1}{1 - x} = 1 + x + x^2 + O(x^3), \tag{D.4}$$

J. Snygg, *A New Approach to Differential Geometry using Clifford's Geometric Algebra*, DOI 10.1007/978-0-8176-8283-5, © Springer Science+Business Media, LLC 2012

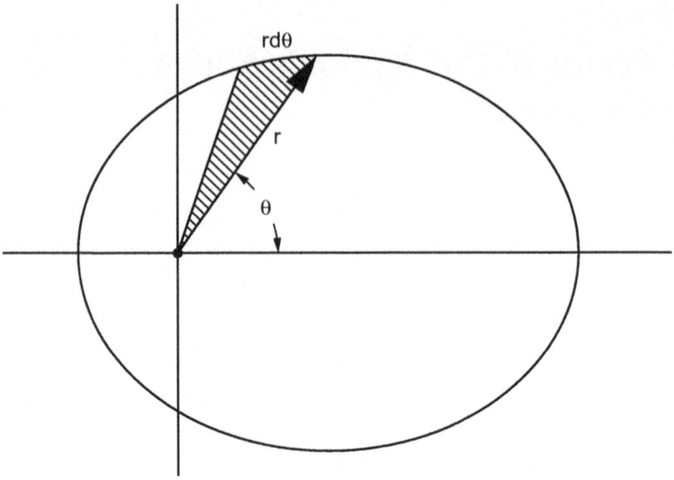

Fig. D.1 According to Kepler's second law, the area swept out by the radial vector from the Sun to a given planet increases at a constant rate

$$\sin(u+x) = \sin u + x\cos u - \frac{x^2}{2}\sin u + O(x^3), \text{ and} \tag{D.5}$$

$$\cos(u+x) = \cos u - x\sin u - \frac{x^2}{2}\cos u + O(x^3). \tag{D.6}$$

According to Kepler's second law, the area swept out by the radial vector r from the Sun to a given planet increases at a constant rate. (See Fig. D.1.) Thus

$$\frac{1}{2}r^2\frac{d\theta}{dt} = h, \text{ where } h \text{ is a constant.} \tag{D.7}$$

To obtain an interpretation for h, we note that

$$\int_0^{2\pi}\frac{1}{2}r^2 d\theta = h\int_0^T dt, \text{ where } T \text{ is the time required for a complete orbit.} \tag{D.8}$$

The integral on the left-hand side of (D.8) is the total area swept out by the radial vector for one revolution which is that of the ellipse. Thus,

$$\pi ab = hT, \text{ and therefore } h = \frac{\pi ab}{T}. \tag{D.9}$$

From (D.2) and (D.4):

$$\begin{aligned} r &= a(1-e^2)(1+e\cos\theta+e^2\cos^2\theta)+O(e^3) \\ &= a(1+e\cos\theta+e^2\cos^2\theta-e^2)+O(e^3). \end{aligned} \tag{D.10}$$

From this result, we have

$$\begin{aligned} r^2 &= a^2(1+2e\cos\theta+3e^2\cos^2\theta-2e^2)+O(e^3) \\ &= a^2\left[1+2e\cos\theta+\frac{3}{2}e^2(1+\cos 2\theta)-2e^2\right]+O(e^3). \\ &= a^2\left[1+2e\cos\theta-\frac{1}{2}e^2+\frac{3}{2}e^2\cos 2\theta\right]+O(e^3). \end{aligned}$$

Using this result along with (D.9), (D.7) becomes

$$\frac{1}{2}a^2\left[1+2e\cos\theta-\frac{1}{2}e^2+\frac{3}{2}e^2\cos 2\theta\right]\frac{d\theta}{dt}=\frac{\pi ab}{T}=\frac{\pi a^2(1-e^2)^{1/2}}{T}+O(e^3).$$

Multiplying both sides of this equation by $2(1+e^2)^{1/2}/a^2$, we have

$$(1+e^2)^{1/2}\left[1+2e\cos\theta-\frac{1}{2}e^2+\frac{3}{2}e^2\cos 2\theta\right]\frac{d\theta}{dt}=\frac{2\pi(1-e^4)^{1/2}}{T}+O(e^3).$$

From (D.3), this becomes

$$\left(1+\frac{1}{2}e^2\right)\left[1+2e\cos\theta-\frac{1}{2}e^2+\frac{3}{2}e^2\cos 2\theta\right]\frac{d\theta}{dt}=\frac{2\pi}{T}+O(e^3),\text{ which implies}$$

$$\left[1+2e\cos\theta+\frac{3}{2}e^2\cos 2\theta\right]d\theta=\frac{2\pi}{T}dt+O(e^3),\text{ and therefore}$$

$$\theta+2e\sin\theta+\frac{3}{4}e^2\sin 2\theta=\frac{2\pi t}{T}+O(e^3). \tag{D.11}$$

It is useful for our purposes the define $\omega = 2\pi/T$. With this convention, we note that, $\theta=\omega t$ if $e=0$. Thus,

$$\theta=\omega t+ef(t)+e^2g(t)+O(e^3). \tag{D.12}$$

To determine $f(t)$ and $g(t)$, we substitute this formula for θ into (D.11) and get

$$\omega t+ef(t)+e^2g(t)+2e\sin(\omega t+ef(t))+\frac{3}{4}e^2\sin 2\omega t=\omega t+O(e^3).$$

Using (D.5), this becomes

$$ef(t) + e^2 g(t) + 2e\,(\sin\omega t + ef(t)\cos\omega t) + \frac{3}{2}e^2 \sin\omega t \cos\omega t = 0.$$

Regrouping terms, this becomes

$$e\,[f(t) + 2\sin\omega t] + e^2\left[g(t) + 2f(t)\cos\omega t + \frac{3}{2}\sin\omega t\cos\omega t\right] = 0.$$

Setting the coefficients of e and e^2 separately equal to zero, we get

$$\begin{aligned} f(t) &= -2\sin\omega t \quad \text{and} \\ g(t) &= \frac{5}{2}\sin\omega t\cos\omega t. \end{aligned}$$

With these results, (D.12) becomes

$$\theta = \omega t - 2e\sin\omega t + \frac{5}{2}e^2 \sin\omega t\cos\omega t + O(e^3). \tag{D.13}$$

Using (D.5) and (D.13), we get

$$\begin{aligned} \sin\theta &= \sin\left(\omega t - 2e\sin\omega t + \frac{5}{2}e^2\sin\omega t\cos\omega t\right) + O(e^3). \\ &= \sin\omega t\left[1 - 2e\cos\omega t - 2e^2\sin^2\omega t + \frac{5}{2}e^2\cos^2\omega t\right] + O(e^3). \end{aligned} \tag{D.14}$$

A similar computation using (D.6) and (D.13), we get

$$\begin{aligned} \cos\theta &= \cos\,(\omega t - 2e\sin\omega t) + O(e^2) \\ &\cos\omega t + 2e\sin^2\omega t + O(e^2). \end{aligned} \tag{D.15}$$

We will use this last equation to get a formula for $r(t)$. From (D.10), we have

$$\begin{aligned} r &= a\left[1 + e\cos\theta + e^2\cos^2\theta - e^2\right] + O(e^3) \\ &= a\left[1 + e\cos\theta - e^2\sin^2\theta\right] + O(e^3) \\ &= a\left[1 + e\left(\cos\omega t + 2e\sin^2\omega t\right) - e^2\sin^2\omega t\right] + O(e^3) \\ &= a\left[1 + e\cos\omega t + e^2\sin^2\omega t\right] + O(e^3). \end{aligned} \tag{D.16}$$

References

Abdali S. Kamal 1997. "The Correct Qibla." http://patriot.net/~abdali/ftp/qibla.pdf

Abers, Ernest S. and Charles F. Kennel 1977. *Matter in Motion: The Spirit and Evolution of Physics*. Boston, Mass. Allyn and Bacon, Inc.

Akivis, M.A. and B.A. Rosenfeld 1993. Élie Cartan (1869-1951) *Translations of Mathematical Monographs* Volume 123. American Mathematical Society, Providence, Rhode Island.

Aleksandrov, Aleksandr D. 1988. "Vladimir Aleksandrovich Fock." (Russian), In Aleksandrov, A.D., *Problemy nauki i pozitsiya uchenogo*. Leningrad: Nauka, pp.489-496.

Armitage, Angus First Printing 1951, Ninth Printing 1963. *The World of Copernicus*. New York, NY: Published as a Signet Science Library Book by The New American Library of World Literature, Inc. Originally published under the title *Sun, Stand Thou Still* in 1947 by Henry Schuman, Inc. New York, NY.

Asprey, Robert B. 1986. *Frederick the Great - The Magnificent Enigma*. New York, NY: Ticknor & Fields.

Auger, P. 1932 *Comptes Rendus* **194**, p.1346.

Averroës 1969. *Tahafut al-Tahafut (The Incoherence of the Incoherence)* Volume 1. Translated from the Arabic with introduction and notes by Simon Van Den Bergh. First Published in 1954, Reprinted 1969. Printed for the trustees of the "E.J.W. Gibb Memorial." and published by Messrs Luzac & Company Ltd., 46, Great Russel Street, London, W.C.I.

Baylis, William E. 1999. *Electrodynamics - A Modern Geometric Approach*. Boston, Massachusetts: Birkhäuser.

Beltrami, Eugenio 1868. "Saggio di interpretazione della geometria non-euclidea." *Giorn. Matem.*, ***6***. 284-312. In his *Opere Matematiche* I: 262-280. Also English translation: Stillwell, John 1996 *Sources of Hyperbolic Geometry*. American Mathematical Society, Providence, RI.

Also: Risoluzione del prolema: "Riportare i punti di una superficie sopra un piano in modo che le linee geodetiche vengano rappresentate da linee rette." *Ann. Mat. pura appl.*, ser. 1,7, 185-204.

Bernoulli, Jakob 1697. "Solutio problematis fraternorum." *Acta Erud.* 16, 211-217.

Bernoulli, Johann 1696. "Problema novum ad cujus solutionem mathematici invitantur." *Acta Erud.*, **15**, 270. In his *Opera Omnio* 1: 179-187.

Blumenthal, Otto 1917. "Karl Schwarzschild." *Sonderabdruck aus dem Jahresberich der Deutschen Mathematiker-Vereiinigun.* XXVI 1917. Heft 1/4. Verlag and Druck von B.G. Teubner in Leipzig. (Available on microfiche in *The Papers of Karl Schwarzschild*, last item in section 2 of reel 27 - - - see Schwarzschild 1975.)

Bonnet, O. 1848. "Memoire sur la théorie générale des surfaces." *Journal Ecole Polytechnique* **19**, H.32, 1-146.

DOI 10.1007/978-0-8176-8283-5,

Bonola, Roberto 1955. *Non-Euclidean Geometry* translated with additional appendices by H.S. Carslaw. (A Dover republication of the first English translation by H.S. Carslaw as published in 1912 by the Open Court Publishing Company.) New York, NY: Dover Publications, Inc.

Boorstin, Daniel J. 1983. *The Discoverers A History of Man's Search to Know his World and Himself.* New York, NY: Random House, Inc.

Bos, H.J.M. 1986. Introduction (pp. xi-xxix) to *Christiaan Huygens' The Pendulum Clock or Geometrical Demonstrations Concerning the Motion of Pendula as Applied to Clocks* (Richard J. Blackwell's translation of Christiaan Huygens' *Horologium oscillatorium sive de moto pendularum ad horologia aptato demonstrationes geometricae.* Paris: F. Muguet 1673.) Ames, Iowa: The Iowa State University Press.

Boyer, Carl B. 1968. *A History of Mathematics.* First Princeton Paperback printing, 1985. Princeton, New Jersey: Princeton University Press.

Brannan, David A., Matthew F. Esplen, and Jeremy J.Gray 1999. *Geometry.* Cambridge, United Kingdom: Cambridge University Press.

Bulgakov, P.G. and A.A. Ahmedov 1977. "Al-Biruni and al-Kindi on the theory of parallels" (Russian) *Obshchestv. Nauki v Uzbek* **8** (cited by JJ O'Connor and EF Robertson www-history.mcs.st-andrews.ac.uk/Biographies/Al-Kindi.html)

Bühler, Walter Kaufmann 1981. *Gauss: A Biographical Study.* New York, NY: Springer-Verlag New York Inc.

Cajori, Florian 1919, 1924. *A History of Mathematics* 2nd Ed. London: Macmillan and Co., Ltd.

Cartan, Élie 1910a. "Sur les développables isotropes et la méthode du trièdre mobile." *C.R. Acad. Sc.* Paris 151, 919-921; *OEuvres complètes:* Partie III, Divers, géométrie différentielle, vols. 1-2, Gauthier-Villars, Paris, 1955, pp. 141-142.

Cartan, Élie 1910b. "La structure des groups de transformations continus et la thèorie du trièdre mobile." *Bull. Sci. Math.* **34**, 250-284; *OEuvres complètes:* Partie III, Divers, géométrie différentielle, vols. 1-2, Gauthier-Villars, Paris, 1955, pp. 145-178.

Cartan, Elie 1966. *The Theory of Spinors.* Cambridge, Massachusetts: The M.I.T. Press.

Casson, Lionel 2001. *Libraries in the Ancient World.* New Haven, Connecticut: Yale University Press.

Charles, Daniel 2005 *Master Mind: The Rise and Fall of Fritz Haber, the Nobel Laureate who Launched the Age of Chemical Warfare.* New York, NY. Ecco.

Chisholm M. 2002. *Such Silver Currents - The Story of William and Lucy Clifford 1845-1929.* Cambridge: The Lutterworth Press.

Clark, Ronald W. 1972. *Einstein - The Life and Times.* New York, NY: Avon Books.

Clarke, C.J.S. 1970. "On the global isometric embedding of pseudo-Riemannian manifolds." *Proc. Roy. Soc. London Series* A 314, pp. 417-428.

Clifford, William Kingdon 1878. "Applications of Grassmann's extensive algebra." *American Journal of Mathematics* vol. 1 350-358. Republished 1882 in *Mathematical Papers by William Kingdon Clifford* (Edited by Robert Tucker) London: Macmillan and Co. 266-276. *Mathematical Papers by William Kingdon Clifford* were reprinted in 196a8 by Chelsea Publishing Company: Bronx, New York.

Clifford, William Kingdon 1882. "On the Classification of Geometric Algebras." Published posthumously in unfinished form as paper XLII in *Mathematical Papers by William Kingdon Clifford* (Edited by Robert Tucker) London: Macmillan and Co. 397-401. *Mathematical Papers by William Kingdon Clifford* were reprinted in 1968 by Chelsea Publishing Company: Bronx, New York.

Copernicus, Nicolas 1959. *Three Copernican Treatises. 2nd Ed.* New York, NY: Dover Publications Inc. translation by Edward Rosen.

Copernicus, Nicolas 1992. *On The Revolutions.* Translation and Commentary by Edward Rosen. Baltimore and London. The Johns Hopkins University Press.

Crowe, Michael J. 1985. *A History of Vector Analysis - The Evolution of the Idea of a Vectorial System.* Mineola, New York: Dover Publications, Inc.

Dalley, Stephanie and John Peter Oleson 2003. "Sennacherib, Archimedes, and the Water Screw: The Context of Invention in the Ancient World", *Technology and Culture* **44**(1).

De Jonge, Alex 1938. *Fire and Water: A Life of Peter the Great.* New York, NY: Coward, McCann, and Geoghegan.

Di Bono, Mario 1995. "Copernicus, Amico, Fracastoro and Tūsī's device: observations of the use and transmission of a model." *Journal for the History of Astronomy* Vol. 26, Issue No. 2, pp.133-154.

Dictionary of Scientific Biography Volume II 1970. Charles Coulston Gillispie Editor in Chief. New York, NY: Charles Scribner's Sons.

do Carmo, Manfredo P. 1976. *Differential Geometry of Curves and Surfaces.* Englewood Cliffs, New Jersey: Prentice-Hall, Inc.

Dühring, Eugen Karl 1873. *Kritische Geschichte der allgemeinen Principien der Mechanik.* Berlin: T. Grieben. Also Micro-opaque New York Readex Microprint 1969 6 cards (Landmarks of Science)

Dunham, William 1999. *Euler: The Master of Us All.* Washington, DC: The Mathematical Association of America.

Dunnington, G. Waldo 1955. *Carl Friedrich Gauss - Titan of Science.* New York: Exposition Press. Also republished with additional material by Jeremy Gray and Fritz-Egbert Dohse 2002 and 2004 by the Mathematical Association of America. (Spectrum Series.)

Durant, Will 1950. *The Age of Faith.* New York: Simon and Schuster.

Durant, Will and Ariel 1965. *The Story of Civilization: Part IX The Age of Voltaire.* New York, NY: Simon and Schuster.

Dzielska, Maria 1995. *Hypatia of Alexandria.* (translated by F. Lyra) Cambridge, Massachusetts Harvard University Press.

Einstein, Albert 1961. *Relativity - The Special and General Theory - A Popular Exposition by Albert Einstein.* (Authorised Translation by Robert W. Lawson) New York, NY: Bonanza Books, Crown Publishers, Inc.

Einstein, Albert 2005. *Relativity -The Special and General Theory* translataed by Robert W. Lawson (Republication of 2nd Edition with Introduction by Roger Penrose, commentary by Robert Geroch, and essay by David C. Cassidy). New York, NY: Pi Press.

Eisenhart, Luther Pfahler 1926. *Riemannian Geometry.* Princeton, NJ: Princeton University Press.

Eisenhart, Luther Pfahler 1947. *An Introduction to Differential Geometry with Use of the Tensor Calculus.* Princeton, NJ: Princeton University Press.

Engel, Friedrich 1911. *Hermann Grassmann's Gesammelte Mathematische und Physikalische Werke.* Dritten Bandes, Zweiten Teil. *Grassmann's Leben Gerschildert* von Friedric Engel. Nebst Einen Verzeichnisse Der Von Grassmann Veröffentlichten Schriften und einer übersicht des Handschriften Nachlasses. Leipzig: Druck and Verlag Von B.G. Teubner. (Also Johnson Reprint Corporation New York, London 1972)

Enneper, Albert 1864. "Analytisch-geometrische Untersuchungen." *Zeitschrift für Mathematik und Physik,* Band 9, pp. 96-125.

Evans, James 1984. "On the function and the probable origin of Ptolemy's equant." American Journal of Physics **52** (12), December 1984, pp. 1080-1088.

Evans, James 1988, "The division of the Martian eccentricity from Hipparchos to Kepler: A history of the approximation to Kepler motion." American Journal of Physics **56** (11), November 1988, pp. 1009-1024.

Evans, James 1998. *The History and Practice of Ancient Astronomy.* Oxford: Oxford University Press.

Eves, Howard 1963. *A Survey of Geometry Vol. I.* Boston, Massachusetts: Allyn and Bacon Inc.

Eves, Howard 1976. *An Introduction to the History of Mathematics* 4th Edition. New York, NY: Holt, Rinehart and Winston.

Faires, Virgil Moring and Robert McArdle Keown 1960. *Mechanism* 5th Edition. New York, NY: McGraw-Hill Book Company, Inc.

Flanders, Harley 1963. *Differential Forms and Applications to the Physical Sciences.* New York, NY: Academic Press.

Fock, Vladimir Alexandrovich 1926. "Zur Schrödingershen Wellenmechanik." *Zeitschrift für Physik,* Band 38, N. 3, pp. 242-250.

Fock, Vladimir Alexandrovich 1929. "Geometrisierung der Diracschen Theorie des Electrons." *Zeitschrift für Physik*, Band 57, pp. 261-277.

Fock, Vladimir Alexandrovich 1930. "Näherungsmethode zur Lösung des quantummechanischen Mehrkörperproblems." *Zeitschrift für Physik*, Band 61, pp. 126-148.

Fock, Vladimir Alexandrovich and Ivanenko (Iwanenko), Dmitri Dmitrievich 1929a. "Géométrie quantique linéaire et déplacement parallèle." *Comptes Rendus des Séances de Ĺ Académie des Sciences*, Tome 188 (Paris) pp.1470-1472.

Fock, Vladimir Alexandrovich and Ivanenko (Iwanenko), Dmitri Dmitrievich 1929b. "Zur Quantengeometrie." *Physikalische Zeitschrift*, Band XXX pp. 648-651. (Published as part of the Proceedings of the Theorectical Physics Conference at the Ukrainian Institute of Physical Technology in Charkow (Kharkow) 1929 pp. 645-655.

Galison, Peter 2003. *Einstein's Clocks, Poincaré's Maps - Empires of Time*. New York, NY: W.W. Norton & Company.

Gauss, Carl Friedrich 1828. "Disquisitiones generales circa superficies curvas." *Commentationes Societatis Regiae Scientiarum Gottingensis Recentiores Göttingen*, Band 6, 1823-1827. Also translation: Gauss, K.F. 1965. "General Investigations of Curved Surfaces." New York, NY: Raven Press.

Gauss, Carl Friedrich 1900. *Carl Friedrich Gauss Werke* Achter Band Herausgegeben von der Königlichen Gesellschaft der Wissenschaten zu Göttingen in commission bei B.G. Teubner in Leipzig.

Gauss, Karl Friedrich 1965. *General Investigations of Curved Surfaces* translated from the Latin and German by Adam Hiltebeitel and James Morehead. Introduction by Richard Courant. Hewlet, New York: Raven Press. Also the same translation 1902, Princeton, NJ: The Princeton University Library.

Gerretsen, Hohan C.H. 1962. *Lectures on Tensor Calculus and Differential Geometry*. Groningen, The Netherlands: P. Noordhoff.

Ginzburg, Vitaly L. 2001. *The Physics of a Lifetime*. Berlin, Heldelberg, New York: Springer-Verlag.

Ginzburg, Vitaly L. Feb. 5, 2003. Three way phone conversation with Sasha Rosenberg acting as translator. Also exchange of e-mail messages before and after.

Goddio, Franck, Andre Bernard, Etienne Bernard, Ibrahim Darwish, Zsolt Kiss 1998. *Alexandria: the submerged royal quarters*. London: Periplus.

Goddio, Franck (Editor) 2008. *Egypt's Sunken Treasures*. Munich, New York: Prestel.

Gorelik, Gennady 1993. "Vladimir Fock: Philosophy of Gravity and Gravity of Philosophy." pp. 308-331. This paper appears in a book entitled: *The Attraction of Gravitation: New Studies in the History of General Relativity* edited by John Earman, Michel Janssen, and John D. Norton. Boston: Birkhäuser.

Gorelik, Gennady 1997. "The Top-Secret Life of Lau Landau." *Scientific American* August 1997, pp. 72-77. New York, NY 10017 (415 Madison Ave.)

Gorelik, Gennady and Victor Ya. Frenkel 1994. *Matvei Petrovich Bronstein and Soviet Theoretical Physics in the Thirties* translated by Valentina M. Levina. Basel, Boston, Berllin: Birkhäuser Verlag.

Graham, Alexander 1981. *Kronecker Products and Matrix Calculus with Applications*. Chichester, West Sussix, England: Ellis Horwood Limited.

Grassmann, Hermann 1995. *A New Branch of Mathematics (*The *Ausdehnungslehre* of 1844, and other works.) translated by Lloyd C. Kannenberg. Chicago and La Salle, Illinois: Open Court.

Green, Peter 1990. *Alexander to Actium: The Historical Evolution of the Hellenistic Age*. Berkeley: University of California Press.

Greene, Robert E. 1970. "Isometric embeddings of Riemannian and pseudo-Riemanniian manifolds" *Memoirs of the American Mathematical Society* Number 97.

Gromov, Mikhail 1986. *Partial Differential Relations*. Berlin: Springer-Verlag.

Haber, Ludwig Fritz 1986. *The Poisonous Cloud: Chemical Warfare in the First World War*. New York, N.Y. Oxford University Press.

Hahn, Otto 1970. *Otto Hahn: My Life The Autobiography of a Scientist.* Translated by Ernst Kaiser and Eithne Wilkins. New York, NY. Herder and Herder New York.

Hall, Tord 1970. *Carl Friedrich Gauss.* A biography by Tord Hall translated by Albert Froderberg. Cambridge, Massachusetts: The M.I.T. Press.

Hargittai, István 2002. *The Road to Stockholm - Nobel Prizes, Science, and Scientists.* Oxford, U.K.: Oxford University Press.

Harris, Michael H. 1995. *History of Libraries in the Western World.* Metuchen, N.J. & London. The Scarecrow Press, Inc.

Hartner, Willy 1973. "Copernicus, the Man, the Work, and Its History." *Proceedings of the American Philosophical Society* 117, no. 6: pp.413-422.

Harvey, L.P. 1992. *Islamic Spain 1250 to 1500* (Paperback Ed.) A hardback edition was published in 1990. Chicago and London: The University of Chicago Press.

Hayes, John R. et al. eds. 1992. *The Genius of Arab Civilization.* New York. New York University Press.

Heath, Sir Thomas 1921. *A History of Greek Mathematics. Volume 1 From Thales to Euclid. Volume 2 From Aristarchus to Diophantus.* Oxford at the Clarendon Press

Hestenes, David and Garret Sobczyk 1984. *Clifford Algebra to Geometric Calculus - A Unified Language for Mathematical Physics.* Dordrecht: D. Reidel Publishing Company.

Hestenes, David 1966. *Space-Time Algebra*, New York: Gordon and Breach.

Holloway, David 1994. "How the Bomb Saved Soviet Physics." The Bulletin of the Atomic Sciientists Vol. 50, No. 6. Nov./Dec. 1994 pp. 46-55.

Hopf, Heinz 1935. "Über die Drehung der Tangenten und Sehnen ebener Kurven." Compositio Math. **2** pp. 50-62.

Hoodbhoy, Perez 1991. *Islam and Science: Religious Orthodoxy and the Battle for Rationality.* Atlantic Highlands, N.J. Zed Books.

Hoyle, Fred 1962. *Astronomy.* London. Crescent Books, Inc.

Hughes, Lindsey 1990. *Sophia, Regent of Russia 1657-1704.* New Haven, Connecticut and London: Yale University Press.

Hughes, Lindsey 1998. *Russia in the Age of Peter the Great.* New Haven, Connecticut and London: Yale University Press.

Huygens, Christiaan 1986. *The Pendulum Clock or Geometrical Demonstrations Concerning the Motion of Pendula as Applied to Clocks.* translated with notes by Richard J. Blackwell with an introduction by H.J.M Bos. Ames, Iowa: The Iowa State University Press.

Ifrah, Georges 2000. *The Universal History of Numbers.* Translated from the French by David Bellos, E.F. Harding, Sophie Wood, and Ian Monk. New York, NY: John Wiley & Sons, Inc.

Inwood, Stephen 1998. *A History of London.* First published by Macmillan Publishers Ltd. Great Britain. New York, NY.: Carroll and Graf Publishers, Inc.

Iwanenko, D. 1932. *Comptes Rendus,* **195** pp. 439-441.

Izotov, G.E. 1993. "Legends and Reality in Lobachevskii's Biography." (Russian) *Piroda* (7) pp. 4-11.

James, Ioan 2002. *Remarkable Mathematicians - From Euler to von Neumann.* Cambridge, U.K.: Cambridge University Press. Washingtion D.C: Mathematical Association of America.

Jancewicz, Bernard 1988. *Multivectors and Clifford Algebra in Electrodynamics.* Singapore: World Scientific Publishing Co. Pte. Ltd.

Jerome, Fred 2002. *The Einstein File: J. Edgar Hoover's Secret War Against the World's Most Famous Scientist.* New York, NY: St. Martin's Press.

Jones, Simon 2007. *World War I Gas Warfare Tactics and Equipment.* Oxford, UK. Osprey Publishing.

Kárteszi, Ferenc (editor) 1987. *János Bolyai Appendix The Theory of Space.* New York, NY: Elesevier Science Publishing Company, Inc.

Katz, Victor J. 1998. *A History of Mathematics - An Introduction.* 2nd. Ed. Reading, Massachusetts: Addison Weslley Longman, Inc.

Kennedy, E.S. and Victor Roberts 1959. *Isis*, Vol. 50, No. 3. (Sept. 1959), pp. 227-235.

King, David A. 1993. *Islamic Mathematical Astronomy* 2nd. Ed. pp. 81-122. Gower House, Croft Road, Aldershot, Hampshire GU11 3HR, Great Britain: Variorum Ashgate Publishing Limited.

Kline, Morris 1981. *Mathematataics and the Physical World.* New York, NY: Dover Publications, Inc.

Klingenberg, Wilhelm 1978. *A Course in Differential Geometry* (translated by David Hoffman). New York, NY: Springer-Verlag.

Koyré, Alexandre 1973. *The Astronomical Revolution.* Translated from the original French text *La révolution astronomique* by Dr. R.E.W. Maddison, F.S.A. Ithaca, NY: Cornell University Press.

Kramer, Edna E. 1981. *The Nature and Growth of Modern Mathematics.* Princeton, New Jersey: Princeton University Press.

Kreyszig, Erwin 1991. *Differential Geometry.* New York, N.Y: Dover Publications, Inc.

Kuhn, Harold W. and Sylvia Nasar, Eds. 2002. *The Essential John Nash.* Princeton, NJ: Princeton University Press.

Kühnel, Wolfgang 2002. *Differential Geometry Curves - Surfaces - Manifolds* (translated from the German by Bruce Hunt) Student Mathematical Library Vol. 16. Providence, RI: American Mathematical Society. Originally published in German Wiesbaden, Germany: Friedr. Vieweg and Sohn as *Wolfgang Kühnel: Differentialgeometrie.* 1. Auglage (1st edition) 1999.

Laptev, B.L. 1992. "Bartels and the shaping of the geometric ideas of Lobachevskii (Russian)" pp. 35-40 in *Dedicated to the memory of Lobachevski* 1 (Kazan, 1992).

Laugwitz, Detlef 1999. *Bernhard Riemann 1826-1866 - Turning Points in the Conception of Mathematics* (translated by Abe Shenitzer). Boston, Massachusetts: Birkhäuser.

Leibniz, G.W. 1697. "Communicatio suae pariter duarumque alienarum ad edendum sibi primum a Dn. Joh. Beroullio." *Acta Erud.* **16**, 205-210. In his *Mathematische Schriften* **5**: 331-336.

Levenson, Thomas 2004. *Einstein in Berlin.* New York, NY: Bantam Dell.

l'Hôpital, G.F.A. d. 1697. "Solutio problematis de linea celerrimi descensus." *Acta Erud.* **16**, 217-220.

Lobachevsky, N. 1837. "Géométrie imaginaire." *Crelle's Journal,* Vol. 17 pp. 295-320. (This journal is sometimes catalogued under the title: *Journal für die reine and angewandte mathematik.)*

Lohne, J. A. 1979. "Essays on Thomas Harriot" *Archive for History of Exact Sciences.* **20**, 189-312.

Lounesto, Pertti 1997. *Clifford Algebras and Spinors.* Cambridge, UK: Cambridge University Press.

MacLeod, Roy (Editor) 2000. *The Library of Alexandria - - - Centre of Learning in the Ancient World.* (The essay by Rober Barnes, entitled "Cloistered Bookworms in the Chicken-Coop of the Muses: The Ancient Library of Alexandria", appears on pp. 61-78.) London: I.B. Tauris Publishers.

Mc Cleary, John 1994. *Geometry from a Differentiable Viewpoint.* New York, NY: Cambridge University Press.

Magnus, Dr. Wilhelm and Dr. Fritz Oberhettinger, and Dr. Raj Pal Soni 1966. *Formulas and Theorems for the Special Functions of Mathematical Physics* 3rd Edition. Springer-Verlag New York Inc.

Margenau, Henry and George Moseley Murphy 1956. *The Mathematics of Physics and Chemistry* 2nd Edition. Princeton, NJ: D. Van Nostrand Company, Inc.

Meschkowski, Herbert 1964. *Non-Euclidean Geometry* (translated by A. Shenitzer). New York, NY: Academic Press Inc. First published in the German language under the title: *Nicheuklidische Geometrie* and copyrighted 1961 by Friedr. Vieweg and Sohn, Braunschweig. Germany.

Minding, Ferdinand 1839. "Wie sich entscheiden läfst, ob zwei gegebene krumme Flächen auf einander abwickelbar sind oder nicht; nebst Bemerkungen über die Flächen von unveranderlichem Krümmungsmaafse." *Crelle's Journal der Mathematik,* Vol. 19 (1839) pp. 370-387. (Sometimes catalogued under the title: *Journal für die reine und angewandte mathematik.*)

Minding, Ferdinand 1840. "Beiträge zur Theorie der kürzesten Linien auf krummen Flächen." *Crelle's Journal der Mathematik,* Vol. 20 (1840) pp. 323-327. (Sometimes catalogued under the title: *Journal für die reine und angewandte mathematik.*)

Minkowski, Hermann 1915a. "Das Relativitätsprinzip." *Jahresbericht der Deutschen Mathematiker - Vereinigung.* **24**, p. 372.

Minkowski, Hermann 1915b. "*Annalen der Physik.*" (Leipzig) p. 927.

Misner, Charles W., Kip S. Thorne, John Archibald Wheeler 1973. *Gravitation.* San Francisco: W.H. Freeman and Company.

Möbius, August Ferdinand 1827. "Der barycentrische Calcul." *Werke 1*, pp. 1-388. (Also R. Baltzer, F. Klein and W. Schiebner (eds.) August Möbius *Gesammelte Werke* (Leipzig, 1885 - 87)

Monastyrsky, Michael 1999. *Riemann, Topology, and Physics 2nd Edition.* With a Foreward by Freeman J. Dyson (Translated by Roger Cooke, James King, and Virginia King.) Boston: Birkhäuser.

Morison, Samuel Eliot 1942. *Admiral of the Ocean Sea: A Life of Christopher Columbus.* Boston: Little Brown and Company. (A one volume condensation of a two volume *Admiral of the Ocean Sea* published at the same time.)

Mosier, John 2001. *The Myth of the Great War.* First Edition. New York, NY. Harper Collins.

Motz, Lloyd and Jefferson Hane Weaver 1993. *The Story of Mathematics.* New York, NY: Avon Books, Inc.

Muir, Jane 1961. *Of Men and Numbers.* New York, NY: Dell Publishing Co., Inc.

Nash, John 1956. "The Imbedding Problem for Riemannian Manifolds." *Annals of Mathematics* **63** (1956), pp. 20-63.

Newcomb, Simon 1898. "Tables of Mercury." *Astronomical Papers 2nd Ed.* vol. VI, prepared for the use of the *Journal of Mathematical Physics,* vol. 3, No. 3. (New York), pp. 566-578.

Newton, Isaac 1697. "The twin problems of Johann Bernoulli's 'Programma' solved." *Phil. Trans.*, **17**, pp. 388-389. In his *Mathematical Papers* **8**: pp. 72-79.

O'Connor, J.J. and E.F. Robertson September 1998. "Leonhard Euler." www.gap-system.org/~history or google "MacTutor History of Mathematics" > Biographies Index > Alphabetical indexes > E > Euler.

O'Connor, J.J. and E.F. Robertson November 1999. "Arabic mathematics: forgotton brilliance?" www.gap-system.org/~history or google "MacTutor History of Mathematics" > History of Topics Index > Arabic mathematics > Arabic mathematics: forgotton brilliance?

O'Connor, J.J. and E.F. Robertson February 2000. "Nikolai Ivanovich Lobachevsky." www.gap-system.org/~history or google "MacTutor History of Mathematics" > Biographies Index > Alphabetical indexes > L > Lobachevsky.

O'Connor, J.J. and E.F. Robertson July 2000. "Eugenio Beltrami." www.gap-system.org/~history or google "MacTutor History of Mathematics" > Biographies Index > Alphabetical indexes > B > Beltrami.

O'Connor, J.J. and E.F. Robertson April 2003. "Pierre Louis Moreau de Maupertuis". www.gap-system.org/~history or google "MacTutor History of Mathematics" > Biographies Index > Alphabetical indexes > M > Maupertuis.

O'Connor, J.J. and E.F. Robertson 2005 "The Size of the Universe." www.gap-system.org/~history or google "MacTutor History of Mathematics" > History of Topics Index > Full Alphabetical List or History Topics > Size of the Universe.

Oleson, John Peter 1984. *Greek and Roman mechanical water-lifting devices. The History of a Technology*, Dordrecht: D. Reidel., ISBN 90-277-1693-5.

Oprea, John 1997. *Differential Geometry and its Applications.* Upper Saddle River, New Jersey: Prentice Hall, Inc.

Osserman, Robert 1995. *Poetry of the Universe: A Mathematical Exploration of the Cosmos.* New York, NY: An Anchor Book published by Doubleday.

Osserman, Robert 2002. *A Survey of Minimal Surfaces* (A republished 1986 Dover edition.) Mineola, NY: Dover Publications, Inc.

Pauli, Wolfgang 1958. *Theory of Relativity.* (translated from German by G. Field.) London: Pergamon Press.
Pais, Abraham 1983. *Subtle is the Lord - Science and the Life of Albert Einstein.* Oxford and New York: Oxford University Press paperback.
Pais, Abraham 1986. *Inward Bound.* New York, NY: Oxford University Press.
Parker, Barry 2003. *Einstein - The Passions of a Scientist.* Amherst, NY: Prometheus Books.
Penrose, Roger and Wolfgang Rindler 1984. *Spinors & Space-Time* Vol. 1 *Two-Spinor Calculus and Relativistic Fields.* Cambridge, UK: Cambridge University Press.
Perrin, F. 1932. *Comptes Rendus* **194** p.2211; **195** p.236.
Pollock, Ethan M. 2000. "The Politics of Knowledges Party Ideology and Soviet Science, 1945 - 1953." Ph.D. Thesis in History. University of California, Berkeley.
Porteus, Ian R. 1981. *Topological Geometry* 2nd Edition. Cambridge, UK: Cambridge University Press.
Porteus, Ian R. 1994. *Geometric Differentiation for the Intelligence of Curves and Surfaces.* Cambridge, UK: Cambridge University Press.
Ptolemy 1998, *Almagest.* (translated by G. J. Toomer) Princeton, New Jersey. Princeton University Press.
Pyenson, Lewis 1985. *The Young Einstein: The Advent of Relativity.* Bristol and Boston: Adam Hilger Ltd.
Qadir, C.A. 1988. *Philosophy and Science in the Islamic World.* New York, NY: Croom Helm.
Ragep, F. Jamil 1993. *Nasīr al-Din al-Tūsī's Memoir on Astronomy:* Springer Verlag (In Volume I, the original Arabic is presented alongside the English translation. Volume II is devoted to commentary by F. Jamil Ragep, the translator.)
Riemann, Bernhard 1868. *Abhandlungen der Königlichen Gesellschaft der Wissenshaften zu Göttingen,* **13**.
Riesz, Marcel 1993. *Clifford Numbers and Spinors,* edited by E. Folke Bolinder and Pertti Lounesto. Dordrecht, The Netherlands: Kluwer Academic Publishers. Also available as Lecture serie Nol 38. 1958. College Park, Maryland: University of Maryland.
Roberts, Victor 1957. "The Solar and Lunar theory of Ibn al-Shātir: A pre-Copernican Copernican Model." *Isis*, Vol. 48, No. 4. (Dec. 1957), pp. 428-432.
Robertson, Priscilla 1960. *Revolutions of 1848 - A Social History.* New York, NY: Harper and Brothers.
Rubenstein, Richard E. 2003. *Aristotle's Children: How Christians, Muslims, and Jews Rediscovered Ancient Wisdom and Illuminated the Dark Ages.* Orlando, FL: Harcourt, Inc.
Saliba, George 2000. "Competition and the Transmission of the Foreign Sciences: Hunayn at the Abbasid Court." *Bulletin of the Royal Institute for Inter-Faith Studies* 2 pp.85-101.
Saliba, George 2007. *Islamic Science and the Making of the European Renaissance.* Cambridge, Mass: The MIT Press.
Schmidt, Franz and Paul Stöckel, editors 1899. *Briefwechsel Zwischen Carl Friedrich Gauss and Wolfgang Bolyai.* Leipzig: B.G. Teubner. Reprinted 1972 by Johnson Reprint Corporation. New York, NY.
Schwarzschild, Karl 1890a. "Zur Bahnbestimmung nach Bruns." *Astronomische Nachrichten,* Band 124.
Schwarzschild, Karl 1890b. "Methode zur Bahnbestimmung der Doppelsterne. *Astronomische Nachrichten,* Band 124.
Schwarzschild, Karl 1915 Letter dated Dec. 22, 1915 addressed to Arnold Sommerfeld published in *Arnold Sommerfeld, Wissenschaftlicher Briefwechsel.* Band I: 1892 - 1918. Ed. M. Eckert and K. Maerker, Berlin et al. Deutsches Museum, Verlag fuer Geschichte der Naturwissenschaften and der Technik, 2000.
Schwarzschild, Karl 1916. "Uber das Gravitationsfeld eines Massen punktes Nach der Einsteinschen Theorie." Sitzber. Deut. Akad. Wiss. Berlin, kl. Math. - Phys. Tech: pp. 189-196.
Schwarzschild, Karl 1975. *The Papers of Karl Schwarzschild (1973 - 1916).* Available on microfiche at the Niels Bohr Library & Archives, AIP, One Physics Ellipse, College Park, MD 20740. (Includes correspondence with his family and fellow scientists.)

Schwarzschild, Karl/author and Hans-Heinrich Voigt/editor 1992. *Gesammelte Werke (Collected Works) von Karl Schwarzschild* Vol.I - III. Springer-Verlag. (Vol. I begins with a short biography of Schwarzschild written by Hans-Heinrich Voigt in English.).

Smith, David Eugene 1923. *History of Mathematics Vol. I: General Survey of the History of Elementary Mathematics.* Ginn and Company. Republished in 1958. New York, NY: Dover Publications, Inc.

Snygg, John 1997. *Clifford Algebra - A Computational Tool for Physicists.* New York, NY: Oxford University Press.

Snygg, John 2002. "Functions of Clifford Numbers or Square Matrices." *Applications of Geometric Algebra in Computer Science and Engineering, pp. 99-107. edited by Leo Durst, Chris Duran, and Joan Lasenby. Boston: Birkhäuser.*

Sobczyk, Garret 1997. "The generalized spectral decomposition of a linear operator." *College Math. J.* **28** (1987), pp. 27-38.

Sobczyk, Garret 2001. "The missing spectral basis in algebra and number theory." The American Mathematical Monthly, **108** (2001).

Spivak, Michael 1970. *A Comprehensive Introduction to Differential Geometry.* Vol. Two. Boston, MA: Publish or Perish, Inc.

Stillwell, John 2001. *Mathematics and its History,* 2nd Ed. New York, NY: Springer.

Struik, Dirk J. 1988. *Lectures on Classical Differential Geometry,* 2nd Ed. New York, NY: Dover Publications.

Sudoplatov, Pavel and Anatoli Pavel with Jerrold L. Schecter and Leona P. Schecter 1995. *Special Tasks.* Boston, MA: Little, Brown and Company.

Swartz, Merlin L. (editor and translator) 1981. *Studies in Islam, Chapter VII* "The Attitude of Orthodox Islam Toward the 'Ancient Sciences.'" by Ignaz Goldziher, translated by Swartz from "Stellung der alten islamischen Orthodie zu den antiken Wissenshaften", Jahrgang 1915, no. 8 (Berlin, 1916), pp. 3-46. New York, Oxford: Oxford University Press.

Swerdlow, N. M. (Noel M.) and O. (Otto) Neugebauer 1984. *Mathematical Astronomy in Copernicus's De Revolutionibus,* Part I and II. (The text appears in the first volume (Part I) and the drawings appear in the second volume (Part II).) New York, NY. Springer-Verlag.

Szenassy, Barna 1987. "Supplement." pp. 220-239. (Article added to *Appendix, the theory of space* by János Bolyai edited by Ferenc Kárteszi (1987).) Amsterdam: Elsevier Sciences Publishers B.V.

Thorne, Kip S. 1994. *Black Holes & Time Warps: Einstein's Outrageous Legacy.* New York, NY: W.W. Norton & Company.

Turnbull, Herbert Westren 1993. *The Great Mathematicians.* New York, NY: Barnes & Nobles Books.

Vacaru, Sergiu 2002, A series of personal conversations May 23-25 at the 6th International Conference on Clifford Algebra and Their Applications; Cookeville, Tennessee, USA May 18-25, 2002.

Vucinich, Alexander 1984. *Empire of Knowledge: The Academy of Sciences of the USSR (1917 - 1970).* Berkeley: University of California Press.

Vucinich, Alexander 2001. *Einstein and Soviet Ideology.* Stanford, California: Stanford University Press.

Weierstrass, Karl 1866. "Untersuchungen über die Flächen, deren mittlere Krümmung überall gleich Null ist." *Monatsberichte der Königlich preussischen Akademie der Wissenshaften zu Berlin.* pp. 612-625.

Westman, Robert editor 1975. *The Copernican Achievement.* Berkeley, California: University of California Press. (This is a collection of papers delivered at a symposium held at the University of California at Los Angeles. The particular paper cited in this book was written by Curtis A. Wilson and appears on pp. 17-39.)

Whitrow, G.J. 1967. *Einstein: The Man and his Achievement.* British Broadcasting Corporation, 35 Marylebone High Street, London W1.

Yoder, Joella G. 1988. *Unrolling Time - Christiaan Huygens and the mathematization of Nature.* Cambridge: Cambridge University Press.

Youschkevitch, A.P. 1971 "Euler, Leonard." Article in *Dictionary of Scientific Biography* Vol. IV, pp. 467-484. Charles Coulston Gillispe, Editor in Chief. New York, NY: Charles Scribner's Sons.

Zackheim, Michele 1991. *Einstein's Daughter: The Search for Lieserl.* New York, NY: Riverhead Books.

Index

Symbols
0-vector, 6, 50
1-vector, 6, 50
2-vector, 6
3-vector, 6

A
Abbasids, 83
Abu Bakr, 82
Adams, John Couch, 42
Akivis, M.A., 23
Alcuin, 81
Aleksey
  son of Peter the Great, 274
algebra, 5, 51
  associative algebra, 51
Ali
  son in law of Mohammad, 82
amplitude, 196, 315
annus mirabilis, 40
aphelion, 99, 417
Apollonius of Perga, 94
applicable surfaces, 298
arc length, 139
arc length parameterization, 138
Archimedes, 255
Archimedes Screw, 255
areal velocity, 95, 415
Argand, Jean Robert, 8
argument, 315
Aristarchus, 90
Aristotle, 90, 94
associative algebra, 51
Augustus, Ernest
  King of Hanover, 222
axial vector, 7, 10
axis of a helix, 239

B
Banu Musa, Jafar Muhammad, 84
Bartels, Martin, 232
Basel Problem, 275
Baylis, William E., xi
Bell, Alan, xii
Beltrami, Eugenio, 216, 302
Beria, Lavrenti Pavlovich, 169
Bernoulli, Daniel, 272
Bernoulli, Jacob, 272
Bernoulli, Johann, 272
Bernoulli, Nikolaus, 272
Bessel, Friedrich Wilhelm, 217
Besso, Michelangelo, 40
binormal vector, 232
biorthogonal, 54
biorthogonal basis, 54
Biruni, al-, 85
bivector, 51
black hole, 408
black holes, 408
Bolker, Ethan, 13
Bolyai, Johann, 216, 218, 299
Bolyai, Wolfgang, 214, 218
Bonaparte, Louis Napoleon, 120
Bonnet, Pierre Ossian, 122, 206
boost, 30
brachistochrone problem, 396
Brahe,Tycho, 91
branch points, 392
Braunschweig, Duke of, 233
Bronstein, Matvei Petrovich, 170

J. Snygg, *A New Approach to Differential Geometry using Clifford's Geometric Algebra*,
DOI 10.1007/978-0-8176-8283-5,

C

Caesar, Julius, 76
calculus of variations, 396
Cartan, Élie, 15, 23
Cartesian frame for a Euclidean space, 49
Cartesian frame for a pseudo-Euclidean space, 49
catenary, 282
catenoid, 270, 385
  isometric to a helicoid, 297
catenoid isometric to helicoid, 388
Catherine I, 273
Catherine the Great
  benefactor of Euler, 280
  enlightened despot, 278
Cauchy-Riemann equations, 381
center of curvature, 193
  curve on 2-dimensional Euclidean plane, 191
central point of a generator, 333, 339
characteristic equation, 264
Christoffel symbols, 134
  coordinate transformation for, 136
  formula for, 135
Cisneros, Archbishop Francisco Jiménez de, 88
Clairaut, Alexis-Claude, 232
Clifford algebra, 4, 5, 52
  real, 52
  universal, 433
Clifford number, 6, 51
Clifford product, 51
Commentariolus, 96
complex conjugate, 379
conformal, 314
connection 1-forms, 151
contravariant, 57
contravariant tensor, 59
coordinate curve, 55
coordinate Dirac vectors, 56
Copernicus, 89, 96
covariant, 57
covariant derivative, 137
  semicolon notation, 136
covariant tensor, 59
Crick, Francis, 167
critical circumference, 408
cross product, 11
cross ratio, 317
Cross-sectional curve, 240
curvature
  curvature for curve in 2-dimensional Euclidean plane, 190
  on 2-dimensional Euclidean plane, 189
  total Gaussian, 211
curvature 2-form, 145
  alternate formula for using Fock–Ivanenko coefficients, 154
  formula for in terms of Fock–Ivanenko coefficients, 153
curvature of a path, 232
curvatures of higher order, 232
curve, 227
curves of constant slope, 240
cuspidal edge, 342
cycloid, 194

D

Dauben, Joseph W., xii
deferent, 91, 94
derivative
  geodesic, 134
  intrinsic derivative, 134
  moving frame derivative, 152
developable helicoid, 254
developable surface, 253, 340
Dickens, Charles, 113
differential forms, 56
Dirac vectors, 6, 49
direct non-Eucldean transformation
  non-Euclidean transformation, 312
direction cosines, 271
directrix, 250, 409
Divine Right to Rule, 398
do Carmo, Manfredo, 376
doubly ruled surface, 251
Drude, Paul, 38
Dühring, Eugen Karl, 217
dummy index, 48
Dyson, Freeman, 175

E

eccentric, 94
eccentricity, 93
edge of regression, 250, 254
eigenvalue, 262
eigenvector, 262
Einstein summation convention, 48
Einstein, Albert, 28, 90, 167
Einstein, Lieserl, 39
ellipsoid
  Gauss curvature, 159
elliptic integral of the first kind, 201
elliptic paraboloid, 161
energy-momentum tensor, 398
Engels, Frederick, 217

Engels, Friedrich, 119
Enneper's surface, 375, 379
epicycle, 91
equant, 95
equivalence principle, 396
escape velocity, 411
ether, 95
Euclidean space, 47
Euler's Theorem (or formula) for convex polyhedrons, 211
Euler's Theorem for Convex Polyhedrons, 210
Euler's Theorem for Normal Curvature
  Version I, 268
  Version II, 352
Euler, Leonhard
  biography, 272
Euler, Paul
  Leonhard's father, 272
Euler-Poincaré characteristic, 212
evolute, 199
  curve in 2-dimensional Euclidean plane, 190
exterior derivative, 181
exterior product, 61, 63

F
Ferdinand, Archduke Franz, 424
first fundamental form, 259
focal point, 409
Fock, Vladimir Alexandrovich, 151, 172
Fock–Ivanenko coefficients, 151
Fracastoro, Girolamo, 106
Frederick the Great, 275
Frenet 2-vector, 233
Frenet equations, 232
Frenet frame, 229
Frenet, Frédéric-Jean, 232
Freundlich, Erwin, 424

G
Galilean transformation, 28
Gamow, George, 166
Gauss and Martin Bartels, 232
Gauss' Theorema Elegantissimum, 206
Gauss, Karl Friedrich, 122
Gauss-Bonnet formula, 206
Gaussian curvature, 125
  formal definition, 142
Gaussian spherical mapping, 286
gear teeth, 282
general linear transformations, 314
General Theory of Relativity, 42
generalized Kronecker delta symbol, 62
generators, 249
genus, 212
geodesic, 121
  equation for, 140
  for a sphere, 123
  formal definition, 140
geodesic curvature, 193
geodesic curvature vector, 260
geodesic derivative, 134
geodesic polar parametrization, 290
geodesics
  calculus of variations, 141
Gerretsen, Johan C.H., xi
Gerulski, Jim, 130
Ghazālī-al, 86
Gibbs, Josiah Willard, 117
Ginzburg, Vitaly, xii
Girard, Albert, 122
Girard, Patrick, xi
Gorelik, Gannady, xii
gradient, 295
Gram-Schmidt process, 68
gravitational lens, 429
Grossmann, Marcel, 39, 40, 42
group
  point group, 16

H
Haller, Friedrich
  Director of the Patent Office in Bern, 39
Hamilton, William Rowan, 8
Hanbal, ibn, 83
harmonic functions, 379
Harriot, Thomas, 122
Hasan
  son of Ali, 82
Heisenberg, Werner, 167
helices
  circular, 228
  generalized, 239
helicoid, 251, 255, 270, 340
  isometric to a catenoid, 297
  lines of curvature for, 358
helicoid isometric to catenoid, 388
Helmholtz, Hermann, 38
Hertz, Heinrich, 38
Hessen, Boris
  phlogiston, 167
Hestenes, David, xi
Hipparchus, 91
Hitler, Adolf, 166
Holliday, Doc, 113

horizon, 407
Hoüel, Guillaume Jules, 302
Husayn
  son of Ali, 82
Huxley, Thomas, 112
Hypatia, 77
hyperbolic paraboloid, 161
hyperboloid
  one sheet, 339
hyperboloid, one sheet
  Gauss curvature, 160
hypersurface, 68, 256

I
Ibn al-Shātir, 88, 97
improper orthogonal transformation, 11
index free Clifford number, 60
indirect non-Euclidean transformation, 312
intrinsic derivative, 134
inverse, 305
inverse of metric tensor, 53
inversion, 305
involute, 280
involutes, 280
Irving, Washington, 109
isochronous, 196
isometric minimal surfaces, 386
isometric surfaces, 289
isometry of catenoid and helicoid, 388
isomorphic, 19
isothermic coordinates, 379
Ivanenko, Dmitrii Dmitrievich, 151, 174

J
Jancewicz, Bernard, xi, 115

K
Kant, Immanuel, 216
Kapitza, Pyotr, 168
Kepler's First Law, 91, 104, 419
Kepler's Second Law, 94, 104, 415
Kepler's Third Law, 104, 423
Kepler, Johannes, 91
Khan, Hulagu, 89
Khwarizimi, al-, 81
Kindi, al-, 82, 84
King George II of England, 277
Kirchoff, Rudolf, 38
Kreyszig, Erwin, 131
Kronecker delta symbol, 54
Kronecker product, 433

L
Lambert, Johann Heinrich, 216
Lancret, Michel-Ange, 232
Landau, Lev Davidovich, 167
Law of Cosines
  for a sphere, 73
  for the Poincaré disk, 326
Law of Sines
  for a sphere, 73
  for the Poincaré disk, 328
Le Verrier, Urbain Jean Joseph, 42
Legendre, Adrien Marie, 122
Leibniz property, 134
Lenard, Philipp, 44
Lieserl, 39
limit rotation, 320
line of curvature, 267, 347
line of stricture, 333, 339
lines of curvature
  for a surface of revolution, 268
Lobachevsky and Martin Bartels, 233
Lobachevsky, Nikolai Ivanovich, 216, 220, 299
Lorentz contraction, 34
Lorentz, Hendrik A., 28
Lotze, Rudolf Hermann, 218
Lounesto, Pertti, xi
lower index coordinate Dirac vectors, 56
lune, 121

M
Mach, Ernst, 38, 114
Mamun, al-, 81, 83
Marić, Mileva, 39, 41
Marx, Karl, 119
Maupertuis, Pierre Louis Moreau de, 276
Maxwell, James Clerk, 27, 38
mean curvature, 375
meridian, 361
meridians, 268
metric tensor, 53
metric, positive definite, 256
Michelson, Albert, 27
Michelson, Albert Abraham, 114
Minding's Theorem, 292
Minding, Ferdinand, 221, 300
minimal polynomial, 347
minimal surface, 375
mixed tensors, 59
Mobius transformations, 314
Möbius, August Ferdinand, 118, 317
Mohammad
  death, 82
  knowledge, 79

Morgan, Frank, xi
Morley, Edward, 27
Morley, Edward William, 114
moving frame, 147
moving frame derivative, 152
Mu'awiya, 82
Mutawakkil, al-, 84

N

Nash embedding theorem
    embedding theorems, 130
Nash, John, 130
Neptune
    discovery of, 42
Neugebauer, Otto, 89
Newton, Isaac, 104
Nguyen, Hieu Duc, xi
non-Riemannian geometry, 69, 130
normal curvature, 260
normal curvature vector, 260
normal variation, 377
normal vector, 232
nuclear plant cooling towers, 253

O

O'Connor, John J., xi, 81
O'Neill, Barrett, xi
one sheet hyperboloid, 251
Oprea, John, 376
order of a group, 19
orthogonal transformation, 11
    improper orthogonal transformation, 11
    proper orthogonal transformation, 11
orthonormal non-coordinate frame, 49, 147
osculating circle, 193
Otis elevator, 395

P

p-form, 181
p-vector, 51
papyrus, 75
parallel, 361
parallel lines, 304
parallel transport, 123, 138, 140
parallels, 268
parchment, 75
patent office, 39
path, 227
Pauli matrices, 12, 52
perihelion, 99, 416
period
    period of a planet, 423
Pernet, Jean, 38
Peter the Great, 273
Phillipe, King Louis, 119
Physcon, 75
Poincaré model, 303
Poincaré, Henri, 28
point group, 16
pole, 393
Porteous, Ian R., 227, 232
Porteus, Ian, 50
positive definite metric tensor, 130
precession, 422
principal curvature, 262, 379
principal curvatures
    n-dimensional surface of revolution, 266
    saddle surface, 264
principal direction, 262
Principle of Least Action, 397
projection operators, 349
proper orthogonal transformation, 11
pseudo-Euclidean space, 47
pseudo-vector, 10
pseudosphere, 208, 271
Ptolemy, 94
Ptolemy, Claudius, 89, 91, 95

Q

qibla, 85
quaternion, 8

R

radius of curvature, 191, 193
Raimundo, Archbishop Don, 88
real Clifford algebra, 52
regular point, 227
regular polyhedrons, 15
Reichenau, Hermann of, 81
retrograde motion, 90
reverse of a Clifford number, 64
Ricci tensor, 398
Riefin, Edgar, 13
Riemann curvature tensor
    definition of, 142
    formula for in terms of Christoffel symbols, 147
    symmetries of, 142
Riemann surfaces, 392
Riemann, Bernhard, 223, 303
Riemannian geometry, 69, 130
right circular cone, 131, 288
right handed curve, 232

Rindler, Wolfgang
  horizon, 407
Robertson, Edmund F., xi, 81
Rosenfeld, B.A., 23
rotation, 320
Rozenberg, Sasha, xii
ruled surface, 249
Rushd, ibn, 87
Rutherford, Ernest, 168

S

Saccheri, Giovanni Girolamo, 216
saddle surface, 251, 339, 353, 359
  lines of curvature for, 355
  principal curvatures of, 264
Saliba, George, xii
scalar, 51
scalar curvature, 398
scalar product, 5
  definition for two Clifford numbers, 64
Schücking, Engelbert, xii, 175, 178, 395
Schwarzschild radius, 407
Schwarzschild, Karl, 399
Schweikart, Ferdinand Karl, 216
second fundamental form, 259
semi colon notation, 136
Senff, Carl Eduard, 232
Serret, Joseph Alfred, 232
Serret-Frenet equations, 232
shape operator, 261
Shiites, 82
signature, 47
signature matrix, 50
singular points, 131
singular points with respect to the representation, 131
Snell's Law, 408
Snygg, Charles, xii
Snygg, Spencer, xii
Snygg, Suzanne, xii
Sobczyk, Garret, xi, 349
special linear transformations, 314
sphere (n-dimensional), 271
spherical excess, 123
spherical triangle, 121
Staples, G. Stacey, xi
Stokes' Theorem, 183
strake, 254
stricture, line of, 333
Struik, Dirk J., xi
summation convention, 48
Sunnis, 82
surface of revolution
  lines of curvature for, 268
Sylvester, James Joseph, 111, 117
symmetric operator, 262
symmetry transformations, 15

T

tangent space, 55
tangent vector, 55, 232
  unit tangent vector, 139
tangential curvature vector, 260
tangential developable, 250, 254, 269, 281
Taurinus, Franz Adolf, 216
tensor
  contravariant, 59
  covariant, 59
tensor components of the shape operator, 261
tensors
  mixed, 59
tessellation, 329
Theorema Egregium, 285
third fundamental form, 287
Thorne, Kip, 408
torsion free, 134
torsions, 232
torus
  Gauss curvature, 157
  lines of curvature for, 362
total Gaussian curvature, 211
trace, 227
tractrix, 282
translation, 320
transpose operator, 262
trivector, 51
Tūsī, Nasīr al-Dīn, al-, 89

U

ultra parallel, 304
Umayyads, 82
unit disk, 303
Unit tangent vector, 139
universal Clifford algebra, 433
upper index coordinate Dirac vectors, 56

V

Vacaru, Sergiu, xii
valence, 59
vector, 51
velocity, 227
Versailles Treaty, 43
Voltaire, 277
Vulcan, 42

W

Watson, James, 167
Weber, Heinrich, 38
Weber, Wilhelm Eduard, 213, 222
wedge product, 5, 11
Weingarten map, 261
Wessel, Casper, 8
Wheeler, John
  black holes, 408
William IV, King Frederick, 119

The manufacturer's authorised representative in the EU is Springer Nature Customer Service Centre GmbH, Europaplatz 3, 69115 Heidelberg, Germany. If you have any concerns regarding our products, please contact ProductSafety@springernature.com

Printed and bound by CPI Group (UK) Ltd, Croydon, CR0 4YY

13/08/2026

02195493-0004